Cities and Nature fosters high-quality multi-disciplinary research addressing the interface between cities and the natural environment. It provides a valuable source of relevant knowledge for researchers, planners and policy-makers. The series welcomes empirically based, cutting-edge and theoretical research in urban geography, urban planning, environmental planning, urban ecology, regional science and economics. It publishes peer-reviewed edited and authored volumes on topics dealing with the urban and the environment nexus, including: spatial dynamics of urban built areas, urban and peri-urban agriculture, urban greening and green infrastructure, environmental planning, urban forests, urban ecology, regional dynamics and landscape fragmentation.

More information about this series at http://www.springer.com/series/10068

Charlie M. Shackleton · Sarel S. Cilliers ·
Elandrie Davoren · Marié J. du Toit
Editors

Urban Ecology in the Global South

Springer

Editors
Charlie M. Shackleton
Rhodes University
Makhanda, Eastern Cape, South Africa

Elandrie Davoren
Rhodes University
Makhanda, Eastern Cape, South Africa

Sarel S. Cilliers
Unit for Environmental Sciences
and Management
North-West University
Potchefstroom, South Africa

Marié J. du Toit
Unit for Environmental Sciences
and Management
North-West University
Potchefstroom, South Africa

ISSN 2520-8306 ISSN 2520-8314 (electronic)
Cities and Nature
ISBN 978-3-030-67649-0 ISBN 978-3-030-67650-6 (eBook)
https://doi.org/10.1007/978-3-030-67650-6

This Springer imprint is published by the registered company Springer Nature Switzerland AG
The registered company address is: Gewerbestrasse 11, 6330 Cham, Switzerland

Contents

The Need for an Urban Ecology of the Global South

Charlie M. Shackleton, Sarel S. Cilliers, Marié J. du Toit, and Elandrie Davoren

Abstract Urban ecology is a key discipline in guiding urban development, sustainability and consequently human wellbeing. However, most urban ecological research has, and continues to be, undertaken in the Global North, and thus urban ecological methods, principles and frameworks are dominated by contributions and understandings from the Global North. However, there are a multitude of local- and national-scale contextual differences between the Global North and the Global South that limit or question the universal application of Global North perspectives and knowledge. This chapter lays the foundation for the rest of the book in two ways. First, it explores the development and definitions of the terms 'urban ecology' and 'Global South'. Second, it presents the major biophysical and socio-economic contextual characteristics of Global South towns and cities that differentiate them from those in the Global North. These contextual differences need to be accounted for in urban ecological research, theory and application towards the development of an urban ecology that is more relevant for the Global South, and when conjoined with understandings from the Global North allow for the exploration and development of truly universal urban ecology principles and frameworks.

Keywords Context · Differences · Global South · Urban ecology · Urban green infrastructure

C. M. Shackleton (✉) · E. Davoren
Department of Environmental Science, Rhodes University, Makhanda 6140, South Africa
e-mail: c.shackleton@ru.ac.za

E. Davoren
e-mail: e.davoren@ru.ac.za

S. S. Cilliers · M. J. du Toit
Unit for Environmental Sciences and Management, North-West University, Potchefstroom, South Africa
e-mail: sarel.cilliers@nwu.ac.za

C. M. Shackleton et al. (eds.), *Urban Ecology in the Global South*, Cities and Nature,
https://doi.org/10.1007/978-3-030-67650-6_1

1 Introduction

With more people globally now living in towns and cities than in rural areas, the future of humanity is increasingly an urban one. Urban areas are the centres of economic growth, innovations, arts and policy development. The supply of basic services and standards of living are frequently, albeit not always, higher amongst urban populations than their rural counterparts. But in many settings these benefits come with costs and trade-offs, such as decreased air quality, water and noise pollution, flooding, high waste loads and little or reduced contact with nature (McMichael 2000; Grimm et al. 2008a; Wu 2014). The prevalence and intensities of these costs and trade-offs demand that urban policy-makers, urban planners, authorities and researchers seek cost-effective, equitable and inclusive ways to promote urban sustainability, resilience and liveability. This quest has been distilled into the various facets of the Sustainable Development Goal number 11, i.e. 'make cities and human settlements inclusive, safe, resilient and sustainable'.

A core discipline to guide the quest for urban sustainability and resilience for the benefit of all urban dwellers is 'urban ecology', which is an interdisciplinary approach to understanding urban social-ecological systems, and hence providing guidance to policy-makers, authorities and planners. However, there are different schools of and approaches to urban ecology, most of which have been developed and advocated within Global North settings, with relatively few contributions from the Global South (Nagendra et al. 2018). However, the histories, dynamics and contexts of most Global South (GS) towns and cities differ quite markedly from those in the Global North (GN). Consequently, many aspects of current principles and frameworks cannot be transferred unquestioningly to GS settings, but rather should be critically examined and tested by researchers, planners and authorities, and, if necessary, adapted or reframed to accommodate the historical and lived realities of the GS. This is the purpose of this book, namely to examine the nature and contributions of urban ecology in GS contexts and to indicate where these may differ from those in the GN.

This opening chapter sets the foundation for the chapters to follow by interrogating and defining the three core concepts of this book, namely urban ecology, Global South and the contextual differences. We also briefly summarise the numbers of research publications on urban ecology from the GS and GN to demonstrate the highly unequal contributions and the dominance of GN perspectives. We end the chapter by presenting the purpose and structure of the book.

2 What Is Urban Ecology?

2.1 Historical Development of the Discipline

The origin of urban ecology can be traced back to the 1920s in the discipline of sociology and is known as the Chicago School of human ecology which studied human living conditions (Weiland and Richter 2011) by investigating the relationships between urban residents and the city (Wu 2014). It was also described by Wu (2008) as a socio-ecological perspective of urban ecology in which cities were regarded as socio-economic systems and humans the primary components. There was no integration with natural sciences but ecological theories such as competition, niche partitioning and succession were used as metaphors to explain spatial differentiation of land uses and people in cities (Wu 2008; Cadenasso and Pickett 2013). Although urban ecology was quite influential in sociology in the United States it lost its impetus by 1950 and the human ecology approach was replaced by a 'contemporary urban ecological approach' which focused on integration of different disciplines within the social sciences (Wu 2014). The social sciences part of urban ecology has, however, been developed further as will be indicated later in this section.

As an 'ecological science' urban ecology started in Europe after World War II through in-depth studies on the distribution and richness of plants and animals in and around cities (Wu 2014) and is often referred to as the 'Berlin School' of urban ecology (Weiland and Richter 2011). The approach used in those studies was bio-ecological (Wu 2008, 2014) also known as the ecology 'in' cities perspective (Sukopp 2008). Pickett et al. (2016) regarded this perspective as the first of the three paradigms in urban ecology. The focus was on the influence of urbanisation on biodiversity, and cities were regarded as disturbed ecosystems with humans as the main drivers of the disturbances (Wu 2008). Despite these earlier bio-ecological studies, urban ecology was only regarded as part of mainstream ecology since the last two decades (Wu 2014). Aspects such as increasing urbanisation and its effect on the environment, as well as a paradigm shift in ecology focusing on non-equilibrium and patch dynamics and an increasing focus on sustainable cities are some of the reasons for the increasing research interest in cities (McDonnell 2011; Wu 2008, 2014; Pickett et al. 2016). Ecology 'in' cities is still a major perspective in urban ecology, focusing on differences in ecological patterns and processes between cities and other environments by studying urban 'analogues' to natural ecosystems such as forest patches, parks, meadows, streams and wetlands (McPhearson et al. 2016a; Pickett et al. 2016). Examples of these types of studies have also been conducted in the GS (e.g. Sheherazade et al. 2017; de Andrade et al. 2020), probably with the exception of China where detailed studies on biodiversity and ecosystem processes of cities are scarce, despite the exponential increase in urban ecological research in that country over the past decade (Wu et al. 2014). Similar approaches and methods as used in natural areas are used in the ecology 'in' cities perspective, and although being small-scale, disciplinary studies, they have resulted in a wealth of information on urban green spaces informing aspects such as biodiversity conservation, and urban planning, design and management (McDonnell 2011; McPhearson et al. 2016a; Pickett et al. 2016; Wu 2014). Within this perspective, comparative studies were often done on regional scales using indirect gradient analyses, namely the urban–rural gradient approach (McDonnell and Hahs 2008), which has also been used in the GS (du Toit and Cilliers 2011; Filloy et al. 2019; Wang et al. 2020).

Another perspective in urban ecology is that of the urban systems perspective, which considers cities as entire ecosystems or landscapes focusing on interactions between socio-economic and biophysical processes, often following interdisciplinary and transdisciplinary approaches (Wu 2008, 2014). This perspective follows an ecosystem ecology approach and is often referred to as the ecology 'of' cities (Grimm et al. 2000), typically incorporating the ecology 'in' cities approach in collecting detailed information on urban biota and urban green spaces (Pickett et al. 2016). Pickett et al. (2016) described this perspective as the second paradigm in urban ecology. From a landscape perspective, cities are regarded as 'spatially complex mosaics of patch types' in which the patches are the result of a combination of anthropogenic and biotic factors (Pickett et al. 2016). The well-known, long-term ecological research (LTER) projects in Baltimore and Phoenix in the USA that started in 1997 helped establish the ecology 'of' cities approach as a standard in 'linking with mainstream ecological ideas' to show the significance of urban ecology to the science of ecology in general. For example, in Baltimore an approach well-known in mainstream ecology, the watershed approach, was tested in an urban area (Pickett et al. 2016). Over the last decade this paradigm has been developed to study cities as complex and interactive systems to investigate the 'interactions and feedbacks between the social and ecological components of those systems' (McPhearson et al. 2016a). Although not explicitly labelled as ecology 'in' cities, the development of urban environmental management and climate adaptation plans in Durban, South Africa (Roberts 2008, 2010; Roberts et al. 2012), aimed for the type of integration of ecological and socio-economic aspects typical of this perspective.

With increasing recognition of the complexity of urban systems and that ecological management can have significant effects on urban sustainability and human well-being, Childers et al. (2015) proposed a move towards an ecology 'for' cities perspective, to address the challenges associated with integrating ecological aspects in urban design to increase urban sustainability. Childers et al. (2015) suggested a 'transformative urban design-ecology nexus' in which the education and training of 'ecologically literate urban designers and engineers, design-literate, engineering-conscious ecologists, broad-thinking and holistically inclined planners and place-aware and activist city residents' is necessary in future. It sounds like a tall order, especially for the GS, but that is probably why it is called 'creative visioning' by Childers et al. (2015). The ecology 'for' cities perspective (the third paradigm in urban ecology) links ecology with social sciences and civic partnerships and highlights the importance of ecological stewardship and ethics that will expand dialogue between all relevant stakeholders (Pickett et al. 2016). According to Wu (2014), sustainability thinking acknowledged cities as 'coupled social-ecological systems, with an increasing emphasis on the relationship between ecosystem services and human well-being in urban areas'. The ecology 'for' cities approach regards resilience as an important goal towards sustainability (Wu 2014; Pickett et al. 2016) which should be included in urban planning (Erixon et al. 2013). Ahern et al. (2014) proposed a transdisciplinary framework for 'safe to fail' adaptive urban planning and design in which urban ecosystem services integrate 'science, professional practice, and stakeholder participation' and 'includes experimental design guidelines, monitoring and assessment protocols and strategies' stimulating a 'learning by doing' perspective that is often a challenge in urban development. The ecology 'for' cities approach still draws on insights generated by the ecology 'in' and 'of' cities approaches, but within the quest for sustainability (Pickett et al. 2016) acknowledges the importance of co-produced research through interactions between scientists and various stakeholders and for governance and policy development (McPhearson et al. 2016a, b).

2.2 *Current Understanding and Definition of Urban Ecology*

It is clear from the different and still evolving perspectives in urban ecology that it is challenging to develop a single comprehensive definition of the discipline. Wu (2014) discussed several definitions from different disciplinary perspectives (e.g. ecology, sociology, geography, planning) and argues that each of these definitions only captures some of the essence of urban ecology. Wu (2014) proposed, therefore, the following broad definition: 'the study of spatiotemporal patterns, environmental impacts, and sustainability of urbanisation with emphasis on biodiversity, ecosystem processes, and ecosystem services'. He acknowledges that socio-economics and planning practices influence patterns of urbanisation but argues that they don't form part of the 'scientific core' of the discipline. For Wu (2014), urban ecology forms part of urban sustainability sciences and ecosystem services form 'a key nexus that links urban ecology and sustainability'. Although McPhearson et al. (2016a) did not offer a definition of urban ecology they argued that for urban ecology to develop into a science of cities the following are needed: 'conceptual synthesis, knowledge and data sharing, cross-city comparative research, new intellectual networks and engagement with additional disciplines'. The debate on the position of urban ecology as a science in its own right (a different ecology with different models) or merely as a sub-discipline of ecology (same ecology but with emphasis on the effects of humans) is ongoing, and Pickett and Cadenasso (2017) emphasise the power of principles to structure a science. Principles are, however, quite flexible and can be described from different viewpoints. Forman (2016) described 90 urban ecological principles from an inductive approach from a large body of cases 'within specified ecological, cultural, and historical contexts', which according to Pickett and Cadenasso (2017), indicates that they are different from those of natural areas. It is, however, argued by Pickett and Cadenasso (2017) that these principles are 'biased towards cities of the Global North'. In their quest to describe urban ecological principles Pickett and Cadenasso (2017) followed a deductive approach to identify 13 principles within five meta-principles which are similar to the ecology of natural areas, namely ecosystems, heterogeneity, dynamic, linked and ecological processes. This emphasises the 'hierarchical nature of urban ecological theory' (Pickett and Cadenasso 2017).

Urban ecology is described as an interdisciplinary and transdisciplinary science drawing on and linking to several other disciplines. There are also several other scientific disciplines which are related to urban ecology and often included in discussions and texts on urban ecology. Acknowledging that other disciplines such as urban geography, urban sociology and anthropology, urban planning and ecological engineering are related to urban ecology, we will focus here only on urban landscape ecology, urban forestry and urban agriculture. The relationship to landscape ecology, the science of studying the relation between spatial patterns (spatial heterogeneity) and processes (ecological and socio-economic) at various scales (Wu 2013), is quite clear as the ecology 'of' cities approach with respect to cities as either ecosystems or landscapes, and therefore Wu et al. (2013) argued that urban ecology has developed into an urban landscape ecology.

Two other closely related disciplines with intimate links to urban ecology are urban agriculture and urban forestry. Various definitions of urban agriculture are in use, such as that of Lin and Egerer (2018) who described urban agriculture as the production of different types of food (e.g. vegetables, fruit, mushrooms, spices, eggs, milk, meat) in a variety of urban green spaces (e.g. community or allotment gardens, private gardens, rooftop gardens, orchards, peri-urban areas) but also acknowledges their importance in providing several other ecosystem services. Urban forestry is a sub-discipline of forestry and refers to the research and management of tree-dominated urban green areas focusing on their 'physiological, sociological, economic and aesthetic benefits' for society (Konijnendijk et al. 2006) but also its ecological importance (Shackleton 2006).

As with urban ecology, there is no unified definition for the concept 'urban', although measures such as population size and density and the amount of impervious surfaces are often used to characterise urban areas (Wu 2014). In defining 'urban' social scientists usually focus on areas with high population densities, whilst ecologists often refer to areas under significant human influence, and therefore McIntyre et al. (2000) proposed that a working definition of 'urban' should be formed for any study of urban ecosystems quantifying different aspects of the urban structure. According to Pickett et al. (2016) the concept of cities refers to 'any urban or urban-influenced settlement'. Urban ecology can therefore also be studied in small- and medium-sized cities and even in rural settlements of which numerous examples exist in the GS (e.g. Molebatsi et al. 2013). Indeed, according to McPhearson et al. (2016b) the most 'pressing urban challenges tend to be found in the GS and in small to medium-sized cities', and these urban areas need to be included in urban ecology. In fact in many developing regions it is often difficult to distinguish between urban and rural areas spatially or by a clear reliance on either urban or a rural income sources but rather a mix (Agergaard et al. 2019). McHale et al. (2013) went as far as arguing that rural can be regarded as the 'new' urban in Africa due to reclassification of rural areas and rural migration to small cities and towns, or the considerable participation in land-based livelihood activities even in what are nominally termed towns or cities

(Shackleton et al. 2020). On the other hand, Forman (2019) highlighted the uniqueness of the ecology of small towns, arguing that they are not merely small cities, as they differ from cities in terms of structure, water availability, biodiversity, their relationships with the surrounding environment and their governance and economic capacities, and therefore warrant separate studies under a new discipline of town ecology.

3 What Is the Global South?

3.1 Global South: An Origin Story

The use of categories or concepts to describe racial or global differences and especially, unequal economic development is not new (Clarke 2018). According to Clarke (2018) the term 'Global South' is the most recent in a long list of catch-all concepts used to group 'diverse economic, social and political experiences and positions into one overarching category'. Previous incarnations included terms such as 'periphery', 'less-developed', 'developing', 'underdeveloped' and 'third world', yet they all served the same purpose, i.e. to identify, cluster and define poorer countries (Clarke 2018). According to Clarke (2018) this tendency to categorise peoples and places became increasingly popular after World War II when many countries achieved independence from colonial rule.

Many scholars credit the US President, Harry Truman, with coining the terms 'developed' and 'underdeveloped' (later 'developing'), when he referred to 'underdeveloped areas' in his 1949 Inaugural Address (Clarke 2018). These terms later became a common way of dividing the world. However, this terminology was problematic for several reasons. Firstly, the focus of economic growth and industrialisation was far too narrow, secondly, it assumed that 'development' would lead to the modernisation of so-called 'traditional societies' and thirdly, implied that there was a universal measurement of 'development', which could be measured against this western standard (Clarke 2018).

In the 1950s and 1960s, the economist Raúl Prebisch campaigned for reform of the world trade system, which moved the terms 'North' and 'South' into the international political lexicon (Dados and Connell 2012). In 1980, the Brandt Commission (more specifically, the Independent Commission on International Development Issues) published their first report, namely 'North–South: A Programme for Survival', in which the so-called 'Brandt line' made its first appearance (Solarz 2012). The Brandt line is a visual depiction of the North–South divide, i.e. the global split in the international community between the generally rich North and poor South (Solarz 2012). However, at the time, the terms 'First, Second and Third World' were popular (Clarke 2018). The term 'Third World' referred to countries that remained non-aligned with either, the United States and their allies ('First World') and the Soviet Union and their allies ('Second World'). After the Cold War ended in the 1990s, these terms became obsolete and the ubiquitous terms 'developed and developing countries' lost some of their theoretical legitimacy, as centres of global economic growth shifted to places outside of the so-called West (Europe, North America) (Pagel et al. 2014) to areas such as Japan, Taiwan and South Korea (Clarke 2018). Simultaneously, the levels of poverty in some countries, such as Brazil and South Africa increased, which made it difficult to categorise countries as a whole (Clarke 2018).

Subsequently, developing countries, predominantly former colonies, began to articulate the idea of a Global South whose interests conflicted with those of developed countries and cut across Cold War divisions (Dados and Connell 2012). According to Clarke (2018), the rise in popularity of the term can be credited to the 'inter-activity among communities, political organisations and other groupings of the "South" to establish their own initiatives and policy responses'. In recent decades, the terms 'Global North' and 'Global South' have become increasingly popular in academic fields such as development studies, international relations, political sciences, natural sciences and medicine (Dados and Connell 2012; Pagel et al. 2014).

3.2 *The Global South Defined*

For many non-governmental organisations (NGOs) and some scholars, the term Global South simply refers to countries classified as low or middle income countries by the World Bank (Clarke 2018). However, the term is far more complicated than that. According to Mahler (2017) 'the Global South as a critical concept has three primary definitions'. The first refers to economically disadvantaged nation-states, predominantly those that originated in the Non-Aligned Movement (Mahler 2017). This definition has traditionally been used within intergovernmental development organisations. However, in recent years, the term 'Global South' has been employed to address peoples and areas that have been negatively impacted by colonisation and subsequently by contemporary capitalist globalisation (Mahler 2017, 2018). In its second definition, Mahler (2017) states that the 'Global South captures a deterritorialised geography of capitalism's externalities and means to account for subjugated peoples within the borders of wealthier countries, such that there are economic Souths in the geographic North and Norths in the geographic South'.

It is from this deterritorialised conceptualisation that the third meaning of the term 'Global South' is deduced. The third definition refers to the 'resistant imaginary of a transnational political subject that results from a shared experience of subjugation under contemporary global capitalism' or colonisation (Mahler 2017, 2018). In this sense, the Global South captures both a political collectivity and ideological formulation that arises from lateral solidarities amongst the world's multiple Souths (Mahler 2017). Consequently, there is no straightforward definition of the term 'Global South' and the debate into its use, meaning, and analytical value is ongoing. As stated by Dados and Connell (2012), the 'term Global South functions as more than a metaphor for underdevelopment. It references an entire history of colonialism, neo-imperialism, and differential economic and social change through which large inequalities in living standard, life expectancy, and access to resources are maintained'.

3.3 An Ongoing Debate

Much like its predecessors, the term 'Global South' is not perfect and has been the subject of ongoing debate. However, as Clarke (2018) contends, 'the term is not static and does not refer to a specific list of countries, groups or communities: it evokes different meanings and is used both descriptively and analytically'. For many scholars, the use of the phrase Global South marked a shift from a cultural or developmental focus towards a geopolitical relation of power (Dados and Connell 2012). This shift is evident in the differing approaches to compiling the Millennium Development Goals (MDGs) in 2000 to the Sustainable Development Goals (SDGs) in 2015 (Horner and Hulme 2019). The MDGs were defined by developed countries and were predominantly aimed at 'developing countries', whilst the SDGs emphasises 'transformation' and sustainability, which require commitments by all countries, i.e. a global approach (Horner and Hulme 2019). The North–South divide is present and increasing, but so too are the inequalities between and within countries (Horner and Hulme 2019; Clarke 2018).

In this book we refer to the development status of the Global South countries and not the geographic framing. Thus, it covers underdeveloped and developing countries (many of which are in the geographic south, but most of Asia is north of the equator but is still considered amongst Global South countries). It broadly categorises the poorer nations from the rich ones, and some argue that it also distinguishes the former colonialised nations from the colonisers. It includes most of Africa, Asia, Latin America and Oceania (Dados and Connell 2012), and excludes Australia, Canada, Europe, Hong Kong, Israel, Japan, Kazakhstan, New Zealand, Russia, Singapore, South Korea, Taiwan, Turkey and USA.

4 Underdevelopment of Urban Ecology Research and Frameworks in the Global South

The discipline of urban ecology has grown in leaps and bounds over the last 2–3 decades, and valuable contributions have been made in understandings of the urban environment. However, numerous authors reveal a relative paucity in studies from the Global South (Table 1). An evaluation of the global distribution of ecological studies analysing 2573 study sites, found that 90% of the studies were carried out

Table 1 The relative proportions of research articles of different aspects of urban ecology from the Global South and the Global North

Subject	No. of studies	Proportion		Source
		Global South	Global North	
Urban nature conservation	787	21	79	Shwartz et al. (2014)
Urban wildlife ecology	571	10	90	Magle et al. (2012)
Urban community gardens	87	9	91	Guitart et al. (2012)
Urban forestry	499	21	79	Shackleton (2012)
Benefits of urban trees	115	26	74	Roy et al. (2012)
Effects of vehicles and roads on wildlife	244	7	93	Taylor and Goldingay (2010)
Urban sustainability research authorship	100	<30	>70	Nagendra et al. (2018)

in countries that ranged in the 70–100th percentile of gross national income (Martin et al. 2012).

Further evidence includes a synthesis by Wilkinson et al. (2013) on urban biodiversity governance where they cautioned about making generalisations because there was a scarcity of studies from Africa, parts of Asia and South America. Furthermore, in a meta-analysis of intra-urban biodiversity variation, representing 75 cities worldwide, 76% of the cities were situated in Europe, North America, Australia and New Zealand (Beninde et al. 2015). Similarly, in the meta-analysis of studies on the influence of socio-economic status on biodiversity by Kuras et al. (2020) only eight of the 34 studies (24%) were from the GS. Kabisch et al. (2015) found that European and the North American research dominated in a review of 219 papers on human–environment interactions in urban green spaces. The disparities between Global North and South were also noted in reviews on ecosystem disservices in cities (von Döhren and Haase 2015), green infrastructure (Bartesaghi Koc et al. 2016; Lindley et al. 2018), perceptions and valuation of urban biodiversity (Botzat et al. 2016), and bee-community responses (De Palma et al. 2016).

To provide a more up to date illustration of the imbalance between Global North and South regarding urban ecological research, we did a survey in Scopus (www.scopus.com) on 8 July 2020 on the entire database for all years. We searched in the titles, abstracts and keywords and limited the results to the subject areas of Environmental Science and Agricultural and Biological Sciences. The specific search criteria were: TITLE-ABS-KEY (urban AND ecology OR 'green infrastructure' OR 'green space' OR 'ecosystem service' OR birds OR vegetation OR biodiversity OR 'species richness' OR soil OR amphibian OR arthropod OR animals OR invertebrates OR mammals) AND (LIMIT-TO (SUBJAREA, 'ENVI') OR LIMIT-TO (SUBJAREA, 'AGRI')).

The search returned 48,057 results, comprising articles (85%), conference papers (7%), reviews (4%), book chapters (3%) and other (1%). The earliest paper recorded in the results was in 1945. We acknowledge that some relevant papers will have been missed in the search, however the large number of returns suggests that these results offer a reasonably representative picture of the situation. The search results indicated that by country, 69% of the documents were authored from the Global North, with 31% represented by the Global South. Analysis of the output from the global top twenty countries indicated that 15 countries were from the Global North and only five countries were from the Global South (Fig. 1a). The country with the highest overall output was the United States, moreover, they had almost double the output of the next highest country (Fig. 1a). However, China, long part of the poorer countries of the Global South, indicated considerable advances in economic improvement and contributions to scientific research by having the second-highest output (Fig. 1a).The top twenty countries in the Global North by output are indicated in Fig. 1b, showing that together with the United States, the United Kingdom, Australia and Germany had the highest outputs. The analysis of the top twenty countries by output in the Global South indicated that China, Brazil and India dominated (Fig. 1c). The top twenty Global North countries produced 38,220 documents, with the top twenty Global South countries producing less than half of that with 16,436 outputs.

This short overview indicates that in terms of research outputs on urban ecology, most come from the Global North. This means that the contexts and understandings of the Global South are inadequately represented and therefore are likely to have less voice and significance in models and policies on urban ecology and development globally. As Nagendra et al. (2018) succinctly state 'urban issues in the Global South are distinctly and statistically different from those in the Global North, but our current urban knowledge is predominantly shaped by research on and from the Global North', echoing the argument of McHale et al. (2013) on the 'Global North bias in urban ecological theory despite acknowledgement of fact that GS cities may represent alternative stable states' (McHale et al. 2013)

5 Why Is Urban Ecology Different in the Global South

It is well appreciated by researchers in almost any discipline that context matters. Context sets the broad- and fine-scale template that shapes what factors come into play, their relative magnitudes, and in what combinations and sequences. In other words, the system characteristics and dynamics operational at any given time and space are dependent to a large degree on the local and wider context. This also includes historical dynamics and legacies when it comes to social-ecological systems. A recognition therefore, that the context of cities, towns and settlements of the GS is in many aspects generally different to that of a typical GN counterpart demands different urban ecological framings, principles and interpretations. However, in making such a claim, we appreciate that there is as much variation between urban settings within the GS as there is between the generalised GS and GN. This is hardly surprising

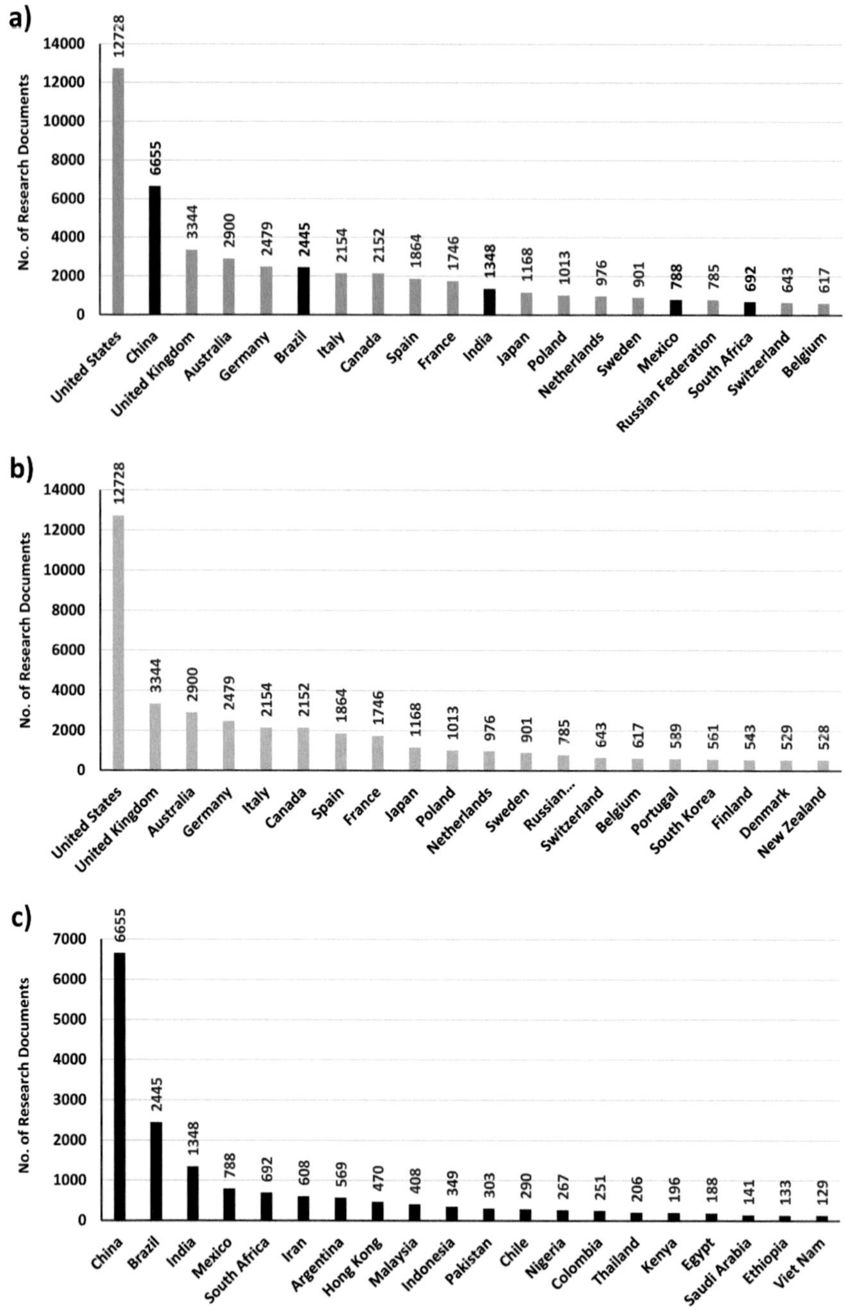

Fig. 1 The top twenty countries **a** globally, **b** in the Global North and **c** in the Global South in terms of research outputs on urban ecology from 1945 to July 2020

given that the GS covers over 60% of the world's terrestrial surface area, houses almost 80% of the world's peoples and includes over a hundred different countries each with its own particular history, political dynamics, cultures, economies and biophysical endowments. Nevertheless, if urban planners, decision-makers, citizens and researchers are to design and implement development plans and policies that are best suited to the needs of the GS, then they need to acknowledge and accommodate the particular characteristics of the GS. Doing so demands an appreciation that GN models, approaches, plans and solutions generally cannot be transferred directly to GS settings and be expected to work in precisely the same way or deliver exactly the same outcomes. Many, if not all, will need to be adapted or replaced to suit the GS conditions and histories. It also demands that GS approaches and innovations be recognised as equally viable and worthy solutions to GS urban ecological and sustainability requirements. Indeed, it is not a case of the GS being 'behind' the GN in terms of development trajectories and will therefore 'catch-up' in time (McHale et al. 2013), but rather the GS offers a myriad of different opportunities, innovations and trajectories than those experienced in the GN. Consequently, individual GS regions, countries and cities can draw from the total pool of GS and GN knowledge and experiences to find potential useful, complementary, hybrid or alternative approaches and outcomes, as shaped and made possible by each one's particular historical pathways and legacies, local biophysical and social endowments and contexts and dynamics, recognising that such will not be static in space or through time.

Whist appreciating the wide diversity of situations across GS settings, we distil several attributes (Table 2) that, when taken together, paint a typical GS context that is sufficiently different from most, albeit not all, GN settings to require a GS urban ecology of, in and for the city. In any given GS city or town some may not apply, but in most GS settings most will apply, and thus present a context that is very different to most towns and cities of the GN. Most are interrelated and dealing with them in isolation cannot do justice to the complexity of the whole, but we do so to highlight core attributes of GS urban contexts.

5.1 Marked Climatic Severity

The majority of GS countries and areas fall in the tropics and subtropics, unlike the GN that mostly falls within the temperate regions. The predominantly tropical and subtropical location is synonymous with more severe climates in terms of being both hotter and having greater extremes in precipitation (amount and seasonality). This is well illustrated by all of the top ten wettest, driest and hottest countries in the world, being in the GS. Additionally, many are marked by strong seasonality in rainfall, which influences water supply to inhabitants and agricultural potential and productivity. The generally higher ambient temperatures may be further exacerbated by the urban heat island effect and climate change, increasing the likelihood of thermal discomfort and heat stress of humans and other organisms (Grimm et al.

Table 2 Generalised characteristics of Global South urban contexts that typically differentiate them from Global North ones

Domain	Attribute
Biophysical context	Marked climatic severity
	High biodiversity
	Invasive species as a colonial legacy
	Urban livestock
	High incidence of ecosystem disservices and geophysical hazards
Socio-economic context	Rapid rates of urban growth • Urban sprawl • Informality in land use and occupation • Land invasions • Many without access to basic services • High indigenous knowledge related to biodiversity due to rural migrants and reliance on locally obtained provisioning services
	High poverty ratios • Significant contribution of informal sector to livelihoods • High levels of vulnerability to economic, social and environmental stresses • Low tax base and per capita budgets for urban planning and ecosystem management • Large numbers of people reliant on local provisioning services
	Low ratios of skilled people • Limited planning and design capacity • Low or no monitoring of urban environment or urban green infrastructure (UGI) • Inadequate waste and pollution prevention or management
	Diversity of belief systems on human–nature relationships
	Prominence of urban agriculture in private and public green spaces
	Insufficiently capacitated urban governance • Inadequate planning, design and management • Low citizen empowerment and participation

2008b). These all contribute to high levels of vulnerability amongst urban populations in the GS, especially of the urban poor.

5.2 *High Native Biodiversity*

The geographical location of most of the GS in the tropics and subtropics means that on average they boast higher native biodiversity at regional scales than do most countries of the GN. Fifteen of the world's 17 megadiverse countries, and nine of the ten most biodiverse countries in the world are in the GS, although size of a country comes into play in such a rankings. Nonetheless, for equivalent areas, most tropical regions have several times more species than temperate ones (Hillebrand 2004) and thus many GS countries feature significantly higher levels of biodiversity than countries in the GN. The higher native biodiversity is also a reflection of lower extents of historical land transformation, although current rates are higher in many GS countries. For example, Europe was once extensively forested but lost much over the last six millennia with the advance of sedentary agriculture (Roberts et al. 2018), whereas rates of forest loss are now greatest in tropical countries. Given that regional-scale levels of biodiversity explain much of the variation at local level (Gaston 2000), these high levels of biodiversity mean that urban planning in the GS needs to be more cognisant of sensitive ecosystems, species and populations within the urban matrix, and for management authorities to ensure their conservation than is the case in the GN.

5.3 *Colonial Species Introductions and Invasions*

Another contribution to high plant species richness in many GS cities is the legacy of species introductions during the colonial period (Shackleton and Gwedla 2021). Colonists and colonial administrations introduced numerous species from Europe to offer familiar reminders of 'home' (Ignativeva and Stewart 2009; Nitoslawski et al. 2016), along with transfers between colonies for aesthetic and economic benefits (Kemp et al. 2020). Since trees are long-lived organisms colonial patterns remain evident in many GS cities to this day, where the older, colonial neighbourhoods have a different tree flora to newer neighbourhoods (Hunte et al. 2019). Urban ecology of and in GS towns and cities must consider these introductions and how they have influenced the provision and value of ecosystem services or nature's contributions to people. It must also address the challenge that dozens have become invasive, thereby posing threats to native biodiversity and ecosystem services in towns and cities, as well as surrounding landscapes (van Wilgen et al. 2020). This is within the context that awareness of invasive species and the capacity of urban authorities to control them is much lower than in the GN.

5.4 *Effects of Urban Livestock*

The presence of different types of domestic livestock (cattle, goats, sheep, pigs, donkeys, poultry) in public and private spaces is common in many GS regions, such as sub-Saharan Africa and SE Asia. For example, in Dar es Salam, the largest city in Tanzania, over 70% of households have livestock (Halloran and Magid 2013), whilst 30% of households in Bengaluru, India, do (Mundoli et al. 2015). The ownership of livestock in urban areas is linked to a combination of cultural, religious and economic factors. The presence of livestock requires owners to secure sufficient feed and water for their animals and to dispose of or use the dung. Some animals free-range and thus feed on vegetation in formal and informal public green spaces, thereby affecting plant vigour and reproduction, and hence species composition. Large species can cause considerable damage to trees and smaller plants in public spaces (Richardson and Shackleton 2014). Additionally, if in sufficient numbers urban livestock can affect soil nutrients and soil compaction (Shackleton et al. 2017), and hence erosion rates. Thus, urban planners must provide sufficient areas for forage provision and landscapers must consider species that are tolerant of herbivory. If the animals are penned, some owners travel out daily to cut fodder by hand to feed to their animals. This needs to be considered when studying plant productivity and species composition in urban settings of the GS. Additionally, the faeces and urine of penned animals may be washed into local water sources. Where small domestic animals are kept in high densities and in close proximity to humans some health risks may arise (Capua and Alexander 2004).

5.5 High Incidence of Ecosystem Disservices and Geophysical Hazards

Any integrated or systemic framework of urban ecology must encapsulate both the positive and the negative dimensions of the interactions between humans and nature and the outcomes for human wellbeing and urban sustainability. This requires consideration of ecosystem disservices alongside ecosystem services (Lyytimäki and Sipilä 2009; Shackleton et al. 2016). This is especially so in the GS, where the diversity of both services and disservices is likely to be greater than in the GN because of the higher levels of biodiversity (Dunn 2010). Indeed, any cursory appreciation of the range and impacts of ecosystem disservices (such as floods, droughts, water-borne diseases, dangerous wildlife and insects) suggests that they are a great deal more common in the GS than the GN. The impacts are also frequently severe, not just because they are more common but because urban populations are more vulnerable and governance institutions have less capacity to prevent or mitigate the impacts. For example, over 4300 people are killed by landslides every year, with the highest number in Asia (Froude and Petley 2018). For many vulnerable households, disservices such as tropical diseases, floods and landslides can mean the loss of part or all of their homestead and belongings (not to mention loss of life) and access to basic services. Thus, ecological design and management of urban green infrastructure in GS towns and cities is not just about optimising ecosystem services and nature's benefits to people, but it must also be about the minimisation or avoidance of ecosystem disservices.

5.6 *Rapid Rates of Urban Growth*

A crucial dynamic affecting most dimensions of urban planning, sustainability and quality of life in the GS is the current large and rapid rates of urban growth. For example, urban growth in North America is approximately 0.8% p.a. and in Europe it is 0.5% p.a., compared to 4.0% in sub-Saharan Africa and 2.5% in SE Asia (World Bank 2018). This is being fuelled mostly by in-migration of rural populations and internal growth. Not only are cities and towns of the GS expanding, the rate at which it has occurred is a lot higher than ever experienced during the development of GN cities (Table 3). This rate and sheer numbers of new people every year can strain even the best urban plans and governance systems (Cohen 2006).

The sheer pace of urban growth in GS cities has several significant implications for urban development, poverty, sustainability, quality of life and hence urban ecology (Cohen 2006; Zhang 2016; Nagendra et al. 2018). The first is the rate of urban sprawl, which is often also associated with development of informal housing or slums. Increasing numbers of the urban poor in GS towns and cities live in homes or shelters located in marginalised and even risky areas (such as within flood lines, on steep or unstable slopes, adjacent to polluting industrial complexes) in homes made of impermanent or scavenged materials. It was estimated in 2003 that approximately one-third of urban dwellers globally live in slums, and nearly all of these are in GS countries and that the rate of growth of slums far exceeded the rate of urban growth (UN Habitat 2003). The last statistic shows the inability of many GS cities and towns to absorb the current rates of urban growth in terms of providing housing and services. This not only leaves many urban citizens vulnerable to environmental, health and economic risks, but it may also result in heavy pressures on urban green infrastructure and the ecosystem services or contributions that it provides to people. Planned and zoned green spaces are taken over for housing or informal businesses (Mensah et al. 2018; Munyati and Drummond 2020), or they may be heavily used for

Table 3 Multiplication rate of selected GS and GN cities in 50-year intervals between 1850 and 2000

Global South cities	1850–2000	1900–2000	1950–2000	Global North cities	1850–2000	1900–2000	1950–2000
Baghdad	40.0x	26.7x	6.9x	London	2.7x	1.1x	0.9x
Dar es Salam	36.5x	24.3x	20.9x	New York	30.2x	4.7x	1.5x
Mexico City	61.0x	36.8x	6.6x	Paris	1.9x	0.8x	0.8x
New Delhi	45.0x	27.0x	7.9x	Tokyo	18.0x	4.2x	2.0x
Mean rate per 50 years	45.6x	28.7x	10.6x		13.2x	2.7x	1.3x

the collection of resources such as firewood, building timber, wild foods, medicines or fodder for livestock (Schlesinger et al. 2015). Many rural migrants have knowledge of useful species and bring that with them when they move to urban centres. High rates of urban growth also strain the provision of urban basic services such as water reticulation and human and solid waste removal, which if overwhelmed, can result in widespread unsanitary conditions and risks of disease.

5.7 High Poverty Ratios

An equally significant characteristic of GS towns and cities is the high poverty ratio (Ravallion 2002; Khan et al. 2016). This applies irrespective of the definitions and measures of poverty that are used (Samanta 2015). The reasons for the high poverty ratios are widely debated and vary between countries and disciplines. But the existence of high levels of urban poverty have implications for all facets of human existence, quality of life and hence urban ecology. High poverty translates into low tax revenues to city authorities, which constrains their abilities to provide and maintain basic and other services (such as urban green infrastructure). The interaction of limited basic services and poverty can be associated with greater reliance on provisioning ecosystem services harvested in and around urban environments, such as water, firewood and building timber, which has ramifications for the supply, maintenance and ecology of UGI. Poverty around the world is also associated with heightened vulnerability to economic, social and environmental stresses and shocks (Hardoy and Pandiella 2009). Not only are poor households and neighbourhoods more vulnerable, they are also less able to cope with shocks such as floods, disease outbreaks and economic structural adjustment, and are more reliant on government or even international assistance when such hardships strike.

5.8 Low Ratios of Skilled People

High poverty rates are also associated with low educational outcomes (and vice versa), and hence low ratios of skilled people. For example, GN countries have, on average, more than 2000 people employed in Research and Development roles per million of population, compared to less than 100 in many GS countries (UNESCO 2018). Corresponding figures regarding expenditure on R and D, is over 2% of GDP for the GN and less than 0.5% for many GN countries (UNESCO 2018). Similar differentials apply across any professional skills group, such as teachers, engineers and medical personnel.

A low skills ratio compounds the constrained ability to plan and provide basic services. Additionally, it means that urban research and urban planning capacities are often insufficient to design and guide sustainable urban development outcomes (Krishnamurthy et al. 2016), which are also linked to low budgets for research and planning. Therefore, models and processes designed elsewhere in different contexts tend to be imported, often without testing or consideration of the contextual and knowledge differences (Khirfan et al. 2013). It also means that there is no, or only limited, resources for the monitoring of key urban development indicators and UGI provision and condition.

5.9 Diversity of Belief Systems on Human–Nature Relationships

That urban green infrastructure and nature experiences are vital for human well-being is no longer in doubt. However, the range of wellbeing measures and nature experiences examined to date have been portrayed almost exclusively from a Euro-centric perspective. Such a perspective presents nature as separate from humanity; humans can use, manage and enjoy nature, but they are not integral to it, and vice versa. In contrast to this dominant worldview in the academic literature and planning practice, many cultures and societies, mostly in GS regions, do not see nature and humanity as separate entities, but rather each are part and one of the same entity (Shackleton and Cocks 2021). Consequently, for some, their relationship with nature can be deeply embedded and take on spiritual or reverential dimensions. For others, nature is the domain where the ancestors reside, which are perceived to have a strong influence over the daily fortunes or misfortunes of their descendants. For yet others, nature is a store to provide plant and animal resources that have deeply held cultural significance (Cocks and Wiersum 2003). Thus, the type of urban nature required moves beyond just formal parks, and the range of activities and practices allowed in urban nature moves beyond relatively passive use for family outings and exercise to far more multifaceted needs and interactions (Cocks et al. 2016; Cocks and Shackleton 2021). Understanding this diversity of currently less formally recognised relationships with nature is vital to developing an urban ecology that recognises how and why particular groups of people view, use and value nature in the way they do, and the 'types' of nature they require, both of which may be very different to the current dominant research and planning discourses in the GN.

5.10 Prominence of Urban Agriculture in Private and Public Green Spaces

In pursuit of the 'Sanitary city' during most of the twentieth century (Pickett et al. 2013) GN urban regulations and planners sought to exclude agriculture from the urban domain. Agriculture was the very antithesis of the modern urban because it was seen as dirty, smelly and potentially unhygienic. Ironically, there are increasing voices to bring agriculture back into urban spaces for a variety of reasons, such as food sovereignty, improved nutrition and health, climate change mitigation and human-nature interactions. Many GS countries sought to emulate this separation of agriculture away from the urban, in policies and regulations at least, but most have failed to do so to the extent achieved in the GN. Thus, arable and livestock agriculture is generally prominent in most GS cities, or particular parts of them (Zezza and Tasciotti 2010; Orsini et al. 2013). This is a consequence of multiple, interacting factors, including the high proportion of rural migrants, poverty, culture, recreational enjoyment and health or environmental awareness. The widespread engagement in urban agriculture in private and public spaces has significant implications for urban ecology in terms of the land required, inputs used, the agricultural wastes and perhaps pollution generated, and the effects on and trade-offs with other ecosystem services and species. These land and resource use patterns are to be considered against the widespread benefits of decreasing food insecurity, vulnerability and carbon emissions and enhancing social cohesion, wellbeing and supporting people's experiences of nature through growing and herding.

5.11 Insufficiently Capacitated Urban Governance

A significant limitation to urban planning and governance in many parts of the GS is the lack of sufficient capacity in terms of adequate numbers of appropriately skilled personnel (Mycoo 2014; Fuseini and Kemp 2015; Satterthwaite 2017). There are diverse historical and context-specific reasons for this, not least being inadequate investment in urban planning and governance sectors. This is further strained by the high rates of urban growth which undermine the opportunities for proactive planning, and thus much has to be done in a reactive or retrospective manner. This means that the capacity to consider and adopt alternative approaches and policies (such as for climate change, disaster management, poverty alleviation, sustainability) is often limited. This shapes how urban systems, including service provision and UGI, are viewed and managed. This capacity constraint is generally greater in small- and medium-sized towns or secondary cities than large cities or megacities (Satterthwaite 2017). The limited governance capacity in many GS settings also applies to citizen participation and involvement. Authorities rarely have the time, capacity or mechanisms to attempt to meaningfully engage with urban residents around land use planning and service provision, and urban citizenry rarely have the agency to proactively engage urban authorities in this regard (Alam and Lovett 2019)

6 Purpose and Structure of This Book

Given that most of the world's land area, human population and biodiversity are found in the GS it is imperative that the ecology in, of, and for GS cities is better researched, understood and communicated in these regions. We believe that in doing so, GS understandings will be better able to contribute and guide urban sustainability solutions applicable to the contexts, conditions and challenges of the GS, as well as those facing cities and urban areas globally. In that way GS contributions and understandings can contribute to the development of truly universal urban ecology principles and frameworks. We also believe that in highlighting the differences and needs of the GS contexts, that researchers working in the GS will be better able to develop framings and contributions that are more suitable for the GS and be empowered to voice and debate them alongside the currently dominant perspectives from the GN. To this end, the purpose of this book is:

- To review and present coherent, current and scientifically sound information and case studies on the urban ecology of cities and towns in the Global South.
- To consider how, where and why urban ecology in the Global South may differ from or parallel that of the Global North and the implications of that.
- To offer a Global South contribution in a field that is predominantly shaped by and reported from the Global North.
- Through each of the above, advance knowledge of and stimulate scholarship around urban ecology in the Global South.

To achieve this aim, the 17 chapters of the book are divided into four sections, along with a concluding chapter which examines some cross-cutting themes and research needs. The first section of the book, spanning four chapters, sets the scene in terms of the rationale for the book (this chapter), the rates and drivers of urbanisation in the GS (Chap. 2), landscape histories in contributing to current urban form (Chap. 3) and urban social ecology in the GS (Chap. 4).

The second section comprises of three chapters around the patterns and processes of urban green infrastructure and biodiversity. Chapter 5 takes a detailed look at urban green infrastructure in the GS, and is followed by Chaps. 6 and 7, which deal with plant biodiversity and animal biodiversity, respectively.

The third section considers the nature, magnitude and use of urban ecosystem services, or nature's contributions to people (Diaz et al. 2018) in GS cities and towns. Chap. 8 does so for provisioning ecosystem services, whilst Chap. 9 considers regulating services and Chap. 10 presents on cultural ecosystem services. There is also a Chap. 11 on ecosystem disservices and one (Chap. 12) on urban agriculture, as two important processes in GS settings that have significantly greater roles in shaping urban ecology and livelihoods in the GS than is typically observed in the Global North.

The last section brings together four chapters dealing with urban planning and management in GS towns and cities. The first (Chap. 13) delves into understandings of urban resilience and sustainability in the GS, which directly links to Chap. 14 on social and ecological connectivity. These two also have relevance to Chap. 15, on ecological planning and design, and Chap. 16, on urban governance. The last and concluding chapter considers the cross-cutting emergent themes from the preceding chapters.

References

Agergaard J, Tacoli C, Steel G, Ørtenblad SB (2019) Revisiting rural-urban transformations and small town development in sub-Saharan Africa. Eur J Dev Res 31:2–11

Ahern J, Cilliers SS, Niemelä J (2014) The concept of ecosystem services in adaptive urban planning and design: a framework for supporting innovation. Landsc Urb Plan 125:254–259

Alam R, Lovett C (2019) Prospects of public participation in the planning and management of urban green spaces in Lahore: a discourse analysis. Sustain 11:3387

Bartesaghi Koc C, Osmond P, Peters A (2016) Towards a comprehensive green infrastructure typology: a systematic review of approaches, methods and typologies. Urban Ecosys 20:15–35

Beninde J, Veith M, Hochkirch A (2015) Biodiversity in cities needs space: a meta-analysis of factors determining intra-urban biodiversity variation. Ecol Lett 18:581–592

Botzat A, Fischer LK, Kowarik I (2016) Unexploited opportunities in understanding liveable and biodiverse cities: a review on urban biodiversity perception and valuation. Glob Environ Change 39:220–233

Cadenasso ML, Pickett STA (2013) Three tides: the development and state of the art of urban ecological science. In: Pickett STA, Cadenasso ML, McGrath B (eds) Resilience in ecology and urban design: linking theory and practice for sustainable cities. Springer, Dordrecht, pp 29–46

Capua I, Alexander DJ (2004) Human health implications of avian influenza viruses and paramyxoviruses. Eur J Clin Microbiol Infect Dis 23:1–6

Childers DL, Cadenasso ML, Grove JM, Marshall V, McGrath B, Pickett ST (2015) An ecology for cities: a transformational nexus of design and ecology to advance climate change resilience and urban sustainability. Sustain 7:3774–3791

Clarke M (2018) Global South: what does it mean and why use the term? https://onlineacademiccommunity.uvic.ca/globalsouthpolitics/2018/08/08/global-south-what-does-it-mean-and-why-use-the-term/. Accessed Sep 2018

Cocks M, Alexander J, Mogano L, Vetter S (2016) Ways of belonging: meanings of "nature" among Xhosa-speaking township residents in South Africa. J Ethnobiol 36:820–841

Cocks ML, Shackleton CM (2021) Situating biocultural relations in city and townscapes: conclusion and recommendations. In: Cocks ML, Shackleton CM (eds) Diversities of nature in cities: enriching belonging, wellbeing and bioculture. Earthscan and Routledge, London, pp 241–268

Cocks ML, Wiersum F (2003) The significance of plant diversity to rural households in the Eastern Cape province of South Africa. Forest Trees Livelih 13:39–58

Cohen B (2006) Urbanisation in developing countries: current trends, future projections and key challenges for sustainability. Technol Soc 28:63–80

Dados N, Connell R (2012) The Global South. Contexts 11:12–13

de Andrade AC, Medeiros S Chiarello AG (2020) City sloths and marmosets in Atlantic forest fragments with contrasting levels of anthropogenic disturbance. Mammal Res 65: 481–491

De Palma A, Abrahamczyk S, Aizen MA, Albrecht M, Basset Y, Bates A, Blake RJ, Boutin C, Bugter R, Connop S, Cruz-Lopez L, Cunningham SA, Darvill B, Diekotter T, Dorn S, Downing N, Entling MH, Farwig N, Felicioli A, Fonte SJ, Fowler R, Franzen M, Goulson D, Grass I, Hanley ME, Hendrix SD, Herrmann F, Herzog F, Holzschuh A, Jauker B, Kessler M, Knight ME, Kruess A, Lavelle P, Le Feon V, Lentini P, Malone LA, Marshall J, Pachon EM, Mcfrederick QS, Morales CL, Mudri-Stojnic S, Nates-Parra G, Nilsson SG, Ockinger E, Osgathorpe L, Parra HA, Peres CA, Persson AS, Petanidou T, Poveda K, Power EF, Quaranta M, Quintero C, Rader R, Richards MH, Roulston T, Rousseau L, Sadler JP, Samnegard U, Schellhorn NA, Schuepp C, Schweiger O, Smith-Pardo AH, Steffan-Dewenter I, Stout JC, Tonietto RK, Tscharntke T, Tylianakis JM, Verboven HA, Vergara CH, VerhulstJ Westphal C, Yoon HJ, Purvis A (2016) Predicting bee community responses to land-use changes: effects of geographic and taxonomic biases. Sci Rep 6:31153

Diaz S, Pacual U, Srenseke M, Martin-López B, Watson RT, Molnár Z, Hill R, Chan KM, Baste IA, Brauman KA, Polasky S, Church A, Lonsdale M, Larigauderie A, Leadley PW, van Oudenhoven AP, van der Plaat F, Schröter M, Lavorel S, Aumeeruddy-Thomas Y, Bukvareva E, Davies K,

Demissew S, Eroul G, Failler P, Guerra CA, Hewitt CL, Keune H, Lindley S, Shirayama Y (2018) Assessing nature's contributions to people. Sci 359:270–272
Dunn RR (2010) Global mapping of ecosystem disservices: the unspoken reality that nature sometimes kills us. Biotropica 42:555–557
du Toit MJ, Cilliers SS (2011) Aspects influencing the selection of representative urbanisation measures to quantify urban-rural gradients. Landsc Ecol 26:169–181
Elmqvist T, Andersson E, Frantzeskaki N, McPhearson T, Olsson P, Gaffney O, Takeuchi K, Folke C (2019) Sustainability and resilience for transformation in the urban century. Nat Sustain 2:267–273
Erixon H, Borgström S, Andersson E (2013) Challenging dichotomies—exploring resilience as an integrative and operative conceptual framework for large-scale urban green structures. Plan Theory Pract 14:349–372
Filloy J, Zurita GA Bellocq MI (2019) Bird diversity in urban ecosystems: the role of the biome and land use along urbanisation gradients. Ecosys 22:213–227
Forman RTT (2016) Urban ecology principles: are urban ecology and natural area ecology really different? Landsc Ecol 31:1653–1662
Forman RTT (2019) Town ecology: for the land of towns and villages. Landsc Ecol 34:2209–2211
Froude MJ, Petley DN (2018) Global fatal landslide occurrence from 2004 to 2016. Nat Haz Earth Syst Sci 18:2161–2181
Fuseini I, Kemp J (2015) A review of spatial planning in Ghana's socio-economic development trajectory: a sustainable development perspective. Land Use Pol 47:309–320
Gaston KJ (2000) Global patterns in biodiversity. Nat 405:220–227
Grimm N, Grove JM, Pickett STA, Redman C (2000a) Integrated approaches to long-term studies of urban ecological systems. BioScience 50: 571–584
Grimm NB, Faeth SH, Golubiewski NE, Redman CL, Wu J, Bai X, Briggs JM (2008b) Global change and the ecology of cities. Sci 319:756–760
Grimm NB, Foster D, Groffman P, Grove MJ, Hopkinson CS, Nadelhoffer KJ, Pataki DE, Peters DP (2008a) The changing landscape: ecosystem responses to urbanisation and pollution across climatic and social gradients. Front Ecol Environ 6:264–272
Guitart D, Pickering C, Byrne J (2012) Past results and future directions in urban community gardens research. Urban For Urban Green 11:364–373
Halloran A, Magid J (2013) Planning the unplanned: incorporating agriculture as an urban land use into the Dares Salaam master plan and beyond. Environ Urban 25:541–558
Hardoy J, Pandiella G (2009) Urban poverty and vulnerability to climate change in Latin America. Environ Urban 21:203–224
Hillebrand H (2004) On the generality of the latitudinal diversity gradient. Am Nat 163:192–211
Horner R, Hulme D (2019) From international to global development: new geographies of 21st century development. Dev Change 50:347–378
Hunte N, Roopsind A, Ansari AA, Caughlin TT (2019) Colonial history impacts urban tree species distribution in a tropical city. Urban For Urban Green 41:313–322
Ignativeva ME, Stewart GH (2009) Homogeneity of urban biotopes and similarity of landscape design language in former colonial cities. In: McDonnell MJ, Hahs AK, Brueste JH (eds) Ecology of cities and towns: a comparative approach. Cambridge University Press, Cambridge, pp 399–421
Kabisch N, Qureshi S, Haase D (2015) Human-environment interactions in urban green spaces: a systematic review of contemporary issues and prospects for future research. Environ Impact Assess Rev 50:25–34
Kemp ME, Mychajliw AM, Wadman J (2020) 7000 years of turnover: historical contingency and human niche construction shape the Caribbean's Anthropocene biota. Proc Roy Soc B 287:20200447
Khan AU, Saboor A, Ali I, Malik WS, Mahmood K (2016) Urbanisation of multidimensional poverty: evidences from Pakistan. Qual Quant 50:439–469
Khirfan L, Momani B, Jaffer Z (2013) Whose authority? Exporting Canadian urban planning expertise to Jordan and Abu Dhabi. Geoforum 50:1–9

Konijnendijk CC, Ricard RM, Kenney A, Randrup TB (2006) Defining urban forestry—a comparative perspective of North America and Europe. Urban For Urban Green 4:93–103
Krishnamurthy R, Mishra R, Desouza KC (2016) City profile: Pune, India. Cities 53:98–109
Kuras ER, Warren PS, Zinda JA, Aronson MFJ, Cilliers S, Goddard MA, Nilon CH, Winkler R (2020) Urban socioeconomic inequality and biodiversity often converge, but not always: a global meta-analysis. Landsc Urb Plan 198:103799
Lin BB, Egerer MH (2018) Urban agriculture: an opportunity for biodiversity and food provision in urban landscapes. In: Ossola A, Niemela J (eds) Urban biodiversity: from research to practice. Routledge, London, pp 71–86
Lindley S, Pauleit S, Yeshitela K, Cilliers S, Shackleton C (2018) Rethinking urban green infrastructure and ecosystem services from the perspective of sub-Saharan African cities. Landsc Urb Plan 180:328–338
Lyytimäki J, Sipilä M (2009) Hopping on one leg: the challenge of ecosystem disservices for urban green management. Urban For Urban Green 8:309–315
Magle SB, Hunt VM, Vernon M, Crooks KR (2012) Urban wildlife research: past, present, and future. Biol Conserv 155:23–32
Mahler AG (2017) Global South. In: O'Brien E (ed.) Oxford bibliographies in literary and critical theory. https://globalsouthstudies.as.virginia.edu/what-is-global-south. Accessed 14 July 2020
Mahler AG (2018) From the tricontinental to the Global South: race, radicalism, and transnational solidarity. Duke University Press, Durham
Martin LJ, Blossey B, Ellis E (2012) Mapping where ecologists work: biases in the global distribution of terrestrial ecological observations. Front Ecol Environ 10:195–201
McDonnell MJ (2011) The history of urban ecology: an ecologlist's perspective. In: Niemelä J, Breuste JH, Guntenspergen G, McIntyre NE, Elmqvist T, James P (eds) Urban ecology: patterns, processes, and applications. Oxford University Press, New York, pp 5–13
McDonnell MJ, Hahs AK (2008) The use of gradient analysis studies in advancing our understanding of the ecology of urbanising landscapes: current status and future directions. Landsc Ecol 23:1143–1155
McHale MR, Bunn DN, Pickett STA, Twine W (2013) Urban ecology in a developing world: why advanced socioecological theory needs Africa. Front Ecol Environ 11:556–564
McIntyre NE, Knowles-Yanez K, Hope D (2000) Urban ecology as an interdisciplinary field: differences in the use of "urban" between the social and natural sciences. Urban Ecosys 4:5–24
McMichael AJ (2000) The urban environment and health in a world of increasing globalisation: issues for developing countries. Bull World Health Organ 78:1117–1126
McPhearson T, Parnell S, Simon D, Gaffney O, Elmqvist T, Bai X, Roberts D, Revi A (2016a) Scientists must have a say in the future of cities. Nat 538:155–156
McPhearson T, Pickett STA, Grimm N, Niemelä J, Alberti M, Elmqvist T, Weber C, Breuste J, Haase D, Qureshi S (2016b) Advancing urban ecology towards a science of cities. BioScience 66:198–212
Mensah CA, Gough KV, Simon D (2018) Urban green spaces in growing oil cities: the case of Sekondi-Takoradi Metropolis, Ghana. Int Dev Plan Rev 40:371–395
Molebatsi LY, Siebert SJ, Cilliers SS (2013) Alien and indigenous plant species diversity of homegardens of a rural settlement in the eastern Kalahari, South Africa. Scr Bot Belg 50:294–303
Mundoli S, Manjunath B, Nagendra H (2015) Effects of urbanisation on the use of lakes as commons in the peri-urban interface of Bengaluru, India. Int J Urban Sustain Dev 7:89–108
Munyati C, Drummond JH (2020) Loss of urban green spaces in Mafikeng, South Africa. World Dev Perspect 19:100226
Mycoo MA (2014) Autonomous household responses and urban governance capacity building for climate change adaptation: Georgetown, Guyana. Urban Clim 9:134–154
Nagendra H, Bai X, Brondizio ES, Lwasa S (2018) The urban south and the predicament of global sustainability. Nat Sustain 1:341–349
Nitoslawski SA, Duinker PN, Bush PG (2016) A review of drivers of tree diversity in suburban areas: research needs for North American cities. Environ Rev 24:471–483

Orsini F, Kahane R, Nono-Wondim R, Gianquinto G (2013) Urban agriculture in the developing world: a review. Agron Sustain Dev 33:695–720
Pagel H, Ranke K, Hempel F, Köhler J (2014) The use of the concept "Global South" in social science and humanities. University of California, Berkeley, 125(13.9)
Pickett ST, Boone CG, McGrath BP, Cadenasso ML, Childers DL, Ogdan LA, McHale M, Grove JM (2013) Ecological science and the transformation to the sustainable city. Cities 32:S10–S20
Pickett STA, Cadenasso ML (2017) How many principles of urban ecology are there? Landsc Ecol 32:699–705
Pickett STA, Cadenasso ML, Childers DL, McDonnell MJ, Zhou W (2016) Evolution and future of urban ecological science: ecology in, of, and for the city. Ecosys Health Sustain 2(7):01229
Ravallion M (2002) On the urbanisation of poverty. J Dev Econ 68:435–442
Richardson E, Shackleton CM (2014) The extent, causes and local perceptions of street tree damage in small towns in the Eastern Cape, South Africa. Urban For Urban Green 13:425–432
Roberts D (2008) Thinking globally, acting locally—institutionalising climate change at the local government level in Durban, South Africa. Environ Urban 20:521–537
Roberts D (2010) Prioritising climate change adaptation and local level resilience in Durban, South Africa. Environ Urban 22:397–413
Roberts D, Boon R, Diederichs N, Douwes E, Govender N, McInnes A, Mclean C, O'Donoghue S, Spires M (2012) Exploring ecosystem-based adaptation in Durban, South Africa: "learning-by-doing" at the local government coal face. Environ Urban 24:167–195
Roberts N, Fyfe RM, Woodbridge J, Gaillard M-J, Davis BA, Kaplan JO, Marquer L, Mazier F, Nielsen, AB, Sugita S, Trondman A-K, Leydet M (2018) Europe's lost forests: a pollen-based synthesis for the last 11,000 years. Sci Rep 8: art 716
Roy S, Byrne J, Pickering C (2012) A systematic quantitative review of urban tree benefits, costs, and assessment methods across cities in different climatic zones. Urban For Urban Green 11:351–363
Samanta D (2015) Urban poverty in developing Asia—dichotomy between the income and non-income dimensions: are we not grossly underestimating its incidence? In: Heshmati A, Maasoumi E, Wan G (eds) Poverty reduction policies and practices in developing Asia: economic studies in inequality, social exclusion and well-being. Springer, Singapore, pp 79–105
Satterthwaite D (2017) The impact of urban development on risk in sub-Saharan Africa's cities with a focus on small and intermediate urban centres. Int J Disaster Risk Reduction 26:16–23
Schlesinger J, Drescher A, Shackleton CM (2015) Socio-spatial dynamics in the use of wild natural resources: evidence from six rapidly growing medium-sized cities in Africa. Appl Geogr 56:107–115
Shackleton CM (2006) Urban forestry South Africa—a Cinderella science? S Afr For J 208:1–4
Shackleton CM (2012) Is there no urban forestry in the developing world? Sci Res Essays 7:3329–3335
Shackleton CM, Cocks ML (2021) Urban nature and biocultural realities. In: Cocks ML, Shackleton CM (eds) Diversities of nature in cities: enriching belonging, wellbeing and bioculture. Earthscan and Routledge, London, pp 1–24
Shackleton CM, Drescher A, Schlesinger J (2020) Urbanisation reshapes gendered engagement in land-based livelihood activities in mid-sized African towns. World Dev 103:104946
Shackleton CM, Guild J, Bromham B, Impey S, Jarrett M, Ngubane M, Steijl K (2017) How compatible are urban livestock and urban green spaces and trees? An assessment in a medium-sized South African town. Int J Sustain Urban Dev 9:243–252
Shackleton CM, Gwedla N (2021) The legacy effects of colonial and apartheid imprints on urban greening in South Africa: spaces, species and suitability. Front Ecol Evol 8: 579813
Shackleton CM, Ruwanza S, Sinasson Sanni GK, Bennett S, De Lacy P, Modipa R, Mtati N, Sachikonye M, Thondhlana G (2016) Unpacking Pandora's Box: understanding and categorising ecosystem disservices for environmental management and human wellbeing. Ecosys 19:587–600
Sheherazade Y, Pradana DH, Tsang SM (2017) The role of fruit bats in plant community changes in an urban forest in Indonesia. Raffles Bull Zool 65:497–505

Shwartz A, Turbé A, Julliard R, Simon L, Prévot A-C (2014) Outstanding challenges for urban conservation research and action. Glob Environ Change 28:39–49
Solarz MW (2012) North-South, commemorating the first Brandt Report: searching for the contemporary spatial picture of the global rift. Third World Q 33:559–569
Sukopp H (2008) On the early history of urban ecology in Europe. In: Marzluff J, Shulenberger E, Endlicher W, Alberti M, Bradley G, Ryan C, ZumBrunnen C, Simon U (eds) Urban ecology: an international perspective on the interaction between humans and nature. Springer, New York, pp 79–97
Taylor BD, Goldingay RL (2010) Roads and wildlife: impacts, mitigation and implications for wildlife management in Australia. Wildl Res 37:320–331
UN Habitat (2003) The challenge of slums: global report on human settlements 2003. Earthscan, London
UNESCO (2018) Unesco Institute for statistics: R and D release. http://uis.unesco.org/en/news/rd-data-release. Accessed 7 July 2020
van Wilgen BW, Measey J, Richardson DM, Wilson JR, Zengeya TA (eds) (2020) Biological invasions in South Africa. Springer, Heidelberg
von Döhren P, Haase D (2015) Ecosystem disservices research: a review of the state of the art with a focus on cities. Ecol Ind 52:490–497
Wang J, Zhou W, Wang J, Yu W (2020) Spatial distribution of urban greenspace in response to urban development from a multi-scale perspective. Environ Res Lett 15:064031
Weiland U, Richter M (2011) Urban ecology—brief history and present challenges. In: Richter M, Weiland U (eds) Applied urban ecology: a global framework. Wiley Blackwell, Chichester, pp 1–11
Wilkinson C, Sendstad M, Parnell S, Schewenius M (2013) Urban governance of biodiversity and ecosystem services. In: Elmqvist T, Fragkias M, Goodness J, Güneralp B, Marcotullio PJ, Mcdonald RI, Parnell S, Schewenius M, Sendstad M, Seto KC, Wilkinson C (eds) Urbanisation, biodiversity and ecosystem services: challenges and opportunities. Springer, Heidelberg, pp 539–588
World Bank (2018) Urban population growth. https://data.worldbank.org/indicator/SP.URB.GROW. Accessed 6 July 2020
Wu J (2008) Making the case for landscape ecology: an effective approach to urban sustainability. Landsc J 27:41–50
Wu J (2013) Key concepts and research topics in landscape ecology revisited: 30 years after the Allerton Park workshop. Landsc Ecol 28:1–11
Wu J (2014) Urban ecology and sustainability: the state-of-the-science and future directions. Landscape Urban Plann 125:209–221
Wu J, He C, Huang G, Yu D (2013) Urban landscape ecology: past, present, and future. In: Fu B, Jones KB (eds) Landscape ecology for sustainable environment and culture. Springer, Dordrecht, pp 37–53
Wu J, Xiang W-N, Zhao J (2014) Urban ecology in China: historical developments and future directions. Landsc Urb Plan 125:222–233
Zezza A, Tasciotti L (2010) Urban agriculture, poverty and food security: empirical evidence from a sample of developing countries. Food Pol 35:265–273
Zhang XQ (2016) The trends, promises and challenges of urbanisation in the world. Habitat Int 54:241–252

Urbanisation in the Global South

Garth Myers

Abstract This chapter argues that the urban world is increasingly a 'southern' urban world. Over the last half-century or more, urbanisation has proceeded far more rapidly in the Global South than the Global North. Analysts of Global South urbanisation critique the frequent use of analytical toolkits developed in the Global North to explain urban processes in the Global South, and therefore this chapter seeks to build from 'southern' understandings of Global South urbanisation. While great diversity exists across the Global South, one can nevertheless identify some common themes. This chapter highlights and discusses variations upon five of these themes across the urban Global South, bringing in examples from Asia, Africa and Latin America: legacies of colonialism and imperialism; rapid processes of urbanisation; sprawling urban form; informal economy and settlement; and variability in governance and service delivery. The chapter's final main segment analyses the implications of the general themes for urban ecologies in the Global South.

Keywords Colonialism · Globalisation · Governance · Industrialisation · Informality · Urban sprawl

1 Introduction

Urbanisation is the multifaceted process creating and adding to towns and cities and the metropolitan regions around them. The study of urbanisation is staggering in breadth and complexity. One broad geographical fact seems evident across this breadth and complexity: over the last half-century or more, the urbanisation of the Global South has proceeded far more rapidly than that of the Global North. Average annual rates of change in the urban population across the world since 1950 bear this out: the rates for 'less developed countries' (often but admittedly not always synonymous with the Global South) were already twice those of 'more developed countries' in the 1950s, but they have been consistently four times as large since 1980. At the

G. Myers (✉)
Centre for Urban and Global Studies, Trinity College, Hartford, CT, USA
e-mail: Garth.Myers@trincoll.edu

C. M. Shackleton et al. (eds.), *Urban Ecology in the Global South*, Cities and Nature,
https://doi.org/10.1007/978-3-030-67650-6_2

most extreme regional ends of these data, Sub-Saharan African urban populations grew by 3.98% annually from 2015 to 2020, while Europe's urban populations only expanded by 0.32% (United Nations 2018). One sees the direct consequences of this global transformation clearly in the rise to prominence (in population rankings) of major urban agglomerations in Asia, Africa and Latin America. By 2020, not only are eight of the 10 largest urban agglomerations in the world located in the Global South, but over 70 of the 100 largest are (United Nations 2018). Indirect consequences are evident in economic rankings. For example, only one Global South city was among the 17 cities which earned an Alpha or Alpha Plus (i.e. the highest ranking) rating of significance to global finance and business management from the Globalisation-and-World-Cities research network in 2000 (Sao Paulo). By 2018, 14 Global South cities were among the 33 cities in this top tier of cities (GaWC 2018).

Thus there is little doubt that the urban world is increasingly a 'southern' urban world (Fox and Goodfellow 2016). How to study or comprehend this shift, on the other hand, is hotly debated. In the last few decades in urban studies, a considerable critique and a vast opening arose, especially out of what is often termed southern urban theory, challenging universal understandings emanating from European and North American cities. This literature has called into question the meaning of cities and the roles, functions and shapes of urban areas, when accepted understandings for these derive from Euro-American contexts and all other cities are measured against them. The literature around planetary urbanisation that sprang from the French urbanist Lefebvre's (1970: 113) hypothesis of 'the planetary nature of the urban phenomenon', for example, has captured the imagination of many scholars (Brenner and Schmid 2014, 2015; Soja and Kanai 2007), along with the relational statistical approach of the globalisation-and-world-cities research network (Derudder et al. 2012). But analysts of urbanisation in the Global South are troubled by the often too-easy transposition of analytical toolkits developed in the Global North, like Lefebvre's or the GaWC's, in order to explain urban processes in the Global South. While accepting that there is some value in understanding urbanisation on a global or planetary scale (Spencer 2015), and that there are merits to empirical comparisons across and between the Global South and Global North (Angel 2012), in this chapter I seek to avoid turning any strictly Northern conceptual lenses onto Southern urbanisation.

One reason for doing so lies in what I view as the first principle that one must adhere to in analysing urbanisation across the Global South, namely respect for diversity. The Global South is an admittedly somewhat fungible delineation originating in the so-called Brandt Line of the 1980s separating out richer and poorer countries on the world map (Fox and Goodfellow 2016). The resultant map was already and remains problematic for many reasons, not least of which is the geographical error of Australia, as a richer nation, being placed in the Global North while China, as a then-poorer nation far to the north of Australia, residing with the Global South. Lumping hundreds of countries ranging in physical size or population or global-historical significance from China or India to Saint Kitts and Nevis or Burundi is ridiculous on the face of it. Simone (2019: 12–13) even calls the Global South a kind of 'science fiction' given its diversity.

When one sifts through the broad data claims in the first paragraph above, for instance, one sees that 49 of the world's 100-largest urban agglomerations are in Asian countries conventionally assigned to the Global South (and 26 are in China alone, when including Taipei and Hong Kong), while only 11 are in Latin America and 10 in Sub-Saharan Africa. Rates of change in the urban population vary incredibly within the Global South, by country and by region. Urban Sub-Saharan Africa is growing at twice the rate of urban South or Southeast Asia, and four times the rate of South America (United Nations 2018). This sort of differentiation is as striking in economic rankings: only one African city (Johannesburg, South Africa) earned an Alpha rating (and even then, an Alpha minus rating) from the GaWC in 2018, as opposed to six Chinese cities and ten other Global South cities in Asia (GaWC 2018).

When one reviews 'the global urban condition with a southern sensibility', however, one does see beyond the massive diversity of the designation (Parnell and Oldfield 2014: 3). If one sticks to the old map of the Global South and thereby includes China, South Korea, India, Brazil and other large middle-income economies, then the prominence of the Global South in twenty-first-century urbanisation is unmistakable, and the 'sensibility' of *what comprises urbanisation now on the planet* emerges from the developing countries of Asia, Africa and Latin America much more than from the minority trends of slow growth or even shrinkage for cities across the Global North. Southern urbanisation is full of vitality and vibrancy, shock and disorder, spontaneity and fluidity. It is also rife with pains from histories of injustices and inequalities, racism and state incapacities, grand dreams and colossal failings (Holston 2008). But there is an imperative for balancing any shared sense of things with the divergences and specific, distinct trajectories of becoming urban. As Sparke (2007: 117) put it, 'the global South is everywhere, but it is also always somewhere, and that somewhere, located at the intersection of entangled political geographies of dispossession and repossession, has to be mapped with persistent geographical sensibility'. Ultimately, there are enough comparable thematic elements to warrant consideration of the Global South as an analytical category in the assessment of urbanisation, as long as one remembers to account for and respect differences across the South.

In the next segment of the chapter, I highlight and discuss variations upon five of the themes for comparability across the urban Global South: legacies of colonialism and imperialism; rapid processes of urbanisation; sprawling urban form; informal economy and settlement; and variability in governance and service delivery. The chapter's final main segment analyses the implications of the general themes for urban ecologies in the Global South.

2 General Settings and Dynamics

Before I examine the five broad themes, there is still a need to 'locate' the Global South, because Global North-dominated urban studies has, until quite recently, left so many elements and places of southern urbanism 'off the map' (Robinson 2002) and largely disconnected from the prevailing story of global urban processes. In the Global North, Robinson (2006) suggested, the assumptions about urbanisation often rest on the connectivity of cities with what it means to be modern; Global South cities are assumed instead to be in the process of *development*. Most discussions of the development of global or world cities in academic urban studies concentrate on links, networks and intertwined economies of the world's major Global North metropolitan centres; non-economic dynamics and spaces of the Global South (beyond a few high-profile megacities) are shortchanged or seen to be effaced within a universal, northern-dominated urban world. Smaller secondary cities and towns in the Global South (which are frequently growing more rapidly than the megacities) gain far less attention, and the broader realities of this globalism and intersection too often fall away (Bunnell 2016, 2017). There is a steadily growing list of exceptions to these general claims, where scholars emphasise relationality of secondary, southern urban areas (e.g. Kanna and Chen 2012; Harrison and Hoyler 2018; Quayson 2014; Schmid et al. 2018; Söderström 2014; Spencer 2015). But given the predominance of Global South urbanisation in the contemporary reality of urbanisation on earth, it is invaluable to attempt to start the discussion from 'southern cities', as 'those that [now] create global connections' (Roy 2014: 17), in terms of what Simone (2001: 17) called 'worlding from below' (see also Ong 2011).

In this way, many recent works of postcolonial and southern urban studies contest the universalising and reductive tendencies of Global North urban theory (e.g. Chattopadhyay 2012; Escobar 2018; Jazeel 2018, 2019; Lawhon and Truelove 2020; Leitner and Sheppard 2016; Mignolo 2002, 2012; Mignolo and Walsh 2018; Parnell and Oldfield 2014; Robinson 2002, 2003, 2006, 2011a, b, 2016; Roy 2011a, b, 2014, 2016; Roy and Ong 2011; Sheppard et al. 2013; Simone 2001, 2016, 2019; Simone and Pieterse 2017). Patel (2014: 45) even argued for the enduring thread of Eurocentrism even in the radical works of Castells (1996), Sassen (1991) and Harvey (1996) that have been so fundamental to global urban studies for the last half-century, and the continuing 'need to deconstruct the provincialism of European universalisms'—while reconstructing the content, concepts and forms of knowledge. Patel (2014) contends that one need not throw out everything that originates in the Global North as 'wrong', but instead situate northern conceptualisations as providing 'only partial and often… flawed understandings'.

Brenner and Schmid (2015: 166) offered seven theses on planetary urbanisation, the most relevant of which here is thesis three: that 'urbanisation involves three mutually constitutive moments', which they called concentrated, extended and differential urbanisation. They argued that in this era, 'the city'—concentrated urbanisation—was no longer the only or the central concern, but one of three processes of urbanisation taking place. Extended urbanisation and differential urbanisation were occurring in under-examined processes across the planet (Murray 2017: 109). Through theses 4–7 that emphasised the 'multidimensional', 'planetary', 'variegated' and 'contested' character of these three processes of urbanisation, respectively, they did invite exploration of the diversity of processes in both Global North and South urban regions.

However, the processes Brenner and Schmid (2015) identified do not seem to be the same in the Global South (Jazeel 2018). As Schmid (2016: 30, 33) acknowledged, the 'classic model of urbanity… has long been overtaken by worldwide urbanisation processes', and 'we are living in a completely different urban world' now. Brenner and Schmid (2015: 160) worked to address that completely different urban world while engaging with postcolonial urban theory, which they saw as highlighting 'the urgency of elaborating alternative categories for understanding the contextually specific patterns and pathways of urbanisation' in Global South contexts. They emphasised the 'equally urgent task of deciphering' the way 'contemporary forms of neoliberal capitalist urbanisation are unfolding across the North/South divide' (Brenner and Schmid 2015: 162). There is nonetheless a potential danger of erasure of 'differences among cities', a tendency to still locate 'the essence of urbanity in the Global North', and the sense that in southern urbanism the sort of 'political economy' critique of neoliberalism upon which this northern work often rests does not provide 'the overriding context within which urban processes unfold' in the Global South (Schindler 2017: 47; Parnell and Robinson 2012).

Moreover, it is clear, as Roy (2014: 13) has written, that 'the urban future' resides 'in the cities of the global south'. Mabin (2014: 24) pointed to the 'overwhelming need for profound and substantial research on what is going on' in southern cities, and on 'attempts at bringing *together* cities across the world'. Even critics of southern urban studies like Storper and Scott (2016: 1121) agree that 'urban theory must now range over the entire world for its sources of data and evidence while remaining fully open to new conceptual insights generated out of the experiences of the cities of the Global South' (see also Scott and Storper 2015). There is still much ranging across southern thinking needed, and openness to its insights for developing global urban studies. When we range around, there are many distinct themes which rise to the surface. In the next segments in succession, I examine five of these—by no means the only themes possible—beginning with the conditions of postcoloniality.

2.1 Historical Impacts of Colonialism and Imperialism

European (and in some cases American or Japanese) colonialism and imperialism in the seventeenth, eighteenth, nineteenth and twentieth centuries had important influences on the shape of urbanisation in Central and South America, the Caribbean, Africa, the Middle East and Southwest Asia, South and Southeast Asia, the Pacific and, to some extent, East Asia (King 1976). Even China, which was never fully colonised, experienced concessionary colonialism in dozens of cities, most famously Hong Kong, Macau, Qingdao, Tianjin, Guangzhou and Shanghai; several 'instant cities' of the contemporary Pearl River Delta, Dongguan and Shenzhen, comprised the landscape and still bear the scars of the Opium War, as the territory between British colonial nodes in Hong Kong and Guangzhou (Al 2014). Even with the generally earlier date of independence from European colonialism for many Latin American territories in comparison to most European-held colonies in Asia or Africa, informal

European imperialism continued to play a prominent role in urban processes—with, for example, the British in Argentina, the French in Mexico, or the US in many countries in the region—into the twentieth century.

The implications of colonial or imperial legacies vary greatly across the Global South (Jacobs 1996; Myers 2003). Yet there are at least three major variations on the same implications. First, the rise of European and then American power across the world generally reoriented urbanisation towards coasts and port cities. In Asia, Africa and Latin America prior to 1500, many of the most important cities politically and economically were in the interiors of the landmasses (McGee 1971). By 2000, coastal or riverine ports near coasts accounted for a considerable number of the major cities of the Global South. Of countries in Africa with a coastline, for instance, in 26 of 35 cases, the largest city is a port; almost every one of these was also the capital city of a colony at one time. Some 32 of the 77 Global South cities among the world's 100-largest urban areas as of 2020 are on or very near the coast.

Second, colonialism and imperialism impacted the urban functions and urban form, to make cities readable in specific ways (Mitchell 1988). In terms of functions, obviously, for port cities that meant cities were predominantly oriented around the extraction of goods for the metropole (Home 1997). In many colonies, industrialisation was stunted, often deliberately. Morphologies were not only shaped to accommodate this extractive role, but also warped by colonial segregationist mentality. Port cities and other urban areas often grew to have separate central business districts by race, or at the very least neighbourhoods and districts defined by racial or ethnic categories (McGee 1971; King 1976). Building codes often used sanitation, health or density as codes for legal enforcement of segregation (Myers 2003).

Third, colonialism shaped the national urban hierarchies in similar ways across many Global South contexts. Colonial rule relied upon spatial control across territories manifested in regional capitals and district headquarters. Colonial urban hierarchies often exhibited a high degree of primacy, where the colonial capital city held an extreme degree of importance and dwarfed other cities in size. The degree to which urban hierarchies were territorially extensive, or where cities of at least modest size came to exist throughout a colony, depended, frankly, on the geography of the resource base. In some extreme cases, such as the former French colony of Mauritania in Northwest Africa, the only railway line the colonists built literally hugged the colony's border, even around a geometric right angle, to ship iron ore from a mine to the colony's only port. In other words—there was no urban hierarchy, just an extractive space economy (Ould-Mey 1996). In larger or more resource-rich colonies, transport networks were extensive, as were investments throughout the hierarchy. Of course, the extent to which colonial regimes re-engineered hierarchies also depended on what sort of level of urbanisation and urban hierarchy existed before colonial rule. Perhaps most famously, British colonial rule linked the already geographically extensive and densely populated urban hierarchy of India through a railway system that became a conduit for its rule over the massive subcontinent, even as it created land to connect the islands that, once joined, became Mumbai, and manufactured what was intended as their largest imperial symbolic statement in the creation of New Delhi as a new colonial capital (replacing Kolkata) in the early twentieth century (Metcalf 1989, 2008).

Unquestionably, every colony varied from the next, as every region's colonial or imperial timeline had distinct beginning and end points. Ferguson (2006) made use of the French colonial cartographic distinction between 'useful' and 'useless' Africa for developing an understanding of how globalisation came to 'hop' onto the continent in some places and ignore vast stretches devoid of resources or geopolitical significance. One might make a similar distinction across the Global South over the 450 years or so of European or Euro-American colonial/imperial domination of the world. Extensive urban hierarchies with dense infrastructures of interconnection going farther back in time existed where colonial and imperial interests and Euro-American settlements were most extensive; where these interests and settlements were least extensive, in many cases to this day urban hierarchies are thinner, as are the infrastructures linking them (Cooper 1987; Fredericks 2018). In much of Sub-Saharan Africa, for instance, colonialism's urban investments in infrastructure barely compare to those colonial regimes made in cities like Mumbai, Kolkata, Hanoi or Sao Paulo (Anand 2017; Gandy 2014; McFarlane 2011). Water, electricity, solid waste management, roads, subways, rails, public transport, health care, schools, airports, land registration, political organisations—it really doesn't matter which way we think of infrastructure: where it was developed at all in colonial urban Africa outside of the white settlement urban enclaves of South Africa, it served the interests of an incredibly small population of colonial administrators, white settlers or elites.

2.2 *Rapid Urbanisation in the Late Twentieth and Early Twenty-First Centuries*

With the broad exception of many settings in Latin America, Global South countries have been at the forefront of demographic urbanisation for at least sixty years. While the rates of change have dropped across the globe, they remain higher for most Global South countries. To some extent, this is just a matter of the world's least urbanised places having the most rapidly growing urban populations—and they have a long way to go to 'catch up' to other settings. For example, Africa's least urbanised country, Burundi (with an annual rate of change in its urban population of 5.68%), was second (and barely so) on the continent to similarly still largely rural Uganda (5.70%) from 2015 to 2020 (United Nations 2018). Africa's most urbanised countries, by contrast, such as South Africa, Egypt, Tunisia or Libya, have among the lowest rates of urban change in the region. In South and Southwest Asia, countries with historically lower urbanisation rates (Afghanistan at 3.37% and Bangladesh at 3.17%) far outpace countries that already had high proportions of their populations in urban areas (such as Sri Lanka at 0.85% or Iran at 1.71%) in rates of urban growth (United Nations 2018). South America is generally considered the Global South region to have had the earliest demographic transition to a majority, urban society. Consequently, its rates of growth for urban populations in 2015–2020 are much lower—from the low of already largely urbanised Uruguay at 0.46% to the high (at least among the continent's independent territories) of comparatively rural Bolivia at 1.97% (United Nations 2018). Nevertheless, the rates of urbanisation have been, and in many Global South countries continue to be, higher than was ever experienced in the Global North even at the peak of its urbanisation surge (Shackleton et al. 2021).

Generally and conventionally in urban studies, it is assumed that urbanisation and economic growth, especially when connected to industrialisation, go together (Fox and Goodfellow 2016). Many Global South cases bear this out, most obviously the cities of China. China's staggering rates of economic growth since the opening up of its economy in 1980 have accompanied staggering rates of urbanisation (Chen 2014). From a growth rate of the urban population of under 2% in the 1970s, China's urbanising trend zoomed above 4% through all of the 1980s, 1990s and 2000s, only slowing again to 2.42% in the most recent five years (2015–2020). Other rapidly industrialising countries have similarly led the way in urbanisation. Botswana, among the poorest and least industrialised countries in the world at independence in the 1960s, even lacked a capital city (the British ran their former colony from an office building in South Africa). From 1965 to 1990, as its diamond industry and other industrial developments took off, Botswana urbanised at more than 10% per year, for 25 years in a row. At the other extreme, countries with low rates of economic growth or industrialisation are often assumed to have correspondingly slow rates of urbanisation, and there are examples in the Global South that support this assumption. Zimbabwe has had one of the weakest economies in Africa for most of the last 20 years, and, sure enough, one of Africa's slowest growth rates for urban population. Central America's lowest rate of growth for the urban population from 2015 to 2020 belonged to the stagnating economy of Nicaragua, at 1.45% (United Nations 2018).

Yet there are also contradictions and puzzles across the southern regions. And, indeed, Sub-Saharan Africa is often highlighted for presenting the main puzzle: numerous countries with very high rates of urbanisation which lack notable rates of economic growth or industrialisation. Urbanisation without growth is evident in many eastern, central and western African countries, where high birth rates and rates of natural increase explain far more of the growth in urban population than a booming economy (which is often quite absent). A number of states in Western, South and Southeast Asia also exhibit the trend of low economic growth with much more rapid urban population growth, such as Afghanistan or Bangladesh. It is in these latter settings—countries with weak economic growth and rapid urbanisation - where the third of this chapter's broad themes for Global South urbanisation is perhaps most obvious as a result, that of sprawling, unplanned urban landscapes.

2.3 Unplanned Sprawl

The physical imprint of urbanism in the Global South tends to be dramatic. The Southeast Asian concept of an extended metropolitan region finds a place on the maps of Sub-Saharan Africa, South Asia and Latin America. The pace of urban growth is often beyond the capacity of states or regulated markets to provide for. This leads to the expanse of the urbanised extents around the edges of cities. Where the Chicago School of urban studies nearly 100 years ago developed a land-use map of concentric zones around a central business district where the farthest circles were ever more elite suburban districts, in many Global South urban regions, the peripheries are often much poorer than the closer-in settlement areas, or at the very least demonstrate stunning juxtapositions of what northerners would call a suburb with slums and squatter settlements.

The Pearl River Delta urban mega-region in southern China shows one pattern for this sprawl, while an examination of Dar es Salaam, Tanzania or Bogota, Colombia, would indicate other variations. For example, the 'instant' city of Shenzhen grew from a scattered zone of more than 200 agricultural and fishing villages with a combined population of around 300,000 people in the late 1970s to a sprawling metropolis of 20 million or so people over scarcely a forty-year period (Altrock and Schoon 2014). Although its official population (12 million) makes it the 26th-largest city in the world, its unofficial or 'floating' population is much larger. Chinese scholars use the terms, *chengzhongcun, chengwaicun* and *chengbiancun* for the villages that became surrounded by city, those straddling the city edge and those in the de facto suburbs, but in all cases today, in English, scholars speak of them as 'urban villages' (De Muelder and Shannon 2014; Wu et al. 2012). Shenzhen's booming economy had such massive demands for labour and thus housing that neither the local state nor the increasingly powerful private sector could provide residential accommodations. The urban villages served as cost-effective homes—literally and figuratively—for the sprawl. Shenzhen's urban planners estimate that more than 80% of the urban area's residents live in urban villages as of 2020, scattered across the entire urban and suburban region (Myers 2020).

Dar es Salaam has grown at a slower but, in the global context, comparable rate to Shenzhen. From a city of a bit more than 150,000 people at independence in 1961, Dar has grown to a metropolis with estimated 6.7 million people by 2020. Unlike Shenzhen, Dar es Salaam's growth is not expected to slow in the near future—the United Nations (2018) estimates that Shenzhen will add three million people by 2035 (a 25% expansion of its official population), while Dar will add nearly six million more residents by then, almost doubling its size in the next fifteen years. And, unlike Shenzhen, neither Dar es Salaam's service sector nor manufacturing sector has been expanding at a rate anywhere near to its population growth. This means that the city's overwhelming majority are poor, and most can only afford to live in the urban region's margins, which lie at ever greater geographical distance from the central business district of the colonial era. A third significant difference lies with the housing stock and residential built environment: where Shenzhen, even in most of its 'urban villages', has very high density in medium-to-high-rise apartment buildings, most of Dar's sprawl consists of single-story and often single-family housing units. But interestingly, as in Shenzhen, in Dar the overwhelming majority, more than three-quarters of the population, resides outside of the formally planned residential areas.

As is typical of many South American urbanisation narratives, Bogota was already a large city of some 600,000 people by 1950, and by 2020 its population is estimated to be nearly 11 million (United Nations 2018). It has therefore grown less rapidly than either Shenzhen or Dar es Salaam—but it is still almost 20 times the size that it was 70 years ago. As in the other two cities, urban growth has outpaced economic growth and the capacity of either the state or the private sector to house the new residents. The urbanised extent (in terms of land use) of Bogota expanded by 30 times in the last fifty years (Planning Secretariat 2013). Most of that extension to the urban fabric occurred around the edges of the city, and in marginal, steeply sloped environments. Large and densely populated sectors of the urban fringe are highly vulnerable to landslides and floods. Yet the city leaders have claimed a place for Bogota as a major model for rethinking urban planning around the world, oriented around tapping into its vibrant culture of innovation (Berney 2019; Planning Secretariat 2013).

The sprawling urban form common among many cities in the Global South is often taken as a negative factor, reinforcing the stereotypes of urban crises. On the other hand, southern urban theorists challenge these stereotypes. Inventive and innovative urban styles and patterns often emerge from these marginalised areas, giving rise to what is variously referred to as Afropolitan development or favela chic or DIY urbanism (Mbembe and Nuttall 2008; Caldeira 2017). If the future of urbanisation is represented in the sprawling metroplexes of the Global South, then powerful creative forces are at work, despite conditions of deprivation and poverty. Exploration of this approach goes together with the next theme of the chapter, the significance of informality.

2.4 *(I)n(f)ormality*

Informal economies and settlements are common across the cities of the Global South—so common, in fact, as to cause me to place the I and F in brackets, to suggest the *normality* at the heart of informality (Myers 2011). In many cities, the informal sector may employ the majority of the workforce, as informal settlements house the majority of the residents. In fact, though, there is great variation across the Global South in how informality is defined and how it relates with formality, in both institutional and built-environment landscapes (Harris 2017).

The phrase, the informal sector, came to prominence in the 1970s, with its origin often credited to the work of Hart (1973) in Ghana on Accra's 'informal economy' and the International Labor Organisation's report on research in Nairobi (ILO 1972). These projects and many since have identified the informal sector as an *autonomous, unregulated, often illegal, small-scale, low technology arena* for jobs many people would use as a stopgap en route into formal sector employment—with the latter meaning *registered, regulated, legal, waged, and often larger-scale, higher-technology, legal work* (ILO 1972:6). Informal-sector activities were marked as including 'ease of entry, reliance on indigenous resources, family ownership of enterprises, small scale of operation, labor-intensive and adapted technology, skills acquired outside the formal school system, and unregulated and competitive markets' (ILO 1972: 6). The distinctions found physical manifestations in informal-settlement structures.

Much of the debate and discussion on urban informality shifted to Latin American urban studies in the ensuing decades (AlSayyad and Roy 2004; AlSayyad 2004; Santos 1979; Portes et al. 1989). Decades of argumentation have shown that informality is a complicated term. Many contemporary scholars argue that it is time to find a new language for this theme. Recent writing on the informal sector in urban Africa for example has argued for 'reconsidering' informality, beyond small business or employment (Hansen and Vaa 2004; Lindell 2010). It is still generally agreed there is a strong trend towards *informalisation* in many Global South cities, meaning an overall widening of informal activity (Harrison et al. 2008; Grant 2009; Konings et al. 2006; AlSayyad 2004, Roy 2005; Harris 2017). The 'new waves of informalisation' in the economy typically rely on 'forms of work beyond the purview of state regulation or lacking legal protection' and on 'casualisation and increased precariousness of work' (Lindell 2010: 4). These new waves are observable in everyday social life in the apparently rising importance of unregistered social networks in the built environment, livelihood strategies, social reproduction, cultural organisation or political mobilisation.

In the last few decades, the drive to formalise informal settlements has become almost a religious movement in development circles (Manji 2006). Because areas of cities dominated by informal arrangements are typically marginalised and poor, it is hardly surprising that planners and urban professionals still abound who are dedicated to the eradication of informality as a way to eliminate poverty. Sometimes eradication is rhetorical, involving adroit re-categorisation or a mantra of formalisation. At other times, the push is quite tangible, as in the bulldozing of informal settlements (Harrison et al. 2008; Kombe and Kreibich 2000).

However, because informal city life is by its very nature 'unregistered, unmonitored' and unruly, there is a 'severe lack of statistical data' regarding its size, form, function or other characteristics, whether one's aim is to champion it, formalise it or bulldoze it out of existence (Bryceson 2006: 9–10). Thus, as difficult as it is to define informality or informal settlements, it is, if anything, more difficult to assess the scope of either. The UN's definition of a slum, for example, does sound something like a typical characterisation of an informal settlement: 'a contiguous settlement where the inhabitants are characterised as having inadequate housing and basic services; a slum is often not recognised and addressed by public authorities as an integral part of the city'. And in estimating slum conditions, UN-Habitat (2010) takes five criteria of household life into account, measuring as a 'deprivation' any lack of: *access to improved water supply; access to improved sanitation; security of tenure; durability of housing; and sufficient living area.* These deprivations are often characteristics of informal settlements. But the variations on these criteria, even in one neighbourhood of one city in the Global South, let alone across southern realms, or between the main regions of the Global South, mean that it is very difficult to make claims for the settlement and housing dynamics of southern urbanisms as being dominated by 'slums' or 'informal settlements' (Myers 2011).

The inexactitude of definition and paucity or inconsistency of data, as well as the apparent pervasiveness of informality within formal ways of doing things in many southern cities, gave rise to thinking that entirely rejects the language shaping the discussion (Mbembe and Nuttall 2008; AlSayyad 2004). The degree to which the city is angled towards a formal vision in housing or land use or towards an informal one varies with each city. The outcomes vary in terms of how formal and informal visions 'work together and how this working together' produces space and the space economy of the city. Where there is more clashing than 'working together', some ascribe this to 'conflicting rationalities' or planning mindsets (Watson 2007). Formal planning, 'grounded in the rationality of Western modernity and development', holds to one notion of 'proper' communities, while the marginalised majority in informal settlements work with a different notion of what a city should be, based on their attempts 'to survive, materially and culturally, in… alien places' (Watson 2007: 69). Some scholars don't see these conflicting rationalities as 'insurmountable', stressing that the capacity of the urban poor to engage with formal planning processes 'is usually limited by the material resources at their disposal rather than "cultural" factors' (Robins 2006: 99).

Perhaps the heart of the concerns about what informality means or what to do about informal settlements centres on land issues, literally the foundation of housing issues as well. The wide influence of Peruvian economist Hernando de Soto's book, *The Mystery of Capital* (2000), and programmes aimed at land titling for the urban poor (Payne 2002) are pertinent here. The key argument of de Soto and others is that formalised and secure property rights reduce poverty. Neoliberal development strategists have argued that when property rights are poorly or vaguely defined, properties are used wastefully, with high transaction costs, and when they are held in common, they are overexploited (Alston et al. 1996). The urban and peri-urban poor's best assets are often their properties, the reasoning goes. Therefore, securing their individual control over them allows the poor, cast as 'heroic entrepreneurs', to gain the greatest value from them (Alston et al. 1999; Manji 2006). The increased market activity created by lower transaction costs and more secure property rights, if they are achieved through 'facilitative institutional arrangements', will, it is claimed, eventually translate into greater economic activity and greater income for the poor (Mooya and Cloete 2007:155). Therefore, the key to bringing to life the 'dead capital' of the informal land holdings of the urban poor and raising them from the depths of poverty is the formalisation of property ownership, with government acting merely as a helpmate in land management (de Soto 2000: 29; Myers 2008, 2011; Carmody and Owusu 2016).

Within informal systems, increasing numbers of urban residents struggle to access land for building houses (Konings et al. 2006: 6). Ordinary residents have very different capacities for exercising informal rights. In many cities, significant tensions arise (or persist) between indigenous residents and migrants, or between classes or genders or races (Lindell 2010: 16; Konings et al. 2006; Grant 2009). Some efforts are made to reach across these divides in organising or deploying informal systems in the interests of the poor majority, but these are still rare.

One emerges from reading the now more than four decades-long debates on informality with the inescapable thought that easy, one-size-fits-all analyses are inadequate. We might need new terms for the apparently fading dichotomy between formal and informal, given how interwoven they are. Yet these terms are still so much part of shaping the discourse that it is hard to throw them away. Global South urbanisation is inseparable from an engagement with informality, especially regarding employment, housing and everyday life, as convoluted and multifaceted as informality may be.

One spinoff from the last two phenomena examined—sprawl and informality—has been the increasing tendency towards new urban enclaves, satellite cities and model suburbs being planned and built throughout the Global South. These new urban worlds are frequently seen by urban planners, states and global capital as alternatives to the failings of existing urban areas. Chief among those failings, one finds questions surrounding governance and the delivery of urban services, to which I now turn.

2.5 Variability in Governance and Service Delivery

There is a gross generalisation that Global South cities lack sufficient capacity for effective governance and the delivery of urban services. To be sure, this is often a valid claim. Kinshasa, in the Democratic Republic of Congo, for instance, is a city now estimated to have more than 10 million people where the municipal government, just a few years ago, had a budget that worked out to about US$ 28 per resident (Myers 2016). Many national governments in the Global South do not allow for substantial local autonomy in fiscal or political terms for municipal authorities. Many cities in South Asia, Sub-Saharan Africa, and Latin America have low rates for the collection and deposit of solid waste in sanitary landfills, inadequate coverage for urban residents for sewage systems or piped water networks, and insufficient public transport.

There is also great uncertainty about the roles of the public, private and popular sectors in governance, governing and politics in Global South cities. The sharpness of meaning and clarity of triumphal vision around the Global North's intertwined ideas of urban democracy and urban neoliberalism are, if not gone entirely, surely lost in the wilderness, even in the Global North. Authoritarian tendencies have found real national government power in the 2010s in societies which had prior commitments to more democratic national and local politics. 'The narrow line between formal democracy and authoritarian rule every day grows thinner as civil rights and liberties are rolled back in the name of security' (Chambers 2017: 26). Yet, substantive social movements for radical democratisation from the grassroots have also taken hold in urban areas, notably in Global South cities. Appadurai (2001: 23) argued nearly two decades ago that 'grassroots movements' were 'finding new ways to combine local activism with horizontal, global networking' and that this phenomenon ought to shift scholars' analytical emphases from 'locations' to 'circulations'. He distinguished rich world cities operating 'like city states in a networked global economy' from 'poorer cities' where the citizens sought 'new ways of claiming space and voice' (Appadurai 2001: 28–29).

However, one must not 'romanticise' the 'global protest wave' of the twenty-first century as championing a common vision of such deeper democracy (Branch and Mampilly 2015: 2–3). Countervailing forces often leave countries and cities in politicised states of instability—in extreme cases (Tripoli, Libya; Port-au-Prince, Haiti; Bamenda, Cameroon; Caracas, Venezuela; Maiduguri, Nigeria; or Mogadishu, Somalia) cities flounder for years without effective governance. Even where urban governance has not reached this semi-permanent meta-crisis mode, many cities are facing 'circumstances of intense volatility and uncertainty' (Simone 2019: 16), looking for any solutions to fiscal, political, social or environmental problems, just when these problems loom larger. Populist or progressive movements for the 'deeper democracy' that Appadurai imagined across the Global South have been flung on the rocks of fiscal crisis; and cultural, religious and racial divides deepen daily in urban areas that now, ironically, have new capacities for cross-cultural communications with internet technologies and social media.

There are always varying limits on the capacity or autonomy of urban governments to affect their own solutions, whether due to the dominance of national governments or the power of the domestic private sector or international capital. Sometimes, from the grassroots up, cities are sharing paths towards 'inventing new visions of democracy and development in which popular interests come first' in addressing urban crises (Branch and Mampilly 2015: 6; Ismail 2014: 269). At other times, militaries and corporate elites share tactics to crush those popular interests. Often, what seems evident is 'some ambiguous mix of the two' tendencies, which Brown (2015: 3) labels as 'progressive' and 'revanchist' forces (see also Stokke 2014: 257). It is remarkable to witness the degree to which policy experiments move around the world's cities filtered through that ambiguous mix—and the resultant politics for ordinary people, whereby 'urban life comes to depend upon improvisation' (Simone 2019: 4).

Neoliberal urban governance frameworks are refracted into a wide variety of forms in cities of the Global South. Borrowing and learning in policy terms, for most southern urbanisms, must be seen through histories of colonialism and imperialism (Robinson 2011c). 'The post-colonial state, having inherited the structure of colonial governance and the project of modernisation, has also inherited its concomitant view of spatial and temporal order' (Chattopadhyay 2012: 89). This inheritance included an 'inability of the state to comprehend the spatial dynamics of the marginalised within its own logic' (Chattopadhyay 2012: 90). This resulted in splintered network infrastructure (Graham and Marvin 2001), but it also splintered cities' political geographies in terms of citizenship, rights or belonging (Carolini 2017). When looked at from some Asian urbanisms, the roles of the state stand out as fundamentally differentiated from Global North experiences; in many East and Southeast Asian countries (like

China), a developmentalist national state facilitating urban development reshaped any inherited Global North agendas (Bae 2012; Chen 2018).

By comparison, in Sub-Saharan Africa, urban governance and neoliberalism are vastly different in theory and practice (Obeng-Odoom 2013: 199). Obeng-Odoom (2013:10) argued for understanding the region's urban governance as having a 'cluster of meanings' linked to 'decentralisation, entrepreneurialism and democratisation'. He noted that 'governments still play a role in governance' (Obeng-Odoom 2013: 13)—as Davidson and Martin (2014: 6) put it, 'city governments still exist!' Their 'existences', on the other hand, come in infinite varieties.

There are also Global South cities that have produced models of effective governance and service delivery. Participatory budgeting (PB) and bus rapid transit (BRT) systems are two prominent examples of Global South-origin urban policy initiatives that have spread into the North. Participatory budgeting strategies started in 1989 in Porto Alegre, Brazil, expanding to other cities in Latin America, Europe, Africa, Asia and North America (Sintomer et al. 2010; Masiya 2012; Ganuza and Baiocchi 2012). Curitiba, Brazil is usually credited with developing the world's first BRT system in 1974. It initially spread into other South American cities before emerging in other Global South urbanisms and northern cities. While Pittsburgh in the US had a BRT that began in 1977, other US cities only began experimenting with BRT in the twenty-first century and with PB in 2009, pushed by non-profit agencies and community activism.

Participatory budgeting ideally 'provides a framework that encourages civic participation at and from the community or grassroots level' (Masiya 2012: 150). When it works, it fosters stronger bonds between ordinary people and municipal governments. For poorer cities, PB can deepen democracy and further urban service delivery. Results on the ground are mixed, with great variation between cities or within one city. Masiya (2012: 157) contrasted the 'robust' and 'progressive' PB of early twenty-first-century Johannesburg with the 'mistrust and tension' for its implementation in Harare, Zimbabwe. Ganuza and Baiocchi (2012) called PB an 'ambivalent… prescription' for deeper democracy and social justice, in part because its radical leftist roots in neighbourhood activism were left behind in the travelling. PB became an example of 'fast policy transfer', but also of 'democratisation… from above'—which is not democratic (Ganuza and Baiocchi 2012: 6; Stokke 2014: 265–266). Like other 'participatory' southern approaches to governance, PB often ends up as a tool 'for the political and economic elite to ensure their interests and profits' (Nastar et al 2018: 504).

BRT policies began with populist fervour in 1974, under Curitiba's long-time mayor, Jaime Lerner. In the mid-1990s, the BRT concept expanded to Quito and then Bogota, where another charismatic mayor, Enrique Penalosa, and his brother Gil developed the Transmilenio system. BRTs are promoted as a subway/train network of public transport—but by bus. Bus-only lanes are built out like an urban subway or light-rail system, thus producing an ostensibly low-cost system serving the masses of poorer cities. The twenty-first century witnessed a rapid globalisation of BRT policies. By 2007, forty cities had BRTs, and by 2016 the number of BRT cities had exploded to more than 200, most in the Global South (Rizzo 2017: 143). Despite the

Southern origins and generally South–South directionality of development, BRTs have ended up as problematic policies in many contexts. The costs of constructing bus-only networks and maintaining fleets of buses proved more expensive than proselytisers claim, and most financing for their construction has come from Global North institutions. BRT policies, like PB, began to appear as technocratic and depoliticising urban solutions, when every step of building, implementing and maintaining a BRT is highly politicised (Rizzo 2017; Wood 2015a, b). Nonetheless, as with PB, BRT expansion does suggest a very tangible way in which the predominant realities of planetary urbanisation in the twenty-first century are increasingly derived from the experiences of Global South cities.

3 Implications for Urban Ecology in the Global South

It should be unsurprising by this point in the chapter to see that there is great variation across the Global South in terms of the implications for urban ecologies of the themes that I have examined. To begin with, Global South cities exist across nearly every climate, biome or elevation found on earth. The variations in the colonial or imperial legacies, pace of urbanisation, degree of sprawl, character of informality and shape to urban governance play out, on top of that biophysical diversity, in infinite variations to ecological issues. Many chapters in this book address key facets in detail, so I will briefly highlight several implications and connections from the previous subsections for ecological and environmental dynamics.

The fact that centuries of colonialism and imperialism helped to redirect urban growth towards coasts—and the fact that contemporary economic globalisation has only increased that factor's significance—means that large proportions of the urban population in the Global South as a whole are on or near oceans as global climate change drives the rise in sea levels. Colonialism and imperialism's legacies of segregation then tend to mean a high degree of inequality and injustice exists spatially in southern cities in terms of the negative externalities of environmental change, including climate change. To take an extreme example: Cape Town, the 'mother city' of white settlement in South Africa, is broadly divided between largely white neighbourhoods at higher elevations and Black and 'Coloured' (mixed-heritage) neighbourhoods that predominate in the low-lying Cape Flats. Climate change has left the Cape Flats more susceptible to flooding, and at the same time, ironically, to chronic potable water shortages. Still-predominantly white areas face less risk from rising sea level and have greater wealth to purchase water for domestic consumption from commercial sources (Myers 2016).

Countries with rapid rates of urbanisation, extensive urban sprawl, significant urban informality and limited urban governance capacity—the last four chapters themes—are likely to experience greater points of environmental crisis with much less resilience and adaptation. Indeed, any one of these can produce stressors on to ecological issues. Rapid rates of urbanisation are directly tied to increased demand for natural resources for construction and infrastructure. The expansion of the urban fabric or footprint over ever more extensive territory likewise increases the need to extend that infrastructure. With rapid urbanisation and sprawl together, the demand will increase for automobile transport—and this is especially acute where poorly funded governance units make public transit investments unlikely. Air pollution will increase as a result. Weak urban governments often go together with weak enforcement (or even an absence) of environmental regulations. Since informal economic activity is often more significant where urban governance is weaker, informality, too, is often linked with environmental crises.

Exceptions or contradictions exist for these generalisations, too. One finds impressive instances of informal-sector or informal-settlement activism to combat ecological crises. Perhaps one of the most widely noted examples exists with the now-international non-governmental organisation, Slum/Shack Dwellers International, which began in India and has now expanded across much of the Global South (Appadurai 2001; Myers 2016). Movements for toilet rights, water rights, environmental justice and many other ecological concerns have emerged among poorer urban areas, often linked with movements for political rights. At the same time, though, many urban environmental policies imposed from hegemonic Global North actors often 'undermine' local actions to combat climate change or ecological crises, thus 'privileging international actors and financial markets' (Ernstson and Swyngedouw 2019: 15). 'Are we not left', Swyngedouw and Ernstson (2019: 37) ask, 'with the gnawing feeling that, despite the elevation of the ecological condition to the dignity of a global public concern, the socio-ecological parameters keep eroding further?'

The literature on climate change mitigation, adaptation and resilience has 'coalesced' with the literature on disaster risk reduction (DRR) over the last decade or so (Simon 2016: 75). Urban fragility in the era of climate change has become the tether connecting cities across the planet and connecting climate change and DRR work (Commins 2018). The 'increasing severity and… frequency of extreme events' join the slow, inexorable rise in sea level as the main impacts of climate change on coastal cities (Simon 2016: 76). The 'structural vulnerabilities' in these cities thoroughly 'constitute a form of unfairness or injustice' (Simon 2016: 77).

Despite such pessimism, ordinary residents have demonstrated direct awareness of climate change realities and other major ecological concerns (Okaka and Odhiambo 2018). Simon and Leck (2014: 624) highlight the role of some southern cities in the C40 learning network, and the Sustainable Urban Development Network (or SUD-Net) out of UN-Habitat in Nairobi for launching the Cities and Climate Change Initiative (CCCI) in 2008. We are a long while from a point at which we might say that cities of the Global South are leading the way in confronting ecological crises in the manner in which one might say they are leading the way in urbanisation in the twenty-first century. But it is not out of the question that southern cities may reconstruct the 'socio-ecological parameters' to provide innovative and even indigenous solutions to environmental crises in an as-yet unforeseen manner.

References

Al S (2014) Introduction: villages in the city: a guide to China's informal settlements. In: Al S (ed) Villages in the city: a guide to South China's informal settlements. Hong Kong University Press, Hong Kong, pp 3–8

AlSayyad N (2004) Urban informality as a "new" way of life. In: Roy A, AlSayyad N (eds) Urban informality: transnational perspectives from the Middle East, Latin America, and South Asia. Lexington Books, Lanham, MD, pp 7–30

AlSayyad N, Roy A (2004) Prologue/dialogue: urban informality: crossing borders. In: Roy A, AlSayyad N (eds) Urban informality: transnational perspectives from the Middle East, Latin America, and South Asia. Lexington Books, Lanham, MD, pp 1–6

Alston L, Eggertsson T, North D (1996) Empirical Studies in Institutional Change. Cambridge University Press, Cambridge

Alston L, Libecap G, Mueller B (1999) Titles, conflict and land use: the development of property rights on the Brazilian frontier. University of Michigan Press, Ann Arbor

Altrock U, Schoon S (eds) (2014) Maturing megacities: The Pearl River Delta in progressive transformation. Springer, Berlin

Anand N (2017) Hydraulic city: water and the infrastructures of citizenship in Mumbai. Duke University Press, Durham NC

Angel S (2012) Planet of cities. Lincoln Institute of Land Policy, Cambridge, MA

Appadurai A (2001) Deep democracy: urban governmentality and the horizon of politics. Environ Urbanis 13:23–43

Bae Y (2012) Governing cities without states? Rethinking urban political theories in Asia. In: Edensor T, Jayne M (eds) Urban theory beyond the west: a world of cities. Routledge, London, pp 95–110

Berney R (2019) Learning from Bogota: pedagogical urbanism and the reshaping of public space. University of Texas Press, Austin

Branch A, Mampilly Z (2015) Africa uprising: popular protest and political change. Zed Books, London

Brenner N, Schmid C (2014) The "urban age" in question. Int J Urban Reg Res 38:731–755

Brenner N, Schmid C (2015) Towards a new epistemology of the urban? City 19:151–182

Brown J (2015) South Africa's insurgent citizens: on dissent and the possibility of politics. Zed Books, London

Bryceson D (2006) Fragile cities: fundamentals of urban life in East and Southern Africa. In: Bryceson D, Potts D (eds) African urban economies: viability, vitality or vitiation? Palgrave Macmillan, New York, pp 3–38

Bunnell T (2016) From world city to the world in one city: Liverpool through Malay lives. Wiley, Chichester

Bunnell T (2017) Locating transnational urban connections beyond world city networks. In: Hannigan J, Richards G (eds) The SAGE handbook of new urban studies. Sage, London, pp 19–29

Caldeira T (2017) Peripheral urbanisation: autoconstruction, transversal logics, and politics in cities of the Global South. Enviro Plan D: Soc Space 35:3–20

Carmody P, Owusu F (2016) Neoliberalism, urbanisation and change in Africa: the political economy of heterotopia. J Afr Dev 18:61–73

Carolini G (2017) Sisyphean dilemmas of development: contrasting urban infrastructure and fiscal policy trends in Maputo, Mozambique. Int J Urb Reg Res 41: 126–144

Castells M (1996) The rise of the network society. Blackwell, Oxford

Chambers I (2017) Postcolonial interruptions: unauthorised modernities. Rowman & Littlefield, London

Chattopadhyay S (2012) Urbanism, colonialism and subalternity. In: Edensor T, Jayne M (eds) Urban theory beyond the west: a world of cities. Routledge, London, pp 75–92

Chen X (2014) Steering, speeding, scaling: China's model of urban growth and its implications for cities of the Global South. In: Parnell S, Oldfield S (eds) The Routledge handbook on cities of the Global South. Routledge, London, pp 155–172
Chen X (2018) Globalisation redux: can China's inside-out strategy catalyse economic development and integration across its Asian borderlands and beyond? Camb J Reg Econ Soc 11:35–58
Commins S (2018) Urban fragility to urban sustainability. African security brief 35, African Center for Strategic Studies, Washington, DC
Cooper F (1987) On the African waterfront: urban disorder and the transformation of work in colonial Mombasa. Yale University Press, New Haven
Davidson M, Martin D (2014) Thinking critically about urban politics. In: Davidson M, Martin D (eds) Urban politics: critical approaches. Sage, London, pp 1–13
De Muelder B, Shannon K (2014) Village in the city as urbanism in China and Elsewhere. In: de Muelder B, Lin Y, Shannon K (eds) Village in the city. Park Books, Zurich, pp 4–19
Derudder B, Hoyler M, Taylor P, Witlox F (2012) Introduction: a relational urban Studies. In: Derudder B, Hoyler M, Taylor P, Witlox F (eds) International handbook of globalisation and world cities. Edward Elgar, Cheltonham, pp 1–4
de Soto H (2000) The mystery of capital: why capitalism triumphs in the west and fails everywhere else. Basic Books, New York
Ernstson H, Swyngedouw E (2019) Politicising the environment in the urban century. In: Ernstson H, Swyngedouw E (eds) Urban political ecology in the Anthropo-obscene: interruptions and possibilities. Routledge, New York, pp 3–21
Escobar A (2018) Designs for the pluriverse: radical interdependence, autonomy, and the making of worlds. Duke University Press, Durham, NC
Ferguson J (2006) Global shadows: Africa in the neoliberal world order. Duke University Press, Durham, NC
Fox S, Goodfellow T (2016) Cities and development, 2nd edn. Routledge, London
Fredericks R (2018) Garbage citizenship: vital infrastructures of labor in Dakar, Senegal. Duke University Press, Durham
Gandy M (2014) The fabric of space. MIT Press, Cambridge, MA
Ganuza E, Baiocchi G (2012) The power of ambiguity: How participatory budgeting travels the globe. J Public Delib 8: 1–14
GaWC (Globalisation and World Cities) Research Network (2018) The world according to the GaWC. Accessed online at https://www.lboro.ac.uk/gawc/world2018.html
Graham S, Marvin S (2001) Splintering urbanism: networked infrastructures, technological mobilities and the urban condition. Routledge, London
Grant R (2009) Globalising city: The urban and economic transformation of Accra, Ghana. Syracuse University Press, Syracuse, NY
Hansen K, Vaa M (eds) (2004) Reconsidering informality: perspectives from urban Africa. Nordic Africa Institute, Uppsala
Harris R (2017) Modes of informal urban development: a global phenomenon. J Plann Lit 33:1–20
Harrison P, Todes A, Watson V (2008) Planning and transformation: learning from the post-apartheid experience. Routledge, New York
Harrison J, Hoyler M (eds) (2018) Doing global urban research. Sage, Los Angeles
Hart K (1973) Informal income opportunities and urban employment in Ghana. J Mod Afr Stud 11:61–89
Harvey D (1996) Justice, nature, and the geography of difference. Blackwell Press, Oxford
Holston J (2008) Insurgent citizenship: disjunctions of democracy and modernity in Brazil. Princeton University Press, Princeton, NJ
Home R (1997) Of planting and planning: The making of British colonial cities. Spon, London
ILO (International Labour Organisation) (1972) Employment, incomes and equality: a strategy for increasing productive employment in Kenya. ILO, Geneva

Ismail S (2014) The politics of the urban everyday in Cairo: infrastructures of oppositional action. In: Parnell S, Oldfield S (eds) The Routledge handbook on cities of the Global South. Routledge, London, pp 269–280
Jacobs JM (1996) Edge of empire: postcolonialism and the city. Routledge, London
Jazeel T (2019) Postcolonialism. Routledge, London
Jazeel T (2018) Urban theory with an outside. Environ Plan D: Soc Space 36:405–419
Kanna A, Chen X (2012) Introduction: bringing the less familiar cities in and together. In: Chen X, Kanna A (eds) Rethinking global urbanism: comparative insights from secondary cities. Routledge, London, pp 1–14
King A (1976) Colonial urban development. Routledge, London
Kombe W, Kreibich V (2000) Informal land management in Tanzania. SPRING Research Series 29, University of Dortmund, Dortmund
Konings P, van Dijk R, Foeken D (2006) The African neighborhood: an introduction. In: Konings P, Foeken D (eds) Crisis and creativity: exploring the wealth of the African neighborhood. Brill, Leiden, pp 1–21
Lawhon M, Truelove Y (2020) Disambiguating the southern critique: propositions, pathways and possibilities for a more global urban studies. Urban Stud 57: 3–20
Lefebvre H (2003 [1970]) The urban revolution. University of Minnesota Press, Minneapolis
Leitner H, Sheppard E (2016) Provincialising critical urban theory: extending the ecosystem of possibilities. Int J Urban Reg Res 40:228–235
Lindell I (2010) Changing landscapes of collective organising in the informal economy. In: Lindell I (ed) Africa's informal workers: collective agency, alliances, and transnational organizing. Nordic Africa Institute, Uppsala, pp 1–30
Mabin A (2014) Grounding southern theory in time and place. In: Parnell S, Oldfield S (eds) The Routledge handbook on cities of the Global South. Routledge, London, pp 21–36
Manji A (2006) The politics of land reform in Africa: from communal tenure to free markets. Zed Books, London
Masiya T (2012) Social accountability in African cities: comparing participatory budgeting in Johannesburg and Harare. In: Chen X, Kanna A (eds) Rethinking global urbanism: comparative insights from secondary cities. Routledge, London, pp 150–165
Mbembe A, Nuttall S (2008) Introduction: Afropolis. In: Nuttall S, Mbembe A (eds) Johannesburg: the elusive metropolis. Duke University Press, Durham, NC, pp 1–33
McFarlane C (2011) Learning the city: knowledge and translocal assemblage. Wiley, Chichester
McGee T (1971) The urbanisation process in the third world. C. Bell, London
Metcalf T (1989) An imperial vision: Indian architecture and Britain's Raj. University of California Press, Berkeley
Metcalf T (2008) Imperial connections: India in the Indian Ocean arena. University of California Press, Berkeley
Mignolo W (2002) The many faces of cosmopolis: border thinking and critical cosmopolitanism. In: Breckinridge C, Pollock S, Bhabha H, Chakrabarty D (eds) Cosmopolitanism. Duke University Press, Durham, NC
Mignolo W (2012) Local histories/global designs: coloniality, subaltern knowledges and border thinking. Princeton University Press, Princeton
Mignolo W, Walsh C (2018) On decoloniality: concepts, analytics, praxis. Duke University Press, Durham, NC
Mitchell T (1988) Colonising Egypt. Cambridge University Press, Cambridge
Mooya M, Cloete C (2007) Informal urban property markets and poverty alleviation: a conceptual framework. Urban Stud 44:147–165
Murray M (2017) The urbanism of exception: the dynamics of global city building in the twenty-first century. Cambridge University Press, Cambridge
Myers G (2003) Verandahs of power: colonialism and space in urban Africa. Syracuse University Press, Syracuse

Myers G (2008) Peri-urban land reform, political-economic reform, and urban political ecology in Zanzibar. Urban Geogr 29:264–288
Myers G (2011) African cities: alternative visions of urban theory and practice. Zed Books, London
Myers G (2016) Urban environments in Africa: a critical analysis of environmental politics. Policy Press, University of Bristol, Bristol
Myers G (2020) Rethinking urbanism: lessons from postcolonialism and the Global South. Bristol University Press, Bristol
Nastar M, Abbas S, Rivero CA, Jenkins S, Kooy M (2018) The emancipatory promise of participatory water governance for the urban poor: reflections on the transition management approach in the cities of Dodowa, Ghana and Arusha, Tanzania. Afr Stud 77:504–525
Obeng-Odoom F (2013) Governance for pro-poor urban development: lessons from Ghana. Routledge, New York
Okaka FO, Odhiambo BDO (2018) Urban residents' awareness of climate change and their autonomous adaptive behavior and mitigation measures in the coastal city of Mombasa, Kenya. S Afr Geogr J 100:378–393
Ong A (2011) Worlding cities: or, the art of being global. In: Roy A, Ong A (eds) Worlding cities: Asian experiments and the art of being global. Wiley, Chichester, pp 1–26
Ould-Mey M (1996) Global restructuring and peripheral states: the carrot and the stick in Mauritania. Rowman & Littlefield, Lanham, MD
Parnell S, Oldfield S (eds) (2014) Routledge handbook to the cities of the Global South. Routledge, London
Parnell S, Robinson J (2012) (Re)theorising cities from the Global South: looking beyond neoliberalism. Urban Geogr 33:593–617
Patel S (2014) Is there a "south" perspective to urban studies? In: Parnell S, Oldfield S (eds) The Routledge handbook on cities of the Global South. Routledge, London, pp 37–47
Payne G (2002) Land, rights, and innovation: improving tenure security for the urban poor. ITDG Publishing, London
Planning Secretariat (Colombia) (2013) Plan de ordenamiento territorial 2013. Office of the Mayor, Bogota
Portes A, Castells M, Benton L (1989) The informal economy: studies in advanced and less developed countries. Johns Hopkins University Press, Baltimore, MD
Quayson A (2014) Oxford street, Accra: city life and the itineraries of transnationalism. Duke University Press, Durham, NC
Rizzo M (2017) Taken for a ride: grounding neoliberalism, precarious labour and public transport in an African metropolis. Oxford University Press, Oxford
Robins S (2006) When shacks ain't chic! Planning for "difference" in post-apartheid Cape Town. In: Bekker S, Leilde A (eds) Reflections on identity in four African cities. African Minds, Johannesburg, pp 97–117
Robinson J (2002) Global and world cities: a view from off the map. Int J Urban Reg Res 26:531–554
Robinson J (2003) Postcolonialising geography: tactics and pitfalls. Singapore J Trop Geogr 24:273–289
Robinson J (2006) Ordinary cities: between modernity and development. Routledge, New York
Robinson J (2011a) Cities in a world of cities: the comparative gesture. Int J Urban Reg Res 35:1–23
Robinson J (2011b) Comparisons: colonial or cosmopolitan? Singapore J Trop Geogr 32:125–140
Robinson J (2011c) The spaces of circulating knowledge: city strategies and global urban Governmentality. In: McCann E, Ward K (eds) Mobile urbanism: cities and policymaking in the global age. University of Minnesota Press, Minneapolis, pp 15–40
Robinson J (2016) Thinking cities through elsewhere: comparative tactics for a more global urban studies. Prog Hum Geogr 40:3–29
Roy A (2005) Urban informality: toward an epistemology of planning. J Am Plan Assoc 71:147–158
Roy A (2011a) Slumdog cities: rethinking subaltern urbanism. Int J Urban Reg Res 35:223–238
Roy A (2011b) Postcolonial urbanism: speed, hysteria, mass dreams. In: Roy A, Ong A (eds) Worlding cities: Asian experiments and the art of being global. Wiley, Chichester, pp 307–335

Roy A (2014) Worlding the south: toward a post-colonial urban theory. In: Parnell S, Oldfield S (eds) The Routledge handbook on cities of the Global South. Routledge, London, pp 9–20
Roy A (2016) Who's afraid of postcolonial theory. Int J Urban Reg Res 40:200–209
Roy A, Ong A (eds) (2011) Worlding cities: Asian experiments and the art of being global. Wiley, Chochester
Santos M (1979) The shared space: the two circuits of the urban economy in underdeveloped countries. Methuen, London
Sassen S (1991) The global city. Princeton University Press, Princeton, NJ
Schindler S (2017) Towards a paradigm of southern urbanism. City 21:47–64
Schmid C (2016) Planetary urbanisation: Henri Lefebvre und das recht auf die stadt. In: Kraft S, Aichinger A, Zhang Z (eds) Planetary urbanism: the transformative power of cities. Arch + Verlag GmbH, Berlin, pp 26–34
Schmid C, Karaman O, Hanakata N (2018) Towards a new vocabulary of urbanization processes: a comparative approach. Urban Stud 55:19–52
Scott AJ, Storper M (2015) The nature of cities: The limits and scope of urban theory. Int J Urban Reg Res 39:1–15
Shackleton CM, Cilliers SS, du Toit MJ, Davoren E (2021) The need for an urban ecology of the Global South. In: Shackleton CM, Cilliers, SS, Davoren E, du Toit MJ (eds). Urban ecology in the Global South. Springer, Cham, pp 1–26
Sheppard E, Leitner H, Maringanti A (2013) Provincialising global urbanism: a manifesto. Urban Geogr 34:893–900
Simon D (2016) Green cities: from tokenism to incrementalism and transformation. In: Simon D (ed) Rethinking sustainable cities: accessible, green and fair. Policy Press, Bristol, pp 61–105
Simon D, Leck H (2014) Urban dynamics and the challenges of global environmental change in the South. In: Parnell S, Oldfield S (eds) The Routledge handbook on cities of the Global South. Routledge, London, pp 613–628
Simone A (2001) On the worlding of African cities. Afr Stud Rev 44:15–41
Simone A (2016) Always something else: urban Asia and Africa as experiment. Basler Afrika Bibliographien, Basel
Simone A (2019) Improvised lives: rhythms of endurance in an urban South. Polity Press, Cambridge
Simone A, Pieterse E (2017) New urban worlds: inhabiting dissonant times. Polity Press, Cambridge
Sintomer Y, Herzberg C, Allegretti G, Röcke A (2010) Learning from the south: participatory budgeting worldwide—an invitation to global cooperation. Capacity Building International, Bonn
Söderström O (2014) Cities in relations: trajectories of urban development in Hanoi and Ouagadougou. Wiley Blackwell, Malden, MA
Soja E, Kanai M (2007) The urbanisation of the world. In: Burdett R, Sudjic D (eds) The endless city: an authoritative and visually rich survey of the contemporary city. Phaidon, New York, pp 54–69
Sparke M (2007) Everywhere but always somewhere: critical geographies of the global South. Glob South 1:117–126
Spencer J (2015) Globalisation and urbanisation: the global urban ecosystem. Rowman & Littlefield, Lanham, MD
Stokke K (2014) Substantiating urban democracy: the importance of popular representation and transformative democratic politics. In: Parnell S, Oldfield S (eds) The Routledge handbook on cities of the Global South. Routledge, London, pp 257–268
Storper M, Scott AJ (2016) Current debates in urban theory: a critical assessment. Urban Stud 53:1114–1136
Swyngedouw E, Ernstson H (2019) O tempora! O mores! Interrupting the Anthropo-obscene. In: Ernstson H, Swyngedouw E (eds) Urban political ecology in the Anthropo-obscene: interruptions and possibilities. Routledge, New York, pp 25–47
United Nations (2018) World Urbanisation prospects: the 2018 revision. (online edition) United Nations Population Division, Bureau of Economic Affairs, New York

UN Habitat (2010) The State of African cities 2010: governance, inequality and urban land markets. UN Habitat, Nairobi

Watson V (2007) Engaging with difference: understanding the limits of multiculturalism in planning in the South African context. In: Murray N, Shepherd N, Hall M (eds) Desire lines: space, memory and identity in the post-apartheid city. Routledge, London, pp 67–79

Wood A (2015a) The politics of policy circulation: unpacking the relationship between South African and South American cities in the adoption of bus rapid transit. Antipode 47:1062–1079

Wood A (2015b) Multiple temporalities of policy circulation: gradual, repetitive and delayed processes of BRT adoption in South African cities. Int J Urban Reg Res 39:568–580

Wu F, Zhang F, Webster C (2012) Informality and the development and demolition of urban villages in the Chinese peri-urban area. Urban Stud 50:1919–1934

The Effect of Landscape History on the Urban Environment: Past Landscapes, Present Patterns

Marié J. du Toit, Amy K. Hahs, and Ian MacGregor-Fors

Abstract Urban environments around the world are the result of their evolutionary and geographic history, and more contemporary influences of the social, cultural, economic, political and technical systems that shaped their construction. In this chapter, we use urban filters as the overarching framework within which specific aspects pertaining to the history of the landscape will be discussed. Legacy effects are discussed under the subthemes of urban form and development history, biodiversity, biological invasions, soils, urban freshwater bodies, natural disasters, the effects of war on urban environments and extinction debts. Lastly, the collective effects of changing urban environments over time manifests in citizens through desensitisation, termed generational amnesia or the extinction of experience, which has important consequences for residents' perceptions towards urban biodiversity. In trying to disseminate patterns and processes in urban areas of the Global South, this overview revealed that more urban ecology research is required to begin untangling the effects of urbanisation and bioclimatic regions from the drivers that are much more context-specific. Growing the number of studies conducted in the Global South is critical to ensure that there is an evidence base that is representative of the full range of conditions for cities around the world.

Keywords Bioinvasions · Colonialism · Extinction debts · Global South · Legacy effects · Urban filters

M. J. du Toit (✉)
Unit for Environmental Sciences and Management, North-West University, Private Bag X6001, Potchefstroom 2520, South Africa

A. K. Hahs
School of Ecosystem and Forest Sciences, University of Melbourne, 500 Yarra Blvd, Richmond, VIC 3121, Australia
e-mail: amy.hahs@unimelb.edu.au

I. MacGregor-Fors
Ecosystems and Environment Research Programme, Faculty of Biological and Environmental Sciences, University of Helsinki, Niemenkatu 73, FI-15140 Lahti, Finland
e-mail: ian.macgregor@helsinki.fi

C. M. Shackleton et al. (eds.), *Urban Ecology in the Global South*, Cities and Nature,
https://doi.org/10.1007/978-3-030-67650-6_3

1 Introduction

The history of a landscape can have a profound impact on currently observed environments. In France, for example, visible differences in plant diversity patterns and soil conditions in forests can be traced to ancient Roman farming practices in the area (Dupouey et al. 2002). Other examples include 200-year-old land-use impacts corresponding to current grassland diversity (Gustavsson et al. 2007); unequal tree cover along socioeconomic gradients in cities (Clarke et al. 2013); and the impact of colonialism on urban vegetation (Abendroth et al. 2012; Hunte et al. 2019). Urban areas are some of the most intensely altered areas on Earth (Grimm et al. 2000) and the history of the landscape can have a lasting legacy on currently observed environments (Clarke et al. 2013). Current ecosystems across the Neotropics are the results of millennia of prehispanic activities, spanning from housing to hunting and farming activities (Piperno 2007). Legacies can be long-term, stretching back centuries, e.g. Caracol, a Maya archaeological site in Belize abandoned in 900 A.D., which influences contemporary tropical rainforest canopy structure (Hightower et al. 2014). Legacies can also be shorter, lasting only a few years, e.g. nitrogen enrichment of soils after the removal of invasive alien nitrogen-fixing plants (Grove et al. 2015). In this way, legacy effects can be defined as "the impacts that previous conditions have on current processes or properties" (Monger et al. 2015). Legacy effects are also referred to as ecological memory (Padisák 1992), biological legacies or antecedent effects, based on different uses by sub-disciplines (Ogle et al. 2015).

One of the distinguishing features and unifying characteristics of the Global South is that it is largely composed of developing countries. This term implies that these regions are in a state of transition and actively working on setting up new trajectories for future legacies but in a space where lessons from the Global North do not always apply. In this chapter, we will look more closely at how historical, geographic, climatic and cultural factors have shaped the interactions between nature, people and the urban form. In particular, we look at which legacies are common across the world, which are shaped by geographic and climatic settings and which legacies are distinctly those of the Global South.

To begin this chapter, we discuss urban filters as the overarching framework within which specific aspects pertaining to the history of the landscape will be discussed. Legacy effects are discussed under the subthemes of urban form and development history, biodiversity, biological invasions, soils, urban freshwater bodies, natural disasters, the effects of war on urban environments and extinction debts. Lastly, the collective effect of changing urban environments over time manifests in citizens through desensitisation, termed generational amnesia or the extinction of experience, which has important consequences for residents' perceptions towards urban biodiversity. The implication of this and other aspects are discussed in the future outlook.

2 Urban Filters

From the broad scale, cities represent a barrier for many biodiversity groups, filtering out a proportion of species from regional species pools (Evans et al. 2009; Williams et al. 2009; Aronson et al. 2014). This 'barrier' is not only given by the physical differences that exist between urban and non-urban systems, but also with the important amount of ecosystemic changes implied by urbanisation, ranging from changes in the biogeochemical cycle due to nutrient enrichment, through to changes in vital resources such as water, food, shelter and conditions for any given species (Grimm et al. 2008). Recently, Aronson et al. (2016) underlined five hierarchical filters that act to determine the species that use, dwell or even thrive in a given urban system: (1) regional climatic and biogeographical factors, (2) human facilitation; (3) urban form and development history, (4) socio-economic and cultural factors and (5) species interactions. Thus, the biological communities that are recorded in a given urban area are the result of a series of species migration processes (both immigration and emigration; Blair and Johnson 2008), together with factors related to the location in which the city was settled and sprawled, as well as the social factors determining its dynamics (Williams et al. 2009).

The magnitude, relative importance and effect of these filters can vary depending upon the conditions of the city they are acting in. For cities in the Global South, the strength, direction and outcomes of these filters may differ from those in the Global North. For example, while much of the urban ecology research to date has been conducted in temperate landscapes, many of the most rapidly urbanising regions are located in areas with higher annual temperatures and lower annual rainfall (Chown et al. 2015). Informal settlements are an urban form that is rarely observed in the Global North, and the pace of urban development in the twenty-first century that is concentrated within the Global South far exceeds anything that has been observed to date (Myers 2021). The socio-economic and cultural factors acting within the Global South are also likely to differ from those in the Global North (Shackleton et al. 2021). Therefore, while the general principle of the filters may hold, it will be important to test them in the Global South in order to more accurately understand the legacies of these different filters on the ecology of the urban systems.

As Aronson et al. (2016) emphasised, there are certain traits of wildlife species attempting to remain in a site that has been urbanised, as well as those colonising urban centres (González-Lagos and Quesada 2017). Such traits range from genetics to morphology, physiology and behaviour (Gil and Brumm 2014). In the long term, evolutionary selection pressures can play crucial roles in the fate of the urban species populations. These evolutionary selection pressures may share similarities with those acting in the Global North, but there may also be distinct selection pressures that arise through the differences in the filters acting on the system. Similarly, the Global South contains a wide range of ecosystems, species and phylogenetic diversity that are not well represented in the Global North. Testing the evolutionary responses of organisms located within the monsoonal systems of southern Asia, the tropical rainforests of the Congo and the subtropical highland systems of the Andes will reveal greater insights into the legacies that cities have on the evolutionary ecology of their local biodiversity.

It has been shown that the filtering process is a semi-permeable one in which species tend to respond punctually (with well-defined thresholds related to the variables that determine its presence) or gradually towards the city's centre (MacGregor-Fors 2010). It is notable that the urban filtering process is not a random one. Recent global assessments have shown that phylogenetic and functional diversity decreases with urbanisation, basically given by species loss, but with the focal loss of evolutionarily distinct species (Sol et al. 2017). Moreover, La Sorte et al. (2018), besides providing evidence that supports the findings from Sol et al. (2017), found that urban avian assemblages contained lower phylogenetic beta diversity. Additionally, these studies showed the loss of functional diversity in urban avian assemblages together with the loss of small and large-bodied, especially broadly distributed species, as well as the loss of herbivores and fewer aquatic feeding species (Reynolds et al. 2021).

Although an important body of knowledge exists for urban wildlife responses across gradients of urbanisation, which could shed light on urban filtering processes, conceptual and practical drawbacks limit comprehension of the phenomenon on a fine spatial scale (McDonnell and Hahs 2008). Specifically, many so-called urban–rural gradients are established in sites within a city, lacking the non-urban component. The diversity of urban forms present in the Global South offer a critical opportunity to extend the characteristics of these urban gradients and reveal new insights into the multiple interacting drivers within them. Other urban–rural gradient studies have included the so-called 'rural' areas which do not necessarily shelter species from regional pools (MacGregor-Fors et al. 2018). However, there have been questions raised about the relevance of a Global North framing of 'rural' in the Global South (Myers 2021). For example, McHale et al. (2013) highlighted that systems of kinship and migration have led to a new form of urban/rural development in a landscape in South Africa, while MacGregor-Fors and Vázquez (2020) have revisited the concept of 'rural' based on their experiences working in South America.

There are not many studies that have specifically addressed the role of the hierarchical filters proposed by Aronson et al. (2016) on wildlife ecology in the Global South. Some studies have shed initial light on how the filtering process occurs in the region. For instance, Filloy et al. (2015) showed that although avian species richness from non-urban areas is susceptible to latitudinal variations in South America, there is a greater similarity in the diversity of birds within the urban centres. However, the effect was greater for cities in a tropical climate compared to those at higher latitudes. In another study investigating bird community responses to urbanisation in three Neotropical cities, Leveau et al. (2017) found that the composition of the surrounding landscape influenced species diversity, with a much more pronounced effect in a city located in the tropics.

Studies have also provided evidence of the importance that cultural factors can have in determining urban biodiversity. One example is that of Jaganmohan et al. (2012), who studied plant diversity in Bangalore (India). Results showed that, on the one hand, single domestic gardens tend to favour variety, with large ones including a high density of large trees. On the other hand, large shared apartment gardens, which also have large trees, are often part of sprawling projects, where pre-existing land-uses are transformed, and thus result in a net biodiversity loss over time. Another example of the importance of cultural, political and historical factors are the differences in the size, management and biodiversity of home gardens in South Africa, where there are differences in the proportion of introduced species and garden styles between cultural groups with signals that these differences may begin to lessen with increasing socio-economic status (Davoren et al. 2016).

Regarding the role of species interactions as a filter, mainly in relation to invasive species in urban ecosystems, studies have indicated the detrimental effect of their presence and abundance on native species in Mexico (e.g. House Sparrow, *Passer domesticus*; MacGregor-Fors et al. 2010). Other studies have assessed space-for-time substitutions, finding an important loss of biodiversity when contrasting urban biodiversity with that of natural ecosystems that resemble those that existed before the urbanisation process (MacGregor-Fors et al. 2012), while others have assessed the ecological patterns occurring along urban–non-urban ecotones (Garaffa et al. 2009; Puga-Caballero et al. 2014). Results of studies focused at the urban fringe showed that urban settings tend to homogenise avian diversity regardless of the adjacent non-urban system, which can be richer when contrasted with some non-urban land-uses (e.g. croplands, Puga-Caballero et al. 2014). Interestingly, the size of urban areas has been shown to influence urban filtering processes. For instance, in the Pampean Region (Argentina), small urban areas do not show filtering effects, but larger urban areas present steep differences (Garaffa et al. 2009).

A vast majority of studies in the Global South focus on the "ecology in the city" paradigm (Pickett et al. 2016), concentrating on patterns mostly within greenspaces or urbanisation intensity gradients. Moreover, the available information is insufficient to ponder the role that the hierarchical filters proposed by Aronson et al. (2016) play in the Global South. Given the high biodiversity sheltered in the Global South (Myers et al. 2000), together with its biogeographic complexity (see Morrone (2014) for the Neotropics as an example), it would not be surprising that the urban filtering processes work differently in the Global South when contrasted with less biodiverse regions. For a numerical perspective, there are examples of cities harbouring hundreds of species, such as Cali (Colombia) (248 bird species; Departamento de Gestión del Medio Ambiente 2010) and Xalapa (Mexico), where ca. 30% of the Mexican avifauna (341 species) has been recorded within the borders of the small-to-medium-sized city (González-García et al. 2014, 2016). It is notable that large urban greenspaces play crucial roles in concentrating most of such biodiversity, and little is known of the filtering processes in highly developed conditions (Escobar-Ibáñez and MacGregor-Fors 2016). Rapidly expanding urban populations and limited resources can result in much lower levels of urban green space provision in developing countries. For example, green spaces cover less than 3% of the land surface area in Lagos city, Nigeria, and less than 10% of the land surface for several towns in South Africa (Mensah 2014). Limited green spaces can have serious implications for biodiversity, as we discuss in more detail later in this chapter.

Information on global trends that include data from the Global South has shown that the filtering processes occur in urban systems. However, based on current knowledge, it is impossible to stipulate the role that the hierarchical filters have on urban species in the Global South, or differences and similarities in relation to the rest of the world. However, we do know that cities from the Global South are highly diverse and that they could be conceived (and used) as urban ecology laboratories due to the rapid urbanisation rates in the region (United Nations 2015). Future research towards better comprehension of the urban filtering processes in the Global South needs to explore variables related to each filter, species pools at different scales, and the traits of the assessed species groups, as outlined by Aronson et al. (2016). Additionally, studies ought to consider and assess the uniqueness of biotic and abiotic traits, and social and cultural diversity of the Global South, as well as the peculiar ways in which cities are established and sprawl, especially within the tropical and subtropical bands.

Urban filter effects can leave legacies, where past impacts are still visible and traceable in current urban environments. The intensity and longevity of these impacts differ based on the processes involved and the specific geographical setting. Below we discuss some of these legacies and how they are similar or different between urban areas in the Global North and South.

3 Legacy Effects

Some urban areas have continuous histories spanning millennia, such as the city of Suzhou in China built in 514 B.C. (Xu 2000). Moreover, in the case of Suzhou, comparison of a 1945 aerial image with a map of the city from 1229 reveals an almost completely unchanged urban form (Mote 1973). The effects of humans on the urban environment in these ancient urban areas vary in intensity, and some can be traced in contemporary landscapes. Archaeological evidence in the remains of Malyan (4000–2000 B.C.), an ancient urban area in Iran, indicates evidence of deforestation of local woodlands along with traces of wood from more distant forests (Miller 1985). Samples taken from contemporary soils, plants and animals indicate enhanced levels of lead and copper in an ancient mining and metalworking site in Southern Jordan (Pyatt et al. 2000).

3.1 Urban Form and Development History

The greatest modern influence on the establishment and development of urban areas in the Global South can be attributed to colonialism by European powers. At the beginning of the Second World War, a third of the world's population, living in two-fifths of the global land area, were in colonies, dependencies or dominions (Ertan et al. 2016). Modern colonisation by Western European powers started as early as the fifteenth century and lasted until the twentieth century (Ertan et al. 2016). It can be argued that the greatest environmental effect of these colonies was the almost ubiquitous deliberate introduction of non-native plant and animal species to the colonial settlements and the disruption to the biocultural knowledge and practices of the area's First Peoples. Colonial administrations were known for their extensive imports of alien species, mostly plants, throughout the colonised world (van Kleunen et al. 2018). It is recorded that Jan van Riebeeck, after landing at the Cape in South Africa in 1652, sent back word to Amsterdam to send them anything that can grow (Huntley 1999).

Unique urban tree species found in Georgetown, Guyana, and not in neighbouring countries reflect its colonial history (Hunte et al. 2019). Other examples of colonial influences on urban tree flora in the Global South include Bangalore, India, where 65% of the total tree flora are exotics (Nagendra and Gopal 2011), with similar patterns in Mendoza, Argentina (Martínez Carretero 2009); Rio de Janeiro, Brazil (dos Santos et al. 2009), Makhanda, South Africa (Kuruneri-Chitepo and Shackleton 2011) and Bandung, Indonesia (Abendroth et al. 2012). A discussion on the establishment of colonies in the United States, Canada, Australia and New Zealand gives a glimpse of the reasoning behind, and scale of impact that nineteenth-century settlers had on their new world (Ignatieva and Stewart 2009).

> …the settlers made themselves at home in these new lands by making it like home. They used European plants and animals and tools of industrial civilisation to transform the countryside with a speed and thoroughness never seen before and on a scale that has never been repeated. The destruction of native ecosystems was a central process, eclipsed only by the subsequent enthusiasm for importing mammals and birds for sentiment and sport. (Ignatieva and Stewart 2009)

Many European colonial cities had uniform grid layouts (Grant 2001; Beverley 2011) with racially separated residential areas (Njoh 2008b; Beverley 2011). In sub-Saharan Africa, racial segregation was a fundamental feature that was present in all colonial cities (Mukoko 1996; Njoh 2008a; Beeckmans 2013). Racial segregation was not invented by modern European colonialism, but it was extensively implemented throughout the colonised world (Raben 2020). This segregation often translated not only into separate areas but also differences in aesthetic planning and provision of green spaces, housing and garden sizes between ethnically delimited residential areas. Moreover, in many areas, segregation legacies persist despite postcolonial planning efforts to desegregate them (Smiley 2010; Muhammad et al. 2015).

Legacies of the 'apartheid' era racial segregation and town planning in South African cities not only manifests as separated areas but also in disparities in the quantity and quality of urban green infrastructure (McConnachie and Shackleton 2010; Venter et al. 2020). Moreover, research indicates significant differences in plant species richness and composition along socio-economic gradients (Lubbe et al. 2010). These legacies of racial segregation are also present in cities of the Global North, most notably in the United States (Andersson et al. 2018), and the legacies of colonisation continue to affect First Nations peoples across the globe. Therefore, it is important to recognise the need to examine and address these issues at a global scale and not frame them as a distinguishing feature unique to the Global South, although they are longer lived in the Global South. Recognising and elevating this as a global issue provides an opportunity to not only acknowledge the global impact but also creates a stronger platform for sharing strategies and examples of efforts to redress the inequities and create trajectories for a more just legacy in the future. In comparing access to green spaces between Global North and Global South cities, Rigolon et al. (2018) found that both contexts have inequalities but that in the Global North inequalities were more strongly related to quality, while inequalities related to proximity were more common in the Global South. Moreover, they found mostly similar patterns regarding proximity, quantity and quality inequities for cities in Africa, Asia and Latin America.

Slums and informal settlements have been part of cities since the start of the concentration of humans in urban areas (Abubakar et al. 2017). It is estimated that one-quarter of the global urban population lives in slums (Kuffer et al. 2016). While slums in former colonies are a legacy of racial segregation and other colonial practices, there are many other types of informal settlements found within the Global South (Dovey and King 2011). Low-income urban realities in regions such as Latin America occur in an array of conditions, basically given by how housing is established. On the one hand, irregular settlement in city peripheries often occurs, with little infrastructure and resources to transform the land, one of the reasons why the environment still resembles the one before the settling (often natural ecosystems or agroecosystems). On the other hand, when planning for low-income housing occurs, construction companies often develop high-density housing units that resemble concrete slabs at the landscape scale, where vegetation components are not only scarce but are basically lawns and isolated small trees and shrubs (Ian MacGregor-Fors, pers.obs.). These patterns starkly expose the repercussions of approaching development with too narrow a focus, as addressing perceived inequities related to urban infrastructure, can result in legacies of inequitable access to nature and biodiversity (McConnachie and Shackleton 2010).

The continued existence and ongoing growth of informal settlements have been linked to low planning capacity and failure of formal land markets (Kuffer et al. 2016). Despite typical negative perceptions and visions of depauperate environments, informal settlements can harbour surprising biodiversity. For example, Gopal and Nagendra (2014) recorded 141 plants species providing several types of ecosystem services to residents in 44 slums of Bangalore (India), while in South Africa Kaoma and Shackleton (2014) reported that 92% of all households in informal settlements of three towns had at least one tree, with an average of 9.7 trees and 4.1 species per household. Studying informal settlements as an urban form has the potential to reveal important lessons for how to temper some of the unsustainable legacies of urban sprawl and architecture that disconnects people from their communities and environments. The recognition of informal settlements as an urban form is not to hold up all informal settlements as a model to aspire to but rather a call to better understand the diversity and lessons that can be learned from including informal settlements as a prevalent urban form. Similar to the argument that every city needs to be contextualised and understood as a product of both its geographic location and social, cultural and political history, the informal areas also need to be considered in this way.

Urban heat islands are a significant legacy of the way urban areas are designed and built. Urban heat islands not only influence the natural environment but can also cause considerable human casualties in cities when combined with heatwaves and higher temperatures (Harlan et al. 2006). For example, heatwaves killed 739 people in Chicago in 1995 within a few days, and 900 people in London in 2003 (Larsen 2003). Moreover, urban heat island effects can vary along socio-economic gradients in a city (Harlan et al. 2006). A study in the United States using 108 cities determined that due to historical housing policies of refusing home loans or insurance based on race; disadvantaged neighbourhoods are up to 7 °C warmer than other residential areas (Hoffman et al. 2020). This pattern was consistent in 94% of the studied areas. There is a paucity of urban heat island (UHI) studies in Africa (Simwanda et al. 2019). Four capital cities with rapid urbanisation in Africa were studied to determine UHI patterns, namely Lagos, Addis Ababa, Lusaka and Nairobi (Simwanda et al. 2019). The results indicated that UHI effects were present in all the cities and that Lagos, with the highest population size and percentage of impervious surfaces, was 3–4 °C warmer than the other cities. Green spaces within a city can offer a cool environment for urban residents, with cooling effects felt up to almost two kilometres from a green space, as measured in Bulawayo, Zimbabwe (Ngulani and Shackleton 2020). A characteristic of most African cities is that much of the core, formal green spaces lie outside the urban area, with a clear separation between high impervious surfaces in the city centre and green spaces in the peripheries (Simwanda et al. 2019). Moreover, they highlight that unplanned and uncontrolled urbanisation worsens this situation and subsequently exacerbates UHI effects. Other research in Global South cities in Southeast Asia also indicated the presence of UHI indicating that temperatures between impervious surfaces and green spaces in the city can vary by as much as 3 °C (Estoque et al. 2017). A study of 5000 cities in Europe indicated that city size and compactness had the biggest influence on UHI intensity, with larger, more compact cities having the highest UHI effects (Estoque et al. 2017).

3.2 Biodiversity

Roman et al. (2018) discussed three important aspects influencing currently observed patterns with respect to the drivers of contemporary urban forest composition and structure. First is the bioregional context which describes the initial conditions in which the city was established; namely, topography, biome, climate, regional species pool and initial vegetation structure, as well as the land use before the establishment of the city. Second, the biophysical legacies represent past ecological disturbances such as pest and disease outbreaks and fires (Roman et al. 2018). Moreover, pests and disease outbreaks can combine with human drivers to co-create legacy effects, for example, the planting of monoculture street trees can have disastrous effects when trees are attacked by pests. Dutch Elm Disease killed 50% of all the elms in

Holland by 1939 and by the 1970s, 50–100 million elms were killed in the United States with some cities losing between 14,000 and 150,000 elms per city (Karnosky 1979). Information on the impact and full extent of urban tree pest outbreaks in the Global South is scarce. Recently, and of grave concern, is the first recording of the presence of the polyphagous shot hole borer tree pest and its fungal symbiont in South Africa (Paap et al. 2018). Current evidence suggests that thousands of urban trees have already died in South Africa, with millions more threatened (Potgieter et al. 2020). Research on 50-year trends of pest outbreaks on the indigenous Camphor tree (*Cinnamomum camphora*) in Chinese urban areas indicates that urbanisation and its associated effects increase urban pest outbreaks, with some areas severely affected (Xiang et al. 2020).

Third are the anthropogenic legacies, which describe the urban development history through specific historical periods such as colonial settlement histories and the establishment of parks and street trees (Roman et al. 2018). In Haikou, China, land-use legacy effects caused significant divergence in functional and taxonomic diversity in woody plants (Yang et al. 2017). Current urban bird density and richness was best described by landscapes of 20 years ago in Sheffield, UK (Dallimer et al. 2015). Additionally, anthropogenic legacies also pertain to socio-economic legacies and the neighbourhood urban form. In Phoenix, United States, historical customs and traditions of preference for mesic garden landscapes with lawns shape present-day urban neighbourhoods with newer neighbourhoods preferring xeric landscapes (Larson et al. 2017). Although described through examples in the United States and Canada, these drivers can be universally applied to any urban area. However, Roman et al. (2018) caution that despite commonalities through shared histories or similar residential layouts, etc. idiosyncrasies can occur highlighting "the crucial role of local sociopolitical history".

3.3 *Biological Invasions*

Since ancient times humans have transported and moved species around (van Kleunen et al. 2018). Urban systems are key components for biological invasions, not only as hubs for domestic animals and plants but by offering potentially invasive species the resources and conditions to establish and further spread outside their limits (Shochat et al. 2010). Additionally, some urban invasive species demonstrate the potential to be responsible for important economic loss (e.g. damage to urban infrastructure; Booy et al. 2017), and even on biodiversity. An example of the latter is the North American invasion of the House Sparrow (*Passer domesticus*), that is not only highly aggressive, but is so numerous in some urban systems that it can exclude up to one-third of the potential local avifauna that could be present in its absence (MacGregor-Fors et al. 2010; García-Arroyo et al. 2020).

Most invasive alien plant introductions originated from agriculture or ornamental horticulture (Hulme et al. 2018). Historical botanical gardens, in particular have been the source of many introductions of ultimately invasive species (Hulme 2015; Guo et al. 2016). Through the sixteenth to twentieth centuries, plant hunters, governments and colonial administrations actively promoted the trade and acquisition of rare and exotic, useful and ornamental species (van Kleunen et al. 2018). The case of *Lantana camara* is an example of the devastating effects of botanical and ornamental introductions, specifically in the colonial era (Bhagwat et al. 2012). In India, the East India Company established a botanical garden in 1786 and subsequent years saw thousands of introductions of alien species to botanical gardens in the subcontinent (Kannan et al. 2012). The first record of a *Lantana* species was in 1807 and by 1874 it was reported as widespread in some areas, its extensive use as a hedge plant helped it escape into the wild (Kannan et al. 2012). By 2010, the area of pasture land invaded in India was estimated at 13.2 million hectares, excluding forests and fallow land (Sharma and Raghubanshi 2010). In Uganda, respondents indicated that *Lantana* invasions reduced potential forage available for livestock by 50%, with similar reductions in crop yields (Shackleton et al. 2017).

Introductions of invasive species in South Africa were both deliberate and accidental. *Pinus pinaster* was planted in Cape Town with the belief that it would increase the water supply from the mountain catchment, while *Acacia* and *Hakea* spp. were introduced to combat soil erosion. Cecil Rhodes (an early colonial capitalist) deliberately introduced, among others, the now widespread European Starling (*Sturnus vulgaris*) to diversify bird fauna, while the massive imports of fodder during colonial settlement and the Anglo-Boer War of 1899–1902 accidentally introduced many invertebrate and forb species to South Africa (Huntley 1999). A recent review specifically details the impacts of biological invasions in South African urban ecosystems (Potgieter et al. 2020). However, not all issues regarding invasive species are clear cut, especially in urban areas. Historically introduced invasive trees grown in plantations in Cape Town significantly threaten native biodiversity, however, residents utilising these plantations for recreation vehemently oppose their removal (van Wilgen 2012). European settlers introduced many alien species which now have strong cultural and heritage values for urban dwellers; examples include *Eucalyptus* spp., *Ailanthus altissima* and *Acacia elata* (Potgieter et al. 2020). Some alien species are so bountiful and characteristic of cities in which they were established that the cities have a nickname based on these plants, e.g. in South Africa, Pretoria is known as Jacaranda City (*Jacaranda mimosifolia*) and Stellenbosch is known as Oak City (*Quercus* spp.) (Potgieter et al. 2020). The usefulness of some invasive species often promotes their distribution and, despite knowledge of invasive status, residents often deliberately cultivate them. In Fijian urban areas and elsewhere on the island, *Psidium guajava* (fruit), *Mikania micrantha* (medicinal) and *Spathodea campanulata* (fencing posts) are deliberately planted for their benefits (Lowry et al. 2020).

Urban freshwater and marine environments are also affected by bioinvasions. The advent of international shipping brought with it huge costs to the environment. Ship ballast, in particular, spread alien species around the world and to urban ports. Carlton (1996) reports several cases of marine bioinvasions through ballast water around the globe. One is the accidental introduction of comb jellyfish from American waters to a Russian port through discharged ballast water. In 1982 they caught one comb jellyfish in the Black Sea, by 1988, they calculated the standing stock of this jellyfish in terms of hundreds of thousands of metric tonnes (Carlton 1996). The proliferation of comb jellyfish had a drastic impact on measured anchovy landings (Carlton 1996). A study measuring phytoplankton species richness in ballast water from nine ships originating from several ports worldwide collectively recorded a total of 342 morphospecies present in ballast water (McCarthy and Crowder 2000). An example in South America is the invasion of the Southeast Asian freshwater bivalve mollusc in the Río de la Plata estuary around 1989 (Boltovskoy et al. 2006). Thirteen years after its introduction, almost the entire estuary was colonised, and it continues to spread up the river waterways through attachment to vessels (Boltovskoy et al. 2006). Several urban areas are located in the estuary and along the river banks. The mollusc has become a major pest for household water supply systems, industries and power plants (Cataldo et al. 2003). In South Africa, a review of marine bioinvasions revealed the ballast water and ship fouling were the dominant pathways for introductions (86 introduced and 39 cryptogenic species) with 53% of the introductions located within urban harbour areas (Mead et al. 2011).

A recent global study on accumulations of invasive alien species, using a novel database of 16,926 established species, indicates that rates of species establishment have increased and that it shows no signs of saturation (Seebens et al. 2017). Therefore, mitigation efforts have not been effective in containing the global spread of invasive species. Overall, Africa has the lowest number of first records of established alien species for all taxa, with Europe and North America having the highest recordings for vascular plants and insects (Seebens et al. 2017).

3.4 Soil

> The thin layer of soil covering the surface of the earth represents the difference between survival and extinction for most land-based life. (Doran and Zeiss 2000)

Urban soil quality and properties are often subject to legacy effects of human disturbances such as earthmoving and pollution, or biological influences such as invasive species. The remnants of past ecosystems or communities can remain present in current systems through ecological memory of soil properties, biological residues and sexual or asexual propagules and these can have a significant effect on the current ecosystem (Schaefer 2010). Even when invasive alien species can be effectively controlled, they often leave a soil legacy that hinders successful restoration; this is particularly true for nitrogen-fixing woody species. A recent review summed up the potential barriers namely: "altered soil microbial communities, depleted native soil seed banks, elevated nitrogen status, secondary invasion and weedy native species dominance, and reinvasion" (Nsikani et al. 2018).

Earthmoving, a daily occurrence in urban areas, can considerably alter soil properties. A study in Japan indicated that it promoted turnover to weeds and that the disturbed soils had higher pH, lower cation exchange capacity, total carbon, total nitrogen and medium-sized particles (Tsuzuki et al. 2020). Moreover, the changes in urban soils are often so pervasive that they can be classified as entirely new soil systems (Effland and Pouyat 1997).

In discussing urban geochemistry, Albanese and Cicchella (2012) state that "toxic metal contamination, organic pollution, smog, acid rain, and greenhouse gas accumulation are the most widespread legacies of an often-uncontrolled growth that has deeply changed the geochemical character of the urban environment". Moreover, "the past is frequently a hidden source of environmental problems with the potential to affect the health of current and future urban residents" (Albanese and Cicchella 2012). A detrimental legacy effect of human activities in urban areas is pollutants in soil. For example, in Copiapó, Chile, urban soils have copper levels above the international guidelines as a result of the city expanding into previously mined areas and its location in the vicinity of a currently active smelter (Carkovic et al. 2016). In China, a review by Wei and Yang (2010) indicated that heavy metal contamination (Cr, Ni, Cu, Pb, Zn and Cd) is common in urban soils and road dust. Legacies of leaded gasoline use and lead-based paints have significantly contaminated urban soils in most countries (Schwarz et al. 2016). Urban soils act as a sink for almost all of the pollutants deposited by atmospheric fallout (Albanese and Cicchella 2012). Urban agriculture practices are also often hampered by urban soils with legacies of degradation and contamination, and novel strategies have had to be implemented to protect growth mediums from contaminated soils (Moskal and Berthrong 2018).

A major problem, with long-lasting legacies, is the bioaccumulation of persistent organic pollutants (POP). A study in Metro Vancouver, British Columbia, Canada researched the bioaccumulation of POPs in urban terrestrial food webs (Fremlin et al. 2020). They found that many legacy POP banned or restricted for more than 40 years were still detected in most of the species groups in the food web. In a study of seven major cities in India, the levels of organochlorine pesticides in urban soils showed varying results based on regional usage patterns (Chakraborty et al. 2015). However, due to the legacies of heavy use in the past, soils are now sources of organochlorine pesticides re-emitting it back into the atmosphere (Chakraborty et al. 2015). Not all Global South cities are equally burdened by organochlorines soil contamination, and in some, the levels were low enough not to pose health risks (Kumar et al. 2018).

3.5 Urban Freshwater Bodies

Urbanisation worldwide has a detrimental impact on urban freshwater sources such as rivers flowing through urban areas. The altered conditions in urban rivers are called the urban stream syndrome, referring to degraded physical, chemical and biological conditions (Walsh et al. 2005; Booth et al. 2016). The Korotoa River, sampled in the urban areas of the Bogra district in Bangladesh, revealed high levels of heavy metal pollution, above the recommended safety levels (Islam et al. 2015). However, the type and degree of ecological responses to urbanisation vary between cities (Utz et al. 2016). Moreover, the role that legacy effects have in influencing urban streams remains understudied (Utz et al. 2016). A study in Maryland, United States, indicated that despite restoration attempts, urbanisation legacies resulted in poor water quality and altered hydrology in urban streams, and aquatic insect communities were also less diverse in comparison to forest sites (Fanelli et al. 2019).

Extreme climates and rapid urbanisation often put a severe strain on urban freshwater bodies in the Global South (Sim and Balamurugan 1991; Wantzen et al. 2019). Residents living near water bodies often have negative perceptions of urban streams and wetlands due to excessive dumping (Beyene et al. 2008), poor water quality and bad smells (Wantzen et al. 2019). In the Global North, policy changes in the United States and Europe allowed similar urban freshwater scenarios to considerably recover in the past few decades (Wantzen et al. 2019). Moreover, the situation in the Global South is compounded by the fact that the urban poor often inhabit riparian zones and lack the capabilities or political influence to effect changes. Additionally, urban planners often overlook or deliberately replace freshwater hydrosystems in urban areas, especially in the Global South (Wantzen et al. 2019). Therefore, Wantzen et al. (2019) introduced the term 'Southern Urban Hydrosystem Syndrome' to describe how the distinctive geographical and socio-political settings in the Global South result in an alternative form compared to the 'Urban Stream Syndrome' that emerged from temperate systems in the Global North. In their paper on the role of urbanisation characteristics on heterogenous stream responses globally, Parr et al. (2016) identify key future research questions among which the importance of understanding legacy effects is central in understanding stream responses. The lack of research on urban streams in developing countries constrains effective management solutions and "if we are going to protect global freshwater resources and ensure the provisioning of freshwater ecosystem services, more research is needed to understand the effects of anthropogenic stressors on fresh waters in lower-income countries" (Capps et al. 2016).

3.6 *Natural Disasters*

> It is clear that the modern city is virtually indestructible. At the same time, there is no question that a catastrophe will profoundly alter a city's fortunes and fate; and therein lays the more compelling matter of resilience and recovery. Indeed, it is possible for a city to be reconstructed, even heroically, without fully recovering. Put another way, resilience involves much more than rebuilding. (Campanella 2006)

Worldwide, urban areas are affected by natural disasters such as flooding, earthquakes, tornadoes, wildfires and hurricanes. These disasters can have lasting impacts on affected communities. Legacies of urban development expansions into floodplains, often unplanned and haphazard, exacerbate the effects of natural disasters such as flooding, moreover, people often resettle in the same high-risk areas after the disaster event (Hardoy and Pandiella 2009). The city of Santa Fe, Argentina, experienced major floods in 2003 and again in 2006–2007 and in both instances the same part of the city was affected, turning one-third of the city situated in a floodplain into a shallow lake (Hardoy and Pandiella 2009). Moreover, socio-demographic legacies can have significant impacts on recovery after disasters. In New Orleans, recovery after Hurricane Katrina was unequal, with historical environmental injustices leading to the most vulnerable people being located in the most high-risk areas, where post-disaster recovery is much slower than in high-income areas (Finch et al. 2010). However, disasters can also allow neighbourhood improvements through redesign and rebuilding (Olshansky et al. 2006).

Disasters are often intensified by legacies of poor urban planning, especially in the Global South. In Freetown, Sierra Leone, a debris slide, debris flow and sediment-laden flood in 2017 caused hundreds of deaths and missing persons and the destruction of many houses (Cui et al. 2019). The disaster was exacerbated by housing construction in dangerous areas, clearing of hillside vegetation, poor quality building materials and low-cost buildings. The scenario was also seen in Haiti, where poor and incorrectly constructed buildings severely exacerbated the destruction caused by the earthquake despite early warnings by seismologists (Laursen 2010). Worldwide, the most vulnerable people are disproportionately affected by disasters, due to poverty and living in the highest risk areas (Collins 2010). In responding to natural disasters, developed countries can often act independently, while developing countries frequently need external help to handle major catastrophes (Lassa 2015). An important consequence of the focus of colonial urban development on port cities for effective resource extraction (Myers 2021), is their vulnerability to future climate-change induced sea-level rises.

3.7 The Ravages of War

War is rarely constrained to human casualties only; the side effects on the environment and urban areas can be devastating. Some of the effects of war on the environment and wildlife include environmental pollution, dramatic habitat alteration, wildlife decimation, overharvesting and severe degradation (Dudley et al. 2002; Hanson et al. 2009; Lawrence et al. 2015). War can also dramatically influence soil environments (Certini et al. 2013). Several countries are still affected by often deadly after-effects of war through the presence of land mines and other unexploded devices that litter old battlegrounds. Some form of landmine related crisis is experienced by one-third of all developing countries (Berhe 2007). War affects urban areas in two main ways, through direct destruction and indirectly through the influx of refugees to urban areas which can be accompanied by increased pressures on local resources.

Many cities were heavily damaged or demolished during the Second World War. Research in Tokyo and Hiroshima (Japan) indicates that in Tokyo, 63,000 street trees were destroyed during the war (Cheng and McBride 2014). Moreover, the atomic bomb dropped on Hiroshima levelled 12 km^2 of the city, with 50% of the trees with broken stems in the blast zone (Cheng and McBride 2014). Hamburg and Dresden (Germany) lost 42 and 51% of their street trees, respectively, during the Second World War (Stilgenbauer and McBride 2010). Desperate residents cut three-quarters of all the urban trees and most of the trees in peri-urban areas for firewood during the Siege of Sarajevo in Bosnia and Herzegovina (1992–1996) (Lacan and McBride 2009). In all five examples, concerted efforts were made to restore the urban forests.

During armed conflicts, the built environment is often intentionally targeted. The "widespread and deliberate destruction of the urban environment" became known as urbicide (Coward 2009). In Palestine, Israeli forces used bulldozers to raze Palestinian settlements and Salmon (2002) observed: "houses are destroyed, olive trees uprooted, orange groves laid waste". In 2007, after a three-and-a-half-month war in northern Lebanon, the Palestinian refugee camp Nahr el-Bared was destroyed. A Palestinian resident stated, "there are no houses left untouched…they removed trees and plants, destroyed buildings…they are against trees and against humans and against stones" (Ramadan 2009). The civil war in Somalia (1988–1990) led to the destruction of several indigenous forests and amenity trees in and around major towns, the most prominent being in Hargeisa, Borama, Berbera and Erigavo; where impoverished residents and internally displaced persons cut trees for firewood, fencing and poles (Candlelight for Health 2006). The political conflicts and poor economic situations of most developing countries, signify that most of these effects have long-term legacies.

A key legacy of war is the major displacement of people. Due to internal conflicts in rural areas of Colombia, refugees seeking asylum in towns put severe pressure on urban forests and surrounding forest remnants (Escobedo et al. 2015). In the Western Area Peninsula Forest Reserve in Sierra Leone, large numbers of displaced people that remained after the war ceased reduced large areas of previously dense forest to "bushes and some areas to bare land" (Conteh et al. 2016). A long-term study spanning 35 years indicated rapid increases in urban built-up areas, decreased forest areas and landscape fragmentation due to resettlement of displaced rural residents in the post-war Western Area in Sierra Leone (Gbanie et al. 2018). In their overview on post-war urbanism, Beall and Goodfellow (2014) state that civil war in Africa caused spectacular rates of urban growth in some areas, for example in Kigali, Rwanda, the annual growth rate was 18% from 1995 to 2000, and Luanda, Angola, grew five times in size during their civil war. Such accelerated growth puts a severe strain on urban areas. The number of internally displaced persons worldwide, as a result of violence and conflict, was 45.7 million at the end of 2019 (Internal Displacement Monitoring Centre 2020). The top ten countries that collectively represent 34.5 million internally displaced persons are all Global South countries.

Not only is the Global South disproportionately affected by contemporary armed conflicts, but also most of the global hotspots of biodiversity are located in the Global South (Shackleton et al. 2021). Globally, 36 regions are recognised as hotspots of biodiversity (Conservation International 2020). Spatial analysis of the hotspots indicates that the boundaries of 170 countries intersect with these regions. Within these countries, 423 cities (with a population larger than 300,000) are located within hotspots, of which 383 show rapid urban growth (Weller et al. 2018). One hundred and forty of these countries (82%) are located in the Global South. Alarmingly, in the period between 1950 and 2000, 90% of the major armed conflicts took place in countries that contain biodiversity hotspots, and more than 80% occurred directly within hotspot areas (Hanson et al. 2009).

3.8 Extinction Debts

The presence and persistence of species in a landscape can be attributed to several factors, including suitable habitat, genetic diversity, presence of food and nutrients, viable offspring that survive and establish, lack of competition or predation and adequately sized habitats. If any of these factors are under pressure or lost, species can decline and even go extinct. Those species most at risk of local extinction are rare, specialist species that inhabit threatened habitats (Zettlemoyer et al. 2019). The major causes of species extinction are invasive species, climate change and habitat loss (Kuussaari et al. 2009). However, not all influences have immediate effects; there are often time delays between cause and effect, termed time lags (Magnuson 1990). These time lags can cause a situation where extinctions can be delayed, and thus, the landscape then incurs a debt, i.e. the extinction debt (Tilman et al. 1994). The presence of extinction debts in a landscape means that extant species can persist in environments which are unsuitable for long-term survival and that they will go extinct in future even when no further changes in habitat conditions or other factors occur (Tilman et al. 1994; Hahs et al. 2009).

Urban environments are highly vulnerable to extinction debts due to intense and extreme changes that typically occur there. Extinction debts have been shown for European cities (Dullinger et al. 2013) and globally (Hahs et al. 2009). Hahs et al. (2009) compared plant extinction rates for 22 cities and found that 67% of the observed trends could be explained by the proportion of native vegetation remaining in the study area and the development history of the city. Cities that had been urbanised for less than 200 years all demonstrated lower than expected extinction rates, while cities that had been urbanised for 200–400 years all showed higher rates of extinction. A couple of cities demonstrated higher than expected rates of extinction, which indicates that additional drivers are acting in the landscape.

Cities in the Global South display a range of development histories; from more recent cities, such as those in sub-Saharan Africa and parts of China, through to cities that are thousands of years old, such as those in Asia and areas of South America. While all of these cities are likely to be carrying an extinction debt, the pace and intensity of urban development in these regions, compounded by limited resources for management and protection, may accelerate the rate at which the extinction debt is realised. Further research is required to fully understand the legacy that this rapid urbanisation will have on the biodiversity of these cities in the future.

Biotic homogenisation is a concept that proposes that the introduction of cosmopolitan species common to many cities, and the loss of locally indigenous species through the extinction debt will result in cities with a reduced species pool that is highly similar around the world (McKinney 2006). Therefore, cities with the highest distinct biodiversity may also be carrying the highest potential extinction debts. For example, changes in urban grassland vegetation in Potchefstroom, South Africa, over 24 years showed that the vegetation was not homogenising, but that the native forb species decreased significantly, an outcome which could partially be explained by below-average rainfall in the preceding years (du Toit et al. 2020). Moreover, vegetation species richness was best predicted by historic landscape measures, indicating the presence of time lags and therefore, potential extinction debts (du Toit et al. 2016, 2020). This research highlights that for cities in the Global South where climate change impacts are predicted to be particularly pronounced, the potential for accelerated rates of extinction may be more likely to be observed than what has previously been recorded in the Global North.

3.9 *Extinction of Experience*

A significant effect influencing attitudes towards nature worldwide is what is coined as "extinction of experience" (Pyle 2003; Miller 2005) or "generational amnesia" (Kahn and Friedman 1995; Kahn 2007). Pyle (2003) explains the extinction of experience as:

> Essentially, the extinction of experience syndrome works as follows: when common species of plants and animals (as well as cultural, architectural, or any other features of diversity) become extirpated in one's everyday environs – within, that is, one's radius of reach – one grows increasingly inured to their absence. (The radius of reach is smaller for the poor, the very old, the very young, and the disabled.) That is, as the richness of the neighbourhood diminishes, the power of the neighbourhood to fascinate, arouse, excite, and stimulate also passes into dullness, ennui, and apathy. Those who know and recognise less, care less, and therefore act less, leading to still more losses. The sequelae of extinction and alienation are further loss and greater detachment, round and round. So the extinction of experience precipitates a cycle of disaffection, degradation, and ultimate separation from nature.

This effect is directly related to the history of the landscape, a person's upbringing and their perceptions of and interaction with nature during childhood. In a classic example Pauly (1995) explains it as the "shifting baseline syndrome". In his discussion on what is seen as acceptable global fish stocks; he explains that each scientist's perception of baselines of fish stocks and species composition is based on what they observed at the beginning of their career; therefore, each new scientist's conceived baseline is based on steadily depleting global fish stocks. This phenomenon causes an underestimation of the true original diversity and abundance of biodiversity, which in turn will have significant ramifications for decision-making on rehabilitation strategies, management and conservation imperatives (Pauly 1995). Acknowledging the existence of the extinction of experience highlights the importance of collecting historical data and incorporating it into contemporary thinking and research frameworks (Pauly 1995).

Since 1500 A.D. there have been 811 documented species extinctions (Dirzo and Raven 2003). In 2020 the IUCN Red List of threatened species (IUCN 2020) stated that the current number of species threatened with extinction numbered 32,000. The Intergovernmental Science-Policy Platform on Biodiversity and Ecosystem Services (IPBES) 2019 global assessment report on biodiversity and ecosystem services predicted that this number is closer to one million species (IPBES 2019). For conservation to be successful support from local communities is essential (Ferketic et al. 2010; Bennett and Dearden 2014). A study in Singapore determined that childhood experience with nature was the most important predictor of their attitude towards wildlife (Ngo et al. 2019). This result emphasises the importance of giving children exposure to nature in order to ensure that as adults, they value urban wildlife (Ngo et al. 2019). In Brazil, students felt differently towards certain species than others and rural children had a higher aversion for some species than urban children due to a history of conflicts (de Oliveira et al. 2018). Moreover, the study found positive correlations between income and schooling level and an affinity for vertebrates, emphasising the importance of education in ensuring future positive attitudes towards urban nature and especially contentious species such as snakes. To combat the negative effects of the extinction of experience, and ensure that future generations care about urban nature conservation, cities need to be designed to afford more opportunities for contact with nature, and environmental education needs to be prioritised.

4 The Future

> Today's land-use decisions will generate tomorrow's legacies. (Ziter et al. 2017)

Urban environments around the world are the result of their evolutionary and geographic history, and more contemporary influences of the social, cultural, economic, political and technical systems that shaped their construction. The impacts of the urban form on the biodiversity and ecosystems of the city are shaped by five broad filters (Aronson et al. 2014). While those filters are likely to be common to all cities, the specifics of the filter and the resulting outcomes are context-dependent. More urban ecology research is required to begin untangling the effects of urbanisation and bioclimatic regions from the drivers that are much more context-specific. Growing the number of studies conducted in the Global South is critical to develop an evidence base that is representative of the full range of conditions for cities around the world.

While the legacies of colonisation have had a prolific effect on cities and First Nations peoples around the world, the problems and challenges of deindustrialisation and the development of sustainable and just cities for the future are exacerbated in the Global South. Of the 7.7 billion people living on the planet in 2019, 86% were living in the Global South. These countries will also be home to the majority of the additional two billion people who will be added in the lead up to the year 2050 (United Nations 2019). The scale and pace of growth in human populations living in cities throughout the Global South are compounded by the lower economic resource base available to support this development and the highest levels of income disparity. This disparity and lower economic resource base results in different financing instruments and infrastructure forms and priorities compared to the Global North, along with a different set of legacies that emerge from them. Understanding these legacies and addressing the challenges that emerge from this model of urban development can only be fully understood by situating sufficient research firmly in the Global South.

The characteristics mentioned above make a case for urban areas of the Global South and Global North sharing information to address similar situations, but also for a need to uniquely address and acknowledge the distinctions displayed in the Global South. There are vast gaps in knowledge of urban ecology in the Global South, with many unresearched countries and urban forms. These need to be addressed urgently to quantify the impacts of the history of these landscapes on contemporary patterns before the patterns *themselves* are history.

While cities of the Global South face their own unique challenges, there are many examples of innovative approaches that emerge from very localised and context-driven projects. There are also strong examples of initiatives and programmes to actively redress existing injustices. These are a unique strength, and highlight the role that the Global South can play in driving the movement to a more sustainable and equitable world. It is only through a genuine and reciprocal dialogue between research arising from both the Global North and the Global South that the trajectories and legacies of urban development can be shifted towards a fairer and more equitable global future.

References

Abendroth S, Kowarik I, Müller N, Von Der Lippe M (2012) The green colonial heritage: woody plants in parks of Bandung, Indonesia. Landsc Urb Plann 106:12–22

Abubakar A, Romice O, Salama AM (2017) Defining slums using multidimensional and relational properties: a dynamic framework for intervention. Int J Archit Res 11:34–54

Albanese S, Cicchella D (2012) Legacy problems in urban geochemistry. Elements 8:423–428

Andersson EK, Lyngstad TH, Sleutjes B (2018) Comparing patterns of segregation in north-western Europe: a multiscalar approach. Eur J Popul 34:151–168

Aronson MF, La Sort FA, Nilon CH, Katti M, Goddard MA, Lepczyk CA, Warren PS, Williams NS, Cilliers S, Clarkson B, Dobbs C, Dolan R, Hedblom M, Klotz S, Kooijmans JL, Kuhn I, Macgregor-Fors I, Mcdonnell M, Mortberg U, Pysek P, Siebert S, Sushinsky J, Werner P, Winter M (2014) A global analysis of the impacts of urbanisation on bird and plant diversity reveals key anthropogenic drivers. Proc Biol Sci 281:20133330

Aronson MF, Nilon CH, Lepczyk CA, Parker TS, Warren PS, Cilliers SS, Goddard MA, Hahs AK, Herzog C, Katti M, La Sorte FA, Williams NS, Zipperer W (2016) Hierarchical filters determine community assembly of urban species pools. Ecol 97:2952–2963

Beall J, Goodfellow T (2014) Conflict and post-war transition in African cities. In: Parnell S, Pieterse E (eds) Africa's urban revolution. Zed Books, London

Beeckmans L (2013) Editing the African city: reading colonial planning in Africa from a comparative perspective. Plan Perspect 28:615–627

Bennett NJ, Dearden P (2014) Why local people do not support conservation: community perceptions of marine protected area livelihood impacts, governance and management in Thailand. Mar Policy 44:107–116

Berhe AA (2007) The contribution of landmines to land degradation. Land Degrad Dev 18:1–15

Beverley EL (2011) Colonial urbanism and South Asian cities. Soc Hist 36:482–497

Beyene A, Legesse W, Triest L, Kloos H (2008) Urban impact on ecological integrity of nearby rivers in developing countries: the Borkena River in highland Ethiopia. Environ Monit Assess 153:461

Bhagwat SA, Breman E, Thekaekara T, Thornton TF, Willis KJ (2012) A battle lost? Report on two centuries of invasion and management of *Lantana camara* L. in Australia, India and South Africa. PLoS One 7: e32407

Blair RB, Johnson EM (2008) Suburban habitats and their role for birds in the urban–rural habitat network: points of local invasion and extinction? Landsc Ecol 23:1157–1169

Boltovskoy D, Correa N, Cataldo D, Sylvester F (2006) Dispersion and ecological impact of the invasive freshwater bivalve *Limnoperna fortunei* in the Río de la Plata watershed and beyond. Biol Inv 8:947–963

Booth DB, Roy AH, Smith B, Capps KA (2016) Global perspectives on the urban stream syndrome. Freshwater Sci 35:412–420

Booy O, Cornwell L, Parrott D, Sutton-Croft M, Williams F (2017) Impact of biological invasions on infrastructure. In: Vilà M, Hulme PE (eds) Impact of biological invasions on ecosystem services. Springer, Cham, pp 235–248

Campanella TJ (2006) Urban resilience and the recovery of New Orleans. J Am Plan Assoc 72:141–146

Candlelight for Health, Education & Environment (2006) Impact of civil war on Natural Resources—a case study for Somaliland [Online]. Available: https://land.igad.int/index.php/documents-1/countries/somalia/conflict-4/878-impact-of-civil-war-on-natural-resources-a-case-study-for-somaliland/file. Accessed 5 Aug 2020

Capps KA, Bentsen CN, Ramírez A (2016) Poverty, urbanisation, and environmental degradation: urban streams in the developing world. Freshwater Sci 35:429–435

Carkovic AB, Calcagni MS, Vega AS, Coquery M, Moya PM, Bonilla CA, Pasten PA (2016) Active and legacy mining in an arid urban environment: challenges and perspectives for Copiapo, Northern Chile. Environ Geochem Health 38:1001–1114

Carlton JT (1996) Marine bioinvasions: the alteration of marine ecosystems by nonindigenous species. Oceanogr 9:36–43

Cataldo D, Boltovskoy D, Pose M (2003) Toxicity of chlorine and three nonoxidising molluscicides to the pest mussel *Limnoperna fortunei*. J Am Water Works Assoc 95:66–78

Certini G, Scalenghe R, Woods WI (2013) The impact of warfare on the soil environment. Earth Sci Rev 127:1–15

Chakraborty P, Zhang G, Li J, Sivakumar A, Jones KC (2015) Occurrence and sources of selected organochlorine pesticides in the soil of seven major Indian cities: assessment of air-soil exchange. Environ Poll 204:74–80

Cheng S, McBride JR (2014) Restoration of the urban forests of Tokyo and Hiroshima following World War II. In: Tidball KG, Krasny ME (eds) Greening in the red zone: disaster, resilience and community greening. Springer, Dordrecht, pp 225–248

Chown SL, Duffy GA, Hahs A (2015) Thermal physiology and urbanisation: perspectives on exit, entry and transformation rules. Funct Ecol 29:902–912

Clarke LW, Jenerette GD, Davila A (2013) The luxury of vegetation and the legacy of tree biodiversity in Los Angeles, CA. Landsc Urb Plan 116:48–59

Collins TW (2010) Marginalisation, facilitation, and the production of unequal risk: the 2006 Paso del Norte floods. Antipode 42:258–288

Conservation International (2020) Explore the biodiversity hotspots [Online]. Available: http://www.cepf.net/resources/hotspots/Pages/default.aspx. Accessed 28 Aug 2020

Conteh A, Gavin MC, McCarter J (2016) Assessing the impacts of war on perceived conservation capacity and threats to biodiversity. Biodivers Conserv 26:983–996

Coward M (2009) Urbicide: the politics of urban destruction. Routledge, New York

Cui Y, Cheng D, Choi CE, Jin W, Lei Y, Kargel JS (2019) The cost of rapid and haphazard urbanisation: lessons learned from the Freetown landslide disaster. Landslides 16:1167–1176

Dallimer M, Davies ZG, Diaz-Porras DF, Irvine KN, Maltby L, Warren PH, Armsworth PR, Gaston KJ (2015) Historical influences on the current provision of multiple ecosystem services. Glob Environ Change 31:307–317
Davoren E, Siebert S, Cilliers S, du Toit MJ (2016) Influence of socioeconomic status on design of Batswana home gardens and associated plant diversity patterns in northern South Africa. Landsc Ecol Eng 12:129–139
de Oliveira JV, Lopes SD, Barboza RR, Alves RR (2018) To preserve, or not to preserve, that is the question: urban and rural student attitudes towards wild vertebrates. Environ Dev Sustain 21:1271–1289
Departamento De Gestión Del Medio Ambiente (2010) Las Aves de Mi Ciudad: Una Guía de Las Aves de Cali. DAGMA, Cali
Dirzo R, Raven PH (2003) Global state of biodiversity and loss. Ann Rev Environ Res 28:137–167
Doran JW, Zeiss MR (2000) Soil health and sustainability: managing the biotic component of soil quality. Appl Soil Ecol 15:3–11
dos Santos AR, Da Rocha CF, Bergallo HG (2009) Native and exotic species in the urban landscape of the city of Rio de Janeiro, Brazil: density, richness, and arboreal deficit. Urban Ecosys 13:209–222
Dovey K, King R (2011) Forms of informality: morphology and visibility of informal settlements. Built Environ 37:11–29
Dudley JP, Ginsberg JR, Plumptre AJ, Hart JA, Campos LC (2002) Effects of war and civil strife on wildlife and wildlife habitats. Conserv Biol 16:319–329
Dullinger S, Essl F, Rabitsch W, Erb K-H, Gingrich S, Haberl H, Hülber K, Jarošík V, Krausmann F, Kühn I, Pergl J, Pyšek P, Hulme PE (2013) Europe's other debt crisis caused by the long legacy of future extinctions. Proc Nat Acad Sci 110:7342–7347
Dupouey JL, Dambrine E, Laffite JD, Moares C (2002) Irreversible impact of past land use on forest soils and biodiversity. Ecol 83:2978–2984
du Toit MJ, Kotze DJ, Cilliers SS (2016) Landscape history, time lags and drivers of change: urban natural grassland remnants in Potchefstroom, South Africa. Landsc Ecol 31:2133–2150
du Toit MJ, Kotze DJ, Cilliers SS (2020) Quantifying long-term urban grassland dynamics: biotic homogenisation and extinction debts. Sustain 12:1989
Effland W, Pouyat R (1997) The genesis, classification, and mapping of soils in urban areas. Urban Ecosys 1:217–228
Ertan A, Fiszbein M, Putterman L (2016) Who was colonised and when? A cross-country analysis of determinants. Eur Econ Rev 83:165–184
Escobar-Ibáñez JF, Macgregor-Fors I (2016) Peeking into the past to plan the future: assessing bird species richness in a neotropical city. Urban Ecosys 19:657–667
Escobedo FJ, Clerici N, Staudhammer CL, Corzo GT (2015) Socio-ecological dynamics and inequality in Bogotá, Colombia's public urban forests and their ecosystem services. Urban For Urban Green 14:1040–1053
Estoque RC, Murayama Y, Myint SW (2017) Effects of landscape composition and pattern on land surface temperature: an urban heat island study in the megacities of Southeast Asia. Sci Total Environ 577:349–359
Evans KL, Newson SE, Gaston KJ (2009) Habitat influences on urban avian assemblages. Ibis 151:19–39
Fanelli RM, Prestegaard KL, Palmer MA (2019) Urban legacies: aquatic stressors and low aquatic biodiversity persist despite implementation of regenerative stormwater conveyance systems. Freshwater Sci 38:818–833
Ferketic JS, Latimer AM, Silander JA (2010) Conservation justice in metropolitan Cape Town: a study at the Macassar Dunes Conservation Area. Biol Conserv 143:1168–1174
Filloy J, Grosso S, Bellocq MI (2015) Urbanisation altered latitudinal patterns of bird diversity-environment relationships in the southern Neotropics. Urban Ecosys 18:777–791
Finch C, Emrich CT, Cutter SL (2010) Disaster disparities and differential recovery in New Orleans. Popul Environ 31:179–202

Fremlin KM, Elliott JE, Green DJ, Drouillard KG, Harner T, Eng A, Gobas F (2020) Trophic magnification of legacy persistent organic pollutants in an urban terrestrial food web. Sci Total Environ 714:136746
Garaffa PI, Filloy J, Bellocq MI (2009) Bird community responses along urban–rural gradients: does the size of the urbanised area matter? Landsc Urb Plan 90:33–41
García-Arroyo M, Santiago-Alarcon D, Quesada J, Macgregor-Fors I (2020) Are invasive house sparrows a nuisance for native avifauna when scarce? Urban Ecosys 23:793–802
Gbanie S, Griffin A, Thornton A (2018) Impacts on the urban environment: land cover change trajectories and landscape fragmentation in post-war Western Area, Sierra Leone. Remote Sens 10:129
Gil D, Brumm H (2014) Avian urban ecology. Oxford University Press, Oxford
González-García F, Straub R, García JA, Macgregor-Fors I (2014) Birds of a neotropical green city: an up-to-date review of the avifauna of the city of Xalapa with additional unpublished records. Urban Ecosys 17:991–1012
González-García F, Straub R, Lobato García JA, Macgregor-Fors I, Santiago Alarcón D (2016) Nuevos registros y notas adicionales comentadas sobre la avifauna de la ciudad de Xalapa, Veracruz, México. Acta Zool Mex 32:253–269
González-Lagos C, Quesada J (2017) Stay or leave? Avian behavioral responses to urbanisation in Latin America. In: Macgregor-Fors I, Escobar-Ibáñez JF (eds) Avian ecology in Latin American cityscapes. Springer, Cham, pp 99–124
Gopal D, Nagendra H (2014) Vegetation in Bangalore's Slums: boosting livelihoods, well-being and social capital. Sustain 6:2459–2473
Grant J (2001) The dark side of the grid: power and urban design. Plan Perspect 16:219–241
Grimm NB, Grove MJ, Pickett STA, Redman CL (2000) Integrated approaches to long-term studies of urban ecological systems. Bioscience 50:571
Grimm NB, Faeth SH, Golubiewski NE, Redman CL, Wu J, Bai X, Briggs JM (2008) Global change and the ecology of cities. Sci 319:756–760
Grove S, Parker IM, Haubensak KA (2015) Persistence of a soil legacy following removal of a nitrogen-fixing invader. Biol Inv 17:2621–2631
Guo X, Ren M, Ding J, Novak S (2016) Do the introductions by botanical gardens facilitate the invasion of *Solidago canadensis* (Asterceae) in China? Weed Res 56:442–451
Gustavsson E, Lennartsson T, Emanuelsson M (2007) Land use more than 200 years ago explains current grassland plant diversity in a Swedish agricultural landscape. Biol Conserv 138:47–59
Hahs AK, Mcdonnell MJ, McCarthy MA, Vesk PA, Corlett RT, Norton BA, Clemants SE, Duncan RP, Thompson K, Schwartz MW, Williams NS (2009) A global synthesis of plant extinction rates in urban areas. Ecol Lett 12:1165–1173
Hanson T, Brooks TM, Da Fonseca GA, Hoffmann M, Lamoreux JF, Machlis G, Mittermeier CG, Mittermeier RA, Pilgrim JD (2009) Warfare in biodiversity hotspots. Conserv Biol 23:578–587
Hardoy J, Pandiella G (2009) Urban poverty and vulnerability to climate change in Latin America. Environ Urban 21:203–224
Harlan SL, Brazel AJ, Prashad L, Stefanov WL, Larsen L (2006) Neighborhood microclimates and vulnerability to heat stress. Soc Sci Med 63:2847–2863
Hightower J, Butterfield A, Weishampel J (2014) Quantifying ancient Maya land use legacy effects on contemporary rainforest canopy structure. Remote Sens 6:10716–10732
Hoffman JS, Shandas V, Pendleton N (2020) The effects of historical housing policies on resident exposure to intra-urban heat: a study of 108 US urban areas. Climate 8:12
Hulme PE (2015) Resolving whether botanic gardens are on the road to conservation or a pathway for plant invasions. Conserv Biol 29:816–824
Hulme PE, Brundu G, Carboni M, Dehnen-Schmutz K, Dullinger S, Early R, Essl F, González-Moreno P, Groom QJ, Kueffer C, Kühn I, Maurel N, Novoa A, Pergl J, Pyšek P, Seebens H, Tanner R, Touza JM, Van Kleunen M, Verbrugge LN, Flory L (2018) Integrating invasive species policies across ornamental horticulture supply chains to prevent plant invasions. J Appl Ecol 55:92–98

Hunte N, Roopsind A, Ansari AA, Caughlin T (2019) Colonial history impacts urban tree species distribution in a tropical city. Urban For Urban Green 41:313–322
Huntley BJ (1999) South Africa's experience regarding alien species: impacts and controls. In: Sandlund OT, Schei PJ, Viken Å (eds) Invasive species and biodiversity management. Kluwer, Dordrecht, pp 363–376
Ignatieva ME, Stewart GH (2009) Homogeneity of urban biotopes and similarity of landscape design language in former colonial cities. In: McDonnell MJ, Hahs AK, Brueste JH (eds) Ecology of cities and towns: a comparative approach. Cambridge University Press, Cambridge, pp 399–421
Internal Displacement Monitoring Centre (2020) 2020 global report on internal displacement. IDMC, Geneva [Online]. Available https://www.internal-displacement.org/sites/default/files/publications/documents/2020-IDMC-GRID.pdf. Accessed 11 Aug 2020
IPBES (2019) Summary for policymakers of the global assessment report on biodiversity and ecosystem services of the Intergovernmental Science-Policy Platform on Biodiversity and Ecosystem Services. In Díaz S, Settele J, Brondízio ES, Ngo HT, Guèze M, Agard J, Arneth A, Balvanera P, Brauman KA, Butchart SHM, Chan KMA, Garibaldi LA, Ichii K, Liu J, Subramanian SM, Midgley GF, Miloslavich P, Molnár Z, Obura D, Pfaff A, Polasky S, Purvis A, Razzaque J, Reyers B, Roy Chowdhury R, Shin YJ, Visseren-Hamakers IJ, Willis KJ, Zayas CN (eds). IPBES secretariat, Bonn, Germany [Online]. Available https://zenodo.org/record/3553579#.X5Laae2xVPY. Accessed 23 Oct 2020
Islam MS, Ahmed MK, Raknuzzaman M, Habibullah Al Mamun M, Islam MK (2015) Heavy metal pollution in surface water and sediment: a preliminary assessment of an urban river in a developing country. Ecol Ind 48:282–291
IUCN (2020) The IUCN red list of threatened species. Version 2020-2 [Online]. Available: https://www.iucnredlist.org/. Accessed 11 Aug 2020
Jaganmohan M, Vailshery LS, Gopal D, Nagendra H (2012) Plant diversity and distribution in urban domestic gardens and apartments in Bangalore, India. Urban Ecosys 15:911–925
Kahn PH (2007) The child's environmental amnesia-it's ours. Children Youth Environ 17:199–207
Kahn PH, Friedman B (1995) Environmental views and values of children in an inner-city black community. Child Dev 66:1403–1417
Kannan R, Shackleton CM, Uma Shaanker R (2012) Reconstructing the history of introduction and spread of the invasive species, *Lantana*, at three spatial scales in India. Biol Inv 15:1287–1302
Kaoma H, Shackleton CM (2014) Homestead greening is widespread amongst the urban poor in three medium-sized South African towns. Urban Ecosys 17:1191–1207
Karnosky DF (1979) Dutch Elm disease: a review of the history, environmental implications, control, and research needs. Environ Conserv 6:311–322
Kuffer M, Pfeffer K, Sliuzas R (2016) Slums from space: 15 years of slum mapping using remote sensing. Remote Sens 8:455
Kumar B, Mishra M, Verma VK, Rai P, Kumar S (2018) Organochlorines in urban soils from Central India: probabilistic health hazard and risk implications to human population. Environ Geochem Health 40:2465–2480
Kuruneri-Chitepo C, Shackleton CM (2011) The distribution, abundance and composition of street trees in selected towns of the Eastern Cape, South Africa. Urban For Urban Green 10:247–254
Kuussaari M, Bommarco R, Heikkinen RK, Helm A, Krauss J, Lindborg R, Ockinger E, Partel M, Pino J, Roda F, Stefanescu C, Teder T, Zobel M, Steffan-Dewenter I (2009) Extinction debt: a challenge for biodiversity conservation. Trends Ecol Evol 24:564–571
La Sorte FA, Lepczyk CA, Aronson MF, Goddard MA, Hedblom M, Katti M, Macgregor-Fors I, Mörtberg U, Nilon CH, Warren PS, Williams NS, Yang J (2018) The phylogenetic and functional diversity of regional breeding bird assemblages is reduced and constricted through urbanisation. Divers Distrib 24:928–938
Lacan I, McBride JR (2009) War and trees: the destruction and replanting of the urban and peri-urban forest of Sarajevo, Bosnia and Herzegovina. Urban For Urban Green 8:133–148

Larsen J (2003) Record heat wave in Europe takes 35,000 lives: far greater losses may lie ahead. Plan B Updates [Online]. Available http://www.earth-policy.org/index.php/plan_b_updates/2003/update29. Accessed 30 Aug 2020

Larson KL, Hoffman J, Ripplinger J (2017) Legacy effects and landscape choices in a desert city. Landsc Urb Plan 165:22–29

Lassa JA (2015) Post disaster governance, complexity and network theory: evidence from Aceh, Indonesia after the Indian Ocean tsunami 2004. PLoS Curr 7: ecurrents.4f7972ecec1b6

Laursen L (2010) Haiti earthquake may have primed nearby faults for failure. Nature. https://doi.org/10.1038/news.2010.51

Lawrence MJ, Stemberger HL, Zolderdo AJ, Struthers DP, Cooke SJ (2015) The effects of modern war and military activities on biodiversity and the environment. Environ Rev 23:443–460

Leveau LM, Leveau CM, Villegas M, Cursach JA, Suazo CG (2017) Bird communities along urbanisation gradients: a comparative analysis among three Neotropical cities. Ornit Neotrop 28:77–87

Lowry BJ, Lowry JH, Jarvis KJ, Keppel G, Thaman RR, Boehmer HJ (2020) Spatial patterns of presence, abundance, and richness of invasive woody plants in relation to urbanisation in a tropical island setting. Urban For Urban Green 48:126516

Lubbe CS, Siebert SJ, Cilliers SS (2010) Political legacy of South Africa affects the plant diversity patterns of urban domestic gardens along a socio-economic gradient. Sci Res Essays 5:2900–2910

MacGregor-Fors I (2010) How to measure the urban-wildland ecotone: redefining 'peri-urban' areas. Ecol Res 25:883–887

MacGregor-Fors I, Morales-Pérez L, Quesada J, Schondube JE (2010) Relationship between the presence of house sparrows (*Passer domesticus*) and Neotropical bird community structure and diversity. Biol Inv 12:126516

MacGregor-Fors I, Morales-Pérez L, Schondube JE (2012) From forests to cities: effects of urbanisation on tropical birds. In: Lepczyk CA, Warren PS (eds) Urban bird ecology and conservation. University of California Press, Berkely, pp 33–48

MacGregor-Fors I, González-García F, Hernández-Lara C, Santiago-Alarcon D (2018) Where are the birds in the matrix? Avian diversity in a Neotropical landscape mosaic. Wilson J Ornithol 130:81–93

MacGregor-Fors I, Vázquez L-B (2020) Revisiting 'rural'. Sci Total Environ 741:132789

Magnuson JJ (1990) Long-term ecological research and the invisible present. BioScience 40:495–501

Martínez Carretero EE (2009) The synanthropic flora in the Mendoza (Argentina) urban area. Urban Ecosys 13:237–242

McCarthy HP, Crowder LB (2000) An overlooked scale of global transport: phytoplankton species richness in ships' ballast water. Biol Inv 2:321–322

McConnachie MM, Shackleton CM (2010) Public green space inequality in small towns in South Africa. Habitat Int 34:244–248

McDonnell MJ, Hahs AK (2008) The use of gradient analysis studies in advancing our understanding of the ecology of urbanising landscapes: current status and future directions. Landsc Ecol 23:1143–1155

McHale MR, Bunn DN, Pickett ST, Twine W (2013) Urban ecology in a developing world: why advanced socioecological theory needs Africa. Front Ecol Environ 11:556–564

McKinney ML (2006) Urbanisation as a major cause of biotic homogenisation. Biol Conserv 127:247–260

Mead A, Carlton JT, Griffiths CL, Rius M (2011) Revealing the scale of marine bioinvasions in developing regions: a South African re-assessment. Biol Inv 13:1991–2008

Mensah CA (2014) Urban green spaces in Africa: nature and challenges. Int J Ecosys 4:1–11

Miller JR (2005) Biodiversity conservation and the extinction of experience. Trends Ecol Evol 20:430–434

Miller NF (1985) Paleoethnobotanical evidence for deforestation in ancient Iran: a case study of urban Malyan. J Ethnobiol 5:1–19

Monger C, Sala OE, Duniway MC, Goldfus H, Meir IA, Poch RM, Throop HL Vivoni ER (2015) Legacy effects in linked ecological–soil–geomorphic systems of drylands. Front Ecol Environ 13: 13–19

Morrone JJ (2014) Biogeographical regionalisation of the Neotropical region. Zootaxa 3782:1–110

Moskal BT, Berthrong ST (2018) Novel soil barrier systems potentially protect urban growing beds from legacy soil contamination and improve soil health. Urban Agric Reg Food Sys 3:180003

Mote FW (1973) A millennium of Chinese urban history: form, time, and space concepts in Soochow. Rice Institute Pamphlet, Rice University Studies, 59

Muhammad MS, Kasim R, Martin D (2015) A review of residential segregation and its consequences in Nigeria. Mediterr J Soc Sci 6:376–384

Mukoko S (1996) On sustainable urban development in sub-Saharan Africa. Cities 13:265–271

Myers G (2021) Urbanisation in the Global South. In: Shackleton CM, Cilliers SS, Davoren E, du Toit MJ (eds). Urban ecology in the Global South. Springer, Cham, pp 27–49

Myers N, Mittermeier RA, Mittermeier CG, Da Fonseca GA, Kent J (2000) Biodiversity hotspots for conservation priorities. Nat 403:853–858

Nagendra H, Gopal D (2011) Tree diversity, distribution, history and change in urban parks: studies in Bangalore, India. Urban Ecosys 14:211–223

Ngo KM, Hosaka T, Numata S (2019) The influence of childhood nature experience on attitudes and tolerance towards problem-causing animals in Singapore. Urban For Urban Green 41:150–157

Ngulani T, Shackleton CM (2020) The degree, extent and value of air temperature amelioration by urban green spaces in Bulawayo, Zimbabwe. S Afr Geogr J 102:344–355

Njoh AJ (2008a) Colonial philosophies, urban space, and racial segregation in British and French colonial Africa. J Black Stud 38:579–599

Njoh AJ (2008b) The segregated city in British and French colonial Africa. Race Class 49:87–95

Nsikani MM, Van Wilgen BW, Gaertner M (2018) Barriers to ecosystem restoration presented by soil legacy effects of invasive alien N_2-fixing woody species: implications for ecological restoration. Rest Ecol 26:235–244

Ogle K, Barber JJ, Barron-Gafford GA, Bentley LP, Young JM, Huxman TE, Loik ME, Tissue DT (2015) Quantifying ecological memory in plant and ecosystem processes. Ecol Lett 18:221–235

Olshansky RB, Johnson LA, Topping KC (2006) Rebuilding communities following disaster: lessons from Kobe and Los Angeles. Built Environ 32:354–374

Paap T, De Beer ZW, Migliorini D, Nel WJ, Wingfield MJ (2018) The polyphagous shot hole borer (PSHB) and its fungal symbiont *Fusarium euwallaceae*: a new invasion in South Africa. Aust Plant Path 47:231–237

Padisák J (1992) Seasonal succession of phytoplankton in a large shallow lake (Balaton, Hungary): a dynamic approach to ecological memory, its possible role and mechanisms. J Ecol 80:217–230

Parr TB, Smucker NJ, Bentsen CN, Neale MW (2016) Potential roles of past, present, and future urbanisation characteristics in producing varied stream responses. Freshwater Sci 35:436–443

Pauly D (1995) Anecdotes and the shifting baseline syndrome of fisheries. Trends Ecol Evol 10:430

Pickett ST, Cadenasso ML, Childers DL, Mcdonnell MJ, Zhou W (2016) Evolution and future of urban ecological science: ecology in, of, and for the city. Ecosys Health Sustain 2:e01229

Piperno DR (2007) Prehistoric human occupation and impacts on Neotropical forest landscapes during the Late Pleistocene and Early/Middle Holocene. In: Bush MB, Flenley JR (eds) Tropical rainforest responses to climate change. Springer, New York, pp 185–212

Potgieter LJ, Douwes E, Gaertner M, Measey J, Paap T, Richardson DM (2020) Biological invasions in South Africa's urban ecosystems: patterns, processes, impacts, and management. In: Van Wilgen BW, Measey J, Richardson DM, Wilson JR, Zengeya TA (eds) Biological invasions in South Africa. Springer, Cham, pp 275–312

Puga-Caballero A, Macgregor-Fors I, Ortega-Álvarez R (2014) Birds at the urban fringe: avian community shifts in different peri-urban ecotones of a megacity. Ecol Res 29:619–628

Pyatt FB, Gilmore G, Grattan JP, Hunt CO, Mclaren S (2000) An imperial legacy? An exploration of the environmental impact of ancient metal mining and smelting in southern Jordan. J Archaeol Sci 27:771–778

Pyle RM (2003) Nature matrix: reconnecting people and nature. Oryx 37:206–214
Raben R (2020) Colonial shorthand and historical knowledge: segregation and localisation in a Dutch colonial society. J Mod Eur Hist 18:177–193
Ramadan A (2009) Destroying Nahr el-Bared: sovereignty and urbicide in the space of exception. Polit Geogr 28:153–163
Reynolds C, Byrne MJ, Chamberlain DE, Howes CG, Seymour CL, Sumasgutner P, Taylor PJ (2021) Urban animal diversity in the Global South. In: Shackleton CM, Cilliers SS, Davoren E, du Toit MJ (eds) Urban ecology in the Global South. Springer, Cham, pp 169–202
Rigolon A, Browning M, Lee K, Shin S (2018) Access to urban green space in cities of the Global South: a systematic literature review. Urban Sci 2:67
Roman LA, Pearsall H, Eisenman TS, Conway TM, Fahey RT, Landry S, Vogt J, Van Doorn NS, Grove JM, Locke DH, Bardekjian AC, Battles JJ, Cadenasso ML, Van Den Bosch CC, Avolio M, Berland A, Jenerette GD, Mincey SK, Pataki DE, Staudhammer C (2018) Human and biophysical legacies shape contemporary urban forests: a literature synthesis. Urban For Urban Green 31:157–168
Salmon C (2002) Palestine notebooks, Sixth Notebook: Sabreen, or patience… . Autodafe, the Censored Library [Online]. Available http://www.mafhoum.com/press3/92Ca61.htm. Accessed 2 Sept 2020
Schaefer VH (2010) Remembering our roots: a possible connection between loss of ecological memory, alien invasions and ecological restoration. Urban Ecosys 14:35–44
Schwarz K, Pouyat RV, Yesilonis I (2016) Legacies of lead in Charm City's soil: lessons from the Baltimore ecosystem study. Int J Environ Res Public Health 13:209
Seebens H, Blackburn TM, Dyer EE, Genovesi P, Hulme PE, Jeschke JM, Pagad S, Pysek P, Winter M, Arianoutsou M, Bacher S, Blasius B, Brundu G, Capinha C, Celesti-Grapow L, Dawson W, Dullinger S, Fuentes N, Jager H, Kartesz J, Kenis M, Kreft H, Kuhn I, Lenzner B, Liebhold A, Mosena A, Moser D, Nishino M, Pearman D, Pergl J, Rabitsch W, Rojas-Sandoval J, Roques A, Rorke S, Rossinelli S, Roy HE, Scalera R, Schindler S, Stajerova K, Tokarska-Guzik B, Van Kleunen M, Walker K, Weigelt P, Yamanaka T, Essl F (2017) No saturation in the accumulation of alien species worldwide. Nat Commun 8:14435
Shackleton RT, Witt AB, Aool W, Pratt CF (2017) Distribution of the invasive alien weed, *Lantana camara*, and its ecological and livelihood impacts in eastern Africa. Afr J Range For Sci 34:1–11
Shackleton CM, Cilliers SS, du Toit MJ, Davoren E (2021) The need for an urban ecology of the Global South. In: Shackleton CM, Cilliers SS, Davoren E, du Toit MJ (eds) Urban ecology in the Global South. Springer, Cham, pp 1–26
Sharma GP, Raghubanshi AS (2010) How Lantana invaded India. Curr Conserv 4:21–22
Shochat E, Lerman SB, Anderies JM, Warren PS, Faeth SH, Nilon CH (2010) Invasion, competition, and biodiversity loss in urban ecosystems. BioScience 60:199–208
Sim LK, Balamurugan G (1991) Urbanisation and urban water problems in Southeast Asia a case of unsustainable development. J Environ Manage 32:195–209
Simwanda M, Ranagalage M, Estoque RC, Murayama Y (2019) Spatial analysis of surface urban heat islands in four rapidly growing African cities. Remote Sens 11:1645
Smiley SL (2010) Expatriate everyday life in Dar es Salaam, Tanzania: colonial origins and contemporary legacies. Soc Cult Geogr 11:327–342
Sol D, Bartomeus I, González-Lagos C, Pavoine S (2017) Urbanisation and the loss of phylogenetic diversity in birds. Ecol Lett 20:721–729
Stilgenbauer J, McBride JR (2010) Reconstruction of urban forests in Hamburg and Dresden after World War II. Landsc J 29:144–160
Tilman D, May RM, Lehman CL, Nowak MA (1994) Habitat destruction and the extinction debt. Nature 371:65–66
Tsuzuki Y, Koyanagi TF, Miyashita T (2020) Plant community assembly in suburban vacant lots depends on earthmoving legacy, habitat connectivity, and current mowing frequency. Ecol Evol 10:1311–1323
United Nations (2015) World urbanisation prospects—2014 revision. United Nations, New York

United Nations (2019) Global issues: population [Online]. Available https://www.un.org/en/sections/issues-depth/population/. Accessed 11 Sept 2020
Utz RM, Hopkins KG, Beesley L, Booth DB, Hawley RJ, Baker ME, Freeman MC, Jones K (2016) Ecological resistance in urban streams: the role of natural and legacy attributes. Freshwater Sci 35:380–397
van Kleunen M, Essl F, Pergl J, Brundu G, Carboni M, Dullinger S, Early R, Gonzalez-Moreno P, Groom QJ, Hulme PE, Kueffer C, Kuhn I, Maguas C, Maurel N, Novoa A, Parepa M, Pysek P, Seebens H, Tanner R, Touza J, Verbrugge L, Weber E, Dawson W, Kreft H, Weigelt P, Winter M, Klonner G, Talluto MV, Dehnen-Schmutz K (2018) The changing role of ornamental horticulture in alien plant invasions. Biol Rev Camb Philos Soc 93:1421–1437
van Wilgen BW (2012) Evidence, perceptions, and trade-offs associated with invasive alien plant control in the Table Mountain National Park, South Africa. Ecol Soc 17:23
Venter ZS, Shackleton CM, Van Staden F, Selomane O, Masterson VA (2020) Green apartheid: urban green infrastructure remains unequally distributed across income and race geographies in South Africa. Landsc Urb Plan 203:103889
Walsh CJ, Roy AH, Feminella JW, Cottingham PD, Groffman PM, Morgan RP (2005) The urban stream syndrome: current knowledge and the search for a cure. J North Am Benth Soc 24:706–723
Wantzen K, Alves CB, Badiane SD, Bala R, Blettler M, Callisto M, Cao Y, Kolb M, Kondolf GM, Leite MF, Macedo DR, Mahdi O, Neves M, Peralta ME, Rotge V, Rueda-Delgado G, Scharager A, Serra-Llobet A, Yengue J-L, Zingraff-Hamed A (2019) Urban stream and wetland restoration in the Global South: a DPSIR analysis. Sustain 11:4975
Wei B, Yang L (2010) A review of heavy metal contaminations in urban soils, urban road dusts and agricultural soils from China. Microchem J 94:99–107
Weller R, Huang C, Kjaersgaard SP, Drozdz Z, Diong N (2018) Hotspot cities: identifying peri-urban conflict zones in the world's biodiversity hotspots. Penn: Current Research on Sustainable Urban Development [Online]. Available https://penniur.upenn.edu/uploads/media/Weller2.pdf. Accessed 11 Sept 2020
Williams NS, Schwartz MW, Vesk PA, Mccarthy MA, Hahs AK, Clemants SE, Corlett RT, Duncan RP, Norton BA, Thompson K, McDonnell MJ (2009) A conceptual framework for predicting the effects of urban environments on floras. J Ecol 97:4–9
Xiang Z, Zhao M, Ogbodo US (2020) Accumulation of urban insect pests in China: 50 years' observations on Camphor Tree (*Cinnamomum camphora*). Sustain 12:1582
Xu Y (2000) The Chinese city in space and time: the development of urban form in Suzhou. University of Hawaii Press, Hawaii
Yang J, Yan P, He R, Song X (2017) Exploring land-use legacy effects on taxonomic and functional diversity of woody plants in a rapidly urbanising landscape. Landsc Urb Plan 162:92–103
Zettlemoyer MA, McKenna DD, Lau JA (2019) Species characteristics affect local extinctions. Am J Bot 106:547–559
Ziter C, Graves RA, Turner MG (2017) How do land-use legacies affect ecosystem services in United States cultural landscapes? Landsc Ecol 32:2205–2218

Urban Social Ecology

Marion Mehring, Katharina Geitmann-Mügge, Fanny Frick-Trzebitzky, and Diana Hummel

Abstract Social Ecology is an emerging scientific field within sustainability science investigating the relationships between society and nature. It is based on the core concept of societal relations to nature. Essential characteristics are: (1) Problem-orientation: The starting point of research is concrete, societal problems with regard to ecological crisis phenomena such as climate change, waste management or protection of biodiversity. (2) Inter- and transdisciplinarity: Research combines academic disciplines with knowledge, values and interests of societal actors. The central aim is achieving learning processes between science and society. (3) Interrelation of theory and empirical research: The concept of societal relations to nature is the theoretical framework, which guides empirical studies. The empirical results in turn contribute to the further development of theories and concepts. (4) System approach: Given the complex interactions between natural and societal processes and structures at different temporal, spatial and social scales, the approach of social-ecological systems is applied. (5) Critical science: Contributing to social-ecological transformations, Social Ecology deals with ignorance, uncertainty and disputed knowledge. Social Ecology critically reflects on the role of researchers and on the limits of scientific knowledge production. Applying Social Ecology this chapter examines three challenges of urban areas in the Global South: rural–urban migration, urban agriculture and green infrastructure.

M. Mehring (✉) · K. Geitmann-Mügge · F. Frick-Trzebitzky · D. Hummel
ISOE—Institute for Social-Ecological Research, Hamburger Allee 45,
60486 Frankfurt am Main, Germany
e-mail: mehring@isoe.de

K. Geitmann-Mügge
e-mail: Geitmann-Muegge@isoe.de

F. Frick-Trzebitzky
e-mail: frick@isoe.de

D. Hummel
e-mail: hummel@isoe.de

M. Mehring · D. Hummel
Senckenberg Biodiversity and Climate Research Centre SBiK-F, Senckenberganlage 25, 60486
Frankfurt am Main, Germany

C. M. Shackleton et al. (eds.), *Urban Ecology in the Global South*, Cities and Nature,
https://doi.org/10.1007/978-3-030-67650-6_4

Keywords Green infrastructure · Rural urban migration · Social-ecological systems · Social ecology · Urban agriculture · Urban social ecology

1 Introduction

Social Ecology is an emerging scientific field within sustainability science investigating the relationships between society and nature. It is based on the core concept of societal relations to nature (SRN) (Hummel et al. 2017; Kramm et al. 2017). In particular, the interrelation and interdependencies between natural and societal structures and processes are considered. With its focus on crisis-driven developments of SRN, Social Ecology provides a problem-oriented, inter- and transdisciplinary perspective on issues which are characteristics of urban ecology in the Global South.

Urban areas face the problem of how to balance the requirements of society with their natural surroundings. Thus, the urban environment is significantly linked to ecosystems (Elmqvist et al. 2018), yet cities are more likely to be seen as the opposite of nature than as a part of environmental systems (Elmqvist et al. 2013). Grimm et al. (2000) argue that conceptualising cities as ecological systems creates opportunities for theoretical advances in planning. For instance, in cities like Accra, Ghana, technical solutions to flood risk are the main paradigm, whereas protection of mangroves could be an efficient and cheaper alternative for buffering against floods in the city (Frick-Trzebitzky and Bruns 2019).

It has to be considered that studies and concepts conducted in the Global North cannot simply be conveyed to the Global South (Robinson 2002; Shackleton et al. 2021). This is because trends in urbanisation in the Global South differ significantly with respect to the historic development of the Global North and it is not expected that the Global South will reach the same state in urban development in the near-term. Instead, cities in the Global South may reach any number of a range of alternative viable states. Consequently, new approaches, strategies and theories of urban development and urban sustainability are required for the Global South (McHale et al. 2013).

Given the intimate nexus of society and the environment in urban settings, it is important that conceptual frameworks explicitly include humans to accurately inform environmental problem-solving (Grimm et al. 2000). Concepts such as ecosystem services (ES) (Millennium Ecosystem Assessment 2005; Fisher et al. 2009), disservices (von Döhren and Haase 2015; Shackleton et al. 2016) and relational values (Chan et al. 2016; Himes and Muraca 2018) can help to address many of the social-ecological challenges of cities. However, recent reviews (Cilliers et al. 2013; Wangai et al. 2016; du Toit et al. 2018) show a deficit in research on social-ecological systems (SES) and interrelations in cities of the Global South.

In this chapter, we will focus on Social Ecology in urban areas in the Global South. First, we will introduce Social Ecology as a critical science of SRN. Second, we will highlight existing challenges and insights from case studies in the Global South. Third, we will conclude with recommendations for future research.

2 Social Ecology as a Critical Science for Understanding Societal Relations to Nature

2.1 Origins and Relationship to Other Fields of Human-Nature Studies

Social Ecology is an emerging inter- and transdisciplinary scientific field in the area of sustainability research. It has a long and multifaceted history with diverse academic roots, as well as relations to other fields of scientific inquiry such as human ecology or political ecology. Since the 1920s, various approaches have emerged, the common feature of which is the scientific study of the reciprocal relationships between humans and their social, biological and physical environment (Becker and Jahn 2006; Fischer-Kowalski and Weeisz 2016). The Chicago School is a key reference for contemporary Human Ecology and Social Ecology. It was founded mainly by Robert E. Park, Ernest Burgess, Roderick D. McKenzie and Louis Wirth in the 1920s and elaborated, in the form of a spatial sociology, to study the principles of organisation of the urban metropolis. From the outset it was closely linked to problems of urban, spatial and regional planning and it placed local communities (and not society as a whole) at the centre of its considerations (Cadenasso and Pickett 2013). The term 'Social Ecology' was suggested by Alihan (1938) to denominate an analytical framework for studies of the relations between humans and their environments, and it was intended to be more integrative than the field of Human Ecology, as put forward by the Chicago School in the 1920s (Alihan 1938). Building on these traditions, several chairs and institutions for social-ecological research have been created in the USA, above all the 'School of Social Ecology' at the University of California in Irvine (Stokols 2018).

In Europe, particularly in Germany and Austria, Social Ecology initially emerged in the 1970s outside universities and across disciplines as interdisciplinary and problem-oriented research and as a scientific reaction to the increasing ecological crises. At the core is the historical insight that nature can no longer be comprehended without society, and society cannot be understood without nature—a hypothesis that also the sociologist Ulrich Beck formulated in the mid-1980s in his book 'Risk Society' (Beck 1992). The European tradition of Social Ecology focuses on the crisis-driven relationships between society and nature, and follows a strong theoretical programme comprising different perspectives (Hummel and Jahn 2014; Kramm et al. 2017). At the Vienna Institute for Social Ecology in Austria, material flow analyses based on historical studies and statistics are combined with the question of the extent to which society needs a material and energetic metabolism based on a 'colonisation' of nature (Haberl et al. 2016). In contrast, at the Institute for Social-Ecological Research in Frankfurt am Main, Germany, ideas of critical theory, ecological critique of technology and science and feminism were initially linked with scientific findings and insights from natural sciences (Becker and Jahn 2006).

Both schools comprise different theoretical traditions and concepts, but they share an inter- and transdisciplinary approach to sustainability research and a common understanding of the challenges posed by the crises of SRN (Becker et al. 2011; Kramm et al. 2017).

Social Ecology partly overlaps with Political Ecology. The latter focuses particularly on the importance of politics and power often associated with a strong criticism of political economy and neoliberalism (Parnell and Robinson 2012; Lawhon et al. 2014). In contrast, Social Ecology emphasises interactions between society and nature (including the role of political regulations and power structures). However, both approaches are mutually compatible with their shared perspective on problems of social-ecological transformations (Görg et al. 2017).

2.2 *Principles of Social Ecology*

Social Ecology is described as the "science of societal relations to nature (SRN)". Essential characteristics are:

- **Problem-orientation**: The starting point of research is concrete, societal problems with regard to ecological crisis phenomena such as climate change, waste management, protection of biodiversity, etc. Scientific analysis is aimed at alternative and more sustainable regulations.
- **Inter- and transdisciplinarity**: Research combines academic disciplines from natural science, social science and engineering science with knowledge, values and interests of societal actors. The central aim is achieving learning processes between science and society towards social-ecological transformations.
- **Interrelation of theory and empirical research**: The concept of 'societal relations to nature' is the theoretical framework which guides empirical studies of, for example water and land management, mobility, biodiversity and consumption. The empirical results in turn contribute to the further development of theory and concepts.
- **System approach**: Given the complex interactions between natural and societal processes and structures of social-ecological problems at different temporal, spatial and social scales, the research framework of SES allows for making SRN proper for empirical analysis.
- **Critical science**: With the objective to contribute to social-ecological transformations, scientific problem-solving deals with challenges of ignorance, uncertainty and disputed knowledge. This implies an attitude in research that critically reflects on one's own role as a researcher and on the limits of scientific knowledge production (Becker and Jahn 2006; Hummel et al. 2017; Jahn et al. 2020).

2.3 Crisis of Societal Relations to Nature

In the scientific investigation of the relationships between society and nature, Social Ecology is based on the core concept of SRN (Görg 1999; Becker and Jahn 2006). The focus is on questions of complex interactions between physical processes and social action on different spatial and temporal scales. The point of departure for the theoretical approach is the fact that changes in these interrelationships can develop into social-ecological crises (Becker and Jahn 2006).

In the Frankfurt approach of Social Ecology, the term SRN marks an epistemological perspective and represents at the same time the central theoretical concept: SRN constitute the 'epistemic object' of research to analyse and understand the relationships between nature and society in their empirical peculiarity (Hummel et al. 2017). SRN emerge from culturally specific and historically variable forms and practices in which individuals, groups and cultures shape and regulate their relations with nature. Thus, they must be considered in their plurality - there is no singular social relationship to nature. The concept comprises an analytical distinction between the physical-material and cultural-symbolic attributes of SRN. This renders it possible to consider their specific materiality, and, at the same time, their embeddedness in symbolic orders, cultural interpretations and power relations. At the normative level, the concept relates to the idea of basic needs; SRN should be regulated in a manner to ensure the satisfaction of needs for all human beings. This perspective explicitly refers to ideas of justice, equity and sustainable development. Hence, basal SRN are essential for individual and societal reproduction and development and comprise, for example, work and production, land use and nutrition, movement and mobility and hygiene and health care. If the regulation of such basal SRN fails, large-scale crises can occur up to the collapse of whole societies (Hummel et al. 2017). The notions of regulation and transformation of SRN play an important role in the concept, both analytically and normatively. SRN are regarded as regulated, i.e. they can be regulated and shaped by societies. Patterns and modes of regulation can be distinguished. While patterns of regulation refer to the interplay of material and symbolic aspects in societies' interaction with nature, modes of regulation represent the power relations, cultural norms and conflicts. Changes in the modes and patterns of regulation can be conceptualised as social-ecological transformations (Görg et al. 2017; Hummel et al. 2017).

2.4 Social-Ecological Systems

One way of making SRN accessible as research objects for the empirical analysis of problems in the context of Urban Ecology of the Global South, is to investigate them as SES. "A social-ecological system consists of a bio-geo-physical unit and its associated social actors and institutions. Social-ecological systems are complex and adaptive and delimited by spatial or functional boundaries surrounding particular

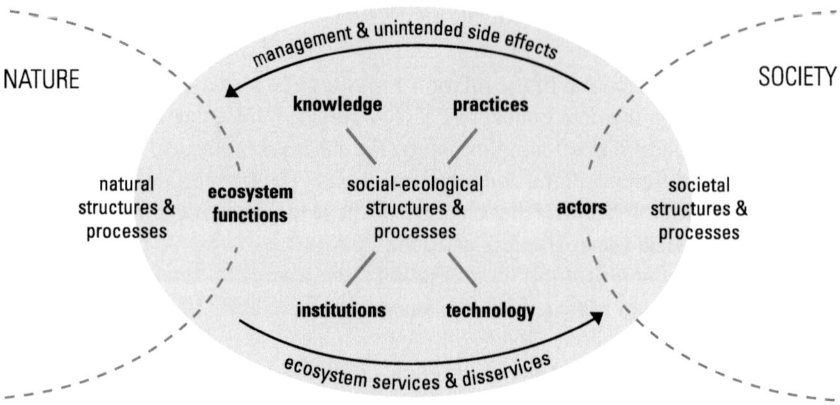

Fig. 1 The concept of social-ecological systems (*Source* Mehring et al. 2017a, pp. 175)

ecosystems and problem contexts" (Glaser et al. 2014). The SES concept can also be applied productively in the framework of Urban Ecology in the Global South. In individual areas of action, such as water supply, urban land use and agriculture, or people's mobility and rural–urban interactions, it can then be examined in which way social and natural structures and processes are linked by certain practices, institutions and other mechanisms and which crisis-prone developments can result from them. Applying the four dimensions of practices, knowledge, institutions and technology (Fig. 1), helps to better understand the particular problems of urban settings in the Global South. When implementing research or projects, for example on new water infrastructure technologies it is essential to also consider the knowledge on how to use and maintain this new technology; the institutional setting needs to make sure that potential new formal regulations meet previous informal regulations; and finally the introduction of new technology might result in or impact current practices of land use. Thus, we see the need for Urban Ecology to shift towards SES integrating the aforementioned societal aspects.

The concept of SES emerged at the end of the 1990s. To date, a large number of different SES approaches have been developed, which are characterised primarily by different concepts of nature and society and their interaction (Binder et al. 2013; Becker 2014). One example is the SES model of the Stockholm Environment Institute and the Resilience Alliance. It focuses on ecosystems which are managed or disturbed by human activities and emphasises the complexity and adaptability of a given SES (Berkes and Folke 1998). Another important SES approach is Elinor Ostrom's model, which particularly emphasises the role of governance and institutions in resource use and management and hence the social elements of the system (Ostrom 2007, 2009). Approaches that are more recent take the concept of ES as a starting point and integrate it into a broader social-ecological context (Reyers et al. 2013; Martín-López et al. 2014; Mehring et al. 2017a). The development of the IPBES (Intergovernmental Platform for Biodiversity and Ecosystem Services)-framework marks a milestone in this discussion (Díaz et al. 2015) and the conceptual development of the NCP—Nature's Contribution to People framework calls for a further development of the ES discussion (Díaz et al. 2018). A more intimate co-directional and co-evolutionary relationship between biodiversity and culture is expressed in the 'biocultural diversity' framing (Maffi 2008; Buizer et al. 2016), which often considers general sustainability issues (including urban ones; see chapters in Cocks and Shackleton 2021), but is rarely applied to environmental crises as per SRN.

The Frankfurt approach of Social Ecology focuses on 'hybrids', i.e. the actual interaction between nature and society (Fig. 1). These social-ecological structures and processes are conceptualised as dynamic interactions where society influences nature and vice versa. In addition to that, both the social and the ecological part are treated in equal depth distinguishing this approach of Social Ecology from others such as Ostrom (2009). In doing so, this conception focuses on the internal dynamics of the SES and thus takes up the complexity of both natural and societal structures and processes (Liehr et al. 2017; Mehring et al. 2017b). Societal actors influence the natural structures, processes and related ecosystem functions via management activities or unintended side effects. Nature provides ES, but also disservices that harm society (Shackleton et al. 2016; Davoren and Shackleton 2021). Knowledge, institutions, technology and practices shape the social-ecological structures and processes as the core part of the SES (Liehr et al. 2017; Mehring et al. 2017b).

Since built, technical and green infrastructures (GI) are of crucial importance in urban areas (e.g. for water provisioning, energy or sewage systems, but also for climate adaptation and mitigation), McPhearson et al. (2016) suggest that urban ecology should move beyond a social-ecological conceptual framing to better address the social-ecological-technical/built system (SETS). The model of SES presented above has the advantage that infrastructures can be considered with 'technology' as one of the four mediating dimensions that addresses all kind of material structures designed, built and controlled by humans for achieving specific purposes and shows the importance of knowledge, institutions and practices related to technology (Hummel et al. 2008). Using several case studies, we will illustrate below how the SRN model of SES can inform Urban Ecology studies in the Global South.

2.5 Social Ecology as Critical Analysis

The central aim of the Frankfurt Social Ecology is a critical analysis of complex interactions between nature and society. In its problem-oriented perspective, consequences are drawn for the process of knowledge production as well as for the problems that drive it. It is not just a matter of better understanding SRN and their crisis-prone developments. The aim is at the same time to generate knowledge with which to intervene in these interactions in a sustainable way. The Frankfurt approach of Social Ecology, therefore, pursues a critical perspective in several respects: on the one hand, the reciprocal constitution of nature through society and of society through nature comes into view. Thus, the supposedly clear distinction between nature and society is criticised, as well as the methodological dualism that has long prevailed in the natural and social sciences. With the procedure of 'double-sided critique' of a naturalisation and a culturalisation of social-ecological problems, one-sided naturalistic or culturalistic descriptions and evaluations can then be identified, related to each other and thus blind spots can be made visible (Becker and Jahn 2006). In particular, the Frankfurt School of Social Ecology attempts to relate social and scientific findings to each other in a theory-driven manner. This involves a critical reflection of the role of science. Science itself is an actor of the SES it sets out to analyse, i.e. SES are self-describing and self-referential systems in which science does not stand outside, but represents a 'participant observer' (Becker 2014). The research problems are understood as expressions of crisis-driven social-ecological conditions, which can only be understood and dealt with in an inter- and transdisciplinary way.

3 Existing Challenges—Insights from Case Studies in the Global South

Following the Frankfurt approach of Social Ecology this chapter presents current research on the complex interactions between nature and society with a particular focus on challenges in the Global South. SRN in the Global South face different challenges compared to other parts of the world. In general, not only social aspects such as power relations (incl. environmental justice), poverty and food and nutrition insecurity play a role, but also the dominance of Western epistemic concepts and perspectives that might have an impact on the development and planning of urban areas. Urban ecology of the Global South needs to address these societal aspects to better capture the drivers and challenges, and be able to better draw conclusions and make recommendations.

In particular, the dichotomy of the physical-material and cultural-symbolic dimensions come to the fore. The relevance and interdependency of both dimensions are more pronounced in the Global South than the Global North (Shackleton et al. 2021). Prominent examples are for instance taboos of using a particular area of land or sacred trees in cities. Besides providing services such as climate regulation, water retention or carbon sequestration, these GI elements have an important cultural-symbolic dimension for the local people that is deeply anchored in their culture as seen, for example, in India (Jaganmohan et al. 2018) and Nigeria (Adedeji 2021). As a result, problems may arise when implementing new formal rules of regulating GI that do not meet the local informal ones. Ignoring this aspect entails the danger that local cultural, historical and natural conditions are not sufficiently taken into account. Along this line, different types of knowledge (traditional ecological knowledge (TEK), indigenous local knowledge (ILK) and scientific) are prevalent and relevant and might have another history and cultural significance compared to the Global North.

This chapter uses Social Ecology to analyse selected urban challenges from countries in Asia, Africa and Latin America. We do so using examples of rural–urban migration, urban agriculture and green infrastructure.

3.1 Rural–Urban Migration: West African Sahel and Eastern Mongolian Steppe

People's mobility and migration movements, but also rural–urban relationships, are of crucial importance for sustainable development of urban areas in the Global South. During the last two decades, there has been a considerable increase of academic literature and empirical studies about the interactions of climate and environmental change and migration in different regions of the world (Hastrup and Fog Olwig 2012; Piguet and Laczko 2014; Gioli et al. 2016). These studies highlight the complexity and context dependency of the migration–environment nexus. Today, there is wide agreement among scholars that migration is multi-causal. Climate change, on its own, does not directly cause migration, but is accompanied by specific environmental changes, that can exacerbate existing vulnerabilities. Against this background, the following examples illustrate the social-ecological conditions of migration and their impacts on the issue of urbanity in the Global South.

3.1.1 Rural–Urban Migration as a Livelihood Strategy—The West African Sahel

The West African Sahel is expected to be one of the regions most affected by climate and environmental change, such as increasing temperatures, oscillations in precipitation patterns and land degradation. The majority of the population relies on small-scale agriculture as their major source of income and livelihood and is thus considerably affected by changing environmental conditions. Migration is both a significant characteristic of the West African culture and a common strategy to deal with the harsh environmental conditions. This can be illustrated by the regions of Linguère in Senegal and Bandiagara in Mali (Hummel 2016; Liehr et al. 2016; van der Land 2018). The temporal and spatial migration patterns in both regions provide interesting information about the processes of urban–rural relationships. Interviews within a standardised survey revealed that around 90% of respondents (residents in rural areas and migrants in urban areas) have personally experienced migration. Besides temporal migration (10 months to five years), seasonal and circular migration (3–9 months) are the dominant patterns. During the dry season, when there is little work to do in farming, people leave their villages and move to urban areas and return to the villages for the harvest season. Regarding spatial patterns, migration within a country was the dominant type, and the majority of respondents moved to the larger urban areas, above all, to the capitals of Bamako and Dakar, respectively

(Hummel 2016; van der Land 2018). Whereas agriculture is the main economic activity for people in the rural areas, migrants in the capitals depend almost exclusively on non-agricultural activities (e.g. trading in commodities, unskilled labour, small business or domestic work [van der Land 2018]). These results highlight the crucial importance of rural–urban relationships (Tacoli 2003). Migrants from both Linguère and Bandiagara maintain strong ties with their home community and visit their village frequently or return home at some point. Thus, migration represents an important livelihood strategy. Moving to urban areas minimises risks of crop failure, contributes to more diversified income sources and food security. However, migration cannot be reduced to a means of adapting to environmental stressors. Economic motivations often overlap with other motives, such as education, family reasons, visits and curiosity—depending on gender, age or educational status (Fig. 2). Particularly, the changing motives and aspirations of the youth are characteristic for overall societal transformations and changing lifestyles: 'Young people often have great aspirations for a better life and a "modern" lifestyle. Life in the rural areas is considered hard and boring and offers little opportunities' (van der Land 2018, p. 75). Cultural-symbolic aspects such as economic independence, the possibility of purchasing consumer goods like mobile phones or fashionable clothing are gaining in importance.

The SES model presented above supports a better understanding of the climate–environment–migration nexus, i.e. the interactions between actors such as migrants and villagers and ecosystem components such as vegetation and soil conditions, which are mediated by specific practices (e.g. seasonal migration), knowledge (e.g.

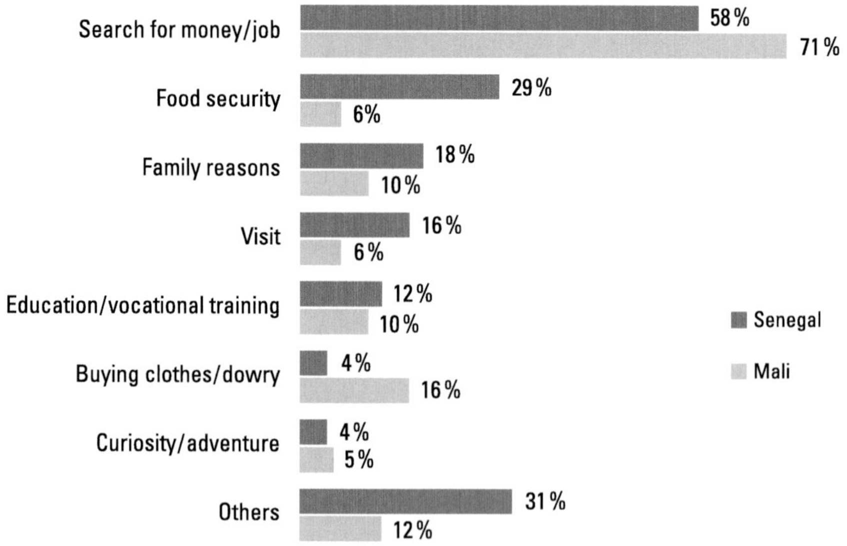

Fig. 2 Motives for rural-urban migration in Senegal and Mali (multiple answers possible) (*Source* micle survey 2012; Senegal n=388, Mali n=398; Hummel and Liehr 2015)

traditional knowledge on agro-biodiversity), technology (e.g., irrigation schemes) and institutions (e.g. social networks or migration policies). Rural–urban migration in West Africa indicates fundamental changes in both physical-material and cultural-symbolic aspects of SRN, as can be illustrated by changing attitudes towards farming. Particularly young people with higher formal education consider rural life no longer desirable because it is regarded as hard and tiring work, and returns are unreliable. Urban living is perceived as alluring, since it promises not only better education and income opportunities, but also a different lifestyle and independence from family and the village community.

For Urban Ecology in the Global South several conclusions can be drawn from the example. First, in terms of urban planning and governance, relationships between rural and urban development processes should be more visibly reflected. For example, infrastructure development in urban areas must consider the promotion of integrated land and water resources management to prevent land degradation in the rural hinterlands. Second, improving the capabilities of the younger generation is of crucial importance. Since education is one of the most frequently mentioned motives for rural–urban migration, and given the attractiveness of urban lifestyles for the youth, formal education must be strengthened (particularly in terms of quality), in addition to the creation of job opportunities outside the agricultural sector. Third, cities are increasingly exposed to the impacts of climate change. This can reinforce migration flows to urban areas which may strain the functionality of urban provisioning systems for water and sanitation, health, education, etc., and can thus increase the population at risk (Adamo 2010). Fourth, in terms of research, more empirical evidence is, therefore, needed about the impacts, challenges and contribution of (environmentally induced) migration to resilient and sustainable cities and social-ecological transformations in urban areas of the Global South (Adger et al. 2019; Serraglio et al. 2019).

3.1.2 Changing Lifestyles and Rural–Urban Migration—The Mongolian Eastern Steppe

The Mongolian Steppe is one of the last remaining intact grassland ecosystems worldwide, characterised by a close coupling of societal and natural processes (Batsaikhan et al. 2014; Wesche et al. 2016). In this SES, mobility is one of the key characteristics of both wildlife and human societies. Flexible herbivore mobility and grazing are essential to cope with the enormous spatio-temporal variabilities in climate conditions (Wesche et al. 2016); droughts and extreme winters occur frequently but are also unpredictable.

At the same time, the Mongolian steppe is characterised by unique long-distance movements of wild ungulates (Nandintsetseg et al. 2019). Recent infrastructure developments of oil exploitation, especially in the Eastern Steppe, caused massive landscape fragmentation. The construction of fenced roads and railway lines hinders free movement of wildlife on the steppe. But, it is critical to avoid any fragmentation of the landscapes and maintain landscape permeability across the entire steppe to protect the globally unique ecosystem, including its wildlife.

This fragmentation process goes along with a profound societal change over the course of the last three decades. Since the 1990s, when Mongolia transitioned from the communist system to a free market and multiparty parliamentary system, the Mongolian economy and society have been altered substantially (Yembuu 2016). The current economic development of Mongolia is accompanied by extensive changes in nomadic lifestyles potentially affecting the unique steppe ecosystem and its biodiversity. The change in lifestyles is mainly characterised by rural–urban migration, resulting in reduced mobility of herders and their livestock causing overgrazing in rural parts of central Mongolia (Mehring et al. 2018), and in an ever-growing population in the capital Ulaanbaatar. Nowadays, almost half of the Mongolian population (1.3 million out of 2.9 million) live in the capital (Central Intelligence Agency 2018). This ongoing rural–urban migration process does not only pose challenges to the rural countryside resulting in a fundamental decline in nomadic pastoralism, but also to the urban capital where the migration process and settlement activities were not controlled over a long time. This resulted in a lack of important urban infrastructure such as roads and health care facilities, and massive air pollution, especially in the winter times from the stoves in the yurt. The question is how nomadic pastoralism can be facilitated in the light of these current societal changes. "Time is pressing as a new generation is born which is growing up in urban environments and with new skill sets, but a potential loss of the tight connection to nature and the nomadic lifestyle" (Mehring et al. 2018, p. 1).

There is evidence that rural–urban migration follows a stepwise process from the rural countryside to the capital (Dore and Nagpal 2006; Save the Children International 2013) associated with a complex interplay between the prevailing social, economic and ecological motives for migration (Tarne 2018). At the same time the cultural-symbolic value and image of the traditional nomadic way of life continues to be important at a personal and societal level (Tarne 2018). Tarne (2018) reports that many middle-aged to older migrants who moved to the capital for economic or ecological reasons have highlighted that they still value these traditional lifestyles and would prefer to return to the rural countryside if it were financially viable. Applying the SES model helps better understanding the potential consequences of these rural–urban migration processes. As described above, especially the herding practices themselves and the traditional ecological knowledge and how to do it and/or to cope with the enormous spatio-temporal variabilities in climate conditions will be lost. For the urban areas the lack of institutions, namely formal regulations and implementation of e.g. health supply alongside the urbanisation process, remain the main challenges.

3.1.3 Regulation and Management of Rural–Urban Migration Processes

With a view to the social-ecological conditions of rural–urban migration illustrated by the case studies of Mongolia and West Africa, it is of crucial importance that policies make use of the positive potential of migration for sustainable development, especially in urban areas. In Mali and Senegal, several national frameworks such as poverty reduction strategy papers (PRSP) take into account the potential benefits of internal migration in terms of development and poverty reduction, e.g. remittances of migrants and transfer of skills, but their implementation is lacking. Moreover, regulation strategies in different policy fields (migration, environment, development) must be better integrated (Hummel 2016). Overall, relationships between rural and urban development processes must be more visibly reflected in policies and development strategies (Tacoli 2003). For example, such strategies should incorporate infrastructure development (in particular health and education) and the promotion of integrated land and water resources management to limit land degradation. Above all, it is important to improve the skills of the younger generation and create more livelihood opportunities for them. The role of education, both in terms of quantity and quality, is one of the most crucial issues, since good education reduces vulnerability by opening new options and capabilities for young people.

In summary, both examples show that it is very important to find strategies on how to enable migration while considering the particularities in both, rural and urban areas. In this sense, infrastructure is key and it is important that the supply of education and health services are guaranteed, e.g. in urban areas, while at the same time the fragmentation of the rural landscape should be reduced. In addition to that, the cultural-symbolic meaning and importance of rural–urban migration needs to be better reflected and integrated into current policies.

3.2 Urban Agriculture: Western and Southern Africa

Under the pressure of basic infrastructure and land for development, securing safe food provision has become a challenge in many fast-growing cities of the Global South (Drescher et al. 2021). In Accra (Ghana), food, including vegetables, sold in supermarkets is largely imported from abroad and sold at relatively high prices. Yet, locally produced vegetables sold on open markets are the main source of nutrients for the majority of Accra's population (Tuholske et al. 2020). Nevertheless, urban farming sites have long been regarded as a 'nuisance' by town planners, and irrigation with the water resources available to urban farmers (mainly untreated water from open drains) potentially poses health risks to both farmers and consumers. Similarly, in North Central Namibia, where urbanisation is a relatively young phenomenon, livestock farming has naturally been moved to towns and is deemed by urban authorities to be creating health risks as the livestock roams freely. Irrigated crop farming is gaining support of city officials as a way to reuse and manage scarce water resources efficiently. This section draws on two examples of urban agriculture (UA) that show how urban vegetable and crop provisioning systems are tightly linked to provision of water, land and health. The analysis draws on literature as well as ongoing research.

3.2.1 Struggling for Access to Food and Land Through Urban Agriculture in Accra, Ghana

Accra is a colonial city established by the German colonisers around a Ga fishermen's village in 1877 (Grant and Yankson 2003). The handwriting of German and British town planners is still visible in town plans with large green spaces allocated for 'recreational activities', as well as for 'nature conservation'. In the light of an extremely vibrant globalised housing market, coupled with population growth, however, green spaces have rapidly become settled on, more or less informally, and land allocated for development often remains vacant due to litigation and speculation (Grant 2014).

These patterns and processes of urban development have led to a patchy and dynamic landscape of built-up areas with pockets of idle lands. Land disputes and litigation, particularly in areas closer the city, have led to traditional fishing villages remaining in seemingly rural, traditional structures, surrounded by faster paced urbanisation across administrative and customary boundaries. These 'peri-urban' structures illustrate the blurredness of what constitutes 'rural' and 'urban', not only in terms of land use and infrastructure, but also in terms of practices in securing access to land, water and food (Allen and Apsan Frediani 2013; Bartels et al. 2018). Agricultural production, for instance, plays a key role in generating livelihoods of the urban poor: "in Ghana, over 70% of the households ranked among the poorest expenditure quintile practice some form of UA. Furthermore, in cities across Ghana, revenue obtained from urban farming appears to contribute to 'over 30% of the income of the poorest quintile'" (Zezza and Tasciotti 2010, p. 268, in Allen and Apsan Frediani 2013, p. 369). It can moreover be considered a strategy for securing access to land (Allen and Apsan Frediani 2013).

Insecure access to food, land and water is managed by the urban poor through engaging in UA, in different forms of organisational structures, and of different degrees of commercialisation. Although not recognised by city authorities as formal land users, urban farmers are organised in cooperatives and similar institutional structures. Their institutions not only help to cope with food insecurity, but also to (re)claim city space and thereby play an active part in urban development (Allen and Apsan Frediani 2013). In the light of poor and insecure water quality, and a lack of resources and support for infrastructure investment, contamination risks from watering with wastewater are reduced by adjusting irrigation practices supported by research conducted at the International Water Management Institute (IWMI) (Drechsel and Keraita 2014). The IWMI approach to safe irrigation with untreated water has been applied in other cities of the Global South where large-scale improvement of irrigation water is not viable (Otoo and Drechsel 2018).

The SES lens brings to the fore the underlying dynamics of UA in Accra by highlighting interactions between the various dimensions. Practices of accessing and securing land through farming emerge from the need to adapt to changing institutions in land allocation (from community-oriented allocation along customary law to individual sales of plots), whereby both tenure security and access to farming land have become uncertain. They are therefore tightly linked to societal processes of urbanisation and globalisation (here in the real estate market). Farming practices require new forms of knowledge about natural processes, such as the transmission of diseases from irrigation with wastewater and technology (e.g. low-tech solutions to water reuse), to reducing health risks. Such knowledge and technology for UA is fostered through the formation of cooperatives and transdisciplinary research programmes on UA in Accra, therein becoming institutionalised. At the same time the UA system draws on deeply entangled institutions of the 'informal sector', namely traditional food markets. The combination of old and new in practices, knowledge, institutions and technology is linked to symbolic shifts, for instance in the meaning of traditional land titles. They materialise in the current urban landscape characterised by large parcels of urban food production amidst the built-up area of Accra.

The underlying dynamics identified call into question dominating (Western) approaches to urban planning, whereby land use is planned strategically and plots are developed—or left idle—accordingly and diverging developments are labelled as 'informal'. Out of the urgency to meet basic needs (e.g. housing and food security), UA has become an important land use in Accra that contributes to local food security. It is nevertheless contested among city authorities and urban farmers face risks from eviction, loss of income opportunities and to environmental health on a daily basis. New approaches in urban planning are needed to balance social-ecological dynamics in planned and unplanned, embedded urban developments.

3.2.2 Securing Water, Sanitation and Food Through Urban Agriculture in North Central Namibia

Transhumance used to be the most common farming practice in North Central Namibia up to the 1980s. Since then, changes in land and water use rights under colonial South African rule and after Independence (1990), as well as rapid population growth, have triggered settlement and urbanisation (Newsham and Thomas 2011; Schnegg and Bollig 2016). While population density is still low (approx. 850,000 inhabitants in an area of close to 35,000 km^2) in comparison to Ghana, the concentration of cattle and people is putting pressure on water and land resources under the prevailing semi-arid conditions (Liehr et al. 2017). Finding adequate water sources for livestock is increasingly challenging, especially during dry years such as 2018 and 2019. Livestock are often brought to drink from the open water canals serving the whole region and can also be seen drinking from untreated wastewater near settlements. This poses risks with respect to the potential spread of water-borne diseases such as cholera or hepatitis E. Investment in infrastructure for water treatment is necessary, especially where towns develop into cities and are obliged to have treatment facilities installed, instead of the less costly but also less efficient wastewater ponds. While the region is becoming more and more urban, people continue to identify with their rural heritage 'in the village' and corresponding status symbols such as herd size.

The population density increases and sedentism in Northern Central Namibia has been paralleled and supported by shifting regimes regulating land use and water rights. These were at first (during South African rule) based on customary systems, and after independence turned into new arrangements irrespective of kinship in the form of land and water use associations (Schnegg and Bollig 2016). Agroecological knowledge and practices that have evolved over more than 400 years were significantly challenged by these developments. As diversification of farming practices is becoming more difficult in the face of increasingly difficult access to land and water, agricultural production becomes more susceptible to drought and environmental hazards, and off-farming diversification is an option in urban areas only (Newsham and Thomas 2011). At the same time, there is an urgent need for food production to become less dependent on imports (Woltersdorf et al. 2015).

Along the SES model (Fig. 1), social processes of urbanisation and demographic change have spurred spatial concentration of farming practice that have now become a stressor on land and water resources (the natural system). Vice versa, environmental change such as climate change-induced droughts have become a burden on agricultural production. The cultural-symbolic dimension of institutions, such as the herd size being a denominator of wealth, persist even in urban areas. At the same time, the need for new knowledge and technology in adaptation to changing conditions is increasingly recognised, as the following example shows.

In North Central Namibia, urban crop production is gaining popularity among city officials as a strategy to manage wastewater by creating produce from the process of evapotranspiration through plants, rather than merely evaporating water from ponds. A pilot for treating wastewater and reusing it in vegetable farming has been developed and set up in Outapi within a German-Namibian research project (Liehr et al. 2017; Woltersdorf et al. 2018). Within the follow-up project 'EpoNa – Enhancing the Effectiveness of Ponds in Northern Namibia', a less cost-intensive system for reusing water from evaporation ponds, instead of setting up a new plant, is being developed and tested. Here the water is used for irrigation of fodder crops, to meet national and WHO standards of food safety. Initial assessments show that both systems offer a great potential for managing urban water and land resources more efficiently, while contributing to urban food security and health (Woltersdorf et al. 2018). What remains a factor of uncertainty in the regulation of urban food provisioning systems, however, is their interaction with the dynamics of institutions regulating urban resource (water and land) availability and crop markets, as lifestyles, food consumption and everyday practices are becoming more and more urban (i.e. individualised). This will also affect the economic viability of commercial UA based on treated water reuse.

Here the SES model (Fig. 1) served to not only analyse but also design UA in the case study region, as new knowledge and technology for adaptation to climate change in UA developed in transdisciplinary research. Specifically, the knowledge generated ranges from scientific understanding and engineering capacity for wastewater treatment to practical knowledge on plant production and wastewater management. Next to knowledge on the water treatment and agricultural production processes themselves, the transdisciplinary process in meetings and workshops revealed interactions within and between relevant actors—such as municipal authorities' dependency on consultants' knowledge in infrastructure planning and management. The knowledge generated enabled proactive formation of new institutions for capacity development, namely a wastewater treatment plant partnership among municipal authorities (Frick-Trzebitzky et al. 2019).

3.2.3 Urban Agriculture: An Approach to Manage Second-Order Problems in Regulation of Land and Water?

The two examples of urban food provisioning systems through UA illustrate the breadth of social-ecological interactions in towns and cities of sub-Saharan Africa. A far greater diversity can be expected when talking about the Global South in general. While in Accra, urban farming has developed rather informally out of people's need for food and land, UA systems in Outapi have been installed rather formally as 'pilot sites' for further implementation in North Central Namibia to manage scarce water resources more efficiently and meet sanitation needs. The intrinsic link between water and land management becomes salient in both cases. The increased individualisation of land and water tenure (Fig. 1: management and unintended side effects) in both Ghana and Namibia has created new environmental and social vulnerabilities (Fig. 1: ecosystem services and disservices) that can be conceptualised as second-order problems. Examples are overstocking, increased exposure to water-borne diseases, insecure tenancy, lack of diversity in farming systems, which altogether lead to greater susceptibility to stresses and shocks such as extreme weather events or diseases. Both UA approaches are societal attempts to regulate these problems. They are, however, inflicted with further uncertainties in the social-ecological relations they produce. For instance, what are the effects of novel food production systems on existing ones? How can they prevail when land and water prices rise (Fig. 1: management and unintended side effects) and under conditions of climate change? How are uneven vulnerabilities in access to land, water, food and health reproduced in UA systems (Fig. 1: ecosystem services and disservices), especially as actors and institutions constantly change and involve both formal and informal ones? Understanding the dynamics of actors, practices, knowledge, ecosystem functions, institutions and technology is a prerequisite for tackling these questions.

3.3 Green Infrastructure: Latin America

Modern cities aim to be resilient to provide present and future inhabitants good, healthy and supportive living and environmental conditions (Elmqvist et al. 2018). In the Global North, a current focus is to expand green spaces in urban areas to preserve and provide a wide range of ES, a process especially prompted by climate change (Haq 2011; du Toit et al. 2018). In 2013, the European Commission issued a corresponding strategy to expand GI in the member states. According to the European Commission GI is defined as "a strategically planned network of natural and semi-natural areas with other environmental features designed and managed to deliver a wide range of ecosystem services in both rural and urban settings" (European Commission 2013). Focusing on the human good as well as ecological aspects, the ongoing implementation of GI is seen as a social-ecological measure. However, in the Global South societal problems related to GI often come to the fore. People are frequently more exposed to economic instability due to insufficient urban planning and higher social inequality (Parves Rana 2009; Barau et al. 2013). In addition, many countries of the Global South experience a higher risk to the consequences of climate change and biodiversity loss (Naciones Unidas and Comisión Económica para América Latina y el Caribe [CEPAL] 2010). At the same time, western concepts of urban planning and development such as GI are not meeting the social and cultural realities and demands in the respective countries of the Global South often resulting in rejection or conflicts. In the following example, we present a case study on GI implementation from Mexico City.

3.3.1 Social Inequality and Green Infrastructure—The Case of Mexico City

Mexico City is the second largest city within Latin America, while Latin America is the most urbanised region in the Global South (Fernández-Álvarez 2017). After a long history of colonisation and a phase of independence, Mexico's government changed with Porfirio Diaz' regime in the late nineteenth century to a structured and planned course of action. He inspired the modernisation of Mexico City, according to the examples of European industrial metropolii. Implementation of GI was considered for the first time and was, along with technical advances, seen as part of a modern city. Nature was largely seen in its decorative value and implemented in an engineered and localised fashion in parks, gardens and avenues (Wakild 2007).

With the population growth of Mexico, starting between 1940 and 1950 and continuing to increase until the 1980s, Mexico turned from a mainly rural country to an urban one (Estados Unidos Mexicanos 2016). Mexico City reached over eight million official inhabitants (Instituto Nacional de Estadística y Geografía 2015). Due to this rapid human population growth including urbanisation, many people live in informal settlements (Torres-Lima and Cruz-Castillo 2019). The Mexican economy is, despite society's orientation towards cities, still based on local natural resources mainly located in rural areas (Naciones Unidas and Comisión Económica para América Latina y el Caribe [CEPAL] 2010; Torres-Lima and Cruz-Castillo 2019). Cities do not provide sufficient work opportunities, leaving two out of three people in poverty (Torres-Lima and Cruz-Castillo 2019).

The difference between the historical development of the Global South and North, as well as the geographical differences, are significant and displayed in the current economic, social and environmental problems (Parves Rana 2009; Myers 2021). Therefore, an alternative state of societal and ecological development is to be expected in the Global South compared to the current and future states of the Global North, requiring new perspectives, theories and new strategies for sustainable development activities (McHale et al. 2013). As a result of this massive and rapid urbanisation process, problems such as environmental injustices exist in Mexican cities (Hardoy et al. 2001; Carruthers 2008). Urban amenities are unevenly distributed and often biased against marginalised populations (Carruthers 2008; Boone et al. 2009). GI in Mexico City appears to be no exception (Fernández-Álvarez 2017).

GI such as parks are supposed to enhance the quality of life (Haq 2011), and it should provide people with additional living space, especially for socially disadvantaged groups (Fernández-Álvarez 2017). High quality GI is efficient in providing ES, when represented locally and regionally in sufficient quantity (Pauleit et al. 2021). It can enhance the resilience of urban areas, reduce environmental risks like floods, and positively affect human health by heat reduction and air quality improvement (Calderón-Contreras and Quiroz-Rosas 2017). Due to the different assessment of priorities during the implementation of GI in the late nineteenth century, GI development and maintenance in Mexico City is now becoming a concern of social inequality. In general, cities became more vulnerable to the risks of environmental disasters in the last decades. Seventy per cent of people living in areas with a high risk of experiencing environmental disaster are city dwellers, with residents of informal settlements often being disproportionately at high risk (Estados Unidos Mexicanos 2016). With low-income communities being more vulnerable to social and environmental problems (Parves Rana 2009), GI is vital to support marginalised social groups, enhance social equality and reduce inequality (Haq 2011). However, poorly planned or implemented GI measures can have the opposite effect, i.e. enhancing social inequality by focusing on services in rich neighbourhoods (Buckingham and Turner 2008; Parves Rana 2009). In Mexico City, GI of higher quantity and quality is mostly located close to conservation zones, leaving other areas of the city without access to the ES provided by GI (Calderón-Contreras and Quiroz-Rosas 2017).

Applying the SRN approach helps understanding the interlinkages between natural and societal processes of GI in urban areas (Calderón-Contreras and Quiroz-Rosas 2017; Torres-Lima and Cruz-Castillo 2019). For example, focusing on everyday practices as part of the SES model helps with better interpretation of the traffic situation in Mexico City and drawing social-ecological conclusions. Due to urban expansion, many Mexico City residents rely on cars to get to their workplace (Estados Unidos Mexicanos 2016; Villa 2018), spending on average two hours per day in transport (UN Environment Programme 2019). Public transportation, where available, is perceived as inconvenient and dangerous. To improve this situation the public institutions for roadways focused on generating higher comfort for individual car traffic, which led to increasing traffic volumes, with attendant negative impacts like high CO_2 emissions (Estados Unidos Mexicanos 2016). The poor air quality, which is mostly due to traffic in Mexico City, leads to health problems. In addition, health risks due to an inactive lifestyle and obesity are increasing. Vulnerable groups, like low-income families, woman and indigenous people, are often more affected by these disadvantages duc to less chances, for example the possibility to choose a close workplace, therefore, spending more time in transport (Bautista-Hernández 2020). With programmes like CiClim (Cities and Climate Change Programme) and 'Muevete en bici', ecological and social problem-solving strategies are combined, implementing bicycle lanes and creating a cycling culture (Villa 2018; UN Environment Programme 2019). Mexican citizens prefer riding a bike in a green and natural environment (Villa 2018). Considering practices and preferences of the local communities, GI as a social-ecological structure can provide new opportunities combing ecological (e.g. place for nature) and social (e.g. health and the promotion of equal opportunities) aspects.

3.3.2 The Importance of the Cultural Meaning of Green Infrastructure

This case study on bicycle lanes and GI is a good example of the relevance of different cultural meanings of a given problem. Besides the physical-material dimension of GI providing ES, such as heat reduction or flood prevention, there is also a cultural-symbolic dimension. This is closely related to the meaning and conceptualisation of nature itself in the Global South. In West Africa, for example in Accra (Ghana), planning of GI or parks is often seen to be in the tradition of postcolonial urban development and neoliberal concepts of urban renewal. These attempts are highly debated as they do not meet the local conditions of urban development (Afenah 2012; Grant 2014). Newer concepts of valuation of nature take these relational aspects (i.e. the relation between nature and society) into account. Chan et al. (2016) and Himes and Muraca (2018) argue that perceptions and valuations of biodiversity and ES, e.g. provided by GI, are particularly associated with social aspects. This concept acknowledges that nature and its relation to society can have a historical background that still has relevance in the current situation. This concept highlights that there are not only ecological benefits of green parks and GI in urban areas in the Global South (such as for climate regulation, water retention), but also a cultural meaning (see chapters in Cocks and Shackleton 2021).

People can have various perspectives on nature due to their culture and belief system. Often the local living conditions and practices are based on local knowledge on the interaction between society and nature. As a result, different conceptualisations and understandings of this relation evolve. This may underlie conflicts, where different people with different cultural backgrounds come together, like in fast growing urban spaces. The concept of SRN can then help disentangle the different perspectives and (local) knowledge systems and align these findings with customised solution strategies in dynamic contexts exacerbated by, for example, climate change and social shifts.

4 Recommendations for Future Research

This chapter has presented some peculiarities of urban Social Ecology in the Global South. Presenting the examples of rural–urban migration, urban agriculture and green infrastructure, the complex interactions and interdependencies between natural and societal structures and processes become clear. This paragraph seeks to highlight how transdisciplinary research is necessary in addressing these issues.

Transdisciplinary research addresses the specific processes and structures of social-ecological problems as described above, and also the processes of scientific knowledge production including the methods and concepts, their consequences and limits. In general, the transdisciplinary research process can be divided into three consecutive steps (Fig. 3): (1) Societal and scientific problems are related to each other to formulate the common research objective; (2) New knowledge is produced via integrating and making use of different disciplines. In this sense interdisciplinarity is a relevant aspect of transdisciplinary research; (3) The knowledge produced is evaluated against the background of its contribution to the societal and scientific progress. This transdisciplinary integration provides results for both the scientific and the societal praxis (Jahn et al. 2012). Ideally, such critical transdisciplinary research meets the following requirements: (1) it is open to theoretical and epistemological pluralism, (2) it provides a strong link between theoretical reflection and empirical analysis of real-world problems and (3) it reflects upon the implications of science as a functional system within society (Jahn 2013; Hummel et al. 2017). In doing

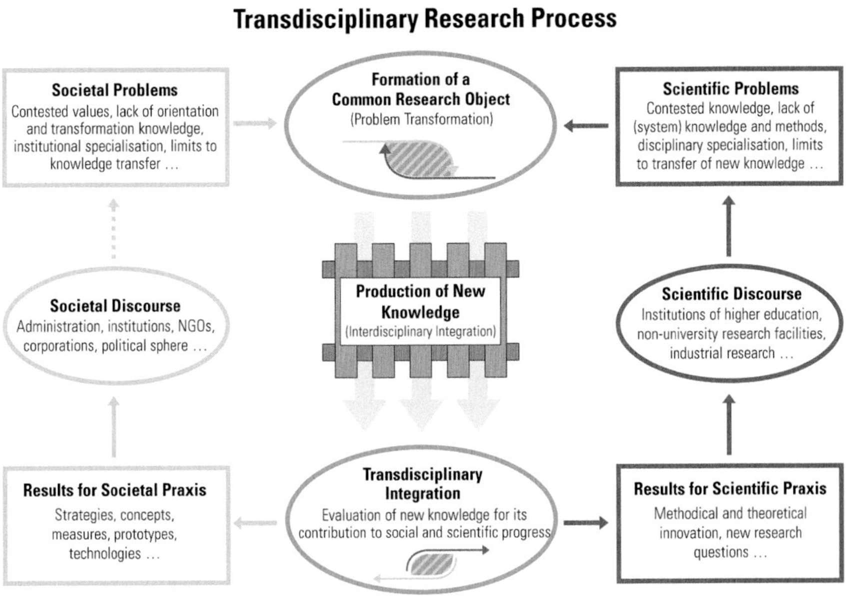

Fig. 3 Model of a transdisciplinary research process (Jahn et al. 2012)

so, transdisciplinarity is an open and reflective research mode. It integrates a diversity of stakeholders as knowledge holders. This is particularly relevant in the Global South to be able to cover the different perspectives of the physical-material and the cultural-symbolic dimensions as exemplified above. Dealing with the different cultural-symbolic dimensions is of major importance, not only the cultural diversity in the Global South, but also the scientific cultures. It is important that researchers consider how they position themselves in the power structure of a given research problem or research space. As a scientist, one is always part of the research process and thereby exert an influence, which is not only as an external observer.

Considering the complexity of social-ecological research problems in the Global South requires a conscious handling of the topics of ignorance, uncertainty and contested knowledge. Thus, the dominance of Western epistemic concepts and perspectives transferred unquestioningly to, and implemented in, the Global South are particularly problematic. This entails the risk that cultural, historical and natural conditions and/or knowledge are not sufficiently taken into account. For example, the Western idea of 'climate refugees' (Berchin et al. 2017): here a linear epistemic idea and dominance is transferred to the Global South with consequences for drawing recommendations. In Western countries, the attitude still dominates that climate change is the main and often only reason for international migration in the Global South and is consequently a phenomenon to be avoided. As a result, efforts are taken to minimise the causes and increase the attractiveness in the given countries. However, in certain areas in the Global South, such as the Sahel region, migration is deeply rooted in the culture and one of several livelihood strategies. In addition to that, causes of migration in these countries are rather social-ecological than only climate change driven (Hummel and Liehr 2015; Hummel 2016). Ignoring these aspects has consequences for recommendations for action if they do not include local peculiarities and context dependency, i.e. draw recommendations on how to shape migration rather than avoiding it. As the case study examples in this chapter illustrate, the transdisciplinary research approach of Social Ecology can help to overcome such shortcomings and develop context-dependent options for action for social-ecological problems in the Global South.

References

Adamo SB (2010) Environmental migration and cities in the context of global environmental change. Curr Opin Environ Sustain 2:161–165

Adedeji JA (2021) Urban biocultural identity of *Yorubas*: intersection of philosophy and nature for wellbeing at Osun Sacred Grove UNESCO Site, Osogbo, Nigeria. In: Cocks ML, Shackleton CM (eds) Diversities of nature in cities: enriching belonging, wellbeing and bioculture. Earthscan and Routledge, London, pp 51–66

Adger WN, Boyd E, Fábos A, Fransen S, Jolivet D, Neville G et al (2019) Migration transforms the conditions for the achievement of the Sustainable Development Goals. Lancet Planet Health 3(11)

Afenah A (2012) Engineering a millennium city in Accra, Ghana: the Old Fadama intractable issue. Urban For 23:527–540
Alihan MA (1938) Social ecology: a critical analyses. Columbia University Press, New York
Allen A, Apsan Frediani A (2013) Farmers, not gardeners. City 17:365–381
Barau AS, Ludin AN, Said I (2013) Socio-ecological systems and biodiversity conservation in African city: insights from Kano Emir's Palace gardens. Urban Ecosys 16:783–800
Bartels LE, Bruns A, Alba R (2018) The production of uneven access to land and water in peri-urban spaces: de facto privatisation in greater Accra. Local Environ 23:1172–1189
Batsaikhan N, Buuveibaatar B, Chimed B, Enkhtuya O, Galbrakh D, Ganbaatar O et al (2014) Conserving the world's finest grassland amidst ambitious national development. Conserv Biol 28:1736–1739
Bautista-Hernández DA (2020) Commuting inequality, role of urban structure, and identification of disadvantaged groups in the Mexico City metropolitan area. J Transp Land Use 13:159–183
Beck U (1992) Risk society: towards a new modernity. Sage, London
Becker E (2014) Social-ecological systems as epistemic objects. In: Glaser M, Krause G, Beate MW, Ratter, Welp M (eds) Human-nature interactions in the Anthropocene: potentials of social-ecological systems analysis. Routledge, New York, pp 37–59
Becker E, Jahn T (2006) Soziale Ökologie: Grundzüge einer Wissenschaft von den gesellschaftlichen Naturverhältnissen. 1. Aufl. Campus Verlag GmbH (Sozialwissenschaften 2001–2008), Frankfurt am Main
Becker E, Hummel D, Jahn T (2011) Gesellschaftliche Naturverhältnisse als Rahmenkonzept. In: Gross M (ed) Handbuch Umweltsoziologie, 1st edn. VS Verlag für Sozialwissenschaften, Wiesbaden
Berchin II, Valuduga IB, Garcia J, de Andrade Guerra JB (2017) Climate change and forced migrations: an effort towards recognising climate refugees. Geoforum 84:147–150
Berkes F, Folke C (eds) (1998) Linking social and ecological systems: management practices and social mechanisms for building resilience. Cambridge University Press, Cambridge
Binder CR, Hinkel J, Bots PW, Pahl-Wostl C (2013) Comparison of frameworks for analysing social-ecological systems. Ecol Soc 18(4)
Boone CG, Buckley GL, Grove JM, Sister C (2009) Parks and people: an environmental justice inquiry in Baltimore, Maryland. Ann Assoc Am Geogr 99:767–787
Buckingham S, Turner M (2008) Understanding environmental issues. Sage, London
Buizer M, Elands B, Vierikko K (2016) Governing cities reflexively: the biocultural diversity concept as an alternative to ecosystem services. Environ Sci Pol 62:7–13
Cadenasso ML, Pickett ST (2013) Three tides: the development and state of the art of urban ecological science. In: Pickett ST, Cadenasso ML, McGrath B (eds) Resilience in ecology and urban design: linking theory and practice for sustainable cities. Springer, Dordrecht, pp 29–46
Calderón-Contreras R, Quiroz-Rosas LE (2017) Analysing scale, quality and diversity of green infrastructure and the provision of urban ecosystem services: a case from Mexico City. Ecosys Serv 23:127–137
Carruthers DV (2008) Environmental justice in Latin America: problems, promise, and practice. MIT Press, Cambridge, MA
Central Intelligence Agency (2018) The world Factbook library. East Asia/Southeast Asia: Mongolia. Washington, DC. Online https://www.cia.gov/library/publications/the-world-factbook/geos/mg.html
Chan K, Balvanera P, Benessaiah K, Chapman M, Díaz S, Gómez-Baggethun E et al (2016) Why protect nature? Rethinking values and the environment. Proc Nat Acad Sci 113:1462–1465
Cilliers S, Cilliers J, Lubbe R, Siebert S (2013) Ecosystem services of urban green spaces in African countries: perspectives and challenges. Urban Ecosys 16:681–702
Cocks ML, Shackleton CM (2021) Diversities of nature in cities: enriching belonging, wellbeing and bioculture. Earthscan and Routledge, London

Davoren E, Shackleton CM (2021) Urban ecosystem disservices in the Global South. In: Shackleton CM, Cilliers SS, Davoren E, du Toit MJ (eds) Urban ecology in the Global South. Springer, Cham, pp 265–292

Díaz S, Demissew S, Joly C, Lonsdale WM, Larigauderie A (2015) A Rosetta Stone for nature's benefits to people. PLoS Biol 13(1):e1002040

Díaz S, Pascual U, Stenseke M, Martín-López B, Watson RT, Molnár Z et al (2018) Assessing nature's contributions to people. Sci 359(6373):270–272

Dore G, Nagpal T (2006) Urban transition in Mongolia: pursuing sustainability in a unique environment. Environ 48:10–24

Drechsel P, Keraita B (2014) Irrigated urban vegetable production in Ghana: characteristics, benefits and risk mitigation. International Water Management Institute, Colombo

Drescher AW, Isendahl C, Caridad Cruz M, Karg H, Menakanit A (2021) Urban and peri-urban agriculture in the Global South. In: Shackleton CM, Cilliers SS, Davoren E, du Toit MJ (eds) Urban ecology in the Global South. Springer, Cham, pp 293 – 324

du Toit MJ, Cilliers SS, Dallimer M, Goddard M, Guenat S, Cornelius SF (2018) Urban green infrastructure and ecosystem services in sub-Saharan Africa. Landsc Urb Plan 180:249–261

Elmqvist T, Fragkias M, Goodness, J, Güneralp B, Marcotullio PJ, McDonald RI et al (eds) (2013) Urbanisation, biodiversity and ecosystem services: challenges and opportunities—a global assessment. Springer, Dordrecht.

Elmqvist T, Bai X, Frantzeskaki N, Griffith C, Maddox D, McPhearson T et al (eds) (2018) Urban planet: knowledge towards sustainable cities. Cambridge University Press, Cambridge

Estados Unidos Mexicanos (2016) Habitat III national report. Mexico

European Commission (2013) Enhancing Europe's natural capital: green infrastructure (GI), 2013. Online https://www.eea.europa.eu/policy-documents/green-infrastructure-gi-2014-enhancing, checked on 7/3/2020.858Z

Fernández-Álvarez R (2017) Inequitable distribution of green public space in the Mexico City: an environmental injustice case. Distribución inequitativa del espacio público verde en la Ciudad de México. https://doi.org/10.22136/est002017697

Fisher B, Turner RK, Morling P (2009) Defining and classifying ecosystem services for decision making. Ecol Econ 68:643–653

Fischer-Kowalski M, Weeisz H (2016) The Archipelago of social ecology and the Island of the Vienna School. In: Haberl H, Fischer-Kowalski M, Krausmann F, Winiwarter V (eds) Social ecology: society-nature relations across time and space. Springer, Heidelberg

Frick-Trzebitzky F, Bruns A (2019) Disparities in the implementation gap: adaptation to flood risk in the Densu Delta, Accra, Ghana. J Environ Pol Plan 21:577–592

Frick-Trzebitzky F, Zimmermann M, Kluge T (2019) Capacity development for wastewater management and water reuse in informal partnerships in northern Namibia. Book of Abstracts of 12th IWA International Conference on Water Reclamation and Reuse, Berlin (16–20 June 2019)

Gioli G, Hugo G, Costa MM, Scheffran J (2016) Human mobility, climate adaptation, and development. Migr Dev 5:165–170

Glaser M, Krause G, Ratter BM, Welp M (2014) New approaches to the analysis of human-nature relations. In: Glaser M, Krause G, Ratter BM, Welp M (eds) Human-nature interactions in the Anthropocene: potentials of social-ecological systems analysis. Routledge, New York, pp 3–12

Görg C (1999) Gesellschaftliche Naturverhältnisse. Westfälisches Dampfboot, Münster

Görg C, Brand U, Haberl H, Hummel D, Jahn T, Liehr S (2017) Challenges for social-ecological transformations: contributions from social and political ecology. Sustain 9:1045

Grant R (2014) Globalising city: the urban and economic transformation of Accra, Ghana. Syracuse University Press, Syracuse

Grant R, Yankson P (2003) Accra. Cities 20:65–74

Grimm N, Grove M, Pickett S, Redmann C (2000) Integrated approaches to long-term studies of urban ecological systems. BioScience 50(7): 571–584

Haberl H, Fischer-Kowalski M, Krausmann F, Winiwarter V (eds) (2016) Social ecology: society-nature relations across time and space. Springer, Geneva

Haq SM (2011) Urban green spaces and an integrative approach to sustainable environment. J Environ Prot 2:601–608
Hardoy J, Mitlin D, Satterthwaite D (2001) Environmental problems in an urbanising world: finding solutions for cities in Africa, Asia, and Latin America. Earthscan, Sterling
Hastrup K, Fog Olwig K (2012) Climate change and human mobility. Cambridge University Press, Cambridge
Himes A, Muraca B (2018) Relational values: the key to pluralistic valuation of ecosystem services. Curr Opin Environ Sustain 35:1–7
Hummel D (2016) Climate change, land degradation and migration in Mali and Senegal: some policy implications. Migr Dev 5:211–233
Hummel D, Jahn T (2014) Sozialökologie. In: Endruweit G, Trommsdorff G, Burzan N (eds.) Wörterbuch der Soziologie. UVK Verlagsgesellschaft mit UVK/Lucius, Konstanz and München, pp 454–457
Hummel D, Liehr S (2015) Migration in the West African Sahel: more than a response to climate change (ISOE Policy Brief, 2/2015). Institute for Social-Ecological Research, Frankfurt am Main
Hummel D, Hertler C, Niemann S, Lux A, Janowicz C (2008) Part A: the analytical framework. In: Hummel D (ed) Population dynamics and supply systems: a transdisciplinary approach. Campus, Frankfurt am Main, pp 11–69
Hummel D, Jahn T, Keil F, Liehr S, Stieß I (2017) Social ecology as critical, transdisciplinary science: conceptualising, analysing and shaping societal relations to nature. Sustain 9(7):1050
Instituto Nacional de Estadística y Geografía (2015) Panorama sociodemográfico de Ciudad de México. INEG, Mexico Ciudad
Jaganmohan M, Vailshery LS, Mundoli S, Nagdendra H (2018) Biodiversity in scared urban spaces of Bengaluru, India. Urban For Urban Green 32:64–70
Jahn T (2013) Sustainability science requires a critical orientation. GAIA 22:29–33
Jahn T, Bergmann M, Keil F (2012) Transdisciplinarity: between mainstreaming and marginalisation. Ecol Econ 79:1–10
Jahn T, Hummel D, Drees L, Liehr S, Lux A, Mehring M, et al (2020) Shaping social-ecological transformations in the Anthropocene. ISOE-Diskussionspapiere 45. ISOE - Institut für sozial-ökologische Forschung, Frankfurt am Main
Kramm J, Pichler M, Schaffartzik A, Zimmermann M (2017) Societal relations to nature in times of crisis: social ecology's contributions to interdisciplinary sustainability studies. Sustain 9(7):1042
Lawhon M, Ernstson H, Silver J (2014) Provincialising urban political ecology: towards a situated UPE through African Urbanism. Antipode 46:497–516
Liehr S, Drees L, Hummel D (2016) Migration as societal response to climate change and land degradation in Mali and Senegal. In: Yaro JA, Hesselberg J (eds) Adaptation to climate change and variability in rural West Africa. Springer, Cham, pp 147–169
Liehr S, Röhrig J, Mehring M, Kluge T (2017) How the social-ecological systems concept can guide transdisciplinary research and implementation: addressing water challenges in Central Northern Namibia. Sustain 9(7):1109
Maffi L (2008) An introduction to biocultural diversity. Terralingua, Salt Spring Island
Martín-López B, Gómez-Baggethun E, García-Llorente M, Montes C (2014) Trade-offs across value-domains in ecosystem services assessment. Ecol Indic 37:220–228
McHale MR, Bunn DN, Pickett ST, Twine W (2013) Urban ecology in a developing world: why advanced socioecological theory needs Africa. Front Ecol Environ 11:556–564
McPhearson T, Pickett ST, Grimm NB, Niemelä J, Alberti M, Elmqvist T et al (2016) Advancing urban ecology toward a science of cities. BioScience 66:198–212
Mehring M, Bernard B, Hummel D, Liehr S, Lux A (2017a) Halting biodiversity loss: how social–ecological biodiversity research makes a difference. Int J Biodivers Sci Ecosyst Serv Manag 13(1):172–180
Mehring M, Zajonz U, Hummel D (2017b) Social-ecological dynamics of ecosystem services: livelihoods and the functional relation between ecosystem service supply and demand: evidence from Socotra Archipelago, Yemen and the Sahel Region, West Africa. Sustain 9(7):1037

Mehring M, Batbuyan B, Bolortsetseg S, Buuveibaatar B, Dashpurev T, Drees L et al (2018) Keep on moving: how to facilitate nomadic pastoralism in Mongolia in the light of current societal transformation processes. ISOE Policy Brief, No. 7, Institute for Social-Ecological Research, Frankfurt am Main
Millennium Ecosystem Assessment (2005) Ecosystems and human wellbeing: synthesis. Island Press, Washington, DC
Myers G (2021) Urbanisation in the Global South. In: Shackleton CM, Cilliers SS, Davoren E, du Toit MJ (eds) Urban ecology in the Global South. Springer, Cham, pp 27–49
Naciones Unidas; Comisión Económica para América Latina y el Caribe (CEPAL) (eds) (2010) Indicadores ambientales de América Latina y el Caribe 2009. With assistance of División de Estadística y Proyecciones Económicas (Cuadernos estadísticos N°, 38). Online http://hdl.handle.net/11362/4316
Nandintsetseg D, Bracis C, Olson KA, Böhning-Gaese K, Calabrese JM, Chimeddorj B et al (2019) Challenges in the conservation of wide-ranging nomadic species. J Appl Ecol 56(8):1916–1926
Newsham A, Thomas DS (2011) Knowing, farming and climate change adaptation in North-Central Namibia. Glob Environ Change 21:761–770
Ostrom E (2007) A diagnostic approach for going beyond panaceas. Proc Nat Acad Sci 104:15181–15187
Ostrom E (2009) A general framework for analysing sustainability of social-ecological systems. Sci 325(5939):419–422
Otoo M, Drechsel P (eds) (2018) Resource recovery from waste: business models for energy, nutrient and water reuse in low- and middle-income countries. Earthscan, London
Parnell S, Robinson J (2012) (Re)theorising cities from the Global South: looking beyond neoliberalism. Urban Geogr 33:593–617
Parves Rana M (2009) Sustainable city in the global North and South: goal or principle? Manage Environ Qual 20:506–521
Pauleit S, Vasquéz A, Maruthaveeran S, Liu L, Cilliers SS (2021) Urban green infrastructure in the Global South. In: Shackleton CM, Cilliers SS, Davoren E, du Toit MJ (eds) Urban ecology in the Global South. Springer, Cham, pp 107–143
Piguet E, Laczko F (2014) People on the move in a changing climate. Springer, Dordrecht
Reyers B, Biggs R, Cumming GS, Elmqvist T, Hejnowicz AP, Polasky S (2013) Getting the measure of ecosystem services: a social–ecological approach. Front Ecol Environ 11:268–273
Robinson J (2002) Global and world cities: a view from off the map. Int J Urban Reg Res 26:531–554
Save the Children International (2013) Shifting livelihoods: trends of pastoralist drop-out and rural to urban migration in Mongolia. Save the Children International, Ulaanbaatar
Schnegg M, Bollig M (2016) Institutions put to the test: community-based water management in Namibia during a drought. J Arid Environ 124:62–71
Serraglio DA, Ferreira HS, Robinson N (2019) Climate-induced migration and resilient cities: a new urban agenda for sustainable development. Seqüência 42:10–46
Shackleton CM, Ruwanza S, Sinasson Sanni GK, Bennett S, De Lacy P, Modipa R, Mtati N, Sachikonye M, Thondhlana G (2016) Unpacking Pandora's Box: understanding and categorising ecosystem disservices for environmental management and human wellbeing. Ecosys 19:587–600
Shackleton CM, Cilliers SS, du Toit MJ, Davoren E (2021) The need for an urban ecology of the Global South. In: Shackleton CM, Cilliers SS, Davoren E, du Toit MJ (eds) Urban ecology in the Global South. Springer, Cham, pp 1–26
Stokols D (2018) Social ecology in the digital age: solving complex problems in a globalised world. Academic Press, London
Tacoli C (2003) The links between urban and rural development. Environ Urban 15:3–12
Tarne A (2018) Analysing the dynamics of rural-urban migration in the light of a changing nomadic. culture in Mongolia. Masters Thesis. University of Vechta, Vechta
Torres-Lima P, Cruz-Castillo JG (2019) Urban processes and socioecological agricultural systems. sustainable trajectories of Chinampa agriculture in Mexico City. Letras Verdes, Revista Latinoamericana de Estudios Socioambientales (25): 168–189

Tuholske C, Andam K, Blekking J, Evans T, Caylor K (2020) Comparing measures of urban food security in Accra, Ghana. Food Secur 12:417–431

UN Environment Programme (2019) Mexico. Online https://www.unenvironment.org/explore-topics/transport/what-we-do/share-road/mexico. Checked on 21 July 2020

van der Land V (2018) Migration and environmental change in the West African Sahel: why capabilities and aspirations matter. Routledge, London

Villa A (2018) Green infrastructure in Mexico: a booster for healthier cities. Deutsche Gesellschaft für Internationale Zusammenarbeit (GIZ). URBANET. Online https://www.urbanet.info/mexico-green-infrastructure/. Checked on 21 July 2020

von Döhren P, Haase D (2015) Ecosystem disservices research: a review of the state of the art with a focus on cities. Ecol Indic 52:490–497

Wakild E (2007) Naturalising modernity: urban parks, public gardens and drainage projects in Porfirian Mexico City. Mexican Stud 23:101–123

Wangai PW, Burkhard B, Müller F (2016) A review of studies on ecosystem services in Africa. Intn J Sustain Built Environ 5:225–245

Wesche K, Ambarlı D, Kamp J, Török P, Treiber J, Dengler J (2016) The Palaearctic steppe biome: a new synthesis. Biodiv Conserv 25:2197–2231

Woltersdorf L, Liehr S, Scheidegger R, Döll P (2015) Small-scale water reuse for urban agriculture in Namibia: modelling water flows and productivity. Urban Water J 12:414–429

Woltersdorf L, Zimmermann M, Deffner J, Gerlach M, Liehr S (2018) Benefits of an integrated water and nutrient reuse system for urban areas in semi-arid developing countries. Res Conserv Recycl 128:382–393

Yembuu B (2016) Mongolian nomads: effects of globalisation and social change. In Robertson M, Tsang PK (eds) Everyday knowledge, education and sustainable futures. Springer, Singapore, pp 89–105

Zezza A, Tasciotti L (2010) Urban agriculture, poverty, and food security: empirical evidence from a sample of developing countries. Food Pol 35:265–273

Urban Green Infrastructure in the Global South

Stephan Pauleit, Alexis Vasquéz, Sreetheran Maruthaveeran, Li Liu, and Sarel S. Cilliers

Abstract Urban green infrastructure (UGI) aims to strategically develop green and blue space networks that contribute to human health and wellbeing, urban sustainability and climate resilience. However, evidence on UGI research and application in the cities of the Global South is limited. An overview of the development of UGI in the three regions of Latin America, Sub-Saharan Africa and Asia is presented. The selective review shows that concepts such as urban greening, urban forestry and urban agriculture are more frequently used. Traditional goals, i.e. city beautification and providing recreational spaces, still prevail, but the reduction of risks (e.g. from flooding and landslides) and securing food and livelihoods are increasingly important drivers for developing UGI. Habitat networks can be another entry point for broadening into multifunctional UGI. However, there are only few examples of integrated approaches that would transcend sector silos. Therefore, strengthening the capacity for strategic planning and collaborative governance

S. Pauleit (✉)
Chair for Strategic Landscape Planning and Management, Faculty of Life Sciences, Technical University of Munich, Emil-Ramann-Str. 6, 85354 Freising, Germany
e-mail: pauleit@tum.de

A. Vasquéz
Department of Geography, Faculty of Architecture and Urban Planning, University of Chile, Portugal 84, Santiago, Chile
e-mail: alexvasq@u.uchile.cl

S. Maruthaveeran
Department of Landscape Architecture, Faculty of Design and Architecture, Universiti Putra Malaysia, UPM, 43400 Serdang, Selangor Darul Ehsan, Malaysia
e-mail: sreetheran@upm.edu.my

L. Liu
Department of Geosciences and Natural Resource Management, University of Copenhagen, Rolighedsvej 2, 1958 Frederiksberg, Denmark
e-mail: liu@ign.ku.dk

S. S. Cilliers
Unit for Environmental Sciences and Management, North-West University, Private Bag X6001, Potchefstroom 2520, South Africa
e-mail: Sarel.Cilliers@nwu.ac.za

C. M. Shackleton et al. (eds.), *Urban Ecology in the Global South*, Cities and Nature,
https://doi.org/10.1007/978-3-030-67650-6_5

of UGI is an important consideration for cities in the Global South and for future research. The potential of local communities for creating and managing UGI should be better recognised in policymaking, along with environmental justice because often the urban poor do not benefit fairly from UGI.

Keywords Africa · Asia · Environmental justice · Green spaces · Informal settlement · Latin America · Planning

1 Introduction

Green infrastructure is considered as a promising concept for the strategic development of networks of green and blue spaces that provide multiple benefits for people (Benedict and McMahon 2002). After its inception in the United States (US) in the 1990s as a response to the pressures created by urbanisation on natural systems, green infrastructure has been particularly adopted in the UK and Australia while more recently the publication of a green infrastructure strategy by the European Commission (EC 2013) has promoted its uptake in the countries of the European Union.

Urban green infrastructure (UGI) has been interpreted in different ways and applied at spatial scales ranging from city-regional greenspace networks to local sustainable urban drainage systems at site level (e.g. Lafortezza et al. 2013; Rouse and Bunster-Ossa 2013; Mell 2014). When understood as a landscape approach, green infrastructure considers the entire diversity of green and blue spaces and even green on built and grey spaces, regardless of their origin, use and ownership (see Cvejić et al. 2015; O'Brien et al. 2017; Bartesaghi Koc et al. 2017).

Research on UGI has been strong in the US (e.g. Ahern 2007; Lovell and Taylor 2013), Australia (e.g. Lin et al. 2018) and Europe (e.g. Mell 2016; Pauleit et al. 2017a, 2019a). The research has provided ample evidence on the value of the UGI concept to address major environmental, social and economic challenges of urbanisation. It has been stressed that for meeting these challenges, the development of UGI should be guided by a number of principles (Fig. 1) among which connectivity and multifunctionality are the key (Benedict and McMahon 2006; Kambites and Owen 2006; Ahern 2007; Mell 2010; Pauleit et al. 2011; EC 2013). Both principles should be understood broadly, including ecological, social and economic dimensions. In this context, ecosystem services have been suggested as a suitable concept to operationalise multifunctionality (Hansen and Pauleit 2014), while embedding UGI into a social-ecological systems perspective is argued to be crucial for enabling the sustainable flow of these services (Andersson et al. 2019).

Among other main principles are the integration of green and grey infrastructures as described above, the coordinated and collaborative development of UGI at multiple scales and in strategic ways, i.e. being based on a long-term vision supplemented by actions and means for implementation (Healey 2010; Davies et al. 2015). UGI requires involvement of a range of disciplines such as planning and landscape

Fig. 1 Core principles for urban green infrastructure. Principles relate to the properties of green infrastructure as well as the process of its development (based on Davies et al. 2015, adapted, picture: Addis Ababa, courtesy of S. Lindley)

architecture, social sciences and the humanities, ecology and engineering, as well as close cooperation between practice and science. Not least, social inclusion has been strongly called for as a key principle, meaning that processes of UGI development are open to all and incorporate the knowledge and needs of diverse parties, emphasising those that are most in need of green space but may be less able to articulate themselves in planning processes (Hansen et al. 2017). Therefore, UGI planning should be conceived as a continuous process of reflection, learning and adaptation that contributes to community building and is able to cope with uncertainty (Kato and Ahrn 2008).

Research has shown that development and implementation of such an ambitious approach is challenging (Davies et al. 2015). It requires strong political support, regulations requesting consideration of green space, available knowledge, and personal and technical capacity. Furthermore, existence of "boundary spanners" (Safford et al. 2017), as well as institutional structures that support collaborative working between different sectors, between science and practice and between the public and the private domains are important enabling factors for the uptake of UGI in urban development. Not least, promotion of UGI requires good understanding of the perception of its values and services by citizens (Miller and Montalto 2019). Many examples of good practices of UGI development have been documented, from top-down planning of green space systems at broader scales (Rouse and Bunster-Ossa 2013; Hansen et al. 2016), to a rich array of citizen-led initiatives at local scales (Buizer et al. 2015; Buijs et al. 2019).

Similar to the related concept of ecosystem services (Haase et al. 2014), most of the green infrastructure research still concentrates on the urban areas of the Global North (Parker and Zingoni de Baro 2019), with some notable exceptions in Asia (Sreetheran and Adnan 2007), Latin America (Vásquez et al. 2019) and in Africa (du Toit et al. 2018; Schäffler and Swilling 2013; Lindley et al. 2018). Based on a selective review of scientific literature, this chapter aims to reduce the knowledge gap on UGI in urban areas of different regions in the Global South by exploring:

- The understanding of what UGI is
- Main functions and goals of UGI
- Approaches for the development of UGI
- Main challenges and enabling factors for UGI.

2 Urban Green Infrastructure in Latin America

Latin America (LA) is one of the most urbanised regions in the world since 80% of its inhabitants live in urban areas and this share is expected to increase to 85% by 2040 (United Nations 2014). This urban expansion has been responsible for an important loss of green spaces, especially through the conversion of agricultural land and natural ecosystems to urban landscapes (Romero et al. 2007; Carrillo Flórez 2011). Impacts on natural ecosystems are particularly critical because LA hosts some of the most biodiversity-rich areas in the world, including several biodiversity hotspots (Myers et al. 2000). However, in semi-arid cities, high-income urban expansion tends to increase the presence of UGI over time (de la Barrera and Henríquez 2017; Hernández and Reyes 2018).

As a result of the rapid urban expansion, historical and socioeconomic factors and weak public policies, LA is characterised by high levels of inequality, exclusion, polarisation and centralisation (Romero-Lankao 2007; Carrión and Dammert 2016). Low-income urban settlements and slums are common on the urban fringe of Latin American cities. They have a low vegetation cover, are poor in biodiversity and vulnerable to floods, landslides and industrial pollution, among others (Vásquez et al. 2017; Dobbs et al. 2019). This starkly contrasts with the construction of real estate megaprojects for high-income households shaping a new urban periphery consisting of territories of corporate control (Lukas et al. 2020). This profound inequitable distribution constitutes one of the main characteristics of UGI in LA, and therefore any related plan or programme should include aspects of environmental justice (Vásquez et al. 2019).

2.1 Types of UGI in Latin America

Urban forests, green spaces and urban parks are the most researched types of UGI. However, only a few articles have used the term UGI (e.g. Vásquez et al. 2016; Castelli et al. 2017), while others have addressed UGI principles without using the term itself (e.g. Andrade et al. 2014). Urban parks, plazas and street trees have received much attention as they are the most common types of public UGI (Reyes and Figueroa 2010), for example, Bogotá's extensive tree inventory counted more than one million trees in public spaces (Andrade et al. 2013; Escobedo et al. 2015). Santiago, Chile's capital, has around 17% tree cover, and only 43% of trees are on public land (Escobedo et al. 2006). In this city, 97% of public green spaces are smaller than a hectare (Reyes and Figueroa 2010).

Semi-natural spaces, and natural spaces mostly located in the urban fringe, are important types of UGI. Notable ones are the forest remnants and wetlands in Latin American cities (Andrade et al. 2013; Vásquez et al. 2016; Doobs et al. 2019). Andrade et al. (2013) reported that there are 800 ha of wetlands remaining in Bogotá that are key for the UGI network. Agricultural lands are common in urban and peri-urban areas in Latin American cities, and offer provisioning ecosystem services (Vásquez et al. 2016; de la Barrera and Henriquez 2017; Dobbs et al. 2019). Notable cases include Quito (1070 urban gardens), Mexico City (2200 ha of agricultural land within the urban boundary) and Havana (90,000 citizens are involved in agriculture) (FAO 2014). Finally, although there are other types of UGI in Latin American cities such as university campuses, streams, garden cemeteries, gardens, military facilities, green roofs and green façades, they are less common and less researched than urban parks and plazas.

2.2 Main Functions of and Goals for UGI in Latin America

The main urban challenges addressed by UGI research in LA are food production, risk reduction and biodiversity conservation. Those that have been particularly emphasised include agriculture in urban and peri-urban areas (Madaleno 2000; Santandreu et al. 2009), governance for food security (Rocha and Lessa 2009), flood and coastal risk assessment and mitigation by using UGI (Zimmermann et al. 2016; Nardini and Gomes Miguez 2016; Silva et al. 2017), the importance of UGI as habitat (MacGregor-Fors and Schondube 2011; García et al. 2014) and citizen reports of wildlife (Gómez and Puerta 2017). Environmental inequalities (Pedlowski et al. 2002; Vásquez and Salgado 2009; Escobedo et al. 2006, 2015), climate regulation (Correa et al. 2006, 2012; Bustos et al. 2016; Inostroza 2014) and air quality (Baumgardner et al. 2012; Reynolds et al. 2017) have received attention in only a few studies.

Vásquez et al. (2019) reviewed public UGI plans in Latin American cities, especially those focused on climate change mitigation and adaptation. They found only four master plans devoted exclusively to UGI development (Mexico City,

Mexico; Campinas, Brazil; Coronel, Chile; Lima, Peru), and seven public policies for combating climate change in cities that incorporate UGI. UGI plans and programmes aim at biodiversity enhancement, urban heat island mitigation and surface run-off control as the main problems in urban environments in Latin America. Other important urban issues such as social and environmental inequality have rarely been incorporated as planning targets. These initiatives have been led by public institutions, in collaboration with public agencies, private partners and NGOs.

2.3 Approaches for the Development of UGI in Latin America

The adoption of the concept of UGI is still rare in governmental and citizen initiatives. Vásquez et al. (2019) pointed out that the concept of UGI is only slowly being incorporated into urban planning policies and initiatives in Latin America. In Colombia, the term Ecological Main Structure, Estructura Ecológica Principal (EMS) tends to be used as an equivalent of UGI. The first is normally preferred over the latter because it is formally included in land use planning instruments (Andrade et al. 2013, 2014). Although the original definition of the Ecological Main Structure emphasises biological connectivity over other environmental and social functions, it seems to be evolving towards a more comprehensive concept more similar to that of UGI (Andrade et al. 2014).

Still, EMS is more focused on natural spaces while UGI also considers green spaces with human management such as parks and tree-lined streets (Remolina 2011). In urban areas, however, some plans for the EMS have also incorporated historical and architectural sites, emphasising its role in the provision of cultural ecosystem services (e.g. Bogotá, Remolina 2011). Already in 2000, the Bogotá Land Use Plan—POT (Plan de Ordenamiento Territorial) incorporated the concept of EMS which in 2007 was introduced as one of the environmental determinants of land use planning by the national environmental legislation (Andrade et al. 2014). Colombia's case can therefore be considered as one of the most successful at incorporating green infrastructure into a mandatory land use planning instrument in LA.

The city of Bogotá has also incorporated green roofs and façades into urban planning regulations for public buildings (Andrade et al. 2013). Other forms for integrating UGI and ecological networks such as Ecological and Recreational Corridors have been included in recent land use plans (Andrade et al. 2013, 2014). The inclusion of UGI has also taken place at local scales in Bogotá, for example, as part of the Plan Fenicia (Plan Parcial Fenicia) that aims at neighbourhood urban regeneration.

In Rio de Janeiro UGI has been used for landslide risk reduction (Lange et al. 2018). The initiative started in the 1980s with the city's reforestation programme Mutirão Reflorestamento (reforestation by voluntary community work) and aims to increase the quality of life in the informal settlements. However, reforestation measures have been mainly implemented in unsettled areas and less around favelas (slums or shantytowns). The initiative is led by municipal authorities and the local communities are included as workers for reforestation activities. Currently, the programme also contracts private companies and uses environmental compensation or tax exemption.

In Chile the tsunami of 2010 has offered an opportunity for UGI planning since many coastal towns in south-central Chile that experienced severe damage had to be reconstructed. For that purpose, special reconstruction plans were launched and the modification of many land use plans were required, where for the first time the concept of UGI for flood and tsunami risk reduction was included. For example, in Pelluhue County the new land use plan incorporates a green buffer zone along the coast where building is restricted in favour of green spaces development (Silva et al. 2017).

The Victor Jara Park in Santiago and the Flood Mitigation Park in Constitución are two examples of UGI for flood mitigation and provision of recreational spaces (Moreno 2013). The first is one of the few examples of intersectoral working between ministries and is aimed at containing the floods that have affected the area for decades. However, this project received much criticism because the design was more a hydraulic work than a park, which disappointed the residents as it did not meet their expectations. The resistance of the neighbours was such that the park had to close for a period to increase the wooded area, add recreational facilities and improve the landscaping.

A further example for UGI is the action research project Aguacruz, in Santa Cruz, Bolivia, financed by the Italian Agency for Cooperation and Development, to improve water availability in the Piraí River basin (Castelli et al. 2017). The modelling of different landscape scenarios showed that the inclusion of green space in urban development cannot compensate for its impacts on water provision, and therefore, UGI should be implemented in the form of agroforestry at the catchment scale. Although the Aguacruz project delivered guidelines and an operational action plan to the city government and stakeholders, their inclusion in formal plans or programmes has not been reported.

The Iguaçu Project developed in the metropolitan region of Rio de Janeiro, Brazil, has included community participation in all stages (Costa et al. 2010). The objective of this project was to improve the housing in the flood areas of the Baixada Fluminense through different interventions, including the development of riverside parks. It was argued that the restoration of the urban river would have positive impacts on security as well as creating job opportunities in the community. However, during project implementation approximately 2000 families had to be relocated and were thereby negatively affected.

Finally, with regard to citizen initiatives, in cities such as Santiago an increasing interest in urban agriculture has been observed. At least 39 initiatives established the Urban Agriculture Network (Red de Agricultura Urbana) in 2011 (Heitmann 2014). These initiatives contribute to UGI development by reclaiming wastelands for community gardens, and are led by different organisations, but mainly by self-organised citizen groups (Contesse et al. 2018). Although some of the municipalities within the Santiago agglomeration offer technical or financial support to urban agriculture initiatives, there is no official policy, programme or plan to promote and sustain them over time.

2.4 Main Challenges and Enabling Factors for UGI in Latin America and the Caribbean

UGI research in LA has focused on quantitative assessments of its spatial structure, provision of ecosystem services, and proposals on how to integrate it into urban planning, neglecting the research on planning and management initiatives. Therefore, important knowledge gaps remain concerning the current status of these initiatives in terms of stakeholder arrangement, degree of success and drivers.

Most of the planning and management initiatives described in scientific literature make only implicit references to the term UGI. This might lead to two important consequences. First, that UGI principles are not fully incorporated into planning and management of urban green and blue spaces, limiting their benefits. Second, the knowledge exchange and dissemination is limited between actors and stakeholders due to the lack of a common concept and vision.

In this sense, adoption of UGI principles in the region has been limited. Most UGI plans developed by public institutions have adopted the idea of multifunctionality, but they still incorporate only a few functions or ecosystem services and tend not to combine different types of ecosystem services, e.g. cultural and regulating (Vásquez et al. 2019). Therefore, there is a need to incorporate all principles to develop multi-purpose and well-integrated UGI solutions in cities affected by multiple problems and characterised by competition for limited urban land (Vásquez et al. 2019).

The main urban challenges addressed by UGI initiatives have been food production, biodiversity protection and mitigation of natural hazards. Although in most cases there is initial recognition of UGI's multiple contributions to these and other problems, normally only one challenge is included as a formal planning objective, missing the opportunity to design multipurpose plans or projects. There are other important urban challenges in LA that have not been sufficiently addressed, such as inequitable distribution of UGI and socio-environmental segregation (Romero et al. 2012; Vásquez et al. 2017; Doobs et al. 2019).

In recent years, new social actors such as NGOs, foundations and community organisations interested in the protection and development of UGI have emerged. For example, in Santiago an academic group from the University of Chile has developed the Stgo+ Green Infrastructure Plan, which is a green infrastructure plan at city scale; the Cerros Islas Foundation promotes the transformation of 26 hills into metropolitan parks; and the Aldea and Mi Parque foundations develop green areas in low-income municipalities, involving neighbours in the design and construction, the municipality in the maintenance and the private sector in funding the works (Vásquez et al. 2017). Furthermore, social movements such as "El Panul" and "Parque La Salle" are self-organised groups aimed at protecting green spaces threatened by urban expansion, through political action and social protests.

The increased number and diversity of actors have been important in improving information and public awareness and taking action against developments that threaten UGI (Vásquez et al. 2017). Civil society and academic actors play active roles in contributing to and mobilising debates and the UGI agenda. However, so far research has not paid enough attention to these initiatives. Thus, producing scientific knowledge about them (e.g. number, type, functioning and issues) is key to promoting understanding of their role as change agents (Vásquez et al. 2017, 2019).

Although neoliberal policies have limited statutory urban planning instruments in LA, several UGI initiatives have been developed as part of formal mandatory plans and programmes, such as land use, reconstruction and ecological restoration plans, and not least, the Ecological Main Structure in Colombia. However, as the private sector is very prominent in urban development in LA, the participation of real estate enterprises and companies should be harnessed to develop parks in disadvantaged areas, green roofs and façades, or a fund should be created to finance UGI initiatives (Vásquez et al. 2017; Contesse et al. 2018).

The main challenges for UGI development are the lack of (1) coordination between sectors, especially between environmental and urban planning institutions, (2) coordination between different administrative jurisdictions, (3) funding both for the construction and the maintenance of UGI and (4) monitoring and control systems that provide relevant data for management as well as to measure the impact of initiatives (Andrade et al. 2013, 2014; Costa et al. 2010; Lange et al. 2018; Vásquez et al. 2016, 2019). In all of this, developing UGI in informal settlements poses particular challenges despite its potential to significantly improve living conditions in risk prone areas. Urban planning regulations, regeneration plans or ordinances seem to have a limited impact due to the very nature of these settlements. Moreover, the restoration or development of UGI can involve the displacement and relocation of vulnerable communities either during implementation of the measures or as a consequence of environmental injustice or green gentrification processes (Anguelovski et al. 2016).

On the other hand, the most important enabling factors are the emergent articulation between public, civil society and private actors, and the rising public awareness of UGI and ecosystem services as key topics in sustainable urban development that promote and scale up UGI as a priority in the agenda of local, regional and national governments (Costa et al. 2010; Contesse et al. 2018; Vásquez et al. 2016, 2017).

3 Urban Green Infrastructure in Africa

Africa is one of the fastest growing continents; its population is estimated to triple by 2050 and will make up 21% of the envisaged urban world population (UN 2014). Most of the growth will be in small and medium-sized cities (Güneralp et al. 2018). Historical and current processes driving urbanisation in Africa are different to other fast-growing regions such as Asia and Latin America (Titz and Chiotha 2019). Unplanned and unregulated growth, the legacy of colonialism, weak urban planning institutions and high levels of informality are important characteristics of urbanisation in Africa (Güneralp et al. 2018). Additionally, the influences of urban populations on natural areas and farmland are immense in terms of aspects such as "extension of transportation networks, transfer of water over long distances and the increased demand for bushmeat" (Güneralp et al. 2018: 7). For instance, Dar es Salaam (Tanzania) lost 30%

of its "green areas" between 2002 and 2008 due to conversion into settlement and farmland (Lindley et al. 2015). The loss of critical ecosystem services through the fragmentation and degradation of natural areas has an especially devastating effect on the low-income urban residents of the continent (Roy et al. 2018), who have to cope with the lack of governmental capacity to provide basic services, lack of adaptive capacity towards climate change, social inequalities, food and nutritional insecurities (Titz and Chiotha 2019).

3.1 Types of UGI in Sub-Saharan Africa

In Sub-Saharan Africa (SSA), UGI is not the only term used to describe urban green and blue spaces. The concept of "urban green structures" (UGS) is used for informal settlements as the green spaces are not planned and do not form a network according to Roy et al. (2018). The term urban forest is also often used for UGI (Chishaleshale et al. 2015; Gashu et al. 2019; Seburanga et al. 2014). Some studies on UGI focused, however, on public open spaces (Shackleton and Blair 2013) or private open spaces such as gardens (Cilliers et al. 2018; Gashu et al. 2019; Kaoma and Shackleton 2014). Up to 74% of the UGI in small and medium-sized towns in South Africa is under private tenure (Shackleton et al. 2018) and gardens and yards cover 53% of the UGI in some Ethiopian cities (Gashu et al. 2019). Agricultural areas, namely farmland which includes field crops and vegetables, are important components of UGI (Abo-El-Wafa et al. 2018), including floodplains and wetlands which are often used for the growing of food but negatively impacted by floods (Douglas 2018). In South African cities which are in biodiversity hot spots, the focus is more on the natural areas in urban open space planning (Roberts et al. 2005; Holmes et al. 2012). One example is Durban, in which the concept of Metropolitan Open Space System (D'MOSS) is used and is defined as a network of natural terrestrial and aquatic ecosystems with high biodiversity value (Boon et al. 2016; eThekwini Municipality 2010). The terms bio-infrastructure, focusing only on indigenous biodiversity (Roberts et al. 2012) and ecological infrastructure that also includes human-modified green areas (Chu et al. 2017), are also used for the D'MOSS. Ecological infrastructure is often regarded as synonymous to GI (da Silva and Wheeler 2017).

3.2 Main Functions of and Goals for UGI in Sub-Saharan Africa

In a recent review of UGI and ecosystem services in SSA, du Toit et al. (2018) found that only 20 out of 53 SSA countries (38%) had at least one article, with the most being from South Africa. Regulating and provisioning ecosystem services were studied the most.

Temperature regulation is seen as an important function of UGI, mainly climate change adaptation and mitigation and risk reduction are the urban challenges addressed by most research on aspects of UGI in SSA (Cavan et al. 2014; Di Leo et al. 2016; Feyisa et al. 2014; Ngulani and Shackleton 2019). Flooding and drainage problems due to urban expansion into flood prone areas such as riverbanks, floodplains and wetlands are a reality in SSA cities (Abo-El-Wafa et al. 2018; Barau et al. 2015; Douglas 2018; Roberts et al. 2012), and some cities have successful flood management schemes with multiple functions (Douglas 2018).

Food provision and security are also important UGI goals as agricultural areas form an important part of UGI (Abo-El-Wafa et al. 2018; de Zeeuw et al. 2011; Magigi 2013; Padgham et al. 2015). Urban and peri-urban agriculture is regarded as important in poverty alleviation and being able to produce their own food can be an "important safety-nets during times when cash or labour to procure the goods elsewhere is limited" according to Shackleton et al. (2018: 279, 280). Several studies on UGI also link climate change adaptation with urban farming (Abo-El-Wafa et al. 2018; Douglas 2018). According to Titz and Chiotha (2019) urban agriculture is a dominant feature of UGI in many urban areas, although it is not recognised as an official land use in many SSA city plans and is even regarded as an illegal practice due to human and environmental health concerns. Of course food provisioning is not just via urban and peri-urban agriculture, but there can also be significant contributions from collecting of wild foods and other products (such as medicines, firewood and building materials), termed urban foraging, from formal and informal UGI (Shackleton et al. 2017). There have been several studies in South Africa showing that collection of wild foods is a widespread practice encompassing multiple spaces and species (e.g. Garekae and Shackleton 2020; Sardeshpande and Shackleton 2020), while Chagomoka et al. (2016) showed that it is an important coping strategy in Ghana during household food shortages. Provisioning ecosystem services and urban agriculture are covered in Chap. 8 (Shackleton 2021) and 12 (Drescher et al. 2021), respectively.

The aesthetic appeal and recreational facilities of public areas such as parks and sports fields are the main management goals (du Toit et al. 2018; Rabare et al. 2009; Shackleton and Blair 2013). Urban residents often perceive recreation as the second most important ES after food provision (Adekunle et al. 2013; Shackleton et al. 2018). Most of the poorer urban residents in South Africa are not satisfied with the availability and management of public green spaces for recreation, showing the social and cultural roles of UGI (Adegun 2018a; Shackleton et al. 2018). Unfortunately, too much emphasis on city beautification may hide the realisation of the importance of UGI as providers of several other ES, as with the management of the urban forests in Johannesburg, South Africa (Schäffler and Swilling 2013) or prevents bottom-up innovative local experiments such as alternative water management practices in Addis Ababa, Ethiopia (Herslund et al. 2018). Biodiversity is generally regarded as important to human health and wellbeing and building the resilience of poor communities (Sutherland et al. 2016), but few UGI studies, except in South Africa (Boon et al. 2016; Holmes et al. 2012; Pfab et al. 2017), address this issue directly. Lastly, job creation is rarely covered as a benefit of UGI in literature but Shackleton and Blair (2013) mentioned its importance in poorer communities, and King and Shackleton (2020) enumerated over 17,000 jobs associated with maintenance of UGI in 12 small towns in South Africa earning over half a billion Rand annually (approx. US$37 million). Schäffler and Swilling (2013) highlighted the importance of job creation in the green industry in Johannesburg, South Africa.

3.3 Approaches for the Development of UGI in Sub-Saharan Africa

Incorporation of UGI is very limited in planning practices in SSA. There are, however, other plans and documents that include urban green spaces, such as systematic conservation plans which have been developed for several large cities in South Africa such as Durban (McLean et al. 2016), Cape Town (Benn 2008; Holmes et al. 2012) and Johannesburg and Pretoria (Pfab et al. 2017) and form part of policies (Spatial Development Frameworks and Integrated Development Plans). In Durban the D'MOSS was integrated in different municipal spatial plans and as a controlled development layer in all municipal schemes (Roberts and O'Donoghue 2013) resulting in disallowing some proposed developments following biodiversity impact assessments (Boon et al. 2016).

In the Gauteng province of South Africa, the green growth concept, that can only materialise through UGI planning, has been included in policy and planning at national, provincial and local levels. At provincial level, the Gauteng Green Strategic Programme (GGSP) follows a green growth approach and includes policies and strategies in terms of growth and development, integrated energy and climate change issues (Bobbins and Culwick 2015). Up to now implementation of the GGSP is limited. Other plans and strategies worth mentioning here, as they include some principles of UGI planning, are the climate adaptation plans or climate resilience plans which have been developed for several cities in SSA such as Addis Ababa (Ethiopia), Dar Es Salaam (Tanzania) and Durban and Cape Town (South Africa) (Roberts et al. 2012; Taylor 2016; Roy et al. 2018).

However, most city planners in South Africa rarely appreciate how green areas can be part of an infrastructure network and how they can provide services that is on par with basic services provided by conventional infrastructure (Bobbins and Culwick 2015). A recent study on the perceptions of professional city planners in South Africa showed that 69% believe that they know what UGI means, but a qualitative investigation revealed that only 23% really comprehend the concept (Cilliers 2019). While principles of multifunctionality, and to a limited degree green-grey integration and social inclusion, were included in planning documents of three cities in Ethiopia, they were not implemented (Girma et al. 2018).

Research on different green areas in the UGI in SSA does mention different UGI principles (Lindley et al. 2018). Urban gardens are multifunctional as they are sources of food and nutrition security, fulfil several regulating and cultural services, and have the potential to increase ecological connectivity (Cilliers et al. 2018). The multifunctional role of floodplains and wetlands are emphasised by Douglas (2018) in that they can alleviate the influence of climate change and are used for urban agricultural practices, indicating the importance of trade-offs between food production and flood reduction (Lindley et al. 2018). Multiple ES provided by UGI are also acknowledged by the urban poor in South Africa, focusing on provisioning services such as food, firewood and medicine (Adegun 2018a, b; Shackleton et al. 2018).

Connectivity is directly mentioned only in a few papers, e.g. the development of an urban biodiversity corridor in Johannesburg that will link the affluent and poorer parts of the city, not only as dispersal corridors for species, but also to focus on urban upliftment as an attempt to alleviate poverty (Burton and Rogerson 2017). However, several papers imply the importance of connectivity between urban green spaces from the perspectives of biodiversity, urban climate or storm water management (e.g. Douglas 2018; Herslund et al. 2018; Roy et al. 2018).

The integration of green infrastructure with the existing grey infrastructure of roads, pavements, buildings and especially water management facilities is only mentioned in a few SSA studies. For instance, in Durban, a green roof pilot project was investigated to determine its role in building the city's "adaptive capacity" towards climate change in terms of aspects such as reducing temperature and storm water run-off (Roberts et al. 2012; Roberts and O'Donoghue 2013), while the potential of UGI for sustainable storm water management was explored for Gauteng by Bobbins and Culwick (2015).

Many papers on urban green spaces and their planning mention the importance of partnerships between different stakeholders (Lindley et al. 2018). In Addis Ababa, Ethiopia, the master plan was revised by local stakeholders from different planning departments of the municipality, namely urban, transport and environment, creating an aura of ownership, while planning in Dar es Salaam was mostly done by foreign experts (Abo-El-Wafa et al. 2018). Several studies have included interviews and workshops with various stakeholders to determine perceptions on UGI and ES (Adekunle et al. 2013; Cilliers et al. 2018; Shackleton et al. 2018), as well as their real uses of urban green spaces (Gashu et al. 2019; Girma et al. 2018; Herslund et al. 2018; Roy et al. 2018). Some of these studies resulted in spatially explicit data of UGI of importance to the stakeholders (Roy et al. 2018) and coverage of different types of UGI (Gashu et al. 2019).

The development of plans and policies focusing on urban green spaces are often driven by researchers at academic institutions. Experts involved in the revision of the structure plan of Addis Ababa, Ethiopia, were mainly university researchers (Herslund et al. 2018). As a result, the terms "ecosystem service provision" and "multifunctional forests" have for the first time been included in the structural plan of Addis Ababa (Herslund et al. 2018: 4). In the three largest cities in South Africa, Johannesburg, Cape Town and Durban, transdisciplinary engagement between researchers, society and the local municipality were developed through City Labs (Anderson et al. 2013; Bobbins and Culwick 2015) and a research action partnership (Cockburn et al. 2016; Taylor et al. 2016). Although the term UGI is not necessarily used in all of these cities, these engagements focus on UGI-related issues and also explore how it can be included in city planning.

There are also several examples of informal non-governmental and citizen-led initiatives in developing and managing UGI that have not developed into UGI plans. Douglas (2018) reported on several community-based, small-scale actions and NGO-driven projects in several SSA cities to improve urban drainage and stress the importance of the "community view" in UGI planning alongside the visions of the municipality, international experts and NGOs. In Dar es Salaam, Tanzania, several innovative methods of adaptation by individuals and community groups were described but their limits were also acknowledged—most of these methods were merely "coping strategies and not progressive or transformative" (Roy et al. 2018: 292).

In urban areas in South Africa, many urban residents have private gardens and trees, which supports a thriving green industry, including gardening services, landscaping businesses and nurseries that resulted in the generation of jobs, revenue and knowledge (King and Shackleton 2020). Forming "service delivery partnerships" with citizen networks can also be responsible for different actions, including ecological knowledge (Schäffler and Swilling 2013). A larger percentage of residents of informal settlements in Johannesburg is willing to pay for green spaces that are developed by entrepreneurial residents than by NGOs or by local government, indicating the importance of non-governmental led actions in UGI development (Adegun 2018a). Practitioners in two Ethiopian cities identified environmental stewardship, spatial planning and economic development as the main drivers of UGI (Gashu et al. 2019).

3.4 Main Challenges and Enabling Factors for UGI in Sub-Saharan Africa

Several studies have indicated that master planning of cities in SSA needs more strategic approaches as they are often vague and outdated even before they can be implemented (Abo-El-Wafa et al. 2018; Herslund et al. 2018). According to Herslund et al. (2015) this will only be achieved if more up to date information about current land use patterns and future urban expansion and how it will influence ES delivery of the UGI, is collected. However, such information is mostly rare.

Mapping and valuation of ES are also limited in SSA, although there are some examples, mainly from South Africa which can be regarded as enabling factors for the implementation of UGI in those cities. Examples of ES mapping are a rapid assessment of ES in Cape Town, South Africa (O'Farrell et al. 2012), the mapping of hotspots for certain ES in Durban, South Africa and their envisaged transformation due to development (Davids et al. 2016, 2018). Examples from outside South Africa included maps focusing on the role of UGI in temperature regulation in Addis Ababa (Ethiopia), Dar es Salaam (Tanzania) and Bobo-Dioulasso (Burkina Faso) (Cavan et al. 2014; Di Leo et al. 2016). Only a limited number of studies determined the monetary value of certain parts of the UGI using different valuation techniques (Roberts et al. 2005; de Wit et al. 2012; Cilliers et al. 2013; Cilliers and Cilliers 2015; Ngulani and Shackleton 2019). Yet, surveys in communities often reveal significant values attached to provisioning services such as firewood, medicinal plants, water and building timber (Kaoma and Shackleton 2015).

There are several other barriers and challenges for the implementation of UGI plans in SSA, such as "socio-cultural values, traditions and perceptions; lack of capacity; poor governance and urban planning; social inequality; and spatial trade-offs and conflicts" as discussed by du Toit et al. (2018). Other studies confirm these challenges as indicated in three towns in Ethiopia where professionals and residents perceived a lack of participation by the community, inefficient skills of the professionals, budget constraints and a lack of coordination between the different sectors in local government as the main issues (Girma et al. 2018). Poor residents in informal settlements are often ignored in planning for the provision of urban green spaces, and therefore fragmentation and transformation of these spaces occur, compromising the important ES that poor people value greatly (Roy et al. 2018). However, several studies have shown there to be marked disparities or inequities in the distribution of UGI, with it being more abundant in more affluent areas and least in the poorest areas (Nero 2017; Venter et al. 2020). In South Africa it was shown that "safety" and "culture" are the two main factors determining the acceptance of UGI, as different cultures value and use green spaces differently, they are often regarded as a luxury or not valued at all because they are perceived to be unsafe (Cilliers 2019).

Although there are several challenges in terms of the implementation of UGI in SSA, a growing awareness of its importance, especially among researchers is an enabling factor. Another enabling factor is the good examples of UGI implementation in SSA, for example in the city of Durban, South Africa. Durban has a long history of ecological open space planning (Roberts et al. 2005) and the application of environmental sustainability principles in planning for biodiversity conservation (Roberts and Diederichs 2002; Boon et al. 2016). Durban is also regarded as a global leader in terms of climate adaptation planning (Roberts et al. 2012) and they have developed municipal adaptation plans focusing on water, health and disaster management (Roberts 2010; Carmin et al. 2012). An important part of ecosystem-based adaptation (EBA) that was followed in Durban is community-based EBA that seeks to increase the adaptive capacity of local communities (Roberts et al. 2012).

4 Urban Green Infrastructure in Asia

The global urban population is projected to grow by 2.5 billion urban dwellers between 2018 and 2050 and nearly 90% of the increase will be concentrated in Asia and Africa (UN 2019), which will bring huge social, economic and environmental transformations (UNFPA 2007). The majority of this development will occur in the 77 low- and middle-income Asian countries including China but not Japan, Hong Kong, Macau, Singapore, South Korea and Taiwan.

Much of the urban growth in Asia has been rapid and poorly planned, with China as an exception (Taubenböck et al. 2009). Ecological and environmental degradation have been a consequence with severe negative impacts on urban green and blue spaces and their ecological, social and economic values (e.g. Nagendra et al. 2012; Sing et al. 2016). Consequently, large-scale destruction of native vegetation has occurred. Trees have been particularly at risk from urban growth and densification (Nagendra and Gopal 2010, 2011). Moreover, rapidly growing cities might not adequately plan for new green spaces, thus leaving large parts of the population underserved in dense settlements in the core areas of the city (Wei 2017; Kumar et al. 2010).

UGI as a term is not much applied in statutory planning practice in Asian countries. Instead, other relevant terms are applied such as urban green space system in China and green network in Malaysia. Scientific publications from China and India have dealt with UGI (e.g. Liu et al. 2019b). Many of them were concerned with green spaces in large cities such as Beijing, Shanghai, Nanjing, Shenzhen and Guangzhou, but there are also studies dealing with middle-sized cities (e.g. Amani-Beni et al. 2018). In studies in Dhaka, Bangladesh, and in Sri Lanka, the term UGI was understood as a strategic approach to address the challenges facing slum dwellers (Birtchnell et al. 2019; Heratha et al. 2018). This section presents a review of the studies relevant to urban green space planning and management in Asia.

4.1 Types of UGI in Asia

Studies conducted in Asia have concentrated on a wide range of components of UGI and related concepts such as forest reserves, urban forests, urban parks, pocket parks, green roofs, vertical greenery systems (e.g. green façades, green walls and living walls), community gardens, home gardens and university campuses (Zinia and McShane 2018; Zhou et al. 2019). Green and blue spaces in informal settlements such as slums were also considered as a potential UGI, particularly for growing food, as well as enhancing social interaction and cohesion (Birtchnell et al. 2019). Within urban areas, urban agriculture includes a diverse range of ground level farming, rooftop farming, as well as community gardens in slums. For example, the slum dwellers in Dhaka (Bangladesh) practise urban agriculture despite facing various difficulties such as land insecurity, space constraints, unemployment, crime, corruption and cultural exclusion (Birtchnell et al. 2019).

Green spaces of religious and cultural significance have been highlighted in many Asian countries. For example, sacred urban areas in India, Thailand and other Asian cities can serve as an important node for nature conservation and cultural services (Gopal et al. 2018). Trees in sacred spaces provide an important buffer against urban environmental stress for the citizen in cities of India and Nepal and they serve as refuges for urban wildlife (Thaiutsa et al. 2008; Adhikari et al. 2015). Water systems on urban surfaces, such as rivers, lakes and ponds, marshes and paddy fields but also fountains and water retention ponds, together with their vegetated shorelines were also considered as components of urban green and blue infrastructure (Liu et al. 2019a).

In China, where statutory planning is strong, five categories of urban green spaces are distinguished: public park, nursery, green buffer, attached green space (such as in industrial and residential areas) and other green space (Ministry of Construction of the P.R. China 2002). Country parks, wetland parks, mining brownfield parks and landfill parks, are some new types of "other green space" that increasingly appear in many Chinese cities (Feng et al. 2019; Gong et al. 2015). In addition, as a response to climate change and the Sponge City programme, blue-green spaces for storm water management and flood mitigation, such as rain gardens, swales, sunken green areas, bio-retention elements, storage ponds and permeable pavements, are some new UGI types that expand quickly in Chinese cities (Jia et al. 2016; Mei et al. 2018).

4.2 Main Functions of and Goals for UGI in Asia

Urban sustainability is an overarching goal that has been often mentioned in scientific studies, especially from China, exploring methods and tools to improve UGI planning and decision-making (e.g. Ma et al. 2018; Wei et al. 2018). Environmental functions of UGI, such as biodiversity enhancement, pollution control, temperature regulation, have been emphasised and most of these studies have been conducted in China (e.g. Wang et al. 2018; Wu et al. 2019; Zhang et al. 2019b) and India (e.g. Jaganmohan et al. 2012; Gopal et al. 2018; Kotharkar et al. 2019). In the light of urban heat island intensification and climate change, improvement of thermal conditions for city dwellers by urban green space has been emphasised (e.g. Liu et al. 2020; Zheng et al. 2020). The regulating ecosystem services of street and park trees have received particular attention. Green roofs and vertical greening systems have also been studied as these forms of urban green infrastructure can be integrated into inner cities with high built densities (e.g. Zhang et al. 2019a; Zhou et al. 2019). Studies from China have frequently applied concepts of continuity and multifunctionality and principles of landscape ecology such as the patch-corridor-matrix to plan UGI, while the term "ecological security" has been used as a goal for UGI planning (e.g. Sun et al. 2018; Wang et al. 2018).

A large number of studies from China, especially after the introduction of the Sponge City initiative in 2015, focus on the role of blue-green infrastructure for flood mitigation and storm water management (e.g. Hu et al. 2019; Qiao et al. 2019). Several other studies in Asia showed the important role of urban green spaces such as parks, gardens, campuses and greenways, and importantly, sacred spaces such as temples and cemeteries, to support biodiversity (e.g. Gopal et al. 2018; Sing et al. 2019). However, much less research has been devoted to the social functions and health benefits from UGI in Asia. Where such studies exist, they clearly demonstrate positive health impacts, e.g. of remnant forests for environmental experience (Foo 2016), stress reduction from park visits (Paul and Nagendra 2017) as well as the promotion of physical activities and social cohesion from neighbourhood green spaces (Liu et al. 2019b; Wang et al. 2019; Zhang and Zhang 2013). For instance, Malaysia emphasises the functions of UGI to provide attractive, comfortable and stimulating urban environment, to support social interaction and to promote active and healthy lifestyles (KLSP 2040 2020). Social benefits of green spaces, including trees, were also observed in informal settlements, e.g. conversing and playing, domestic activities such as cooking and washing dishes, manufacturing of broomsticks, and tyre repair in Bengaluru, India (Gopal and Nagendra 2014). The wooded groves (*thopes*) in Bengaluru, India, not only provide food and timber but are also used as places of worship of family deities at tree shrines (Mundoli et al. 2017).

Environmental justice and social equality increasingly appear as topics in UGI studies from China (Zhu et al. 2019; Wang et al. 2019). On the other hand, the economic values of urban green space have only been addressed in few of the reviewed papers. For instance, economic valuation of urban green spaces in Kuala Lumpur (Malaysia), and Guangzhou (China) showed the distance, accessibility and size effects of different green space types as well as riverbank greening on the property prices of residential areas (Chen and Li 2018; Samad et al. 2020). Attention is also drawn to the cost-effectiveness of UGI implementation, e.g. for flood mitigation and storm water management, in comparison to conventional engineering approaches (Mei et al. 2018).

4.3 Approaches for the Development of UGI in Asia

Although UGI has been widely applied as a term in internationally published papers by Chinese scholars in the recent decade, in statutory planning of urban green spaces in China the term "urban green space system planning (UGSSP)" has been used since the late 1980s. The Chinese planning system is characterised as a top-down planning approach with comprehensive policies, regulations, criteria and guidelines at the national level, which are passed down to the local levels (Liu 2008). The planning of urban green space systems followed detailed regulations concerning, among others, green space types, protection of ecologically sensitive areas and ancient trees. In addition to UGSSP, further national regulations and guidelines exist in China, e.g. to define "green lines" where no construction is allowed in certain areas. The Opinions of the State Council on Strengthening Urban Infrastructure Construction (The State Council of the P.R. China 2013) also suggest cities enforce the ecological landscape construction. It is accompanied by detailed guidelines for the planning, design and management of various green space types.

With the rising awareness of climate change impacts, in 2014 China introduced a nation-wide programme of "Sponge City Construction" (SCC) to use (at least partially) blue-green infrastructure for dealing with storm water challenges such as pluvial flooding, water pollution and water resources scarcity, as well as ecosystem services (MOHURD 2014). The goals for SCC are that 70–85% of annual precipitation should be managed onsite, that 20% of urban built areas should achieve this goal by 2020 and that 80% of urban built areas should achieve it by 2030. In 2014 and 2015,

14 and 16 Chinese cities, respectively, were selected as pilot cities to implement the SCC within three years, each receiving a large amount of national level funding (Qiao et al. 2019). Since 2015, a growing number of scientific papers on UGI have examined the "Sponge City" concept (e.g. Xia et al. 2017; Chan et al. 2018). Furthermore, the National Ecological Garden City nomination in China has also greatly promoted the development of UGI since the 1990s. The National Garden City Series Standards (MOHURD 2016) mandated sectorial plans such as UGSSP, Greenway Plan and Sponge City Plan, and requires quantitative and qualitative criteria applied in, for example, the implementation of the so-called "green lines", wetland protection, brownfield restoration and urban water system restoration.

Some papers, especially from China, explored UGI and similar planning concepts such as greenways, greenbelts, green networks and green corridors (Yu et al. 2006; Wang et al. 2014). The development of greenbelt and river corridors in cities have been related with urban forest and tree planting programmes since 2000, where ecological and recreational functions have gained more attention (Yu et al. 2006). In Malaysian cities, such as Kuala Lumpur green network is applied as a strategic concept for green space planning to create soft traffic corridors, connect green spaces in and outside of the city and link green space with water networks (KLSP 2040 2020).

Urban and peri-urban agriculture is another approach to the development of UGI in and around urbanising regions in Asia, with particular importance for providing food security and generating income for the poor in a mostly informal economy (Birtchnell et al. 2019). In Indonesia, urban and peri-urban agriculture was promoted during times of economic crises in 1998 and 2007 by temporarily making vacant land available for farming and hence improve food security (Pribadi and Pauleit 2016). However, there is still a lack of recognition of urban and peri-urban agriculture in planning in this region (Zinia and McShane 2018). Vertical farming has also been explored as an option for promoting food supply in densely built urban areas such as in China and India. Green roofs development in China has just started and there is no relevant policy (Xiao et al. 2014). A fuller account of urban agriculture in the Global South is given in Chap. 12 (Drescher et al. 2021).

4.4 *Main Challenges and Enabling Factors for UGI in Asia*

Diversity of the Asian region and the scarcity of research outside of China and India does not allow a comprehensive overview of the state of UGI planning. However, from our review it emerges that research has widely demonstrated the services and values of UGI in this region. Ecosystem services have received particular attention, while cultural and economic services of urban green spaces are still less understood. Much less is known about the planning and governance of the UGI. Research on instruments and regulations for the development of UGI is still very limited in the Asian region. Uy and Nakagoshi (2008) reported that 18 m^2 of urban green space per capita is required by the Hanoi Master Plan for the capital of Vietnam, which the authors considered as insufficient for maintaining ecological processes in the city. Yet, in other cities actual urban green space provision can be much lower, as is the case for Colombo, Sri Lanka (Li and Pussella 2017). Also, in many Asian cities, private gardens are often prioritised over the creation of accessible public green spaces (Jim and Zhang 2015). In addition, blue space (surface waters) were not as popular as green spaces in considering ecosystem services and their utility in most Asian cities because blue spaces are mostly polluted (Zinia and McShane 2018). However, in China, public green spaces are of more concern, and blue spaces are regarded highly important as part of UGI through the Sponge City programme.

Densely populated cities such as Dhaka, and like many other mega-cities in Asia, present challenges associated with the affordability and practicality of maintaining green adaption strategies and related ecosystem services given widespread poverty and the relatively high-priced land (for green and/or blue spaces) (Zinia and McShane 2018). According to a study on the implementation of the green network plan of Kuala Lumpur, although Kuala Lumpur has established parks, small parks and gardens, connecting these green areas is often hindered by obstacles, including private buildings and illegal hawkers along the pedestrian walkways which block free access of the public (Rasli and Kanniah 2018). Another issue is the limited protection from rain and sunlight for the pedestrians in a hot and humid climate. Therefore, pedestrian protection is suggested to be one of the important parameters for the quality of green corridors (Rasli and Kanniah 2018). In addition, in terms of green space in Kuala Lumpur, there is also usually conflict in terms of policies from several authorities in charge of green spaces. Areas of UGI must also compete with other physical developments and are constantly under risk of land acquisition, changes and modification. They often need to make way for new developments that are seen as more beneficial monetarily. This was also observed in Jakarta where, between 2001 and 2014 Jakarta Metropolitan City exhibited an increase in built-up area up to 81%, followed by a decrease in vegetated land and water bodies (Murayama et al. 2015).

China stands out with its strong government-led planning approach. Major principles of UGI by content—as was mentioned before—have been gradually adopted in statutory planning in this country. However, there is little research on the uptake of these principles in practice. Process principles of UGI—"strategic" as well as "inter- and transdisciplinary"—have received attention in statutory planning. Strategic planning of UGI is facilitated by the top-down system of China and its national policies, regulations and guidelines. On the other hand, China's sectorial planning and institutional structure make the adoption of an inter- and transdisciplinary approach to the planning of green infrastructure difficult. It seems that some strong national regulatory and institutional measures such as the "Sponge City Leader office", involving a mayor and the leaders of relevant sectors of the public administration, can overcome sectorial boundaries. However, of all principles of UGI, the principle of socially inclusive planning may have been the least considered in a full scale, despite having gained attention in academia (e.g. Yu et al. 2006; Yao et al. 2019).

Only a limited number of studies from China focused on the challenges and barriers of UGI implementation. In a study on the implementation of a large-scale project of planning 50 million new urban trees around Beijing, Yao et al. (2019) identified challenges and barriers such as the lack of civic participation, conflict with farmland conservation by extensive use of farmland, discrepancies between plans and outcomes, inadequate rule-making in a short period of planning, and the dominance of an authoritative discourse. Moreover, challenges and barriers of recent Sponge City construction have been identified by several studies (Liu et al. 2019a; Nguyen et al. 2019; Qiao et al. 2019), such as the scarcity of urban land in dense cities for UGI, the lack of quantification of UGI's cost-effectiveness for storm water management, the lack of cooperation among sectors and the lack of clear guidance for adopting a systematic approach to the planning and design of sustainable storm water management systems that provide multiple benefits.

5 Discussion

This chapter has examined the state of urban green infrastructure (UGI) as a planning approach in the three main regions of the Global South, i.e. Latin America (LA), Sub-Saharan Africa (SSA) and Asia. Specifically we explored the understanding of UGI in these different regions, approaches for its development, as well as main challenges and enabling factors.

5.1 Understanding of UGI

In Latin America, Sub-Saharan Africa and Asia, urbanisation has exerted tremendous pressure on natural systems and their biodiversity and related ecosystem services (McDonald et al. 2013). Farmland near and in urban areas has also experienced dramatic declines (e.g. Pauleit et al. 2019b) while there often is a lack and unjust distribution of green spaces within the built environment (e.g. Pham et al. 2012; Vásquez et al. 2017). These problems are likely to increase under ongoing future urbanisation (Angel et al. 2011; Seto et al. 2012; UN 2019) if decisive action is not taken to better protect and integrate green and blue spaces in urban development processes as UGI (Fohlmeister et al. 2015).

In all three regions, UGI thinking has rarely been used in research and practice. Still, in Latin America and Sub-Saharan Africa (specifically South Africa) the term has been applied in some cases. Other terms such as Ecological Main Structure in Colombia as well as urban green structure, urban forest and metropolitan open space systems, bio-infrastructure and ecological infrastructure in SSA have their specific meaning in their respective contexts and appear to be more biased towards addressing ecological issues. But they provide ground for strategic development of green and blue networks. Yet, in all three regions, there still is a strong focus both in research and in practice on single green space types such as urban parks, gardens and forests. This has become particularly evident in Asia where a large quantity of literature has dealt with a diversity of green and blue space types, including some unique types such as sacred sites.

An exception to this focus on individual green spaces in Asia is China, where "urban green space systems" are planned at urban scale and share some of the characteristics of the UGI concept. Also, China distinguished itself with the "Sponge City" concept as a strategic approach for storm water management, flood mitigation and water resource management, introducing some new UGI elements such as swales and rain gardens. Despite these and further notable exceptions found in all three regions, overall, traditional views of a primarily aesthetic and recreational role of urban green still appear to predominate in planning (du Toit et al. 2018). A multifunctional landscape approach is mostly lacking that would also consider the larger amount of green spaces on private or community land, remnants of natural areas as well as farmland with their diversity of ecological, social and economic functions being a part of UGI (see, e.g. Escobedo et al. 2006; Pribadi and Pauleit 2016). In line with this, understanding of the UGI concept by planners is still limited (Cilliers 2019). Consequently, it may not come as a surprise that strategic approaches to the development of urban green and blue spaces have been almost exclusively reported from the bigger, thriving cities in the GS, whether it is Kuala Lumpur (Malaysia), Johannesburg and Durban (South Africa) or Bogotá (Columbia), where administrative capacity for planning is well-developed. On the other hand, there is a dearth of information from smaller urban areas with few exceptions (e.g. Shackleton et al. 2018).

5.2 *Main Functions and Goals of UGI*

The limited volume of UGI research and application in the Global South provides evidence on its potential to successfully address major challenges of urbanisation. Apart from the traditional focus on recreation and aesthetics, risk reduction from natural hazards exacerbated by climate change (such as riverine and pluvial flooding, heat, and landslides) emerged as a major entry point for developing UGI in the cities of all three regions (Cavan et al. 2014; Inostroza 2014; Xia et al. 2017). Ecosystem based adaptation in Durban is an example of addressing climate change issues (Roberts et al. 2012). These findings are also supported by the research on regulating ecosystem services of UGI in all three regions (Escobedo 2021). Such research is particularly strong in China (Hu et al. 2019) where the cooling functions of urban green and its role for flood mitigation in the Sponge City concept are emphasised.

The provision of food by urban agriculture can be highlighted as another important goal for UGI in the Global South (see also Drescher et al. 2021). Yet, urban agriculture and the multifunctional benefits it can provide are not yet sufficiently recognised in the realms of policymaking (Pribadi and Pauleit 2016; Contesse et al. 2018). Other objectives of UGI such as enhancing human health, nature experiences, social cohesion and promoting the shift towards a greener economy (Pauleit et al. 2017a) are still hardly considered in policies for UGI of the Global South, despite some supporting scientific evidence and few exceptions such as the Green Growth concept of the Gauteng region (South Africa) (Bobbins and Culwick 2015). Therefore, there is still much scope for multifunctional UGI planning to simultaneously address basic needs in the Global South such as sanitation issues, food security and poverty alleviation.

Connectivity seems to be the principle most developed and applied, as examples of Colombia's Ecological Main Structure with Bogotá (Remolina 2011) as one case for its urban application and the green space corridors for the Gauteng region in South Africa (Burton and Rogerson 2017). Ecological connectivity is emphasised but offers potential for the broadening into the development of multifunctional corridors as the above examples demonstrate (Andrade et al. 2014; Burton and Rogerson 2017).

Overall, based on the available literature it is difficult to establish whether there are notable differences between and within the three regions concerning the functions and goals of UGI planning. Differences may be rather dependent on local contexts and priorities. The above does suggest, however, that in all of them, there is a gap between the evidence provided by research on a variety of ecological, and a much lesser degree social and economic functions of UGI, and their consideration and adoption in the planning realm. Notable exceptions such as Bogotá (Columbia), Gauteng (South Africa) and Kuala Lumpur (Malaysia) may inspire future development of UGI elsewhere in the GS.

5.3 Approaches for the Development of UGI

China clearly is a special case in the Asian region, and the Global South, with its top-down planning system with comprehensive policies, regulations, criteria and guidelines at the national level, which are passed down to the local level. Moreover, China is distinct by its strong economic growth and high administrative capacity that enables large-scale projects for the development of UGI. In most other countries of the GS, neoliberal policies predominate and statutory urban planning has limited influence on urban development and green spaces, respectively. Nevertheless, UGI initiatives have been developed in many countries as part of formal mandatory plans and programmes, such as land use, reconstruction and ecological restoration plans, and not least, the Ecological Main Structure in Colombia. In SSA, systematic conservation plans and climate adaptation plans can be an entry point to the development of UGI. Gauteng's Green Growth concept is an example of all three regions where economic goals drive the development of UGI.

Additionally, in all three regions, a number of bottom-up initiatives have been observed where local communities organise themselves to establish and manage urban green spaces for recreation and other purposes, not least producing food (e.g. Contesse et al. 2018). The importance of such bottom-up initiatives in UGI development at urban scale is still little understood. Some studies on, e.g. community gardens, rooftop gardens and urban agriculture indicate their potential (e.g. Birtchnell et al. 2019; Contesse et al. 2018; Zinia and McShane 2018).

5.4 Main Challenges and Enabling Factors for UGI

A severe lack of information on green spaces was reported for all three GS regions to reliably monitor UGI types, their distribution, dynamics and multiple benefits that would compare to recent research from Australia and Europe (Bartesaghi Koc et al. 2017; Cvejić et al. 2015; Wolff and Haase 2019). Therefore, the basis for evidence based UGI planning is limited in the cities of the GS.

China certainly stands apart with its tradition of urban green system planning and the Sponge City concept in over 30 cities (Liu 2008; Qiao et al. 2019). This is in stark contrast to most other cities in the Global South, with the exception of Durban in South Africa, where the capacity for planning and plan implementation is often very weak whereas informality is high (UN-Habitat 2015). Lack of skilled staff and financial resources pose particular constraints for the establishment and management of green spaces as was reported, for instance, for Ethiopian cities (Girma et al. 2018).

However, cooperation between different sectors, such as hydrological engineering, urban planning, as well social and environmental sectors, also still appears to be weak in practice in China (Xia et al. 2017), and a gap between UGI policy and practice has also been noted more generally for this country. In LA, lack of coordination between sectors, especially between environmental and urban planning institutions, as well as the lack of coordination between different administrative jurisdictions have been equally highlighted. Developing platforms such as the City Labs and research action partnerships in South Africa may help to reduce these barriers and support the mainstreaming of UGI into other urban policies, e.g. for city wide transport networks, water supply and waste water treatment and other technical infrastructures. Also, the example of the new Master Plan for Addis Ababa (Ethiopia) has shown that scientists can make an important impact by providing relevant information on UGI (Abo-El-Wafa et al. 2018).

Elsewhere, small-scale projects to manage storm water run-off via UGI, such as Durban's green roof pilot project (Roberts et al. 2012), were observed, while the difficulties of intersectoral working have been well illustrated by the case of the Victor Jara Park in Santiago (Moreno 2013). Nevertheless, such projects may provide opportunities for government to develop capacity for UGI planning at broader scales as well as to better engage with citizens. Socially inclusive approaches to planning pose particular challenges in the Global South where large parts of urban areas are developed in informal ways and poverty levels are high. In China, on the other hand, the top-down planning tradition undermines (or conceals) some opportunities for citizen involvement. The disconnect between city level government and administration, on the one hand, and local communities, which are struggling with manifold problems, on the other hand, is common to all three regions (Herslund et al. 2015; Vásquez et al. 2016).

In this situation, the development of UGI in socio-economically disadvantaged communities appears to receive little attention in urban policies despite their great appreciation and use by the residents (Roy et al. 2018). Where government-led greening efforts in deprived communities are made, community needs and wishes may not always be well understood, as indicated by the examples of the Victor Jara Park, Santiago, and the reforestation initiative in Rio de Janeiro (Lange et al. 2018).

Therefore, the lack of social inclusion is closely connected to issues of social and environmental injustice which have been raised in this chapter with regard to green space provision, access and quality (Rigolon et al. 2018) and the provision of trees in public and private spaces (e.g. Gwedla and Shackleton 2019). The latter is especially true for indigenous communities located in cities and peri-urban areas of the GS as they tend to be closely attached to some types of UGI. Cases such as the Iguaçu project also show that the implementation of UGI projects can negatively affect the livelihoods of local communities. Eviction of informal settlers from flood prone areas disrupts their lives socially and economically (Douglas 2018). Involvement of local communities and NGOs from early on is key for avoiding or reducing such conflicts (Adegun 2018a, b; Douglas 2018). On the other hand, the potential of community-initiated activities to create and maintain UGI such as community gardens (Cilliers et al. 2018; Contesse et al. 2018) or reforestation of hazard prone slopes (Lange et al. 2018) has been demonstrated in all three regions.

6 Conclusions

Although we can draw some conclusions from this review of UGI in the Global South for advancing its development, the diversity of the urban areas with their specific processes of urbanisation and the scarcity of scientific studies, limits the possibility to generalise. In this light, the following conclusions shall only be pointers for further debate:

1. Considerable efforts are still needed to raise strong political support for UGI development, whether under this name or related terms of urban greening, urban forestry, ecosystem-based adaption or the more recent concept of nature-based solutions (Pauleit et al. 2017b). Finding the locally relevant urban challenges for a broad range of actors in a collaborative process is key to the success of UGI planning as a strategic approach. For this purpose, the importance of green and blue spaces to address basic health and safety needs such as fresh water supply, food security, sanitation and reduction of risks can provide a main political driver. Habitat corridor planning for biodiversity conservation can be another good starting point for the development of UGI from where to broaden into ecosystem-based adaptation to climate change, green mobility, socio-spatial integration and the inclusion of other themes. Importantly, issues of social and environmental injustice deserve much more attention to address the blatant inequalities in the provision of UGI and access to its services in urban areas of the Global South.
2. Small projects probably offer the greatest chances to kick off UGI development as they are less abstract and complex than city wide strategies. They may provide learning arenas for the stakeholders involved and, in the case of success, create momentum for transfer and upscaling into wider strategies. This review has

revealed a broader range of such projects. Further efforts are, however, required to give them more visibility and to facilitate knowledge exchange and dissemination among GS researchers and practitioners.

3. As the capacity of authorities to cater for urban green spaces is often weak, bottom-up initiatives are important to conserve and manage green spaces and trees, and where possible even develop them at neighbourhood level. Local governments should take a supportive role to realise this potential and reduce conflicts between authorities and local residents by decentralising mandates and resources to city- and sub-city levels and establish a close cooperation across public and private actors, including land owners and the estate sector, across levels of the decision-making hierarchy.
4. UGI needs to be supported by solid evidence on the types of UGI present, the different bundles of ecosystem services they can provide and who benefits from access to those, including economic valuation as was attempted in some South African cities (e.g. de Wit et al. 2012; Kaoma and Shackleton 2015). Also, evidence on the effects of UGI on human health and wellbeing are still much in need while exploring at the same time how potentially negative effects of UGI, such as the spread of diseases or crime, can be avoided.
5. There is even a greater need for systematic studies on planning and governance of UGI in the GS that would provide valuable information on successful approaches, but also help to identify barriers to the development of UGI, whether for individual cities, countries or larger regions. Most of all, the transmission of research findings into practice and collaboration between different sectors of the administration should be strengthened, e.g. be creating stakeholder platforms such as City Labs and Research Action Partnerships. Collaboration between authorities may also help to reduce resource constraints to some degree by pooling information and human resources. Yet, urban planning authorities need to be strengthened and more green space planners are required at all administrative levels that are well versed with UGI thinking.

In conclusion, UGI as multipurpose networks of green and blue spaces has great potential to address major challenges of urbanisation in the Global South. Concepts of UGI planning developed in the Global North cannot be blindly copied but they provide a useful lens to critically review the current state of UGI policy, research and implementation in the Global South. Core ideas and principles of UGI such as connectivity, multifunctionality, green-grey integration and social inclusion need to be tailored to the specific situations of the urban areas in Latin America, Africa and Asia for this purpose with risk management, dealing with poverty and informality being key priorities.

References

Abo-El-Wafa H, Yeshitela K, Pauleit S (2018) The use of urban spatial scenario design model as a strategic planning tool for Addis Ababa. Landsc Urb Plan 180:308–318

Adegun OB (2018a) Resident's relationship with green infrastructure in Cosmo City, Johannesburg. J Urban 11:329–346
Adegun OB (2018b) When green is grievous: downsides in human-nature interactions in informal settlements. J Urban 11:347–361
Adekunle MF, Agbaje BM, Kolade VO (2013) Public perception of ecosystem service functions of peri—urban forest for sustainable management in Ogun State. Afr J Environ Sci Technol 7:410–416
Adhikari YP, Fischer A, Pauleit S, Ticktin T (2015) Sustainable conservation perspectives for epiphytic orchids in the Central Himalayas, Nepal. Appl Ecol Environ Res 133:753–767
Ahern J (2007) Green infrastructure, a spatial solution for cities. In: Novotny V, Brown P (eds) Cities of the future. IWA Publishing, London, pp 267–283
Amani-Beni M, Zhang B, Xie G, Xu J (2018) Impact of urban park's tree, grass and waterbody on microclimate in hot summer days: a case study of Olympic Park in Beijing, China. Urban For Urban Green 32:1–6
Anderson PML, Brown-Luthango M, Cartwright A, Farouk I, Smit W (2013) Brokering communities of knowledge and practice: reflections on the African Centre for Cities' CityLab programme. Cities 32:1–10
Andersson E, Langemeyer J, Borgström S, McPhearson T, Haase D, Kronenberg J, Barton DN, Davis M, Naumann S, Röschel L, Baró F (2019) Enabling green and blue infrastructure to improve contributions to human well-being and equity in urban systems. Bioscience 69:566–574
Andrade G, Remolina F, Wiesner D (2013) Assembling the pieces: a framework for the integration of multi-functional ecological main structure in the emerging urban region of Bogotá, Colombia. Urban Ecosys 16:723–739
Andrade G, Remolina F, Wiesner D, Montenegro F (2014) La estructura ecológica principal en lo local. propuesta de aplicación en la renovación urbana de Fenicia, Las Aguas, Bogotá. Revista Nodo 8:43–54
Angel S, Parent J Civco DL, Blei AM (2011) Making room for a planet of cities. Lincoln Institute of Land Policy, Cambridge, MA
Anguelovski I, Shi L, Chu E, Gallagher D, Goh K, Lamb Z, Reeve K Teicher H (2016) Equity impacts of urban land use planning for climate adaptation: critical perspectives from the Global North and South. J Plann Educ Res 36: 333–348
Barau AS, Maconachie R, Ludin AN, Abdulhamid A (2015) Urban morphology dynamics and environmental change in Kano, Nigeria. Land Use Pol 42:307–317
Bartesaghi Koc C, Osmond P, Peters A (2017) Towards a comprehensive green infrastructure typology: a systematic review of approaches, methods and typologies. Urban Ecosys 20:15–35
Baumgardner D, Varela S, Escobedo FJ, Chacalo A, Ochoa C (2012) The role of a peri-urban forest on air quality improvement in the Mexico City megalopolis. Environ Pollut 163:174–183
Benedict MA, McMahon ET (2002) Green infrastructure: smart conservation for the 21st century. Renew Res J 20:12–17
Benedict MA, McMahon ET (2006) Green infrastructure: linking landscapes and communities. Island Press, Washington, DC
Benn G (2008) City of Cape Town BioNet terrestrial systematic conservation plan re-analysis: methods and results. GeoCline Consulting CC. http://cpu.uwc.ac.za/CapeTown/CCTBioNet_MethodReport.pdf. Accessed 26 Apr 2019
Birtchnell T, Gill N, Sultana R (2019) Sleeper cells for urban green infrastructure: harnessing latent competence in greening Dhaka's slums. Urban For Urban Green 40:93–104
Bobbins K, Culwick C (2015) Green growth transitions through a green infrastructure approach at the local government level: case study for the Gauteng city region. J Public Admin 50:32–49
Boon R, Cockburn J, Douwes E, Govender N, Ground L, Mclean C, Roberts D, Rouget M, Slotow R (2016) Managing a threatened savanna ecosystem (KwaZulu-Natal Sandstone Sourveld) in an urban biodiversity hotspot: Durban, South Africa. Bothalia 46: a2112
Buijs A, Hansen R, Van der Jagt S, Ambrose-Oji B, Elands B, Rall EL, Mattijssen T, Pauleit S, Runhaar H, Olafsson AS, Møller MS (2019) Mosaic governance for urban green infrastructure:

upscaling active citizenship from a local government perspective. Urban For Urban Green 40:53–62
Buizer IM, Elands BH, Mattijssen TJ, Jagt AP, Ambrose, B, Geroházi E, Santos E (2015) The governance of urban green spaces in selected EU-cities: policies, practices, actors, topics. GREEN SURGE Deliverable 6.1, Available via https://greensurge.eu/working-packages/wp6/. Accessed 9 May 2018
Burton C, Rogerson JM (2017) The making of green urban infrastructure: the Klipriviersberg Urban Biodiversity Corridor. Afr J Hospit Tour Leis 6:3
Bustos ML, Ferrelli F, Piccolo MC (2016) El rol del arbolado urbano sobre la temperatura invernal de la villa balnearia de Pehuen Co, Argentina. Revista Univers Geogr 25:57–72
Carmin J, Anguelovski I, Roberts DC (2012) Urban climate adaptation in the Global South: planning in an emerging policy domain. J Plann Educ Res 32:18–32
Carrillo Flórez MJ (2011) La dinámica de crecimiento del borde urbano sobre los Cerros Orientales de Bogotá. Posibilidades de gestión de ciudad en zonas de ladera. Doctoral dissertation, Universidad del Rosario, Rosario
Carrión F, Dammert M (2016) Los estudios urbanos en América Latina: un espejo donde mirarse. In: Metzger P, Rebotier J, Robert J, Urquieta P, Vega P (eds) La cuestión urbana en la región andina: miradas sobre la investigación y la formación. Babel, Quito, pp 241–280
Castelli G, Foderi C, Guzman B, Ossoli L, Kempff Y, Bresci E, Salbitano F (2017) Planting waterscapes: green infrastructures, landscape and hydrological modeling for the future of Santa Cruz de la Sierra, Bolivia. Forests 8:437
Cavan G, Lindley S, Jalayer F, Yeshitela K, Pauleit S, Renner F, Gill S, Capuano P, Nebebe A, Woldegerima T, Kibassa D, Shemdoe R (2014) Urban morphological determinants of temperature regulating ecosystem services in two African cities. Ecol Indic 42:43–57
Chagomoka T, Unger S, Drescher A, Glaser R, Marschner B, Schlesinge J (2016) Food coping strategies in northern Ghana, A socio-spatial analysis along the urban-rural continuum. Agric Food Sec 5:4
Chan FK, Griffiths JA, Higgitt D, Xu S, Zhu F, Tang Y-T, Xu Y, Thorne CR (2018) Sponge City in China: a breakthrough of planning and flood risk management in the urban context. Land Use Pol 76:772–778
Chen WY, Li X (2018) Impact of urban stream pollution: a comparative spatial hedonic study of high-rise residential buildings in Guangzhou, south China. Geogr J 184:283–297
Chishaleshale M, Shackleton CM, Gambiza J. Gumbo D (2015) The prevalence of planning and management frameworks for trees and green spaces in urban areas of South Africa. Urban For Urban Green 14:817–825
Chu E, Anguelovski I, Roberts D (2017) Climate adaptation as strategic urbanism: assessing opportunities and uncertainties for equity and inclusive development in cities. Cities 60:378–387
Cilliers EJ (2019) Reflecting on Green infrastructure and spatial planning in Africa: the complexities, perceptions, and way forward. Sustain 11:455
Cilliers EJ, Cilliers SS (2015) From green to gold: a South African example of valuing urban green spaces in some residential areas in Potchefstroom. Town Reg Plann 67:1–12
Cilliers SS, Cilliers EJ, Lubbe R, Siebert S (2013) Ecosystem services of urban green spaces in African countries: perspectives and challenges. Urban Ecosys 16:681–702
Cilliers SS, Siebert SJ, du Toit MJ, Barthel S, Mishra S, Cornelius A, Davoren E (2018) Health clinic gardens as nodes of social-ecological innovation to promote garden ecosystem services in Sub-Saharan Africa. Landsc Urb Plan 180:294–307
Cockburn J, Rouget M, Slotow R, Roberts D, Boon R, Douwes E, O'Donoghue S, Downs CT, Mukherjee S, Musakwa W, Mutanga O, Mwabvu T, Odindi J, Odindo A, Procheş S, Ramdhani S, Ray-Mukherjee J, Sershen T, Schoeman MC, Smit AJ, Wale E, Willows-Munro S (2016) How to build science-action partnerships for local land use planning and management: lessons from Durban, South Africa. Ecol Soc 21:28

Contesse M, van Vliet BJ, Lenhart J (2018) Is urban agriculture urban green space? A comparison of policy arrangements for urban green space and urban agriculture in Santiago de Chile. Land Use Pol 71:566–577
Correa E, Martínez C, Lesino G, De Rosa C, Cantón A (2006) Impact of Urban parks on the climatic pattern of Mendoza's Metropolitan Area, in Argentina. PLEA2006—23rd Conference on Passive and Low Energy Architecture, Geneva, pp 6–8
Correa E, Ruiz MA, Canton A, Lesino G (2012) Thermal comfort in forested urban canyons of low building density: an assessment for the city of Mendoza, Argentina. Build Environ 58:219–230
Costa L, Vescina L, Barcellos D (2010) Environmental restoration of urban rivers in the metropolitan region of Rio de Janeiro, Brazil. Urban Environ 4:13–26
Cvejić R, Eler K, Pintar M, Železnikar Š, Haase D, Kabisch N, Strohbach M (2015) A typology of urban green spaces, ecosystem provisioning services and demands. GREEN SURGE Deliverable D3.1. 19 Mar 2016. http://greensurge.eu/workingpackages/wp3/files/D3.1_Typology_of_urban_green_spaces_1_.pdf/D3.1_Typology_of_urban_green_spaces_v2_.pdf. Accessed 25 Jan 2019
da Silva JM, Wheeler E (2017) Ecosystems as infrastructure. Perspect Ecol Conserv 15:3235
Davids R, Rouget M, Boon R, Roberts D (2016) Identifying ecosystem service hotspots for environmental management in Durban, South Africa. Bothalia 46:2118
Davids R, Rouget M, Boon R, Roberts D (2018) Spatial analyses of threats to ecosystem service hotspots in Greater Durban, South Africa. Peer J 6:e5723
Davies C, Hansen R, Rall E, Pauleit S, Lafortezza R, De Bellis Y, Santos A, Tosics I (2015) The status of European green space planning and implementation based on an analysis of selected European city-regions. EU FP7 project GREEN SURGE, Deliverable D5.1, https://greensurge.eu/working-packages/wp5/files/D_5.1_Davies_et_al_2015_Green_Infrastructure_Planning_and_Implementation_v2.pdf. Accessed 22 Mar 2020
de la Barrera F, Henríquez C (2017) Vegetation cover change in growing urban agglomerations in Chile. Ecol Indic 81:265–273
de Wit M, van Zyl H, Crookes D, Blignaut J, Jayiya T, Goiset V, Mahumani B (2012) Including the economic value of well-functioning urban ecosystems in financial decisions: evidence from a process in Cape Town. Ecosys Serv 2:38–44
de Zeeuw H, Van Veenhuizen R, Dubbeling M (2011) The role of urban agriculture in building resilient cities in developing countries. J Agric Sci 149:153–163
Di Leo N, Escobedo FJ, Dubbeling M (2016) The role of urban green infrastructure in mitigating land surface temperature in Bobo-Dioulasso, Burkina Faso. Environ Dev Sustain 18:373–392
Dobbs C, Escobedo FJ, Clerici N, De la Barrera F, Eleuterio AA, MacGregor-Fors I, Reyes-Paecke S, Vásquez A, Zea Çamaño JD, Hernández HJ (2019) Urban ecosystem services in Latin America: mismatch between global concepts and regional realities? Urban Ecosys 22:173–187
Douglas I (2018) The challenge of urban poverty for the use of green infrastructure on floodplains and wetlands to reduce flood impacts in intertropical Africa. Landsc Urb Plan 180:262–272
Drescher AW, Isendahl C, Caridad Cruz M, Menokamit A, Karg H (2021) In Shackleton CM, Cilliers SS, Davoren E, du Toit MJ (eds) Urban Ecology in the Global South. Springer, Cham, pp 293–324
du Toit MJ, Cilliers SS, Dallimer M, Goddard M, Guenat S, Cornelius SF (2018) Urban green infrastructure and ecosystem services in sub-Saharan Africa. Landsc Urb Plan 180:249–261
EC (European Commission) (2013) Green infrastructure (GI): enhancing Europe's natural capital. Communication from the Commission to the European Parliament, The Council, The European Economic and Social Committee and The Committee of the Regions, Com(2013), 249 final, Brussels
Escobedo FJ (2021) Understanding urban regulating ecosystem services in the Global South. In: Shackleton CM, Cilliers SS, Davoren E, du Toit MJ (eds) Urban ecology in the Global South. Springer, Cham, pp 227–244

Escobedo FJ, Nowak DJ, Wagner JE, De la Maza CL, Rodríguez M, Crane DE, Hernández J (2006) The socioeconomics and management of Santiago de Chile's public urban forests. Urban For Urban Green 4:105–114
Escobedo FJ, Clerici N, Staudhammer CL, Corzo GT (2015) Socio-ecological dynamics and inequality in Bogotá, Colombia's public urban forests and their ecosystem services. Urban For Urban Green 14:1040–1053
eThekwini Municipality (2010) Development assessment guidelines, DMOSS local action for biodiversity. Environmental Planning and Climate Protection Dept. https://www.emdlotiuip.co.za/wp-content/uploads/2018/02/Development-Assesment-Guidelines-DMOSS-1FEB2018.pdf. Accessed 1 May 2019
Food and Agriculture Organization of the United Nations (FAO) (2014) Ciudades más verdes en América Latina y el Caribe. Un informe de la FAO sobre agricultura urbana y periurban en la region, http://www.fao.org/3/a-i3696s.pdf. Accessed 15 Mar 2019
Feng S, Hou W, Chang J (2019) Changing coal mining brownfields into green infrastructure based on ecological potential assessment in Xuzhou, Eastern China. Sustain 11:2252
Feyisa GL, Dons K, Meilby H (2014) Efficiency of parks in mitigating urban heat island effect: an example from Addis Ababa. Landsc Urb Plan 123:87–95
Fohlmeister S, Pauleit S, Coly A, Touré H, Yeshitela K (2015) The way forward: climate resilient cities for Africa's future. In: Pauleit S, Coly A, Fohlmeister S, Gasparini P, Jørgensen G, Kabisch S, Kombe WJ, Lindley S, Simonis I, Yeshitela K (eds) Urban vulnerability and climate change in Africa: a multidisciplinary approach. Springer, Dordrecht, pp 369–399
Foo CH (2016) Linking forest naturalness and human well-being: a study on public's experiential connection to remnant forests within a highly urbanised region in Malaysia. Urban For Urban Green 16:13–24
García A, Lorenzo C, Vazquez LB, Reyna-Hurtado R (2014) Roedores y murciélagos en espacios verdes en San Cristóbal de Las Casas, Chiapas, México. Therya 5:615–632
Garekae H, Shackleton CM (2020) Foraging wild food in urban spaces: the contribution of wild foods to urban dietary diversity in South Africa. Sustain 12:678
Gashu K, Gebre-Egziabher T, Maru M (2019) Drivers for urban green infrastructure development and planning in two Ethiopian cities: Bahir Dar and Hawassa. Arboricult J 41:48–63
Girma Y, Terefe H, Pauleit S, Kindu M (2018) Urban green infrastructure planning in Ethiopia: the case of emerging towns of Oromia special zone surrounding Finfinne. J Urban Manage 8:75–88
Gómez M, Puerta M (2017) Citizen biodiversity: daily reports of wildlife in urban environments. In: Mejía M (ed) Urban Nature, Instituto de Investigación de Recursos Biológicos Alexander von Humboldt. Bogotá D. C., p 60–63
Gong L, Mao B, Qi Y, Xu C (2015) A satisfaction analysis of the infrastructure of country parks in Beijing. Urban For Urban Green 14:480–489
Gopal D, Nagendra H (2014) Vegetation in Bangalore's slums: boosting livelihoods, well-being and social capital. Sustain 6:2459–2473
Gopal D, von der Lippe M, Kowarik I (2018) Sacred sites as habitats of culturally important plant species in an Indian megacity. Urban For Urban Green 32:113–122
Güneralp B, Lwasa S, Masundire H, Parnell S, Seto KC (2018) Urbanisation in Africa: challenges and opportunities for conservation. Environ Res Lett 13:015002
Gwedla N, Shackleton C (2019) Perceptions and preferences for urban trees across multiple socio-economic T contexts in the Eastern Cape, South Africa. Landsc Urb Plan 189:225–234
Haase D, Larondelle N, McPhearson T, Schwarz N, Hamstead Z, Kremer P, Artmann M, Wurster D, Breuste J, Borgström S, Jansson A, Elmqvist T, Andersson E, Langemeyer J, Gomez-Baggethun E, Kabisch N, Rall EL, Pauleit S, Hansen R, Voigt A, Qureshi S (2014) A quantitative review of urban ecosystem services assessment: concepts, models and implementation. Ambio 43:413–433
Hansen R, Pauleit S (2014) From multifunctionality to multiple ecosystem services? A conceptual framework for multifunctionality in green infrastructure planning for urban areas. Ambio 43:516–529

Hansen, R, Werner R, Santos A, Luz A, Száraz L, Tosics I, Vierikko K, Davies C, Rall E, Pauleit S (2016) Advanced urban green infrastructure planning and implementation. EU FP7 project GREEN SURGE, Deliverable D5.2, www.greensurge.eu. Accessed 25 Jan 2019
Hansen R, Rall E, Chapman E, Rolf W, Pauleit S (2017) Urban green infrastructure planning: a guide for practitioners. GREEN SURGE Deliverable 5.3. http://greensurge.eu/working-packages/wp5. Accessed 22 Mar 2020
Healey P (2010) In search for the "strategic" in spatial strategy making. Plann Theory Prac 10:439–457
Heitmann J (2014) Red de agricultura urbana: identificando la agricultura urbana en Santiago de Chile. In: Fuentes A (ed) Traduciendo el zumbido del enjambre. Ediciones CU, Santiago, pp 26–35
Heratha HM, Halwatura RU, Jayasinghe GY (2018) Evaluation of green infrastructure effects on tropical Sri Lankan urban context as an urban heat island adaptation strategy. Urban For Urban Green 29:212–222
Hernández Á, Reyes S (2018) The effects of urban expansion on green infrastructure along an extended latitudinal gradient in Chile over the last thirty years. Land Use Pol 79:725–733
Herslund L, Lund DH, Jørgensen G, Mguni P, Kombe W, Yeshitela K (2015) Towards climate change resilient cities in Africa: initiating adaptation in Dar es Salam and Addis Ababa. In: Pauleit S, Coly A, Fohlmeister S, Gasparini P, Jørgensen G, Kabisch S, Kombe WJ, Lindley S, Simonis I, Yeshitela K (eds) Urban vulnerability and climate change in Africa: a multidisciplinary approach. Springer, Dordrecht, pp 319–348
Herslund L, Backhaus A, Fryd O, Jørgensen G, Liu L, Mguni P, Mkupasi M, Workalemahu L, Yeshitela K (2018) Challenges and opportunities for developing water resilient green cities in Addis Ababa and Dar es Salaam: in search of champions and paths for urban transition. Landsc Urb Plan 180:319–327
Holmes PM, Rebelo AG, Dorse C, Wood J (2012) Can Cape Town's unique biodiversity be saved? Balancing conservation imperatives and development needs. Ecol Soc 17:28
Hu H, Tian Z, Sun L, Wen J, Liang Z, Dong G, Liu J (2019) Synthesised trade-off analysis of flood control solutions under future deep uncertainty: An application to the central business district of Shanghai. Water Res 166:11506
Inostroza L (2014) Open spaces and urban ecosystem services: cooling effect towards urban planning in South American cities. Tema (June 2014): 523–524
Jaganmohan M, Vailshery LS, Gopal D, Nagendra H (2012) Plant diversity and distribution in urban domestic gardens and apartments in Bangalore, India. Urban Ecosys 15:911–925
Jia Z, Tang S, Luo W, Li S, Zhou M (2016) Small scale green infrastructure design to meet different urban hydrological criteria. J Environ Manage 171:92–100
Jim CY, Zhang H (2015) Effect of habitat traits on tree structure and growth in private gardens. Landsc Ecol 30:1207–1223
Kambites C, Owen S (2006) Renewed prospects for green infrastructure planning in the UK. Plann Pract Res 21:483–496
Kaoma H, Shackleton CM (2014) Homestead greening is widespread amongst the urban poor in three medium-sized South African towns. Urban Ecosys 17:1191–1207
Kaoma H, Shackleton CM (2015) The direct use value of urban tree non-timber forest products to household income in poorer suburbs in South African towns. For Policy Econ 61:104–112
Kato S, Ahrn J (2008) Learning by doing: adaptive planning as a strategy to address uncertainty in planning. J Environ Plann Manage 51:543–559
King A, Shackleton CM (2020) Maintenance of private and public urban green space provides significant employment in Eastern Cape towns, South Africa. Urban For Urban Green 54:126740
KLSP (2020) Kuala Lumpur Structure Plan 2040. Kuala Lumpur City Hall (DBKL), Kuala Lumpur
Kotharkar R, Bagade A, Ramesh A (2019) Assessing urban drivers of canopy layer urban heat island: a numerical modeling approach. Landsc Urb Plan 190:103586
Kumar M, Mukherjee N, Sharma G, Raghubanshi AS (2010) Land use patterns and urbanisation in the holy city of Varanasi, India: a scenario. Environ Monit Assess 167:417–422

Lafortezza R, Davies C, Sanesi G, Konijnendijk CC (2013) Green infrastructure as a tool to support spatial planning in European urban regions. iForest Biogeosci For 6: 102–108
Lange W, Sandholz S, Nehren U (2018) Strengthening urban resilience through nature: the potential of ecosystem-based measures for reduction of landslide risk in Rio de Janeiro. Lincoln Institute of Land Policy, Washington, DC
Li L, Pussella P (2017) Is Colombo city, Sri Lanka secured for urban green space standards? Appl Ecol Environ Res 15:1789–1799
Lin BB, Meyers JA, Barnett GB (2018) Establishing priorities for urban green infrastructure research in Australia. Urb Pol Res 37:30–44
Lindley SJ, Gill SE, Cavan G, Yeshitela K, Nebebe A, Woldegerima T, Kibassa D, Shemdoe R, Renner F, Buchta K, Abo-El-Wafa H, Printz A, Sall F, Coly A, Marème Ndour N, Feumba RA, Zogning MO, Tonyé E, Ouédraogo Y, Bani Samari S, Sankara BT (2015) Green infrastructure for climate adaptation in African Cities. In: Pauleit S, Coly A, Fohlmeister S, Gasparini P, Jørgensen G, Kabisch S, Kombe WJ, Lindley S, Simonis I, Yeshitela K (eds) Urban vulnerability and climate change in Africa: a multidisciplinary approach. Springer, Dordrecht, pp 107–152
Lindley S, Pauleit S, Yeshitela K, Cilliers S, Shackleton C (2018) Rethinking urban green infrastructure and ecosystem services from the perspective of sub-Saharan African cities. Landsc Urb Plan 180:328–338
Liu L (2008) Status and prospects for urban green structure planning in China—Weihai city as a case study. Forest and Landscape Research, University of Copenhagen, Copenhagen
Liu L, Fryd O, Zhang S (2019a) Blue-green infrastructure for sustainable urban stormwater management: lessons from six municipality-led pilot projects in Beijing and Copenhagen. Water 11:2024
Liu Z, Lin Y, de Meulder B, Wang S (2019b) Can greenways perform as a new planning strategy in the Pearl River Delta, China. Landsc Urb Plan 187:81–95
Liu Z, Brown RD, Zheng S, Jiang Y, Zhao L (2020) An in-depth analysis of the effect of trees on human energy fluxes. Urban For Urban Green 50:126646
Lovell ST, Taylor JR (2013) Supplying urban ecosystem services through multifunctional green infrastructure in the United States. Landsc Ecol 28:1447–1463
Lukas M, Fragkou M, Vásquez A (2020) Hacia una ecología política de las nuevas periferias urbanas: suelo, agua y poder en Santiago de Chile. Revista Geog Norte Grande 76:100–123
Ma Y, Rong K, Mangalagiu D, Thornton TF, Zhu D (2018) Co-evolution between urban sustainability and business ecosystem innovation: Evidence from the sharing mobility sector in Shanghai. J Clean Prod 188:942–953
MacGregor-Fors I, Schondube JE (2011) Gray vs. green urbanisation: relative importance of urban features for urban bird communities. Basic Appl Ecol 12:372–381
Madaleno I (2000) Urban agriculture in Belém, Brazil. Cities 17:73–77
Magigi W (2013) Urbanisation and its impacts to food systems and environmental sustainability in urban space: Evidence from urban agriculture livelihoods in Dar es Salaam, Tanzania. J Environ Prot 4:1137–1148
McDonald RI, Marcotullio PJ, Güneralp B (2013) Urbanisation and global trends in biodiversity and ecosystem services. In: Elmqvist T, Fragkias M, Goodness J, Güneralp B, Marcotullio PJ, McDonald RI, Parnell S, Schewenius M, Sendstad M, Seto KC, Wilkinson C (eds) Urbanisation, biodiversity and ecosystem services: challenges and opportunities: a global assessment. Springer, Dordrecht, pp 31–52
McLean CT, Ground LE, Boon RG, Roberts DC, Govender N, McInnes A (2016) Durban's systematic conservation assessment. EThekwini Municipality, Environmental Planning and Climate Protection Department, Durban http://www.durban.gov.za/City_Services/development_planning_management/environmental_planning_climate_protection/Publications/Documents/DurbansSystematicConservation.pdf. Accessed 26 Apr 2019
Mei C, Liu J, Wang H, Yang Z, Ding X, Shao W (2018) Integrated assessments of green infrastructure for flood mitigation to support robust decision-making for sponge city construction in an urbanised watershed. Sci Tot Environ 639:1394–1407

Mell IC (2010) Green infrastructure: concepts, perceptions and its use in spatial planning. PhD thesis, University of Newcastle, Newcastle

Mell IC (2014) Aligning fragmented planning structures through a green infrastructure approach to urban development in the UK and USA. Urban For Urban Green 13:612–620

Mell IC (2016) Global green infrastructure: lessons for successful policy-making, investment, and management. Taylor & Francis, London

Miller SM, Montalto FA (2019) Stakeholder perceptions of the ecosystem services provided by green infrastructure in New York City. Ecosys Serv 37:100928

Ministry of Construction of the P.R. China (2002) Standard for classification of urban green space, CJJ/T 85-2002. Ministry of Construction of the P.R. China, Beijing (in Chinese)

Ministry of Housing and Urban-Rural Development (MOHRUD) (2014) Sponge city construction technical guide—construction of low-impact development rainwater system (trial). http://www.mohurd.gov.cn/wjfb/201411/W020141102041225.pdf. Accessed 16 Oct 2020

Ministry of Housing and Urban-Rural Development (MOHURD) (2016) Notice of the Ministry of Housing and Urban-Rural Development on Issuing the National Garden City Series Standards and the Management Measures for Application and Evaluation, Jiancheng (2016) No. 235 (in Chinese). http://www.mohurd.gov.cn. Accessed 18 Apr 2019

Moreno O (2013) Paisajes resilientes: reflexiones en torno a la reconstrucción de territorios desde el manejo y diseño de infraestructuras verdes, en el marco de las estrategias de gestión de riesgo ante desastres. Revista Electrón Geog Aust 5:1–19

Mundoli S, Manjunatha B, Nagendra H (2017) Commons that provide: the importance of Bengaluru's wooded groves for urban resilience. Int J Urban Sustain Dev 9:184–206

Murayama Y, Estoque RC, Subasinghe S, Hou H, Gong H (2015) Land-use/land-cover changes in major Asian and African cities. Annual report on the multi-use social and economy data bank, vol 92. http://giswin.geo.tsukuba.ac.jp/capital-cities/data/2015_capital_project_report.pdf. Accessed 12 Feb 2020

Myers N, Mittermeier RA, Mittermeier CG, Da Fonseca GA, Kent J (2000) Biodiversity hotspots for conservation priorities. Nat 403(6772):853–858

Nagendra H, Gopal D (2010) Street trees in Bangalore: density, diversity, composition and distribution. Urban For Urban Green 9:129–137

Nagendra H, Gopal D (2011) Tree diversity, distribution, history and change in urban parks: studies in Bangalore, India. Urban Ecosys 14:211–223

Nagendra H, Nagendran S, Paul S, Pareeth S (2012) Graying, greening and fragmentation in the rapidly expanding Indian city of Bangalore. Landsc Urb Plan 105:400–406

Nardini A, Gomes Miguez M (2016) An integrated plan to sustainably enable the City of Riohacha, (Colombia) to cope with increasing urban flooding, while improving its environmental setting. Sustain 8:198

Nero BF (2017) Urban green space dynamics and socio-environmental inequity: multi-resolution and spatiotemporal data analysis of Kumasi, Ghana. Int J Rem Sens 38:6993–7020

Ngulani T, Shackleton CM (2019) The degree, extent and value of air temperature amelioration by urban green spaces in Bulawayo, Zimbabwe. S Afr Geogr 102:344–355

Nguyen TT, Ngo HH, Guo W, Wang XC, Ren N, Li G, Ding J, Liang H (2019) Implementation of a specific urban water management-Sponge City. Sci Total Environ 652:147–162

O'Brien L, de Vreese R, Kern M, Sievänen T, Stojanova B, Atmis E (2017) Cultural ecosystem benefits of urban and peri-urban green. Urban For Urban Green 24:236–248

O'Farrell PJ, Anderson PM, Le Maitre DC, Holmes PM (2012) Insights and opportunities offered by a rapid ecosystem service assessment in promoting a conservation agenda in an urban biodiversity hotspot. Ecol Soc 17:27

Padgham J, Jabbour J, Dietrich K (2015) Managing change and building resilience: a multi-stressor analysis of urban and peri-urban agriculture in Africa and Asia. Urban Clim 12:183–204

Parker J, Zingoni de Baro ME (2019) Green infrastructure in the urban environment: a systematic quantitative review. Sustain 11:3182

Paul S, Nagendra H (2017) Factors influencing perceptions and use of urban nature: surveys of park visitors in Delhi. Land 6:1–23
Pauleit S, Liu L, Ahern J, Kazmierczak A (2011) Multifunctional green infrastructure planning to promote ecological services in the city. In: Niemelä J (ed) Handbook of urban ecology. Oxford University Press, Oxford, pp 272–285
Pauleit S, Hansen R, Rall EL, Zölch T, Andersson E, Luz A, Santos A, Szaraz L, Tosics I, Vierikko K (2017a) Urban landscapes and green infrastructure. In: Shugart H (ed) Oxford research encyclopedia of environmental science. Online publication at https://doi.org/10.1093/acrefore/9780199389414.013.23
Pauleit S, Zölch T, Hansen R, Randrup TB, Konijnendijk van den Bosch C (2017b) Nature-based solutions and climate change: four shades of green. In: Kabisch N, Korn H, Stadler J, Bonn A (eds) Nature-based solutions to climate change adaptation in urban areas: linkages between science, policy and practice. Springer Nature, Cham, pp 29–50
Pauleit S, Ambrose-Oji B, Andersson E, Anton B, Buijs A, Haase D, Elands B, Hansen R, Kowarik I, Kronenberg J, Mattijssen T, Stahl Olafsson A, Ral lE, van der Jagt AP, Konijnendijk van den Bosch C (2019a) Advancing urban green infrastructure in Europe: outcomes and reflections from the GREEN SURGE project. Urban For Urban Green 40: 4–16
Pauleit S, Pribadi DO, El Wafa H (2019b) Peri-urban agriculture: lessons learnt from Jakarta and Addis Ababa. Field Act Sci Rep 20:18–25
Pedlowski MA, Da Silva VA, Adell JJ, Heynen NC (2002) Urban forest and environmental inequality in Campos dos Goytacazes, Rio de Janeiro, Brazil. Urban Ecosys 6:9–20
Pfab MF, Compaan PC, Whittington-Jones CA, Engelbrecht I, Dumalisile L, Mills L, West SD, Muller PJ, Masterson GP, Nevhutala LS, Holness S, Hoare DB (2017) The Gauteng conservation planp Planning for biodiversity in a rapidly urbanising province. Bothalia 47:a2182
Pham T-T, Apparicio P, Seguin A-M, Landry S, Gagnon M (2012) Spatial distribution of vegetation in Montreal: an uneven distribution or environmental inequity? Landsc Urb Plan 107:214–224
Pribadi DO, Pauleit S (2016) Peri-urban agriculture in Jabodetabek Metropolitan Area and its relationship with the urban socioeconomic system. Land Use Pol 55:265–274
Qiao XJ, Liu L, Kristoffersson A, Randrup TB (2019) Governance factors of sustainable stormwater management: a study of case cities in China and Sweden. J Environ Manage 248:109249
Rabare RS, Okech R, Onyango GM (2009) The role of urban parks and socioeconomic development: case study of Kisumu Kenya. Theor Empir Res Urban Manage 3:22–36
Rasli FN, Kanniah KD (2018) Green corridors for liveable and walkable city: a case of Kuala Lumpur. Chem Eng Trans 63:391–396
Remolina F (2011) Figuras municipales de conservación ambiental en Colombia: ¿áreas protegidas, redes ecológicas o infraestructuras verdes? Nodo: Arquitectura Ciudad. Medio Ambiente 6:65–76
Reyes S, Figueroa I (2010) Distribución, superficie y accesibilidad de las áreas verdes en Santiago de Chile. Revista Latinoamer Est Urban Region 36:89–110
Reynolds C, Escobedo F, Clerici N, Zea-Camaño J (2017) Does "greening" of neotropical cities considerably mitigate carbon dioxide emissions? The case of Medellin, Colombia. Sustain 9:785
Rigolon A, Browning MH, Lee K, Shin S (2018) access to urban green space in cities of the Global South: a systematic literature review. Urban Sci 2:67
Roberts D (2010) Prioritising climate change adaptation and local level resilience in Durban, South Africa. Environ Urbanis 22:397–413
Roberts D, Diederichs N (2002) Durban's Local Agenda 21 programme: tackling sustainable development in a post-apartheid city. Enviro Urbanis 14:189–201
Roberts D, O'Donoghue S (2013) Urban environmental challenges and climate change action in Durban, South Africa. Environ Urbanis 25:299–319
Roberts D, Boon R, Croucamp P, Mander M (2005) Resource economics as a tool for open space planning in Durban, South Africa. In: Tryzna T (ed) The urban imperative. California Institute of Public Affairs, Sacramento, pp 44–48

Roberts D, Boon R, Diederichs N, Douwes E, Govender N, McInnes A, Mclean C, O'Donoghue S, Spires M (2012) Exploring ecosystem-based adaptation in Durban, South Africa: "Learning-by-doing" at the local government coal face. Environ Urbanis 24:167–195

Rocha C, Lessa I (2009) Urban governance for food security: the alternative food system in Belo Horizonte, Brazil. Int Plann Stud 14:389–400

Romero H, Molina M, Moscoso C, Sarricolea P, Smith P, Vásquez A (2007) Caracterización de los cambios de usos y coberturas de suelos causados por la expansión urbana de Santiago: análisis estadístico de sus factores explicativos e inferencias ambientales. In: De Mattos C, Hidalgo R (eds) Santiago de Chile: movilidad espacial y reconfiguración metropolitana. Instituto de Estudios Urbanos y Territoriales e Instituto de Geografía, Pontificia Universidad Católica de Chile, pp 251–270

Romero H, Vásquez A, Fuentes C, Salgado M, Schmidt A, Banzhaf E (2012) Assessing urban environmental segregation, UES: the case of Santiago de Chile. Ecol Indic 23:76–87

Romero-Lankao P (2007) Are we missing the point? Particularities of urbanisation, sustainability and carbon emissions in Latin American cities. Environ Urbanis 19:159–175

Rouse DC, Bunster-Ossa IF (2013) Green infrastructure, a landscape approach. American Planning Association, Washington, DC

Roy M, Shemdoe R, Hulme D, Mwageni N, Gough A (2018) Climate change and declining levels of green structures: Life in informal settlements of Dar es Salaam, Tanzania. Landsc Urb Plan 180:282–293

Safford HD, Sawyer SC, Kocher SD, Hiers JK, Cross M (2017) Linking knowledge to action: the role of boundary spanners in translating ecology. Front Ecol Environ 15:560–568

Samad AN, Abdul-Rahim AS, Mohd Yusof MJ, Tanak K (2020) Assessing the economic value of urban green spaces in Kuala Lumpur. Environ Sci Pollut Res 27:10367–10390

Santandreu A, Perazzoli AG, Terrile R, Ponce M (2009) Urban agriculture in Montevideo and Rosario: a response to crisis or a stable component of the urban landscape. Urban Agric Mag 22:12–13

Sardeshpande M, Shackleton CM (2020) Urban foraging: land management policy, perspectives, and potential. PLoS ONE 15(4):e0230693

Schäffler A, Swilling M (2013) Valuing green infrastructure in an urban environment under pressure: the Johannesburg case. Ecol Econ 86:246–257

Seburanga JL, Kaplin BA, Zhang QX, Gatesire T (2014) Amenity trees and green space structure in urban settlements of Kigali, Rwanda. Urban For Urban Green 13:84–93

Seto KC, Güneralp B, Hutyra LR (2012) Global forecasts of urban expansion to 2030 and direct impacts on biodiversity and carbon pools. Proc Nat Acad Sci 109:16083–16088

Shackleton CM (2021) Ecosystem provisioning services in Global South cities. In: Shackleton CM, Cilliers, SS, Davoren E, du Toit MJ (eds) Urban ecology in the Global South. Springer, Cham, pp 203–226

Shackleton CM, Blair A (2013) Perceptions and use of public green space is influenced by its relative abundance in two small towns in South Africa. Landsc Urb Plan 113:104–112

Shackleton CM, Hurley PT, Dahlberg A, Emery MR, Nagendra H (2017) Urban foraging: a ubiquitous human practice but overlooked by urban planners, policy and research. Sustain 9:1884

Shackleton CM, Blair A, de Lacy P, Kaoma H, Mugwagwa N, Dalu MT, Walton W (2018) How important is green infrastructure in small and medium sized towns? Lessons from South Africa. Landsc Urb Plan 180:273–281

Silva R, Lithgow D, Esteves LS, Martínez ML, Moreno-Casasola P, Martell R, Pereira P, Mendoza E, Campos-Cascaredo A, Winckler Grez P, Osorio AF (2017) Coastal risk mitigation by green infrastructure in Latin America. Proc Inst Civil Eng Marit Eng 170:39–54

Sing KW, Dong H, Wang WZ, Wilson JJ (2016) Can butterflies cope with city life? Butterfly diversity in a young megacity in Southern China. Genome 59:751–761

Sing KW, Luo J, Wang W, Jaturas N, Soga M, Yang X, Dong H, Wilson JJ (2019) Ring roads and urban biodiversity: distribution of butterflies in urban parks in Beijing city and correlations with other indicator species. Sci Rep 9:7653

Sreetheran M, Adnan M (2007) Green network development of Kuala Lumpur city: from the perspective of Kuala Lumpur structure plan. Forum 7:38–42
Sun J, Li YP, Gao PP, Xia BC (2018) A Mamdani fuzzy inference approach for assessing ecological security in the Pearl River Delta urban agglomeration, China. Ecol Indic 94:86–396
Sutherland C, Sim V, Buthelezi S, Khumalo D (2016) Social constructions of environmental services in a rapidly densifying peri-urban area under dual governance in Durban, South Africa. Bothalia 46:a2128
Taubenböck H, Wegmann M, Roth A, Mehl H, Dech S (2009) Urbanisation in India: spatiotemporal analysis using remote sensing data. Comp Environ Urban Syst 33:179–188
Taylor A (2016) Institutional inertia in a changing climate: climate adaptation planning in Cape Town, South Africa. Intn J Clim Change Strat Manage 8:194–211
Taylor C, Cockburn J, Rouget M, Ray-Mukherjee J, Mukherjee S, Slotow R, Roberts D, Boon R, O'Donoghue S, Douwes E (2016) Evaluating the outcomes and processes of a research-action partnership: the need for continuous reflective evaluation. Bothalia 46:a2154
Thaiutsa B, Puangchit L, Kjelgren R, Arunpraparut W (2008) Urban green space, street tree and heritage large tree assessment in Bangkok, Thailand. Urban For Urban Green 7(3):219–229
The State Council of the P.R. China (2013) Opinions of the State Council on strengthening urban infrastructure construction. Ministry of Construction of the P.R. China, Beijing Guofa [2013] No. 36 (in Chinese)
Titz A, Chiotha SS (2019) Pathways for sustainable and inclusive cities in Southern and Eastern Africa through urban green infrastructure. Sustain 11:2729
UNFPA (2007) The state of world population 2007: unleashing the potential of urban growth. UNFPA, New York
UN (United Nations) (2014) World urbanisation prospects: the 2014 revision. United Nations, New York, http://esa.un.org/unpd/wup/. Accessed 27 Apr 2019
UN (United Nations) (2019) World urbanisation prospects: the 2018 revision. United Nations, New York
UN-Habitat (2015) HABITAT III Issue Papers 22: informal settlements. http://habitat3.org/wp-content/uploads/Habitat-III-Issue-Paper-22_Informal-Settlements-2.0.pdf. Accessed 15 Aug 2019
Uy PD, Nakagoshi N (2008) Application of land suitability analysis and landscape ecology to urban greenspace planning in Hanoi, Vietnam. Urban For Urban Green 7:25–40
Vásquez A, Salgado M (2009) Desigualdades socioeconómicas y distribución inequitativa de los riesgos ambientales en la comunas de Peñalolén y San Pedro de la Paz. Revista Geog Norte Grande 43:95–110
Vásquez A, Devoto C, Giannotti E, Velásquez P (2016) Green infrastructure systems to face fragmented cities in Latin America: the case of Santiago, Chile. Proc Eng 161:1410–1416
Vásquez A, Lukas M, Salgado M, Mayorga J (2017) Urban environmental (in)justice in Latin America: the case of Chile. In: Holifield R, Chakraborty J, Walker G (eds) The Routledge handbook of environmental justice. Routledge, London, pp 556–566
Vásquez A, Giannotti E, Galdámez E, Velásquez P, Devoto C (2019) Green infrastructure planning to tackle climate change in Latin American cities. In: Henriquez C, Romero H (eds) Urban climate in Latin-American cities. Springer, Dordrecht, pp 329–354
Venter ZS, Shackleton CM, van Staden F, Selomane O, Masterson VA (2020) Green apartheid: urban green space remains unequally distributed across income and race geographies in South Africa. Landsc Urb Plan 203:103889
Wang F, McShea WJ, Wang D, Li S, Zhao Q, Wang H, Lu S (2014) Evaluating landscape options for corridor restoration between giant panda reserves. PLoS ONE 9:e105086
Wang H, Dai X, Wu J, Wu X, Nie X (2019) Influence of urban green open space on residents' physical activity in China. BMC Public Heal 19:1093
Wang YC, Shen JK, Xiang WN (2018) Ecosystem service of green infrastructure for adaptation to urban growth: function and configuration. Ecosys Heal Sustain 4:132–143
Wei F (2017) Greener urbanisation? Changing accessibility to parks in China. Landsc Urb Plan 157:542–552

Wei J, Qian J, Tao Y, Hu F, Ou W (2018) Evaluating spatial priority of urban green infrastructure for urban sustainability in areas of rapid urbanisation: a case study of Pukou in China. Sustain 10:327

Wolff M, Haase D (2019) Mediating sustainability and liveability: turning points of green space supply in European cities. Front Environ Sci 7:61

Wu C, Li J, Wang C, Song C, Chen Y, Finka M, Rosa DL (2019) Understanding the relationship between urban blue infrastructure and land surface temperature. Sci Total Environ 694:133742

Xia J, Zhang Y, Xiong L, He S, Wang L, Yu Z (2017) Opportunities and challenges of the sponge city construction related to urban water issues in China. Sci China Earth Sci 60:652–657

Xiao M, Lin Y, Han J, Zhang G (2014) A review of green roof research and development in China. Renew Sustain Energy Rev 40:633–648

Yao N, van den Bosch CC, Yang J, Devisscher T, Wirtz Z, Jia L, Duan J, Ma L (2019) Beijing's 50 million new urban trees: strategic governance for large-scale urban afforestation. Urban For Urban Green 44:126392

Yu K, Li D, Li N (2006) The evolution of greenways in China. Landsc Urb Plan 76:223–239

Zhang K, Zhang J (2013) Health function evaluation and exploring its mechanisms in the Shanghai Green Belt, China. Acta Ecol Sinica 33:4189–4198

Zhang F, Chung CK, Yin Z (2019a) Green infrastructure for China's new urbanisation: a case study of greenway development in Maanshan. Urban Stud 57: 508–524

Zhang G, He BJ, Zhu Z, Dewancker B (2019b) Impact of morphological characteristics of green roofs on pedestrian cooling in subtropical climates. Int J Environ Res Public Heal 16: 179

Zheng S, Guldmann JM, Liu Z, Zhao L, Wang J, Pan X (2020) Modeling of shade creation and radiation modification by four tree species in hot and humid areas: case study of Guangzhou, China. Urban For Urban Green 45:126545

Zhou D, Liu Y, Hu S, Hu D, Neto S, Zhang Y (2019) Assessing the hydrological behaviour of large-scape potential green roofs retrofitting scenarios in Beijing. Urban For Urban Green 40:105–113

Zhu Z, Lang W, Tao X, Feng J, Liu K (2019) Exploring the quality of urban green spaces based on urban neighborhood green index: a case study of Guangzhou City. Sustain 11(19):5507

Zimmermann E, Bracalenti L, Piacentini R, Inostroza L (2016) Urban flood risk reduction by increasing green areas for adaptation to climate change. Procedia Eng 161:2241–2246

Zinia NJ, McShane P (2018) Ecosystem services management: an evaluation of green adaptations for urban development in Dhaka, Bangladesh. Landsc Urb Plan 173:23–32

Urban Plant Diversity: Understanding Informing Processes and Emerging Trends

Pippin M. L. Anderson, Luke J. Potgieter, Lena Chan, Sarel S. Cilliers, and Harini Nagendra

Abstract Plant diversity in cities is important for numerous social and ecological reasons. In this chapter we use Aronson et al.'s (Ecology 97:2952–2963, 2016) hierarchical filters to explore urban biodiversity trends and driving processes in cities of the Global South. Modelled expansion in the Global South shows a high degree of coincidence with biodiversity hotspots with sprawling, informal, settlement rapidly transforming landscapes. Alien plants are frequently introduced through human facilitation and this has been linked to aspirational aesthetics and livelihoods. The relative contribution of indigenous versus alien plant species varies and histories, in particular of colonisation, are important. Like cities in the Global North, invasive aliens are a threat to urban biodiversity in and beyond cities in the Global South. Cities here differ in having less resources to tackle these threats. Biodiversity in cities of the Global South further differs from the Global North in the role it plays in people's lives where social and cultural practices make for diverse engagements. Emerging work suggests these are means through which to engage the public in biodiversity conservation. Individual garden management, combined with remnant vegetation,

P. M. L. Anderson (✉)
Department of Environmental and Geographical Science, University of Cape Town, Cape Town, South Africa
e-mail: pippin.anderson@uct.ac.za

L. J. Potgieter
Centre for Invasion Biology, Department of Botany and Zoology, Stellenbosch University, Stellenbosch, South Africa

L. Chan
International Biodiversity Conservation Division, National Parks Board of Singapore, Singapore, Singapore
e-mail: Lena_chan@nparks.gov.sg

S. S. Cilliers
Unit for Environmental Sciences and Management, North-West University, Potchefstroom, South Africa
e-mail: Sarel.Cilliers@nwu.ac.za

H. Nagendra
Centre for Climate Change and Sustainability, Azim Premji University, Bangalore, India
e-mail: harini.nagendra@apu.edu.in

C. M. Shackleton et al. (eds.), *Urban Ecology in the Global South*, Cities and Nature,
https://doi.org/10.1007/978-3-030-67650-6_6

makes for diverse biodiversity outcomes and refutes the blanket application of ideas developed in the Global North such as the luxury effect. The role of urban agriculture and provisioning services plays an important role in garden biodiversity. The current amount of remnant biodiversity in cities in the Global South presents an opportunity to conserve these elements.

Keywords Aliens · Biodiversity · Human facilitation · Remnant biodiversity

1 Introduction

The world is currently urbanising at an unprecedented rate, especially in the Global South (UN DESA 2018). This rapid growth is placing significant pressure on natural ecosystems around and within the urban matrix through an increasing demand for land and ecosystem services. Plant diversity in cities, and the associated plant communities, are important for several reasons, not least being that the supply of a full range of ecosystem services is dependent on plant diversity. For example, at a local scale plants are often important for livelihood benefits such as fuelwood or medicinal plants (Shackleton 2021), and at a larger scale for their role in underpinning the ecological processes that regulate some of earth's life-support systems (Escobedo 2021), such as nutrient cycling, for example, and most relevant at present, carbon sequestration (Reiss et al. 2009). Biodiversity is also critical to system resilience where ecological systems with higher species richness are more persistent in the face of change (Elmqvist et al. 2003; Oliver et al. 2015). This identified resilience function is important in cities in the Global South where change is rapid (Sitas et al. 2021), and cities have been identified as particularly susceptible to the impacts of global change and are home to the most vulnerable populations of the world (Seto et al. 2011). Miller and Hobbs (2002) remind us that we cannot afford to focus conservation efforts solely on pristine sites but that we must also turn our attention to biodiversity conservation "where people live and work". It is here in the emerging cities of the world that we will achieve important conservation goals, and also meet biodiversity-related well-being and educational goals.

Plant diversity can be measured in several ways. While species are the most common metric (Purvis and Hector 2000), these can be presented according to counts at any given scale: giving measures of plant species richness, or on their relative contribution to a community; giving measures of evenness, or a combination of these two measures through a myriad of ever emerging diversity metrics. Each species will also have its own status according, for example, to their dominance or rarity, whether they are indigenous, endemic, alien or invasive, or as relates to their conservation IUCN Red List status which categorises species according to their level of threat. For example, the application of the IUCN Red List at even a city-state level has been useful as demonstrated in the case of Singapore (Davison et al. 2008). Plants are also allocated traits, often informed by human interests (such as leaf phenology where a homeowner may wish to avoid planting a deciduous species which requires more maintenance with seasonal leaf-shed) or ecological questions that may relate to their functional or ecosystem service contribution (such as plant height or biomass which might be used to test vegetation response to climate variability associated with global change) (Williams et al. 2015). It is important to remember when discussing diversity, that different measures, taken at different scales, will result in the presentation of variable patterns in plant diversity (McDonnell and Hahs 2015).

While we know rapid urbanisation is having a significant impact on biodiversity, how this plays out is not yet well understood (Silva et al. 2015; Aronson et al. 2016). A general synthesis of urban biodiversity remains in a fledging state (McDonnell and Hahs 2015), and this lack of detailed data and knowledge is even more pronounced in the Global South (Silva et al. 2015; Pena et al. 2016; Lindley et al. 2018). In this chapter we use Aronson et al.'s (2016) hierarchical filters (Fig. 1) to explore trends and processes governing urban biodiversity in cities of the Global South. Any plant community is the result of a series of hierarchically imposed filters and much of the study of ecology has been dedicated to establishing the rules that underpin plant community assembly. In cities, where overall a much greater proportion of the vegetation is planted, the filters determining which plant traits remain in the system, or which colonise the system, are different to those informing rural, more natural contexts. In addition to the more traditional historical filters established for rural plant communities, in cities the filters are heavily influenced by social factors and these can vary significantly across time and space. Aronson et al.'s (2016) model presents this journey, where the emergent plant community is a function of its original regional species pool and human contributions following which filters of urban form and history, and social and cultural considerations, will select, or filter for certain

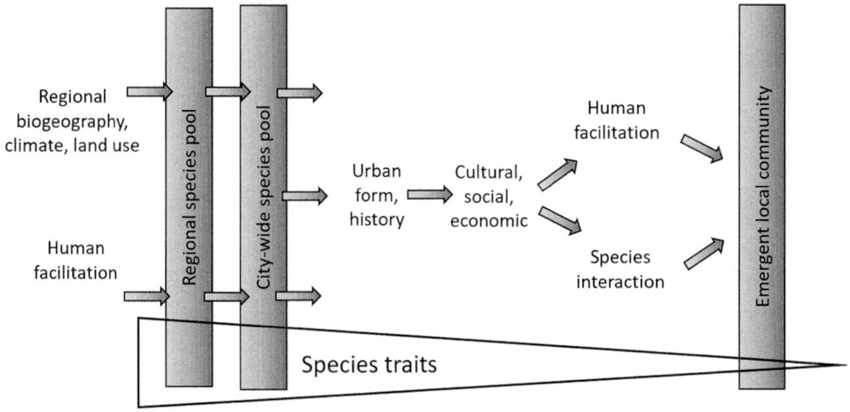

Fig. 1 Hierarchical filters operating in cities that guide species selection and direct community assembly (adapted from Aronson et al. [2016]). Here we use their hierarchical filters (in horizontal text) to explore filtering and biodiversity outcome in cities of the Global South. Resultant species pools are presented in the vertical boxes. Life histories and functional traits, which play a role throughout, are represented by the triangle

species. Along the way species, traits themselves, and human facilitation continue to filter the species pool with an outcome that is both directed by human engagement and the plant communities themselves (Fig. 1). We present examples from some Global South cities to demonstrate regional cases or trends, and how these contrast to in particular circumstances, to the Global North, and in doing so also show that while we look for general understandings around community assembly in the Global South, every city will differ to some degree (Silva et al. 2015).

Aside from the more socially influenced filters, which will be explored in detail below, every city's regional climate and geography will set it apart from others (Aronson et al. 2016). Yet in their study of the factors driving city diversity, Aronson et al. (2016) found anthropogenic factors as more important than traditional filters such as climate. Here we examine the anthropogenic factors that define the diversity of cities of the Global South, and not the more traditional filters of climate and geography. Similarly, we elect not to focus specifically on the biotic engagements that relate to species interactions, unless through examples where relevant, as these are not particular per se to cities of the Global South. We also look at original species pools and then explore the filters of human facilitation (the role of invasive alien flora), emergent urban morphologies, socio-economic and cultural factors and the role of small-scale individual land management. We conclude by considering the most pressing concerns and reflect on the opportunities apparent in securing urban biodiversity in cities of the Global South.

2 Species Pools and Traits

The pool of species in any city will be a filtered subset of the regional pool and other species introduced through human facilitation (Akinnifesi et al. 2010; Aronson et al. 2016). The species are filtered through a set of selective pressures—in the case of cities these are primarily anthropogenic in origin. Species will be selected based on their response traits and those best suited to the city context will persist or thrive. These response traits have been well studied in urban ecology, particularly in the Global North (Duncan et al. 2011), but with some work emerging in the Global South (Goodness et al. 2016; Anderson et al. 2020). The species themselves then become part of the selective filters in operation through effect traits. Effect traits are those that inform system functioning and in turn inform community assembly processes (Suding et al. 2008). These effect traits have received far less attention in the urban ecology literature. The process of establishing urban plant diversity is complex and iterative in nature, and while not examined here, the emergent plant communities themselves will become an additional filter.

There is considerable work still to be done in this area, especially in the cities of the Global South where no large-scale systematic comparative study has been done on explored indigenous plant extinctions and associated traits (Pauchard et al. 2006; Silva et al. 2015). The original species pool however is crucial and in the case of urban development in the Global South, is recognised as significantly biodiverse. For example, the 2018 IPBES Regional Assessment for Africa highlights the substantial extant nature in Africa and, while noting deeply concerning declines, describes extraordinary species richness and biodiversity that persists on the continent in remnant nature (IPBES 2018).

3 Urban Plant Diversity Filters: Regional Climate Biogeography and Land Use

Güneralp and Seto (2013) developed models of likely urban expansion with a view to understanding biodiversity impacts, and demonstrate a close association between urban settlement, critical habitats and high biodiversity. Their models show future urban expansion in the rapidly urbanising Global South as placing considerable pressure on critical biodiversity. Regions with the most pronounced urban growth, namely SouthEast Asia, Central America, Mid Asia, Southern America, Western Asia and North Africa, are all home to biodiversity hotspots and numerous protected areas (Güneralp et al. 2017; Seto et al. 2012). While Asia has the most variable patterns of urban settlement (home to more than 60% of the world's population), urban expansion in this region will be critical to regional biodiversity (Seto 2013; Silva et al. 2015). Dramatic economic growth, and a shift from production to a more consumption-oriented economy is matched by dramatic urbanisation in the region, particularly in China, which has four of the 34 global biodiversity hotspots (Güneralp and Seto 2013). Predicted growth in the east of the country and across the western-corridor between Hangzhou and Shenyang will coincide with significant biodiversity. Here land use dynamics are primarily informed by economic drivers with little consideration for environmental effects and ecosystem services (Güneralp and Seto 2013). India by comparison, has been slower to urbanise, but the recent development of a government strategy for the establishment of cities acknowledges the growing trend of urbanisation (Nagendra et al. 2012). Here urbanisation is also flagged as coinciding with noteworthy biodiversity, and research demonstrates negative impacts on biodiversity with the clearing of land for settlement and atmospheric and freshwater pollution. While ecosystem impacts and losses are recorded, some environmental wins are noted with, for example, shifts in energy options in cities (Nagendra et al. 2012). Similarly, in biodiversity conservation, new species, rediscoveries and new records of plants continue to be recorded in highly densely populated cities like Singapore (Leong-Skornickova et al. 2014; Lim et al. 2018).

Urbanisation in Africa tends not to be tied to economic growth and as a result, urban populations are characteristically poor, and settlement has a high degree of informality (Myers 2021). Persistent ties to rural landscapes post-urbanisation mean urban impacts on biodiversity are widely felt (Anderson et al. 2013). Some urban dwellers may forage off the natural environment in the city in a manner more typical of rural lifestyles (Anderson et al. 2013; Kaoma and Shackleton

2015). Here, existing biodiversity hotspots are already significantly impacted (Rebelo et al. 2011) and future predictions show negative biodiversity impacts, especially in the north of the region (Güneralp and Seto 2013). South America by contrast, is one of the most urbanised regions of the world (Pauchard and Barbosa 2013). Urban settlement here is characterised by highly unequal settlement with informal, and high-cost housing in the city. Biodiversity hotspots that coincide with urban settlement receive little attention in planning, and informal settlement tends to encroach on sensitive habitats of water courses and steep slopes, resulting in significant environmental degradation (Akinnifesi et al. 2010).

4 Urban Plant Diversity Filters: Global Human Facilitation

Cities are defined by their high population density and human facilitation is naturally one of the most significant factors determining biodiversity in a city. Urban areas are foci for the introduction of species and as a result, their biota may comprise a high proportion of alien species, many of which were intentionally introduced to deliver, augment or restore particular ecosystem services. Urban areas worldwide share certain features that facilitate the proliferation of some alien species. Intensive cultivation and repeated introductions of many alien species have resulted in the availability of large numbers of propagules, increasing the likelihood of their establishment and persistence (Pyšek 1998; Kowarik et al. 2013). The complex networks of dispersal pathways and vectors in cities (networks of roads, railways and other conduits for the transport of people and commodities) offer abundant opportunities for rapid dissemination of propagules both within urban settings and outwards (Alston and Richardson 2006; von der Lippe and Kowarik 2008; McLean et al. 2017). Urbanisation has been suggested as a major cause of biotic homogenisation (McKinney 2006) driven by species invasions and indigenous species losses. However, several studies, mostly from the Global North, suggest that alien species can drive within city floral differentiation (e.g. Olden and Poff 2003; Kühn and Klotz 2006). Further work is required to elucidate the effects of urbanisation and the underlying drivers of biotic homogenisation in the Global South.

A recent example of facilitated introduction comes from India where returning technology professionals from Silicon Valley in California, USA, brought with them an aesthetic desire for alien palm trees, a novel and conspicuous feature of the Indian landscape (Nagendra 2016). These trees, introduced for social and cultural reasons to do with globalisation, economic and migratory relationships, will in turn serve as a selective force in the ecology they occupy. Other research from Bangalore, India, suggests that while Indian residents favoured trees with scented flowers, preferred for their use in local rituals of worship, British colonists changed the tree-scape by planting trees with showy, scentless, spectacularly coloured flowers, bringing in an aesthetic that favoured the visual over the olfactory (Nagendra 2016).

Silva et al. (2015) point out that described urban biodiversity patterns that hold for the Global North, with large volumes of work from European and North American cities, may not necessarily be true in the Global South. They go on to say that it is difficult to generalise, and what limited work there is, does not always show consistent trends and more studies are needed to understand common responses. In their study of urban bird diversity in Valdivia in Chile they found urban vegetation to be mostly alien, with less than 30% of urban plant species indigenous to the region, while bird species were almost all indigenous (Silva et al. 2015). By contrast in Sao Luis in north-eastern Brazil over 60% of all urban garden plant species were indigenous to the region while only half the bird species held the same status (Akinnifesi et al. 2010). In Bangalore, India, 77% of trees in landscaped recreational spaces, such as city parks, are introduced species, in contrast in informal settlements, there were few trees because of the limited space availability, but 52% of these trees are indigenous (Nagendra and Gopal 2011; Gopal et al. 2015). While some trends hold true, in particular with the general degradation and loss of indigenous diversity and environment, each city and region will have exceptions and its own unique trends, and the filters will vary in type or intensity. The full understanding of urban biodiversity trends in the Global South is still to be revealed and is better described for some regions or cities, than others.

Invasive alien species in urban areas can negatively affect ecosystem services (Pyšek and Richardson 2010) or create ecosystem disservices (Hobbs et al. 2013; Potgieter et al. 2017; Vaz et al. 2017). For example, invasive species can reduce local biodiversity, damage property and infrastructure, cause toxicity and allergic reactions, increase the frequency and severity of fires at the urban–wildland interface and compromise the safety and security of urban residents (Reis et al. 2008; van Wilgen et al. 2012; Shochat et al. 2010; Nentwig et al. 2017; Potgieter et al. 2018). The high number of people in urban areas, and vulnerability of the urban poor of the Global South, means these impacts are amplified and felt more acutely here (Seto et al. 2012; Davoren and Shackleton 2021). Complex human–nature interactions may also give rise to unique threats associated with invasive alien species (such as crime, see Potgieter et al. 2019a). Some invasive alien species can be both beneficial and harmful. Such species generate conflicts around their use and management (Dickie et al. 2014; Potgieter et al. 2019b). While not particular to cities of the Global South (Ives and Kelly 2016), the high dependency on the local natural environment in developing world cities can make these conflicts more intense (Anderson et al. 2013). For example, an invasive plant species may serve as an important source of fuelwood, but displace indigenous biodiversity, leading to conflicts over which should be prioritised for control (Petersen et al. 2012; Potgieter et al. 2018).

While there is some "overflow" of natural-area invaders into urban settings, cities around the world share a similar suite of invasive species from all taxonomic groups, showing that general features common to urban environments are key drivers of these invasions (Potgieter et al. 2017). Management of invasive species however differs markedly in different parts of the world. This is often determined by the availability of funding and city planning prioritisation approaches—some cities prioritise urban green space, while others direct limited funding allocated for environmental issues to other obligations more closely aligned with socio-economic imperatives

(Irlich et al. 2017; de Satgé and Watson 2018). Developing countries face many operational challenges associated with managing urban invasions. Funding insecurity, inadequate management capacity and ineffective monitoring (Shackleton et al. 2021) hamper invasive species control efforts (Irlich et al. 2017). Scaling up, major socioeconomic pressures such as addressing the need for education, housing, nutrition and healthcare and transport infrastructure (Goodness and Anderson 2013) further complicate attempts to manage urban invasions (Fig. 2). Challenges in addressing rising levels of poverty and unemployment, and the formal housing deficit, all place considerable strain on remaining vegetation patches, which are in high demand for conversion to housing or industrial development (Goodness and Anderson 2013). Cities of the Global South have levels of prosperity, social and physical infrastructure,

Fig. 2 Without adequate budgets to manage them, invasive alien pines run rampant and degrade the natural environment in north-eastern Madagascar with high biodiversity loss (Photograph: P. O'Farrell)

environmental sustainability and consequent quality of life levels that are overall significantly lower than cities of the Global North, thus presenting a challenge of an altogether different magnitude for biodiversity management (Nagendra et al. 2018).

Understanding the social dimensions mediating responses to biological invasions is becoming increasingly important in invasive species management (Shackleton et al. 2019b, c). Urban areas typically have a greater number and diversity of stakeholders compared to rural areas (Gaston et al. 2013) and this leads to a larger diversity of views about invasive alien species that may disproportionately complicate management but can also represent an opportunity for effective citizen engagement for rapid response. Effective stakeholder engagement is therefore a crucial component in managing urban biological invasions (Novoa et al. 2018; Shackleton et al. 2019a).

Knowledge of the patterns and processes of invasions in urban settings of the Global South is poor, despite long histories of alien species introductions into many urban centres (e.g. Potgieter et al. 2020). Urban areas are surrogates for global change (Lahr et al. 2018) and we need further understanding of how invasive species spread, their effects on biodiversity, ecosystem services and human well-being, and the challenges associated with managing urban biological invasions.

5 Urban Plant Diversity Filters: Emerging Urban Morphology

A common feature of many cities in the Global South is their colonial legacy (Myers 2003), which has resulted in certain cities being exposed to species through historical global colonial networks. Urban morphology, the shape and form of a city, how this emerges through time, how much greenspace remains, and in what configuration, is a critical filter of urban plant biodiversity (Aronson et al. 2016). In Guyana, urban tree diversity is dominated by 57 alien species, the origins and spatial distribution of which have been linked directly to colonial engagements in city development (Hunte et al. 2019). The particular circumstances of the Guyana case effectively demonstrate the role of colonial development and teleconnections. An environmental history of the city of Cape Town, South Africa, shows the introduction of specific species in association with broader colonial connections, in particular through contact with other fire-driven, Mediterranean-type ecologies (Anderson and O'Farrell 2012). While introduced species add to the collective richness of a flora, they can have negative consequences. In the case of the city of Cape Town, a biodiversity hotspot of global significance with very high endemic diversity, the introduction of alien species has commonly been at the expense of the indigenous flora through competitive suppression of indigenous species and altered fire regimes (Rebelo et al. 2011). However, not all colonial urban ecology legacies are detrimental to overall biodiversity, and both Bangalore and Delhi in India—cities situated in semi-arid regions—are cited as being significantly improved in terms of habitability as a result of extensive colonial tree planting which has greened these cities (Nagenda and Mundoli 2019). A study of Bandung in Indonesia by Abendroth et al. (2012) shows a tree legacy in urban parks which reflects a colonial policy which promoted both alien and indigenous planting, a legacy they look to continue. When planning and designing the ecology of emerging cities, planners will need to engage in and consider the role of useful alien (and sometimes even invasive) species in conjunction with biodiversity considerations. In some instances, alien species may need to be tolerated for the social benefits they provide, particularly in the developing world context where the state may not be in a position to provide alternatives. In Cape Town, an invasive species management framework was proposed to address conflicts in terms of control of alien species with the aim to tolerate some invasive species with important ecosystem services, but with the premise that choices made are "logic and defensible" (Gaertner et al. 2016).

Plant community data, and accurate data, for cities in the Global South is lacking. Table 1 gives some insights where non-native flora for three Global South cities compared to three Global North cities shows highly variable numbers of recorded non-native species. This table also shows the percentage of each pool that is shared with at least one other of these cities. This shows variable results flagging the role of distinct geographies, histories and contemporary practices filtering emergent plant communities. This data hints at a more distinctive non-native plant communities in the Global South, but while data are slim, firm conclusions are hard to draw. When comparing non-native plant species between each of the three Global South cities, and the same for the three Global North cities, there were only five species shared between the Global South cities, while there are 76 species shared between the Global North cities. This points to the highly interconnected nature of cities in the Global North, compared to those in the Global South. Here the role of longer histories of connectivity and trade outweigh shared colonial pasts.

Many developing world cities still have remnant patches of natural vegetation in the city and these are under considerable threat from the pace and nature of urban settlement (Lindley et al. 2018). Greenspaces and patches are important repositories of plant diversity and provide supporting green infrastructure and ecosystem

Table 1 Numbers of non-native plant species were recorded in six cities, three from the Global South, and three from the Global North

	City	Country	Number of non-native plant species	Percentage shared non-native plant species
Global South	Bujumbura	Burundi	197	38.6
	Cape Town	South Africa	910	25.7
	Niamey	Niger	83	59.0
Global North	Berlin	Germany	511	55.0
	Brno	Belgium	307	70.0
	Toronto	Canada	851	43.8

Included are percentages of species shared with at least one other of these six cities shown here. Data source: Global Urban Biological Invasions Consortium (GUBIC) data repository services. The shape and spatial arrangement of these patches, their size and connectivity, determine the space available for biodiversity to persist or establish (Aronson et al. 2016). Exactly how this finer-scale urban morphology promotes and selects for biodiversity is still not fully understood (McDonnell and Hahs 2013). For instance, while large parks have more biodiversity compared to small and medium parks in Indian cities, the presence of small urban green spaces such as home gardens, avenue trees and other pockets of green spaces within a 1–5 km radius increases the biodiversity of specific mobile taxa such as birds, butterflies and insects, at varying spatial scales (Swamy et al. 2019). Benchimol et al. (2017) describe urban settlement in South America as disorderly, with little consideration to securing or establishing greenspaces in the city development process. They note that while the benefits of these spaces are increasingly being acknowledged, they tend to exist as small, ad hoc spaces in an otherwise dominant hard urban form. In acknowledgement of their value however Indonesia passed a law that cities must secure at least 30% of their footprint for urban greenspace. While this is difficult to achieve in older, more dense cities in the country, new urban settlements have been earmarked to secure at least this amount of original land in the city as they move towards more ecologically functional and resilient cities (Arifin and Nakagoshi 2011). The current growth and development of new urban settlements, and expansion of urban settlement in the Global South presents an opportunity to secure indigenous biodiversity. Pauchard et al. (2006) make a strong case for the inclusion of biological considerations in city development and planning of cities in the Global South. Sing et al. (2016), presenting case studies of cities in South Korea, are candid in their reminder that urban planning in the Global South is not always guaranteed when they say, "Urban planning, when feasible". The rate of urbanisation combined with factors such as weak governance, lack of capacity and political instability often mean cities are settled without planning (Myers 2021).

Many cities in the Global South are characterised by high informality (de Satgé and Watson 2018; Myers 2021). This has various implications for biodiversity. Informal settlements may be established in sensitive habitats, as demonstrated in cities such as Cape Town, Bangalore, and Sao Paulo (Goodness and Anderson 2013; Nagendra et al. 2012; Pauchard and Barbosa 2013) with negative biodiversity impacts and the exposure of urban residents to higher environmental risks. Conversely, residents of informal settlements, typically the urban poor, are highly dependent on locally procured ecosystem services, and use the natural environment for foraging, livelihoods and as an extension of their own shelters (Petersen et al. 2012; Shackleton et al. 2017). These aspects mean elements of biodiversity and associated plant communities are sometimes fostered in these environments. Anderson et al. (2020), Nagendra (2016) and Pauchard and Barbosa (2013) note that informal settlements hold more indigenous biodiversity than the more homogenised wealthier suburban gardens. This biodiversity resource is potentially a benefit to those communities who suffer informality and often lack easy access to the more traditionally designed nature typical of cities of the Global North. While the concept of green infrastructure has been best described and explored in the urban ecology literature emanating from the Global North, Lindley et al. (2018) note that it is in fact the direct benefits of urban green infrastructure that are most evidently felt in the Global South where poverty and lack of formal infrastructure and service delivery mean the demand on local urban nature and biodiversity is greater.

5.1 Socio-Economic and Cultural Factors

Socio-economic gradients (Pickett and Cadenasso 2006), and cultural factors (Cook et al. 2012) have been demonstrated as determinants of urban biodiversity in numerous cities around the world. Many cities in the Global South are defined, among other factors, by inequality and poverty (de Satgé and Watson 2018; Nagendra et al. 2018), presenting much more pronounced socio-economic gradients, or generally a poorer socio-economic setting (Shackleton et al. 2021). While gradients tend to show a decline in diversity with socio-economic status (Pickett and Cadenasso 2006; Anderson et al. 2020), the so-called luxury effect (Hope et al. 2006), this is not always the case (Leong et al. 2018). Cilliers et al. (2012) for example, show that some poorer households living on city margins retain a higher number of indigenous plant species in their yards for food and medicine, contributing to higher species richness. The natural environment in developing world cities is often recognised as important for livelihood support, for example in the collection and sale of fuelwood or natural medicinal products (Petersen et al. 2012; Schlesinger et al. 2015), and as foraging sites for food security (Anderson et al. 2013; Garekae and Shackleton 2020). In remote hilly areas such as the trans-Himalayan mountains, where communities are dependent on local biodiversity to supply much of their needs, even a relatively minor increase in urbanisation can lead to a decrease in the dependence on local species, and an increased use of imported ecosystem services from distant areas (Murali et al. 2019). These diverse practices can be dependent on extant indigenous plants or introduced species that are grown and fostered in the city (Petersen et al. 2012; Nagendra 2016), with variable consequences for biodiversity.

Naturally, cities in the Global South have different, and diverse, cultural practices that set them apart from the Global North (Cocks and Shackleton 2021). Many spiritual and religious practices in India and countries in Africa require natural or semi-natural spaces for worship (Okyerefo and Fiaveh 2017; Ngulani and Shackleton 2019), and this can contribute to the protection of these spaces to the benefit of biodiversity. In the city of Bangalore in India, sacred sites in cities have higher indigenous plant diversity, in particular tree diversity, than other green areas or open spaces of the city (Gopal et al. 2018; Jaganmohan et al. 2018) (Fig. 3). Remnant forest patches serve as important sacred sites in many African cities. For example, in Ghana, natural forests in cities are important for worship, and in seeking solace and support for mental health issues and healing (Okyerefo and Fiaveh 2017). In South Africa, ancestors are believed to reside in forests, and remnant patches are critical sites for communicating with ancestors (Cocks et al. 2016). In their paper exploring perceptions and value for biodiversity in cities, Chen and Jim (2010) found only cultural aspects to be a useful predictor of willingness to pay.

Fig. 3 Sacred trees in the city of Bangalore, India (Photograph: H Nagendra)

In their work in Oaxaca and Mexico, Robson and Berkes (2011) make a strong case in support of the view that indigenous cultural worldviews and practices should be seen as an agent of landscape renewal that fosters biological and cultural diversity (see also chapters in Cocks and Shackleton 2021). There is a push to embrace cultural practices as part of national and international conservation agendas where these should be formally recognised in their potential contribution to securing urban biodiversity.

5.2 *Local Land Management*

In urban areas biodiversity patterns are also manipulated at the scale of the individual parcels, i.e. gardens. Urban gardens can be a significant contributor to plant diversity as they cover a large part of the urban green infrastructure in most cities and provide multiple ecosystem services (Cameron et al. 2012; Cilliers et al. 2018b). Although a larger focus has initially been placed on biodiversity studies of gardens in the Global North (e.g. Smith et al. 2006) several studies have recently been completed in the Global South, for example in Brazil (e.g. Akinnifesi et al. 2010), China (e.g. Clarke et al. 2014), India (e.g. Jaganmohan et al. 2018) and sub-Saharan Africa (e.g. Bigirimana et al. 2012; Lubbe et al. 2011) to name a few. Gardens in cities can be government-owned (e.g. food, botanical and health clinic gardens) but are mostly privately owned (e.g. home, domestic, medicinal, community) (Cilliers et al. 2018b). Most private gardens are not explicitly managed to increase biodiversity as they are needs-driven (Clarke et al. 2014; Molebatsi et al. 2010) and provide various provisioning, regulating, cultural and supporting ecosystem services (see Cilliers et al. 2018b for an overview of ecosystem services provided by gardens in sub-Saharan Africa).

One reason why the importance of plant diversity in gardens is often overlooked, is the perception of low species diversity and a large number of alien plant species that may become invasive. Private gardens could be a source for plant invasions (Alston and Richardson 2006). Lubbe et al. (2011) found that 10% of the garden species in Potchefstroom, South Africa, are declared invaders, while Bigirimana et al. (2012) reported that 27% of the garden species in Bujumbura, Burundi, are invasive in other countries, for example 11 of these species are invasive in the Kruger National Park in South Africa (Foxcroft et al. 2008). However, several studies from the Global South have also shown that gardens may have a high plant diversity (e.g. Clarke et al. 2014; Jaganmohan et al. 2012; Lubbe et al. 2011), mainly due to the large pool available for species selection and the elimination of environmental stress due to maintenance and management practices (Lubbe et al. 2010). Private gardens often have a higher plant species diversity than other land-uses (Lubbe et al. 2010) and include many indigenous plant species (Akinnifesi et al. 2010; Albuquerque et al. 2005). Indigenous and even endangered plant species can therefore be protected in private gardens, mostly unintentionally. A total number of 25 indigenous tree species which are representative of the surrounding natural vegetation ("caatinga") occur frequently in most home gardens studied in Alagoinha, Brazil (Albuquerque et al. 2005). Although only 30% of the 835 plant species in 100 gardens in Potchefstroom were indigenous, 61 species were endemic to South Africa and 18 species had a threat status according to the South African National Red Data list (Lubbe et al. 2011).

Urban garden plant diversity is determined by specific garden management practices that can be driven by various social, cultural and socio-economic factors (Cilliers et al. 2018a; Lubbe et al. 2010; Davoren et al. 2016). As indicated in the previous section there is often a positive correlation ("luxury effect") between plant diversity (indigenous and aliens) and the socio-economic status of the garden owners (Bernholt et al. 2009; Clarke et al. 2014; González-Ball et al. 2017; Lubbe et al. 2010). There may also be no correlation (neutral) as was indicated by garden studies in Brazil (Eichemberg et al. 2009) and Puerto Rico (Melendez-Ackerman et al. 2014), especially for alien plant species. The "luxury effect" is often eliminated by differences in the context of the natural environment and economics across cities (Leong et al. 2018; Melendez-Ackerman et al. 2014). All these correlations also exist in cities in the Global North. The plant species diversity-socio-economics correlation can even sometimes be negative (see Cilliers et al. 2012). In terms of cultural aspects the layout and structure of the gardens may determine plant diversity. Molebatsi et al. (2010) described a unique layout of the home gardens of the Batswana people in South Africa ("Tswana tshimo") in which various micro-gardens (e.g. ornamental, vegetable, fruit trees, medicinal, natural remnant) are increasing the plant species richness and have more indigenous and utilitarian species. The large plant species richness in home gardens in Rio Claro in Brazil was also ascribed to the cultural background of the homeowners who came mostly from rural areas (Eichemberg et al. 2009). However, home gardens studies in South Africa have shown that socio-economic status overrides cultural preferences in Batswana home gardens, as an increase in socio-economic status has led to a switch from tshimo garden design to a more Westernised design type (also with more alien and ornamental species) (Davoren et al. 2016) (Fig. 4). In Niamey, Niger and Hawassa City, Ethiopia, high plant species richness of gardens was determined by the presence of specific ethnic groups as well as factors such as garden size, gender of the gardener and socio-economic factors (Bernholt et al. 2009; Regassa 2016). In Beijing, China, the "hierarchy of needs", local agricultural traditions and yard size were described as the main drivers for garden plant diversity (Clarke et al. 2014).

Interestingly, a review of domestic garden studies in the Global North focuses on a variety of ecosystem services such as temperature regulation, carbon sequestration, water runoff reduction, air pollutant removal, increase of human health and wellbeing, increasing of social capital, etc. (Cameron et al. 2012). Although these ecosystem services are also important for private gardens in the Global South, many garden studies have focused more on urban agriculture and agroforestry in urban and rural areas (e.g. Albuquerque et al. 2005; Bernholt et al. 2009; Pandey et al. 2007). Several of these and other garden studies have also indicated that the large numbers of alien species could be the result of a large food plant diversity, for example 55% of the garden plant species in Niamey, Niger (Bernholt et al. 2009) and 79% in Săo Luis City in Brazil (Akinnifesi et al. 2010) are food plants, including fruit and nut trees and arable crops such as grains and vegetables. Although this situation could be challenging for the conservation of indigenous plant diversity, these gardens "could provide adaptive capacity to food shortages" according to Vila-Ruiz et al. (2014). Multi-layered tropical home gardens in Africa and Latin America are therefore often regarded as agroforestry systems in which a large genetic diversity of different food species and local cultivars can be preserved (Agbogidi and Adolor 2013; Akinnifesi et al. 2010).

Another local, though the less bounded influence on urban biodiversity is urban agriculture, in particular livestock husbandry, which is common practice in many cities of the Global South (Drescher et al. 2021). Opitz et al. (2016) explore the re-emergence of urban agriculture in the Global North and note that where food security concerns and associated urban agricultural responses are becoming increasingly relevant in the Global North, these are features of Global South cities that "never went away". The practice is often shunned, or considered illegal, by officials where the global view of what constitutes a city does not allow for the presence of livestock (Schiere et al. 2006). However, small-scale farming for livelihoods, food security and culture and identity purposes are well described in cities of the Global South (Lerner and Eakin 2011; Drescher et al. 2021). Global South cities are often described as peri-urban with densification of previously rural settlements (McHale et al. 2013), or historical villages are "swallowed" by city expansion (Shah 2012) and often agriculture is a remnant practice. Lerner and Eakin

Fig. 4 **a** A typical "Tswana tshimo" garden in South Africa with characteristic bare areas and several different micro gardens—small flower beds can be seen here with a vegetable garden in the background (Photograph: S. S. Cilliers). **b** A more westernized garden design is used by Batswana gardeners with a higher socioeconomic status in South Africa deviating from the "Tswana tshimo" garden design type. Several exotic ornamental species are included in these gardens as well as a lawn (Photograph: S. S. Cilliers)

(2011) describe a spatial configuration in cities of the Global South of a fusion of livelihoods with agriculture practised in the backyard by the older generations, while the youth exit through the front door in search of more modern careers or pursuits. While their intent is somewhat metaphorical, the truth is urban livestock keeping in the Global South can take place in a number of spaces ranging from private yards, to more public spaces such as commons or remnant spaces like road verges. The implications of livestock in cities, while mostly recorded for the Global North, are wide-ranging where cattle must be fed and watered, impact local soils through trampling, damage local trees and produce large amounts of waste (Opitz et al. 2016). Increasingly there is a concern for the wildlife–livestock–human interface as a complex multi-host community for novel pathogen transmission in cities (Hassell et al. 2017) that warrants closer attention.

6 Forward Glancing: Urban Biodiversity Management and Conservation

Cities of the Global South, in most cases, have the advantage of more intact natural remnant plant biodiversity than their Global North counterparts. However, this is under considerable development pressure with rapid, and often unplanned, development. In addition, social imperatives relating to systemic poverty, informality and unemployment pose further threats. These same features, of poverty and informality, mean natural remnants often present a livelihood service and this could in turn guarantee their conservation. However, as cities move from a production to a consumption mode, and achieve their development goals, and before an established urban middle-class emerges, urban biodiversity is often not prioritised. Cultural and religious use of natural spaces present an opportunity to ensure the conservation of features of urban plant diversity and these should be fostered. Although private gardens are also important in the Global North, research on gardens in the Global South does indicate their importance as part of the urban green infrastructure, not only in terms of conservation of indigenous plant diversity and the protection of genetic and crop diversity which is important for food resilience, but also as a source of plant invasions which are detrimental to the conservation of natural habitats.

Effective environmental governance is critical, but not a strength of the Global South, and certainly is not likely to look the same as it does in the Global North. City planners and managers in the Global South are under considerable pressure to address immediate and pressing social pressures and development goals (Shackleton et al. 2021). While there is a call for the systematic inclusion of biological considerations in planning, Sing et al. (2016) and Cilliers (2018) note the challenges, and frequent absence, of ecological and green infrastructure planning in the Global South. What will be required is an augmentation of traditional conservation approaches, and likely a novel, blended approach to securing biodiversity, and one that embraces the high utility of the local natural environment to many urban citizens in the Global South. Climate change poses significant risks to city sustainability and resilience in the Global South, and this presents a further useful opportunity to motivate for the conservation or restoration of biodiversity in these cities. In Indian cities, urban commoning approaches appear to hold promise for ecosystem restoration, with urban communities coming together to protect and restore ecosystems ranging from indigenous forests to wetlands, lakes and landscaped city parks, using approaches from street protests and street theatre to civic awareness campaigns, public interest litigations in court and collaboration with municipal governments for restoration (Nagendra 2016). Although sometimes criticised for their focus on the recreational use of ecosystems, and on landscaping that promotes alien biodiversity and excludes the poor, such citizen movements hold promise for an alternative, bottom-up approach towards conservation in cities of the Global South (Nagendra 2016). Another positive trend is emerging in China, one of the megadiverse countries, which has recently made some major governance changes, including the establishment of a new ministry named Ministry of Ecology and Environment, in taking definitive steps towards green development and "ecological civilisation" (Xu et al. 2019). These cases are useful examples of successful engagements, and shifts in engaging society and government and other countries in the Global South should look to these as useful successful examples. Biodiversity does not recognise boundaries and given the evident influence of wider ecologies it is important that a regional approach be adopted where sharing of experiences and knowledge must be encouraged and facilitated. There are numerous useful organisations in different regions, and across regions, that can be used towards knowledge sharing such as the Association of SouthEast Asian Nations (ASEAN) Centre for Biodiversity, an intergovernmental regional centre that facilitates co-operation among members in the region towards biodiversity conservation, ICLEI who work with local governments globally on sustainability agendas who have a Cities with Nature branch and are active in cities of the Global South, and URBIS who are a global network promoting collaboration in improving urban form and design for biodiversity potential and ecosystem services and have an active Global South membership (ICLEI, n.d.). The 15th Meeting of the Conference of the Parties to the Convention on Biological Diversity is poised to adopt a post-2020 global biodiversity framework that will guide the road towards the 2050 Vision of "Living in Harmony with Nature". With more than 55% of the world's population living in urban areas and with a projection of close to 90% of the increase in urban population taking place in Asia and Africa (UN DESA 2018), it is imperative that biodiversity conservation must be fostered in these growth areas. While we must still build case studies and generate more inventory data, it is evident both through how biodiversity currently manifests in cities in the Global South, and the very different nature of the filters when compared to how these play out in the Global North, that simply applying biodiversity conservation solutions generated in the Global North is not a useful solution in this very different context. Novel, context-specific, solutions will be needed to secure urban biodiversity in the Global South.

References

Abendroth S, Kowarik I, Müller N, von der Lippe M (2012) The green colonial heritage: woody plants in parks of Bandung, Indonesia. Landsc Urb Plan 106:12–22

Agbogidi OM, Adolor EB (2013) Home gardens in the maintenance of biological diversity. Appl Sci Rep 1:19–25

Akinnifesi FK, Sileshi G, Da Costa J, De Moura EG, Da Silva RF, Ajayi OC et al (2010) Floristic composition and canopy structure of home-gardens inSão Luís city, Maranhão State, Brazil. J Hortic For 2:72–86

Albuquerque UP, Andrade LH, Caballero J (2005) Structure and floristics of homegardens in Northeastern Brazil. J Arid Environ 62:491–506

Alston KP, Richardson DM (2006) The roles of habitat features, disturbance, and distance from putative source populations in structuring alien plant invasions at the urban/wildland interface on the Cape Peninsula, South Africa. Biol Conserv 132:183–198

Anderson P, O'Farrell P (2012) An ecological view of the history of the City of Cape Town. Ecol Soc 17(3)

Anderson PM, Okereke C, Rudd A, Parnell S (2013) Regional assessment of Africa. In: Elmqvist T, Fragkias M, Goodness J, Guneralp J, Marcotullio PJ, McDonald RI, Parnell S, Sendstad M, Schewenius M Seto KC, Wilkinson C (eds) Urbanisation, biodiversity and ecosystem services: challenges and opportunities. Springer, New York, pp 453–459

Anderson P, Charles-Dominique T, Ernstson H, Andersson E, Goodness J, Elmqvist T (2020) Post-apartheid ecologies in the city of Cape Town: an examination of plant functional traits in relation to urban gradients. Landsc Urb Plan 193:103662

Arifin HS, Nakagoshi N (2011) Landscape ecology and urban biodiversity in tropical Indonesian cities. Landsc Ecol Eng 7:33–43

Aronson MF, Nilon CH, Lepczyk CA, Parker TS, Warren PS, Cilliers SS, Goddard MA, Hahs AK, Herzog C, Katti M, La Sorte FA (2016) Hierarchical filters determine community assembly of urban species pools. Ecol 97:2952–2963

Benchimol JF, do Nascimento Lamano-Ferreira AP, Ferreira ML, Cortese TT, Ramos HR (2017) Decentralised management of public squares in the city of São Paulo, Brazil: implications for urban green spaces. Land Use Pol 63:418–427

Bernholt H, Kehlenbeck K, Gebauer J, Buerkert A (2009) Plant species richness and diversity in urban and peri-urban gardens of Niamey, Niger. Agrofor Sys 77:159–179

Bigirimana J, Bogaert J, De Canniere C, Bigendako M-J, Parmentier I (2012) Domestic garden plant diversity in Bujumbura, Burundi: role of the socio-economical status of the neighborhood and alien species invasion risk. Landsc Urb Plan 107:118–126

Cameron RW, Blanuša T, Taylor JE, Salisbury A, Halstead AJ, Henrico B, Thompson K (2012) The domestic garden: its contribution to urban green infrastructure. Urban For Urban Green 11:129–137

Chen WY, Jim CY (2010) Resident motivations and willingness-to-pay for urban biodiversity conservation in Guangzhou (China). Environ Manage 45:1052–1064

Cilliers EJ (2018) Reflecting on green infrastructure and spatial planning in Africa: the complexities, perceptions and way forward. Sustain 11:455

Cilliers SS, Siebert SJ, Davoren E, Lubbe CS (2012) Social aspects of urban ecology in developing countries, with an emphasis on urban domestic gardens. In: Richter M, Weiland U (eds) Urban ecology: a global framework. Blackwell, New York, pp 123–138

Cilliers SS, Siebert SJ, Du Toit MJ, Barthel S, Mishra S, Cornelius A, Davoren E (2018a) Garden ecosystem services of Sub-Saharan Africa and the role of health clinic gardens as social-ecological systems. Landsc Urb Plan 180:294–307

Cilliers SS, Siebert SJ, Du Toit MJ, Davoren E (2018b) Managing urban green spaces for biodiversity conservation: An African perspective. In: Ossola A, Niemelä J (eds) Urban biodiversity: from research to practice. Earthscan, London, pp 167–185

Clarke LW, Li L, Jenerette GD, Yu Z (2014) Drivers of plant diversity and ecosystem services production in home gardens across the Beijing Municipality of China. Urban Ecosys 17:741–760

Cocks ML, Shackleton CM (2021) Diversities of nature in cities: enriching belonging, wellbeing and bioculture. Earthscan and Routledge, London

Cocks M, Alexander J, Mogano L, Vetter S (2016) Ways of belonging: meanings of "nature" among Xhosa-speaking township residents in South Africa. J Ethnobiol 36:820–842

Cook EM, Hall SJ, Larson KL (2012) Residential landscapes as social-ecological systems: a synthesis of multi-scalar interactions between people and their home environment. Urban Ecosys 15:19–52

Davison GW, Ng PK, Ho HC (eds) (2008) The Singapore red data book: threatened plants and animals of Singapore, 2nd edn. Nature Society, Singapore

Davoren E, Shackleton CM (2021) Urban ecosystem disservices in the Global South. In: Shackleton CM, Cilliers SS, Davoren E, du Toit MJ (eds). Urban ecology in the Global South. Springer, Cham, pp 265–292

Davoren E, Siebert SJ, Cilliers SS, Du Toit MJ (2016) Influence of socioeconomic status on design of Batswana home gardens and associated plant diversity patterns in northern South Africa. Landsc Ecol Eng 12:129–139

de Satgé R, Watson V (2018) Urban planning in the Global South: conflicting rationalities in contested urban space. Springer, Berlin

Dickie IA, Bennett BM, Burrows LE, Nunez MA, Peltzer DA, Porté A, Richardson DM, Rejmánek M, Rundel PW, van Wilgen BW (2014) Conflicting values: ecosystem services and invasive tree management. Biol Inv 16:705–719

Drescher AW, Isendahl C, Caridad Cruz M, Karg H, Menakanit A (2021) Urban and peri-urban agriculture in the Global South. In: Shackleton CM, Cilliers SS, Davoren E, du Toit MJ (eds). Urban ecology in the Global South. Springer, Cham, pp 293–324

Duncan RP, Clemants SE, Corlett RT, Hahs AK, McCarthy MA, McDonnell MJ, Schwartz MW, Thompson K, Vesk PA, Williams NS (2011) Plant traits and extinction in urban areas: a meta-analysis of 11 cities. Glob Ecol Biogeog 20:509–519

Eichemberg MT, Amorozo MC, Moura LC (2009) Species composition and plant use in old urban homegardens in Rio Claro, Southeast of Brazil. Acta Botanica Brasilica 23:1057–1075

Elmqvist T, Folke C, Nyström M, Peterson G, Bengtsson J, Walker B, Norberg J (2003) Response diversity, ecosystem change, and resilience. Front Ecol Environ 1:488–494

Escobedo FJ (2021) Understanding urban regulating ecosystem services in the Global South. In: Shackleton CM, Cilliers SS, Davoren E, du Toit MJ (eds). Urban ecology in the Global South. Springer, Cham, pp 227–244

Foxcroft LC, Richardson DM, Wilson JR (2008) Ornamental plants as invasive aliens: Problems and solutions in Kruger National Park, South Africa. Environ Manage 41:32–51

Gaertner M, Larson BM, Irlich UM, Holmes PM, Stafford L, Van Wilgen BW, Richardson DM (2016) Managing invasive species in cities: a framework from Cape Town, South Africa. Landsc Urb Plan 93:184–193

Garekae H, Shackleton CM (2020) Foraging wild food in urban spaces: the contribution of wild foods to urban dietary diversity in South Africa. Sustain 12:678

Gaston KJ, Ávila-Jiménez ML, Edmondson JL (2013) Managing urban ecosystems for goods and services. J Appl Ecol 50:830–840

González-Ball R, Bermúdez-Rojas T, Romero-Vargas M (2017) Floristic composition and richness of urban domestic gardens in three urban socioeconomic stratifications in the city Heredia, Costa Rica. Urban Ecosys 20:51–63

Goodness J, Anderson PM (2013) Local assessment of Cape Town: navigating the management complexities of urbanisation, biodiversity, and ecosystem services in the cape floristic region. In: Elmqvist T, Fragkias M, Goodness J, Guneralp J, Marcotullio PJ, McDonald RI, Parnell S, Sendstad M, Schewenius M Seto KC, Wilkinson C (eds) Urbanisation, biodiversity and ecosystem services: challenges and opportunities. Springer, Dordrecht, pp 461–484

Goodness J, Andersson E, Anderson PM, Elmqvist T (2016) Exploring the links between functional traits and cultural ecosystem services to enhance urban ecosystem management. Ecol Indic 70:597–605

Gopal D, Manthey M, Nagendra H (2015) Vegetation in Bangalore's slums: composition, species distribution, density, diversity and history. Environ Manage 55:1390–1401

Gopal D, von der Lippe M, Kowarik I (2018) Sacred sites as habitats of culturally important plant species in an Indian megacity. Urban For Urban Green 32:113–122
Güneralp B, Seto KC (2013) Futures of global urban expansion: uncertainties and implications for biodiversity conservation. Environ Res Lett 8:014025
Güneralp B, Lwasa S, Masundire H, Parnell S, Seto KC (2017) Urbanisation in Africa: challenges and opportunities for conservation. Environ Res Lett 13:015002
Hassell JM, Begon M, Ward MJ, Fèvre EM (2017) Urbanisation and disease emergence: dynamics at the wildlife-livestock-human interface. Trends Ecol Evol 32:55–67
Hobbs RJ, Higgs ES, Hall CM (2013) Defining novel ecosystems. In: Hobbs RJ, Higgs ES, Hall CM (eds) Novel ecosystems: intervening in the new ecological world order. Wiley, Chichester, pp 58–60
Hope D, Gries C, Zhu W, Fagan WF, Redman CL, Grimm NB, Nelson AL, Martin C, Kinzig A (2006) Socioeconomics drive urban plant diversity. Proc Nat Acad Sci 100:8788–8792
Hunte N, Roopsind A, Ansari AA, Caughlin TT (2019) Colonial history impacts urban tree species distribution in a tropical city. Urban For Urban Green 41:313–322
ICLEI Local Governments for Sustainability (n.d.) Urban agendas and biodivercity. http://old.iclei.org/index.php
IPBES (2018) Summary for policymakers of the regional assessment report on biodiversity and ecosystem services for Africa of the Intergovernmental Science-Policy Platform on Biodiversity and Ecosystem Services. IPBES secretariat, Bonn, Germany, 49 p. ISBN No: 978-3-947851-00-3
Irlich UM, Potgieter LJ, Stafford L, Gaertner M (2017) Recommendations for municipalities to become compliant with national legislation on biological invasions. Bothalia 47:a2156
Ives CD, Kelly AH (2016) The coexistence of amenity and biodiversity in urban landscapes. Landsc Res 41:495–509
Jaganmohan M, Vailshery LS, Gopal D, Nagendra H (2012) Plant diversity and distribution in urban domestic gardens and apartments in Bangalore, India. Urban Ecosys 15:911–925
Jaganmohan M, Vailshery LS, Mundoli S, Nagendra H (2018) Biodiversity in sacred urban spaces of Bengaluru, India. Urban For Urban Green 32:64–70
Kaoma H, Shackleton CM (2015) The direct use value of urban tree non-timber forest products to household income in poorer suburbs in South African towns. For Pol Econ 61:104–112
Kowarik I, von der Lippe M, Cierjacks A (2013) Prevalence of alien versus native species of woody plants in Berlin differs between habitats and at different scales. Preslia 85:113–132
Kühn I, Klotz S (2006) Urbanisation and homogenisation—comparing the floras of urban and rural areas in Germany. Biol Conserv 127:292–300
Lahr E, Dunn RR, Frank SD (2018) Getting ahead of the curve: cities as surrogates for global change. Proc R Soc B 285:20180643
Leong M, Dunn RR, Trautwein MD (2018) Biodiversity and socioeconomics in the city: a review of the luxury effect. Biol Lett 14:20180082
Leong-Skornickova J, Thame A, Chew PT (2014) Notes on Singapore native Zingiberales I: a new species of Zingiber and notes on the identities of two further Zingiber taxa. Gard Bull Singapore 66:133–167
Lerner AM, Eakin H (2011) An obsolete dichotomy? Rethinking the rural-urban interface in terms of food security and production in the global south. Geogr J 177:311–320
Lim RC, Lindsay J, Middleton D, Ho BC, Leong PK, Niissalo MA, van Welzen PC, Esser H-J, Ganesan SK, Lua HK, Johnson DM, Murray NA, Leong-Skornickova J, Thomas DC, Ibrahim A (2018) New records and rediscoveries of plants in Singapore. Gard Bull Singapore 70:67–90
Lindley S, Pauleit S, Yeshitela K, Cilliers SS, Shackleton C (2018) Rethinking urban green infrastructure and ecosystem services from the perspective of sub-Saharan African cities. Landsc Urb Plan 180:328–338
Lubbe CS, Siebert SJ, Cilliers SS (2010) Political legacy of South Africa affects the plant diversity patterns of urban domestic gardens along a socioeconomic gradient. Sci Res Essays 5:2900–2910
Lubbe CS, Siebert SJ, Cilliers SS (2011) A floristic analysis of domestic gardens in the Tlokwe Municipal area, South Africa. Bothalia 41:353–363
McDonnell MJ, Hahs AK (2013) The future of urban biodiversity research: Moving beyond the 'low-hanging fruit'. Urban Ecosys 16(3): 397–409

McDonnell MJ, Hahs AK (2015) Adaptation and adaptedness of organisms to urban environments. Ann Rev Ecol Evol Syst 46:261–280
McHale MR, Bunn DN, Pickett ST, Twine W (2013) Urban ecology in a developing world: why advanced socioecological theory needs Africa. Front Ecol Environ 11:556–564
McKinney ML (2006) Urbanisation as a major cause of biotic homogenisation. Biol Conserv 127:247–260
McLean P, Gallien L, Wilson JR, Gaertner M, Richardson DM (2017) Small urban centres as launching sites for plant invasions in natural areas: insights from South Africa. Biol Inv 1(19):3541–3555
Miller JR, Hobbs RJ (2002) Conservation where people live and work. Conserv Biol 16:330–337
Melendez-Ackerman EJ, Santiago-Bartolomei R, Vila-Ruiz CP, Santiago LE, Garcia-Montiel D, Verdejo-Ortiz JC, Manrique-Hernández H, Hernandez-Calo E (2014) Socioeconomic drivers of yard sustainable practices in a tropical city. Ecol Soc 19(3):20
Molebatsi LY, Siebert SJ, Cilliers SS, Lubbe CS, Davoren E (2010) TheTswana tshimo: a homegarden system of useful plants with a particular lay out and function. Afr J Agric Res 5:2952–2963
Murali R, Suryawanshi K, Redpath S, Nagendra H, Mishra C (2019) Changing use of ecosystem services along a rural-urban continuum in the Indian Trans-Himalayas. Ecosys Serv 40:101030
Myers G (2003) Verandahs of power: colonialism and space in urban Africa. Syracuse University Press, Syracuse
Myers G (2021) Urbanisation in the Global South. In: Shackleton CM, Cilliers SS, Davoren E, du Toit MJ (eds). Urban ecology in the Global South. Springer, Cham, pp 27–49
Nagendra H (2016) Nature in the city: Bengaluru in the past, present, and future. Oxford University Press, Oxford
Nagendra H, Gopal D (2011) Tree diversity, distribution, history and change in urban parks. Urban Ecosys 14:211–223
Nagenda H, Mundoli S (2019) Cities and canopies: trees in Indian cities. Penguin Random House, New Delhi
Nagendra H, Sudhira HS, Katti M, Tengö M, Schewenius M (2012) Urbanisation, ecosystems and biodiversity: assessments of India and Bangalore. Cities for Life Summit, parallel to the eleventh meeting of the Conference of the Parties to the Convention on Biological Diversity (CBD), 15 October 2012, Hyderabad, India. https://www.econstor.eu/bitstream/10419/191589/1/978-94-007-7088-1.pdf
Nagendra H, Bai X, Brondizio ES, Lwasa S (2018) The urban south and the predicament of global sustainability. Nat Sustain 1:341–349
Nentwig W, Mebs D, Vilà M (2017) Impact of non-native animals and plants on human health. In: Vilà M, Hulme P (eds) Impact of biological invasions on ecosystem services. Springer, Cham, pp 277–294
Ngulani T, Shackleton CM (2019) Use of public urban green spaces for spiritual services in Bulawayo, Zimbabwe. Urban For Urban Green 38:97–104
Novoa A, Shackleton R, Canavan S, Cybele C, Davies SJ, Dehnen-Schmutz K, Fried J, Gaertner M, Geerts S, Griffiths CL, Kaplan H (2018) A framework for engaging stakeholders on the management of alien species. J Environ Manage 205:286–297
Okyerefo MP, Fiaveh DY (2017) Prayer and health-seeking beliefs in Ghana: understanding the 'religious space' of the urban forest. Health Sociol Rev 26:308–320
Olden JD, Poff NL (2003) Towards a mechanistic understanding and prediction of biotic homogenisation. Am Nat 162:442–460
Oliver TH, Heard MS, Isaac NJ, Roy DB, Procter D, Eigenbrod F, Freckleton R, Hector A, Orme CD, Petchey OL, Proença V (2015) Biodiversity and resilience of ecosystem functions. Trends Ecol Evol 30:673–684
Opitz I, Berges R, Piorr A, Krikser T (2016) Contributing to food security in urban areas: differences between urban agriculture and peri-urban agriculture in the Global North. Agric Hum Values 33:341–358
Pandey CB, Rai RB, Singh L, Singh AK (2007) Home gardens of Andaman and Nicobar, India. Agric Syst 92:1–22

Pauchard A, Barbosa O (2013) Regional assessment of Latin America: rapid urban development and social economic inequity threaten biodiversity hotspots. In: Elmqvist T, Fragkias M, Goodness J, Guneralp J, Marcotullio PJ, McDonald RI, Parnell S, Sendstad M, Schewenius M Seto KC, Wilkinson C (eds) Urbanisation, biodiversity and ecosystem services: challenges and opportunities. Springer, Dordrecht, pp 589–608

Pauchard A, Aguayo M, Peña E, Urrutia R (2006) Multiple effects of urbanisation on the biodiversity of developing countries: the case of a fast-growing metropolitan area (Concepción, Chile). Biol Conserv 127:272–281

Pena JC, Magalhães DM, Mourao AC, Young RJ, Rodrigues M (2016) The green infrastructure of a highly-urbanised Neotropical city: the role of the urban vegetation in preserving native biodiversity. Capa 11:66–78

Petersen LM, Moll EJ, Collins R, Hockings MT (2012) Development of a compendium of local, wild-harvested species used in the informal economy trade, Cape Town, South Africa. Ecol Soc 17:26

Pickett ST, Cadenasso ML (2006) Advancing urban ecological studies: frameworks, concepts, and results from the Baltimore Ecosystem Study. Aust Ecol 31:114–125

Potgieter LJ, Gaertner M, Kueffer C, Larson BM, Livingstone SW, O'Farrell PJ, Richardson DM (2017) Alien plants as mediators of ecosystem services and disservices in urban systems: a global review. Biol Inv 19:3571–3588

Potgieter LJ, Gaertner M, Irlich UM, O'Farrell PJ, Stafford L, Vogt H, Richardson DM (2018) Managing urban plant invasions: a multi-criteria prioritisation approach. Environ Manage 62:1168–1185

Potgieter LJ, Gaertner M, O'Farrell PJ, Richardson DM (2019a) Does vegetation structure influence criminal activity? Insights from Cape Town, South Africa. Front Biogeogr 11:e42035

Potgieter LJ, Gaertner M, O'Farrell PJ, Richardson DM (2019b) Perceptions of impact: invasive alien plants in the urban environment. J Environ Manage 229:76–87

Potgieter LJ, Douwes E, Gaertner M, Measey GJ, Paap T, Richardson DM (2020) Biological invasions in South Africa's urban ecosystems: patterns, processes, impacts and management. In: van Wilgen BW, Measey GJ, Wilson JR, Zengeya T (eds) Biological invasions in South Africa. Springer, Cham, pp 275–309

Purvis A, Hector A (2000) Getting the measure of biodiversity. Nat 405:6783

Pyšek P (1998) Alien and native species in Central European urban floras: a quantitative comparison. J Biogeogr 25:155–163

Pyšek P, Richardson DM (2010) Invasive species, environmental change and management, and health. Ann Rev Environ Res 35:25–55

Rebelo AG, Holmes PM, Dorse C, Wood J (2011) Impacts of urbanisation in a biodiversity hotspot: conservation challenges in Metropolitan Cape Town. S Afr J Bot 77:20–35

Regassa R (2016) Useful plant species diversity in home gardens and its contribution to household food security in Hawassa city, Ethiopia. Afr J Plant Sci 10:211–233

Reis JN, Palma T, Ribeiro GS, Pinheiro RM, Ribeiro CT, Cordeiro SM, da Silva Filho HP, Moschioni M, Thompson TA, Spratt B, Riley LW (2008) Transmission of *Streptococcus pneumoniae* in an urban slum community. J Infect 57:204–213

Reiss J, Bridle JR, Montoya JM, Woodward G (2009) Emerging horizons in biodiversity and ecosystem functioning research. Trends Ecol Evol 24:505–514

Robson JP, Berkes F (2011) Exploring some of the myths of land use change: can rural to urban migration drive declines in biodiversity? Glob Environ Change 21:844–854

Schiere H, Rischkowsky B, Thys E, Schiere J, Matthys F (2006) Livestock keeping in urbanised areas, does history repeat itself. Cities Farming for the Future. Online publication by Resource Centre on Urban Agriculture and Food Security, pp 350–366

Schlesinger J, Drescher A, Shackleton CM (2015) Socio-spatial dynamics in the use of wild natural resources: evidence from six rapidly growing medium-sized cities in Africa. Appl Geogr 56:107–115

Seto KC (2013) Regional assessment of Asia. In: Elmqvist T, Fragkias M, Goodness J, Guneralp J, Marcotullio PJ, McDonald RI, Parnell S, Sendstad M, Schewenius M Seto KC, Wilkinson C (eds). Urbanisation, biodiversity and ecosystem services: challenges and opportunities. Springer, Dordrecht, pp 53–56

Seto KC, Fragkias M, Güneralp B, Reilly MK (2011) A meta-analysis of global urban land expansion. PLoS ONE 6:23777
Seto KC, Güneralp B, Hutyra LR (2012) Global forecasts of urban expansion to 2030 and direct impacts on biodiversity and carbon pools. Proc Nat Acad Sci 109:16083–16088
Shackleton C, Hurley P, Dahlberg A, Emery M, Nagendra H (2017) Urban foraging: ubiquitous but overlooked by urban planners and policy. Sustain 9: 1884
Shackleton CM (2021) Provisioning services in Global South cities. In: Shackleton CM, Cilliers SS, Davoren E, du Toit MJ (eds). Urban ecology in the Global South. Springer, Cham, pp 203–226
Shackleton CM, Cilliers SS, du Toit MJ, Davoren E (2021) The need for an urban ecology of the Global South. In: Shackleton CM, Cilliers SS, Davoren E, du Toit MJ (eds). Urban ecology in the Global South. Springer, Cham, pp 1–26
Shackleton RT, Adriaens T, Brundu G, Dehnen-Schmutz K, Estévez RA, Fried J, Larson BM, Liu S, Marchante E, Marchante H, Moshobane MC (2019a) Stakeholder engagement in the study and management of invasive alien species. J Environ Manage 229:88–101
Shackleton RT, Larson BM, Novoa A, Richardson DM, Kull CA (2019b) The human and social dimensions of invasion science and management. J Environ Manage 229:1–9
Shackleton RT, Richardson DM, Shackleton CM, Bennett B, Crowley S, Dehnen-Schmutz K, Estévez R, Fischer A, Kueffer C, Kull C, Marchante E, Novoa A, Potgieter LJ, Vaas J, Vaz AS, Larson BH (2019c) Explaining people's perceptions of invasive alien species: a conceptual framework. J Environ Manage 229:10–26
Shah AM (2012) The village in the city, the city in the village. Econ Politic Weekly, 17–19
Shochat E, Lerman SB, Anderies JM, Warren PS, Faeth SH, Nilon CH (2010) Invasion, competition, and biodiversity loss in urban ecosystems. BioScience 60:199–208
Silva CP, García CE, Estay SA, Barbosa O (2015) Bird richness and abundance in response to urban form in a Latin American city: Valdivia, Chile as a case study. PLoS ONE 10(9):e0138120
Sing KW, Jusoh WF, Hashim NR, Wilson JJ (2016) Urban parks: refuges for tropical butterflies in Southeast Asia? Urban Ecosys 19:1131–1147
Sitas N, Selomane O, Hamann M, Gajjar SP (2021) Towards equitable urban resilience in the Global South within a context of planning and management. In: Shackleton CM, Cilliers SS, Davoren E, du Toit MJ (eds). Urban ecology in the Global South. Springer, Cham, pp 325–345
Smith RM, Thompson K, Hodgson JG, Warren PH, Gaston KJ (2006) Urban domestic gardens (IX): composition and richness of the vascular plant flora, and implications for native biodiversity. Biol Conserv 129:312–322
Suding KN, Lavorel S, Chapin FS III, Cornelissen JH, Díaz S, Garnier E, Goldberg D, Hooper DU, Jackson ST, Navas M-L (2008) Scaling environmental change through the community-level: a trait-based response-and-effect framework for plants. Glob Change Biol 14:1125–1140
Swamy S, Devy S, Nagendra H (2019) Building biodiversity in neighbourhood parks in Bangalore city, India: ordinary yet essential. PLoS ONE 14(5):e0215525
United Nations Department of Economic and Social Affairs (2018) The 2018 revision of world urbanisation prospects. https://www.un.org/development/desa/publications/2018-revision-of-world-urbanization-prospects.html
van Wilgen BW, Forsyth GG, Le Maitre DC, Wannenburgh A, Kotzé JD, van den Berg E, Henderson L (2012) An assessment of the effectiveness of a large, national-scale invasive alien plant control strategy in South Africa. Biol Conserv 148:28–38
Vaz AS, Kueffer C, Kull CA, Richardson DM, Vicente JR, Kühn I, Schröter M, Hauck J, Bonn A, Honrado JP (2017) Integrating ecosystem services and disservices: insights from plant invasions. Ecosys Serv 23:94–107
Vila-Ruiz CP, Meléndez-Ackerman E, Santiago-Bartolomei R, Garcia-Montiel D, Lastra L, Figuerola CE, Fumero-Caban J (2014) Plant species richness and abundance in residential yards across a tropical watershed: implications for urban sustainability. Ecol Soc 19(3)
von der Lippe M, Kowarik I (2008) Do cities export biodiversity? Traffic as dispersal vector across urban–rural gradients. Div Distrib 14:18–25
Williams NS, Hahs AK, Vesk PA (2015) Urbanisation, plant traits and the composition of urban floras. Persp Plant Ecol Evol Syst 17:78–86
Xu W, Pimm SL, Du A, Su Y, Fan X, Li A, Liu J, Ouyang Z (2019) Transforming protected area management in China. Trends Ecol Evol. https://doi.org/10.1016/j.tree.2019.05.009

Urban Animal Diversity in the Global South

Chevonne Reynolds, Marcus J. Byrne, Dan E. Chamberlain, Caroline G. Howes, Colleen L. Seymour, Petra Sumasgutner, and Peter J. Taylor

Abstract Urban animal ecology is a rapidly growing research area, yielding fascinating insights into the patterns and processes that shape biodiversity in the city. However, much of this research has focused on cities in the developed world, where the mechanisms affecting biodiversity might be very different to those in the developing cities of the Global South. Here we detail how the contemporary cities of the

C. Reynolds (✉) · M. J. Byrne · C. G. Howes
School of Animal, Plant and Environmental Sciences, University of the Witwatersrand, Johannesburg 2001, South Africa
e-mail: Chevonne.reynolds@wits.ac.za

M. J. Byrne
e-mail: Marcus.byrne@wits.ac.za

C. Reynolds · C. L. Seymour · P. Sumasgutner
Fitzpatrick Institute of African Ornithology, DST-NRF Centre of Excellence, University of Cape Town, Rondebosch 7701, South Africa
e-mail: C.Seymour@sanbi.org.za

M. J. Byrne · P. J. Taylor
DST-NRF Centre of Excellence for Invasion Biology, School of Animal, Plant and Environmental Sciences, University of the Witwatersrand, Johannesburg 2001, South Africa
e-mail: Peter.taylor@univen.ac.za

D. E. Chamberlain
Department of Life Sciences and Systems Biology, University of Turin, Via Accademia Albertina 13, 10123 Turin, Italy

C. L. Seymour
South African National Biodiversity Institute, Kirstenbosch Gardens, Private Bag X7, Claremont 7735, South Africa

P. Sumasgutner
Core Facility Konrad Lorenz Research Centre for Behaviour and Cognition, University of Vienna, Fischerau 11, 4645 Grünau Im Almtal, Austria

P. J. Taylor
South African Research Chair on Biodiversity Value and Change, School of Mathematical and Natural Science, University of Venda, Thohoyandou 0950, South Africa

Zoology and Entomology Department and Afromontane Research Unit, University of the Free State, QwaQwa Campus, Private Bag X13, Phuthaditjhaba 9866, South Africa

C. M. Shackleton et al. (eds.), *Urban Ecology in the Global South*, Cities and Nature,
https://doi.org/10.1007/978-3-030-67650-6_7

Global South diverge from their Global North counterparts and explain how several key differences in pattern can have important consequences for urban animal diversity and ultimately ecological function and ecosystem services. Our focus in this chapter is on several key taxonomic groups, including, birds, mammals, herpetofauna and invertebrates, and incorporates a case study on urban predators, as well as some views on novel insights that can be gained from studying urban animal diversity in the Global South. Additionally, we synthesise the available urban animal diversity research from the Global South and explore how varying landscape patterns, distinct abiotic conditions, and vastly different socio-economic contexts can lead to greatly different outcomes for biodiversity in Global South cities.

Keywords Assemblage · Biodiversity · Invertebrate · Pattern and process · Urbanisation · Vertebrate

1 Introduction

Ecological research on animal diversity in urban ecosystems is rapidly expanding (e.g. Faeth et al. 2011; McPhearson et al. 2016; Marzluff 2017; Ramírez-Restrepo and MacGregor-Fors 2017), and interfaces with many other disciplines, including sociology, human health, and economics (Wu 2014). The current surge of interest in urban animal ecology has been driven by several factors, including rising concerns of the environmental impacts of expanding urbanisation on animal diversity (both in cities and surrounding landscapes) (Grimm et al. 2008; Mcdonald et al. 2008; McDonnell and MacGregor-Fors 2016), opportunities to use urban systems to understand how animal communities respond to changes in landscape composition and configuration (Faeth et al. 2005, 2011; Breuste et al. 2008; Soga et al. 2014), climate change, as animals respond to urban "heat islands" (Oke 1982; Cregg and Dix 2001; Kaiser et al. 2016), and finally, the chance to explore how novel selection pressures in urban environments can affect the evolutionary and behavioural responses of organisms that allow them to thrive in the city (Lowry et al. 2013; Johnson and Munshi-South 2017). Thus, once viewed as highly degraded habitats of little ecological interest (Wu 2014), cities are now seen as complex ecosystems supporting important biodiversity and providing an excellent experimental backdrop for the advancement of ecological and evolutionary theory.

Animal species can play several key roles in urban environments and for some, urban environments may hold important populations. For example, some species are able to thrive in urban environments ["urban exploiters", e.g. the bat species, *Chaerephon pumilus*, *Tadarida aegyptiaca*, *Otomops martiensseni* and *Scotophilus dinganii*, where hunting success is enhanced by artificial light (Schoeman 2016)], and can reach densities far higher than those in natural or semi-natural habitats. Some generally declining species may be largely confined to urban areas (Mason 2000; Ives et al. 2016), for example, the endangered Western Leopard Toad (*Amietophrynus pantherinus*), where a significant proportion of the species' population occurs on the highly urbanised "Cape Flats" in Cape Town, South Africa (Measey and Tolley 2011). Such urban biodiversity may also contribute to ecosystem functions like pollination, and urban green spaces can support a variety of pollinating insect species (Baldock et al. 2019). For the most part, however, urban diversity tends not to be of conservation concern, often being fairly common, or frequently non-native, and can regularly be perceived as problematic (McKinney 2006; Kark et al. 2007). Nevertheless, these species may be a city dweller's only contact with nature. This in itself may have value in that such experiences, while "artificial", enhance the wider appreciation of nature and thus also reinforce the value of protecting biodiversity more generally [the so-called "pigeon paradox"; (Dunn et al. 2006)]. Finally, there is increasing evidence that urban green spaces, and their associated animal and plant communities, enhance the quality of human lives through psychological, cultural, and health benefits (e.g. Fuller et al. 2007; Dearborn and Kark 2010; Belaire et al. 2015).

2 Contemporary Cities and the Consequences for Animal Diversity

While a great deal is known about the diversity, and in some cases ecology and evolution, of several animal taxa in the cities of the Global North—e.g. birds (McKinney 2008; Faeth et al. 2011; Beninde et al. 2015), cities of the Global South are not as well researched. A recent meta-analysis exploring the factors determining variation in intra-urban biodiversity (both plants and animals) showed that of the 75 studies included in the analysis, less than 20% were conducted in cities of the Global South, with only one study from the African continent (Beninde et al. 2015). The need to focus on biodiversity in cities of the Global South is more pressing than ever, given that by 2030 these countries will contain 80% of the world's urban population (Montgomery 2008; Ramalho and Hobbs 2012). Whereas the first global urban transition (1750–1950) took place in Europe and North America and happened relatively gradually, the second urban transition (1950–2030) is happening largely in Africa and Asia and at an unprecedented rate and scale (Montgomery 2008; Myers 2021). Urban land cover is expected to triple from its current extent by 2030, with considerable loss of habitat in the biodiverse Global South [e.g. the Guinean forests in West Africa (7% loss), the Philippines (4% loss) and the Western Ghats of India and Sri Lanka (4% loss)] (Seto et al. 2012). Unfortunately, few studies have addressed the ecological effects and biodiversity impacts of this rapid urban growth in these biodiverse regions (McDonnell and MacGregor-Fors 2016). Indeed, Magle et al. (2012) identified the rapidly urbanising areas in South America, Africa, and Asia as one of the most critical gaps in urban wildlife research.

The shape of these "new" cities is also very different to their historical predecessors (Seto et al. 2010). There has been a shift from geographically well-defined cities, consisting of a dense urban core surrounded by approximately concentric circles of suburban and peri-urban habitat, to more sprawling mosaics of mixed urban and rural land-covers (Redman and Jones 2005; Seto et al. 2010). Furthermore, rapidly urbanising rural landscapes distant from large cities, such as the sprawling settlements in rural Apartheid "homelands" in South Africa, represent novel modes of urbanisation not adequately captured by theory developed in the Global North (McHale et al. 2013). Indeed, the extensive rural regions with urban-level population densities that typify parts of Asia and Africa have been referred to as "ruralopolises" (Qadeer 2000). This is relevant for two reasons. Firstly, an urban-to-rural gradient approach has been the traditional approach for exploring how biodiversity changes with increasing degrees of urbanisation in the Global North. For example, it has shown how diversity (usually species richness) of different taxonomic groups varies in response to indicators of urbanisation, such as the percentage of sealed surfaces, population density, and socio-economic status (Hope et al. 2003; McKinney 2008; Ramalho and Hobbs 2012; Chamberlain et al. 2019). However, these gradient approaches, which are usually based on simple measures, oversimplify the processes that affect biodiversity in urban areas, and are being replaced by methods more pertinent to developing cities and that explicitly view the city as a mosaic (Seto et al. 2010). Increasingly, urban studies use spatial pattern metrics from the field of landscape ecology to test the effects of urbanisation on diversity (McGarigal et al. 2012). Many of the indices developed to characterise the complexity, shape, and configuration of landscapes are particularly useful to the study of urban landscapes (Seto et al. 2010). Secondly, the contrast between contemporary and historical urban configurations bears a remarkable similarity to the debate in the agricultural literature on the implications of land-sharing versus land-sparing for biodiversity (Green et al. 2005), an approach that has recently been considered for urban development in Australia (Geschke et al. 2018). Given current and future projections of demand, land-sparing is the preferred strategy in many agricultural landscapes (Phalan et al. 2011; Hulme et al. 2013; Kamp et al. 2015). Similarly, this may be the case for cities of the Global South where there is an urgent need to implement the best development strategy to minimise the negative impacts of urbanisation on native biodiversity (Lin and Fuller 2013; Soga et al. 2014).

It is not only differences in landscape structure that may affect patterns of animal diversity in the cities of the Global South, but also vastly differing abiotic and socio-economic templates (Shackleton et al. 2021). Firstly, there are critical climatic differences. Cities of the Global South are generally located in tropical regions (Gupta 2002), in contrast to those of the temperate North. Thus, in the temperate North, animal breeding seasons are short and synchronised, whereas in the tropical South they tend to be long and asynchronous (Hau 2001) and create very different opportunities for urban wildlife. For example, the urban heat island effect (Oke 1995; Arnfield 2003; Streutker 2003) is one of the most consistent differences in the abiotic environment between urban areas and adjacent non-urban areas, with higher ambient temperatures and a lower fluctuation in diel temperatures in cities, leading to earlier springs and longer frost-free seasons with reduced snowfall that advances dates of egg-laying by birds, as well as leaf emergence and flowering in plants (Wilby 2007). However, although elevated temperatures in the urban environment can buffer the effects of cold stress in cold places it could add to heat stress in the already hot cities of the Global South (Grimm et al. 2008).

Second, socio-economic differences vary dramatically between cities of the Global North and South, but only a few studies on urban wildlife in the Global South have incorporated socio-economic aspects (Chamberlain et al. 2020). Human activities are often spatially clustered for socio-economic and historical reasons (Kinzig et al. 2005). These practices can structure the urban landscape as significantly as the more classically referred to urban physical structures (e.g. building density) and ultimately can create ecologically relevant social gradients that can be independent of an urban–rural transition (Kumar et al. 2019). In the densely populated “ruralopolises” of the Global South, residents of settlements retain strong economic and cultural ties with the transformed rural matrix in which they are embedded (Qadeer 2000; McHale et al. 2013). Human behaviour and culture are increasingly recognised as essential components of urban areas, underlining the urgent need for more insight into their ecological consequences (Alberti et al. 2003). While some studies have related biodiversity, or individual-level behavioural responses by urban animals, to human socio-cultural factors (e.g. Kinzig et al. 2005; Kumar et al. 2018), it is largely unknown whether community or individual responses translate into population-level consequences (but see Kumar et al. 2019).

Taken together, these important differences between perhaps more traditional cities and the rapidly expanding urban mosaics that represent cities of the Global South present very different challenges for urban biodiversity and opportunities for urban biodiversity research and conservation. Here, our viewpoint is largely of animal diversity in the city (McPhearson et al. 2016) and we focus on primary ecological research from urban areas addressing how patterns of urbanisation and development drive patterns of animal diversity and affect the ecology of organisms in urban habitats. Where possible, we address the consequences of these changes in pattern and process for ecosystem function and services, and how this may vary between the Global North and Global South.

3 Animal Diversity in the City

The patterns of biodiversity change in relation to urbanisation gradients in cities are well established, although these patterns are much better studied in cities of the Global North, and indeed, many of these gradients do not apply in cities of the Global South. For animals, there is generally a decline in species diversity and richness with increasing urbanisation (Faeth et al. 2011). Changes in species abundance often vary across taxa, with some, such as birds, increasing in abundance in more urbanised areas, usually because of increases in non-native species such as House Sparrow (*Passer domesticus*) and Feral Pigeon (*Columba livia domestica*) (Bhatt and Joshi 2011; MacGregor-Fors et al. 2012). In fact, cities tend to be hotspots of biological invasions (Gaertner et al. 2017), and this is particularly true of cities in the Global South that have a lower capacity to prevent the introduction of invasive alien species

(Early et al. 2016). However, in contrast to patterns of plant diversity in the city, which is usually under human control, Faeth et al. (2011) proposed that patterns of animal diversity are more of a response to these altered plant communities or to landscape pattern. For the most part humans have relatively little direct control over the abundance or diversity of animal taxa in the urban environment compared to plants. Nevertheless, human activities certainly do influence animal communities, either directly, for example through provision of resources (Robb et al. 2008; Galbraith et al. 2015), intentional or unintentional introduction of vertebrate species (Gaertner et al. 2017), or indirectly through the many environmental changes linked to urbanisation (Grimm et al. 2008). This complex situation makes understanding the mechanisms driving patterns of animal diversity more difficult to untangle, and the effects of drivers may vary dramatically with landscape context. In general, though, increasing patch size and landscape connectivity correlate with increasing biodiversity in cities (Beninde et al. 2015).

Green infrastructure (e.g. parks, private gardens, allotments) is generally regarded as being a key driver of urban biodiversity, providing critical habitat, and in the Global North there is good evidence that the extent of these habitats dictates urban animal diversity, at least for birds and insects (see Beninde et al. 2015). While some studies from the Global South support the positive effect of green infrastructure on urban biodiversity (Purkayastha et al. 2011; MacGregor-Fors et al. 2012; Filloy et al. 2015; Entiauspe-Neto et al. 2016), others have suggested that habitat quality, rather than the extent of green infrastructure per se, is more important (Dures and Cumming 2010; Silva et al. 2015; Kruger et al. 2015). These results suggest that there may be greater variation in the quality of green infrastructure, including private gardens and public green spaces, in the Global South compared to the Global North (Silva et al. 2015). Alternatively, the contrast between urban spaces and green spaces in the Global South is not as marked as in the North, given that many cities in the South occur within a transformed rural matrix rather than the planned, built-up tarmac/asphalt and concrete urban environments of the rural North.

In the subsequent sections, we detail how diversity and abundance of different animal taxa change in relation to processes of urbanisation and discuss the mechanisms that may be responsible for these patterns. Unfortunately, as already stated much of what is known is from studies conducted in developed cities of the Global North. Despite several attempts to highlight the urgent need for more studies to take place in cities of the Global South (Magle et al. 2012; Marzluff 2017), little progress has been made (see study location data in Faeth et al. 2011; Haase et al. 2014; Beninde et al. 2015). Where possible, we present studies from Global South countries and discuss how results may vary based on the different landscape, abiotic and socio-economic contexts presented by Global South cities.

4 Urban Birds

For birds, urbanisation acts as a filter such that species richness is reduced, but individual species may occur at higher abundance, at least at intermediate levels of urbanisation (Batáry et al. 2018). This general pattern is also reflected in responses of bird communities in the Global South, in that there is much-existing evidence of a decline in bird species richness as urban cover increases (e.g. van Rensburg et al. 2009; Bhatt and Joshi 2011; MacGregor-Fors and Schondube 2011; MacGregor-Fors et al. 2012; Silva et al. 2015; Chamberlain et al. 2017). There are, however, some notable exceptions. Leveau and Leveau (2005) found that species richness was greater in peri-urban and urban compared to rural areas in Argentina, the latter being habitats dominated by agriculture and monoculture plantations. Dures and Cumming (2010) found a positive correlation between species richness and building density in Cape Town, South Africa, a pattern that was driven by the prevalence of the invasive tree *Acacia saligna* in the surrounding landscape, a monoculture-forming species that is damaging to the native Fynbos vegetation. Filloy et al. (2015) found opposing patterns in terms of species richness and urbanisation in South America, whereby species richness was greater in natural compared to urban habitat in a city located in a humid tropical biome, but that this pattern was reversed for a city in the arid biome. This is probably unsurprising, as in an arid setting gardens in urban areas are likely to be far more productive than the natural surroundings. These examples illustrate how the type of surrounding habitat can influence conclusions about impacts of urbanisation on bird communities.

Several studies have reported that species richness of both plants and animals (although most of the latter group have only considered birds) correlate positively with wealth status within urban areas, the so-called Luxury Effect (Hope et al. 2003; Leong et al. 2018). The existence of such an effect suggests environmental injustice in that poorer sections of society do not have access to higher levels of biodiversity and hence the benefits with which it is associated. However, the Luxury Effect has mostly been considered in the developed world (Chamberlain et al. 2020). We are aware of only four studies that have tested for the existence of the Luxury Effect in the Global South. There was support for the Luxury Effect in both Morelia, Mexico (MacGregor-Fors and Schondube 2011) and Belo Horizonte, Brazil (Perillo et al. 2017), but there was no significant association between species richness and socio-economic status in Valdivia, Chile (Silva et al. 2015). Chamberlain et al. (2019), however, found that support for the Luxury Effect depended on the degree of urbanisation across a large sample of urbanised landscapes in South Africa (Fig. 1). At low levels of urban land cover, there was support for the Luxury Effect, but in highly urbanised centres, the relationship flipped, higher wealth status being linked to lower species richness. Unfortunately, the mechanism underpinning this result was not resolved as increased tree cover, which although significantly correlated with bird species richness, did not explain the Luxury Effect. While these studies give some general support to the Luxury Effect in the Global South, it remains largely a phenomenon of the developed world. Given that wealth inequality is generally higher in the Global

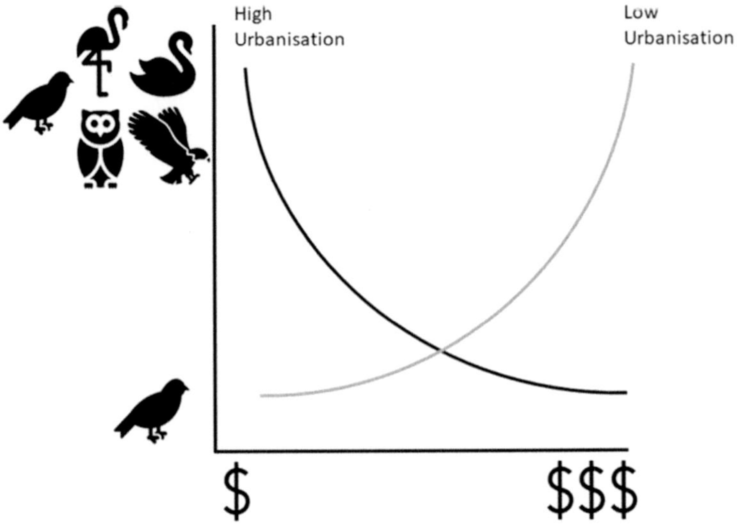

Fig. 1 In analysing the relationship between bird species richness and income level of urban dwellers in South Africa, Chamberlain et al. (2019) tested the "luxury effect", which posits a positive correlation between wealth and biodiversity. They found that the relationship between species richness and income differed across an urbanisation gradient. The "luxury effect" was evident in areas of lower urbanisation (grey curve), while the opposite pattern was found in more urbanised areas (black curve) (Adapted from Chamberlain et al. [2019])

South than elsewhere (Alvaredo et al. 2018), the existence of the Luxury Effect, and strategies to minimise it and hence address environmental injustice issues, should be considered a high research priority. This may not necessarily equate to tree planting, as areas of natural or invasive vegetation can be viewed with suspicion in some contexts, as hiding places from which criminals may attack people, as in Cape Town, South Africa (Potgieter et al. 2019).

There is a less consistent response of bird abundance to urbanisation gradients in cities of the Global South, with evidence of both negative (Bhatt and Joshi 2011; MacGregor-Fors and Schondube 2011) and positive (van Rensburg et al. 2009; MacGregor-Fors et al. 2012; Silva et al. 2015; Chamberlain et al. 2017) influences of increasing urban land cover. These results may to some extent be due to the length and position of the urbanisation gradient in each study, given that there appears to be a general non-linear response of bird abundance to urbanisation gradients (Batáry et al. 2018). In such a situation, a gradient from the urban centre to suburban areas may show a negative effect of urbanisation, whereas a gradient from suburban to rural habitats might show a positive effect. Nevertheless, such a diversity of responses may also arise because many cities of the Global South do not fit the classic urban–rural gradient typical of many Global North cities (Seto et al. 2010).

Urbanisation also acts as a filter on species traits, and hence has important consequences for ecological function. However, most empirical data confirming species traits linked to urbanisation stem from the Global North. While there are a number of inconsistencies between the findings of different studies (e.g. Møller 2009; Evans et al. 2011), urban bird species of the Global North tend to be generalists in their use of habitats, and are more likely to nest above the ground and to exploit a plant-based diet (Chace and Walsh 2006; Møller 2009; Evans et al. 2011). The latter effect is likely associated with the exploitation of anthropogenic food (e.g. bird feeders and bird tables). Such food sources are likely to have profound effects on the ecology of urban birds (Robb et al. 2008), but we are only aware of a single study that has considered anthropogenic food subsidies of urban birds in the Global South (Stofberg et al. 2019). The study detailed how behavioural flexibility in the Red-winged Starling (*Onychognathus morio*) acted as a mechanism to cope with dramatic fluctuations in anthropogenic food availability due to change in the presence of humans. Red-winged Starlings on the University of Cape Town campus consumed more anthropogenic food on high-human presence days (e.g. weekdays) and there was a strong indication that daily mass gain was greater on these high-human presence days (Stofberg et al. 2019). One study that considered bird traits in an urban context in the Global South showed that insectivores and frugivores were the most negatively affected by urbanisation, whereas omnivores and granivores were fairly tolerant (Sanz and Caula 2015). Scavengers may also benefit from urban habitats in the Global South, which is likely linked to poverty and the lack of investment in municipal waste management (Chamberlain et al. 2017; Pomeroy and Kibuule 2017). Nesting habitat may also be an important component of how birds respond to urbanisation—Leveau and Leveau (2005) found lower negative responses of ground and shrub nesters to urbanisation, but positive responses of tree nesters.

One feature that seems particularly marked in cities of the Global South is that non-native species dominate in highly urbanised areas. Two species with European origins, House Sparrow (*Passer domesticus*) and Feral Pigeon (*Columba livia domestica*), are almost ubiquitous in the exotic avifauna (e.g. Bhatt and Joshi 2011; MacGregor-Fors et al. 2012), but other species such as mynas (*Acridotheres* spp) also seem to be common exotic species in the Global South (Fig. 2a) (van Rensburg et al. 2009; Bhatt and Joshi 2011). The populations of exotic species may be enhanced by cultural practices such as releasing prayer animals (Gilbert et al. 2012; Su et al. 2017). Additionally, the vulnerability of transformed ecosystems such as urban areas to invasive species appears to be a global trend (Godefroid and Ricotta 2018), and unlike the temperate zone, there is a lack of data on what kinds of decorative ornamental plants are being used in urban green areas in tropical countries (Müller et al. 2013) and how the species composition might change in the face of a global temperature increase.

Native bird species diversity declined with an increase in exotic plant species in Delhi, India (Khera et al. 2009) and were in lower abundance in exotic street trees in Grahamstown, South Africa (Shackleton 2016). However, the prevalence of a non-native tree species has led to higher species richness in urban Cape Town compared to outlying areas (Dures and Cumming 2010; see above). This might be because bird species richness is positively correlated with vertical habitat heterogeneity (Karr and Roth 1971): non-native tree species increase habitat heterogeneity in a vegetation type dominated by shrubs. Also, birds can profit from alien plants in urban green space, as

Fig. 2 (a) Indian Myna (*Acridotheres tristis*) (Koshy Sahab); (b) bats in a roost (Bat Interest Group Gauteng); (c) Southern Tree Agama (*Acanthocercus atricollis*) (Blair Cowie); (d) Parktown Prawn or King Cricket (*Libanasidus vittatus*) (Johan Heyns); (e) Trees in suburban Johannesburg, South Africa (Shivan Parusnath)

seen in the recent colonisation by Black Sparrowhawks (*Accipiter melanoleucus*) of Cape Town, South Africa (Martin et al. 2014), or the Crowned Eagle (*Stephanoaetus coronatus*) population in Durban, South Africa (McPherson et al. 2016a) that is found at higher breeding densities than those reported in natural (non-urban) habitats. Both raptor populations nest in exotic pine (*Pinus* spp.) and eucalyptus trees (*Eucalyptus* spp.) (Malan and Robinson 2001; Tarboton 2001; Malan and Shultz 2002), that are currently the subject of invasive plant removal programs. This might affect other species such as African Fish-Eagles (*Haläeetus vocifer*) in the Western Cape of

South Africa where eucalyptus trees provide the only suitable nesting sites (Welz and Jenkins 2005). Such invasive plant removal is an area of conflict in many urban systems, particularly where they provide habitat or food for wildlife and where they affect charismatic species such as birds and butterflies (Dickie et al. 2014).

5 Urban Mammals

Next to birds, mammals have been the most studied group of urban vertebrates (Magle et al. 2012). Although most larger mammals are absent from cities, apart from domesticated species such as pets (dogs, cats, and horses) or livestock (goats, sheep, and cattle), several orders of smaller mammals have adapted to urban landscapes, including bats (Chiroptera), rodents (Rodentia), insectivores (Soricomorpha: shrews, moles, and hedgehogs), golden moles (Afrosoricida), small and medium-sized carnivores (Carnivora), wild pigs (Perrisodactyla), small and medium-sized ungulates (Cetartiodactyla) and certain primates (Order Primates) (Feldhamer et al. 1988; Taylor et al. 1999; Ramkissoon 2006; Picker and Griffiths 2011; Nowak et al. 2017; Foord et al. 2018). Even in some of the world's biggest cities, small mammals can persist in parks and gardens. For example, the "Wildlife Garden" at the Natural History Museum in central London supports at least eight species of mammal (https://www.nhm.ac.uk/visit/galleries-and-museum-map/wildlife-garden.html). However, while the mechanisms driving mammal diversity are less well-known in the Global South, it is likely that urban green infrastructure such as green corridors will have a positive influence on mammalian biodiversity and ecosystem services in general, as exemplified by the Durban Metropolitan Open Space System, DMOSS (Davids et al. 2016).

Although most large mammals are absent from metropolitan areas, some larger mammals such as mesocarnivores (e.g. coyotes, *Canis latrans*) and baboons (*Papio ursinus*) can increase their populations in cities resulting in increasing conflicts with humans (Gehrt et al. 2010; Hoffman and O'Riain 2012). Several mesocarnivore species have become synanthropic (living close to humans and benefiting from them) in urban areas, including raccoons (*Procyon lotor*), kit foxes (*Vulpes macrotis*), and red foxes (*Vulpes vulpes*) (Gehrt et al. 2010; Scott et al. 2014). In urban areas, due to increasing resources, mesocarnivores often decrease their home ranges. In the Chicago metropolitan area, coyote home ranges are usually associated with natural areas within the urban matrix (Gehrt et al. 2009). Although they have been known to attack pets and even humans, the study found that coyotes avoid areas and times of high-human activity, including a shift to more nocturnal habits. In expanding urban settlements and associated rangelands and agricultural areas in Limpopo Province of South Africa, occupancy by small and medium-sized carnivores (mesocarnivores) was highest in crop fields adjacent to villages and was influenced negatively by the relative abundance of domestic dogs and positively by the relative abundance of livestock (Williams et al. 2018; Foord et al. 2018). Primates can also become nuisance species that raid food in urban areas. A notable example concerns Chacma baboons

(*Papio ursinus*) in Cape Town where human populations have encroached on 16 baboon colonies resulting in increased human–wildlife conflict due to food raiding by baboons and baboons subsequently associating humans with food (Hoffman and O'Riain 2012).

Bats can thrive in urban settings, both in cities of the Global North, e.g. Chicago, USA (Gehrt and Chelsvig 2003), and the Global South, e.g. Durban, South Africa (Taylor et al. 1999). A meta-analysis of 180 bat species from Europe, Asia, Australia, and North and South America showed that species-species traits accounted for differences in urban tolerance of bats. Bats that thrive in urban areas tend to be open and edge space foraging, as well as in bats with flexible roosting strategies (Jung and Threlfall 2018). Open space foraging, free-tailed bats of the Family Molossidae are most notable in this regard and are frequently commensal with humans in towns and cities, particularly in Africa and Madagascar (Goodman 2011; Happold and Happold 2013). In the USA, Mexican free-tailed bats (*Tadarida brasiliensis*) typically roost in caves but colonies of this species can also occupy human-made structures like bridges. A colony of 1.5 million individuals occupies the Congress Avenue Bridge in central Austin, Texas, where it is a major tourist attraction during summer months. In the Midwest United States, Gehrt and Chelsvig (2003) found that heterogenous urban landscapes within cities represented islands of habitat for several bat species, and bat activity was higher in patches of intact woodland in the urban matrix, than in the surrounding depauperate landscapes. Similarly, an example from the Global South, i.e. Durban, South Africa, recorded a particularly high diversity of bats within the urban boundaries. This included 16 bat species out of an estimated regional total of 19 from three types of roosts; including buildings, where they roosted either in attics (several species of molossid bats as well as house bats, *Scotophilus dingani*), or on the outer walls of houses (Mauritian tomb bats, *Taphozous mauritianus*); tall exotic trees such as various palms species (especially Royal Palm), where fruit bats (*Epomophorus wahlbergi*) roosted on the midribs on the under-surfaces of leaves, and underground roosts including artificial structures such as an abandoned ammunition dump and an overflow tunnel constructed for the Shongweni Dam (Taylor et al. 1999).

Colonies of bats living in proximity with humans have raised concerns about the possible transmission of zoonotic diseases like Ebola and rabies-like viruses (Lyssavirus) from bats to humans. However, such fears are often exaggerated. Among other wildlife species implicated, only three fruit bat species limited to West and Central Africa have been implicated as possible reservoirs of Ebola (Leroy et al. 2005). Likewise, the incidence of rabies-like viruses and other potentially zoonotic viruses in bats is very low in North America, Europe, Australia, and Africa (Markotter et al. 2006; Schneeberger and Voigt 2016). At the same time, the white-nose syndrome fungus (*Geomyces destructans*) probably transported by humans from Europe to USA has killed more than one million cave-dwelling bats across the USA (Hallam and McCracken 2011; Frick et al. 2016).

Light pollution associated with cities is known to negatively impact on biodiversity, especially nocturnal species including bats (see Box 1). Light pollution is a global threat, with the accelerated use of electric lighting increasing at 6%

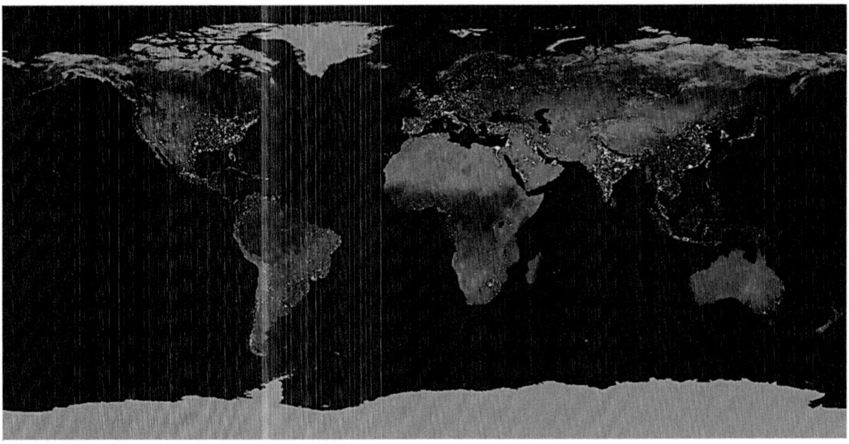

Fig. 3 Global view of artificial light at night (ALAN) for the year 2016 (NASA Earth Observatory https://earthobservatory.nasa.gov/features/NightLights)

per year (Hölker et al. 2010). While some open-air foraging bats prey on insects attracted by mercury vapour street lights, slow-flying bat species (e.g. horseshoe bats, *Rhinolophus* species) avoid brightly lit areas, including areas lit by both high-pressure sodium and energy-conserving bright light-emitting diode (LED) lamps (Stone et al. 2009, 2012; Ives et al. 2016). Additionally, Schoeman (2016) showed that brightly lit stadiums in Durban favoured urban species and may homogenise urban bat diversity. This was because bright lights greatly favoured urban exploiter (light-tolerant) species such as yellow house bats (*Scotophilus* spp.) and little free-tailed bats (*Chaerephon pumilus*) in favour of urban-avoiders (light-averse species), which included some of the rarer species. Neotropical fruit-eating bats disperse the seeds of forest plants including pioneer species. Such bats often avoid lighted areas when foraging, which can result in reduced ecosystem services (forest regeneration) due to light pollution (Lewanzik and Voigt 2014).

Box 1: Novel insights on the effect of light pollution on animal diversity from the Global South

The dark night sky, under which all biota evolved, has rapidly disappeared within the span of four human generations. Twenty-three per cent of land is considered to be light polluted (Bennie et al. 2015, 2018) (Fig. 3). The biological effect of Artificial Light at Night (ALAN) is only now being teased apart from the normal fluctuations of animal night-time behaviours. Many vertebrate nocturnal predators such as bats, owls, and toads appear to benefit from increased light, while their vertebrate prey species reduce their nocturnal activity or show reduced immune responses (Gaston et al. 2013). Nocturnal invertebrates are well known to fly to lights at night, but despite some insects showing increased flight activity to light as expected (Manfrin et al. 2017), unexpectedly moths, which we often observe at light sources, showed no population effects of ALAN (Spoelstra et al. 2015). This surprising result may point to the complexity of what appears to be a simple situation. However, we do not know what each insect uses light at night for, or under what circumstances that may change.

Nocturnal dung beetles may reveal this complexity. All dung beetles will fly to lights (Houston et al. 1982; Howden et al. 1991; Hill 1996), and ball rolling species are well known to use nocturnal celestial cues such as the moon, polarised light, and even the stars to orientate in the savanna at night (Dacke et al. 2003, 2004, 2013), suggesting they are sensitive to the quality and intensity of light for finding their way. Nevertheless, those natural sources can be replaced by a variety of artificial lights, to which the beetles respond by continuing to steadfastly roll onwards in search of a ball burial spot (Dacke et al. 2013; el Jundi et al. 2015; Smolka et al. 2016; Foster et al. 2017). Full blown city light is likely to be so much more disruptive than experimental lights, and it may have contributed to the catastrophic loss of insects seen in the last 40 years (Sánchez-Bayo and Wyckhuys 2019). African and particularly South African cities offer a brief period in which to test this hypothesis. This is because the continent remains dark on global maps of light pollution, with the exception of the Nile in the North, and the influence of cities such as Johannesburg, Durban, and Cape Town (Bennie et al. 2015) that lie in three very different climatic regions in one country, South Africa. These cities and their environs offer an ideal testing ground for impacts of ALAN on insects.

Urban mammals include some of the world's top 100 most invasive mammals (Lowe et al. 2000), such as feral domestic cats (*Felis catus*), the house mouse (*Mus musculus*), and rats (*Rattus* spp.) and have had a huge negative impact on urban landscapes and biodiversity, not to mention the impacts on human health and food security (Meerburg et al. 2009a, b). Domestic cats have been involved in the extinction of 26% of the world's bird, mammal, and reptile species listed as threatened/extinct on the International Union for Conservation of Nature (IUCN) Red List, and have contributed to the threat status of a further 25% (Doherty et al. 2016). Sterilisation campaigns are being practised in some South African towns (Picker and Griffiths 2011), although caught, sterilised and released feral cats, which occur at densities far higher than natural predators, continue to prey on wildlife after release and transmit diseases (Longcore et al. 2009). Over 60 known zoonotic diseases are spread by rodents, such as plague, leptospirosis, and toxoplasmosis (Taylor et al. 2008; Meerburg et al. 2009a). Some 280 million undernourished people would benefit from increased crop production if greater attention were paid to rodent control (Meerburg et al. 2009b). Based on a successful rodent-control campaign applied in the city of Boston, USA, the "Boston Model" was developed as a suggested method to control rodents in cities (Colvin and Jackson 1999). The success of the model relied on an integrated approach involving a strong health department that provided a coordinating role and communicated well with several other key players: (1) code enforcement/legal; (2) pest control operators; (3) public works departments; (4) community groups; (5) neighbourhood services; and (6) sewer department. As shown by Taylor et al. (2008) based on a case study at Cato Crest, an informal settlement outside Durban, South Africa, the Boston Model is only partially applicable to developing cities where some services such as sewer departments and rodent pest control services are often absent. An integrated rodent-control campaign involving public education, disease-testing of rodents and residents, environmental clean-up efforts, and vector control with both snap traps and live traps was successful at Cato Crest, at least in the medium-term. This was because it was led by an interdisciplinary and international research team having biological, socio-economic, and zoonotic disease expertise (www.nri.org/ratzooman) that replaced some of the functions of the Boston Model and also enlisted commitment and support from a local community group (involving councillors, residents and the Cato Manor Area-Based Management) as well as the eThekwini Municipality of Durban, which provided vector control and environmental hygiene services to the informal settlement as well as employing three local residents as rat-catchers on a continual basis (Taylor et al. 2008).

Finally, in some cases, closely related exotic species have replaced native species. An example is the case of the North American grey squirrel (*Sciurus carolinensis*), which was introduced into Europe (and elsewhere) where it has replaced the native red squirrel (*Sciurus vulgaris*) through competition and by transmitting the squirrel poxvirus (Gurnell et al. 2004). In the Western Cape province of South Africa, where grey squirrels are a tourist attraction in central Cape Town, the impacts are less severe and include damage to seedlings and ornamental plants and consumption of fruits and vegetables (Picker and Griffiths 2011). In other cases, unrelated species such as eutherian mammals (e.g. rabbits, goats, dingos, and other predators) have invaded Australian cities and rural areas, displacing and threatening native marsupials. In the city of Cape Town and the slopes of Table Mountain in South Africa, introduced Himalayan tahr (*Hemitragus jemlahicus*) and European fallow deer (*Dama dama*) were common until recently on Table Mountain where they threatened local plant and animal communities. Numbers of Himalayan tahr reached > 600 by the 1970s and they damaged mountain fynbos plant communities, as well as causing erosion and displacing local klipspringer populations (Picker and Griffiths 2011). Culling was introduced as a control measure and only occasional reports suggest that a few may have survived. Already present in Cape Town in 1869 and widely translocated to all provinces except Limpopo, European fallow deer have proliferated. Culling programmes are underway in Cape Town where they have damaged local fynbos plants through trampling and overgrazing (Picker and Griffiths 2011).

6 Urban Herpetofauna

Reptiles and amphibians are among the least studied taxa in urban wildlife research (Magle et al. 2012), and even more so in the Global South. Like other non-avian vertebrates, reptiles and amphibians tend to decrease in species diversity and abundance with increasing urbanisation (Germaine and Wakeling 2001; McKinney 2008; Hamer and McDonnell 2008, 2009; Faeth et al. 2011; Ackley et al. 2015). This trend is even more marked for large and venomous reptiles that are considered nuisance species, for example *Python* spp., and are, therefore, more likely to be removed from both urban and suburban areas (Garden et al. 2006; Hamer and McDonnell 2009). Additionally, for amphibians, urbanisation results in a significant erosion of genetic variability (Hamer and McDonnell 2008). This is because patterns of urbanisation present a unique challenge to amphibian populations. For example, many pond-breeding amphibian species have complex lifecycles and require different types of habitat for breeding and non-breeding life stages (Pope et al. 2000). Thus, the preservation of amphibian populations in urban landscapes often requires a delicate balance between maintaining both suitable aquatic habitats and intact and/or permeable terrestrial habitats (Wells 2007; Pillsbury and Miller 2008; Hamer and McDonnell 2008; Bickford et al. 2010).

Reptile diversity in urban spaces is related to variables at both landscape scale (e.g. mean natural patch size or percent road cover) and local scale (e.g. number of vegetation strata or number of fallen logs) (Germaine and Wakeling 2001; Hamer and McDonnell 2009; Garden et al. 2010). However, as with other vertebrate taxa, habitat structure is a major driver of reptile species richness in urban areas. A decline in reptile species richness is often linked to a concurrent loss of habitat structure in more heavily urbanised areas (Germaine and Wakeling 2001; Hamer and McDonnell 2009; Garden et al. 2010). This highlights the importance of remnant habitat patches, such as urban parks that preserve habitat heterogeneity (Purkayastha et al. 2011; Entiauspe-Neto et al. 2016). In urban parks, large reptiles may persist and even thrive, for example, the predatory Tegu lizard (*Salvator merianae*) in São Paulo, Brazil (Sazima and D'Angelo 2013). Additionally, the urban–rural matrix of densely populated "ruralopolises" across Asia and Africa, degraded rangelands and woodlands may still harbour high reptile diversity, even in comparison to nearby conservation areas, again suggesting that habitat heterogeneity is paramount in supporting urban reptile diversity (Smart et al. 2005).

Similarly, amphibian diversity in urban areas also responds to both local and landscape-scale features but of both the aquatic and terrestrial habitats (Hamer and McDonnell 2008). A recent study, conducted in Potchefstroom, South Africa, showed that at the local-scale larger pond size and increased pond vegetation heterogeneity increased amphibian diversity, while at the landscape scale increasing amounts of urban infrastructure in the surrounding landscape and lack of connectivity had negative effects on amphibian diversity (Kruger et al. 2015). These results are in line with what has been shown from studies conducted in Australia, USA, and Europe (see Hamer and McDonnell 2008 for reviews) and suggests common mechanisms between the Global North and South that may drive amphibian diversity in cities. For example, larger pond size is synonymous with increased species richness, diversity, and population persistence because of the species area relationship (MacArthur and Wilson 1967; Fahrig 2013). Additionally, vegetation heterogeneity, and the increased vertical and horizontal structure in the ponds, provides much-needed mate calling locations and subsequent oviposition sites for frogs and provides improved predator avoidance for larvae (Babbit and Tarr 2002; Egan and Paton 2004; Kruger et al. 2015). However, as these ponds become more isolated, immigration and emigration rates decrease causing amphibian diversity to decline (Marsh and Trenham 2001; Hamer and McDonnell 2008).

The quality of available habitat is another important factor affecting amphibian diversity in cities (Hamer and McDonnell 2008), with water quality, in particular, varying considerably between Global North and Global South cities (Walsh 2000; Vörösmarty et al. 2010). Today, many urban areas of the Global South remain inadequately served by sewage treatment and industrial effluent infrastructure and are considered ecologically degraded (Walsh 2000). Amphibians are highly sensitive to changes in water quality, largely due to their aquatic larval stage and physiological requirements (Phillips 1990; Blaustein et al. 1994). While studies of how frog populations respond to polluted aquatic environments in cities of the Global South are lacking, evidence from the Global North suggests that increased levels of nitrogen and heavy metals reduce survivorship, growth, and development rates in amphibians (Casey et al. 2005; Snodgrass et al. 2007). It has been suggested that these urban ponds might represent ecological traps by presenting attractive cues, such as vegetation and water, for example, to the North American Wood Frog (*Rana sylvatica*), which is highly sensitive to urbanisation. These ponds however are not viable breeding habitats because of accumulated pollutants and toxins (Snodgrass et al. 2008). Studies on amphibian diversity are desperately needed in the Global South. Research from rapidly expanding urban landscapes in Shanghai, China, clearly demonstrates that habitat fragmentation and deteriorating pond quality decrease amphibian abundance and community diversity (Zhang et al. 2016). Given that amphibians are the most threatened vertebrate taxon globally, with approximately 2,200 amphibians explicitly threatened by urbanisation, basic research on amphibian ecology in urbanising landscapes of the Global South is a much-needed research avenue (Hamer and McDonnell 2008).

Urban reptiles face a host of threats, many of which may be exacerbated in the Global South. For example, cat predation is a major threat to urban reptiles, which make up a sizable proportion of the organisms caught by domestic and feral cats (Woods et al. 2003; Loyd et al. 2013; Morling 2014). In the Global South, this may be exacerbated since the majority of cats in this region are free-ranging and are less likely to be receiving adequate nutrition from their owners, resulting in increased predation of wild prey (Silva-Rodríguez and Sieving 2011). Snakes, and other reptiles that resemble snakes, are also at risk of being deliberately killed by people. Snakes are one of the most feared and disliked group of animals globally, and this is particularly true in the parts of the Global South where snakes are surrounded by superstition and myths (Burkhardt et al. 2003; Kaltenborn et al. 2006; Özel et al.

2009; Ballouard et al. 2013). This may be linked to there being far higher diversities of highly poisonous snakes in the Global South compared to the Global North. In South Africa, snakes were the top-ranked fear for 9–13-year-old children, with over 45% of children reporting that they were very afraid of them, while in Tanzania, 84% of people were very afraid (Burkhardt et al. 2003; Kaltenborn et al. 2006). This fear puts urban snakes in Global South countries at a greater risk of being killed by humans or even selectively run over by vehicles on the road (Ashley et al. 2007; Ballouard et al. 2013). While many snake species are actively targeted out of fear or superstition, others such as certain species of turtles, frogs, snakes, and lizards are exploited as a food source (Purkayastha et al. 2011). Additionally, reptile species are also killed to be used in traditional medicine, contributing to the decline in abundance of highly sought-after species in densely populated areas (Smart et al. 2005).

Finally, while herpetofaunal species richness generally declines with urbanisation, some groups have been able to successfully adapt to urban environments. As with other taxa, small and generalist species are more likely to survive in urbanised environments (Garden et al. 2006; Hamer and McDonnell 2009). Germaine and Wakeling (2001) also documented arboreal lizard species that were able to remain in even the most urbanised environments, because they can adapt to human-made structures. In South Africa, the arboreal blue-headed tree agama (*Acanthocercus atricollis*) thrives in peri-urban settings despite persecution (Fig. 2c), due to factors such as the presence of large trees, lower densities of predators, abundant food sources, and less competition (Smart et al. 2005). Additionally, some urban nocturnal gecko species regularly utilise artificial night lighting in human-landscapes to improve their hunting success (Perry et al. 2008).

7 Urban Invertebrates

Invertebrates, most notably the insects, are one of the most diverse group of animals on earth (Farrell 1998; Forbes et al. 2018). Despite this, or perhaps because of it, relatively little is known of how invertebrates respond to urbanisation, particularly in the Global South. Those studies that have been conducted have tended to focus on the "pest" aspect of invertebrates, and how to manage, or simply get rid of, them (e.g. Dreistadt et al. 1990; Rust and Su 2012). There has been far less research on how urbanisation influences invertebrate diversity and community structure (McIntyre 2000). Yet invertebrates, by virtue of their diversity, short lifecycles and therefore relatively short response times to disturbance, variation in tolerances of environmental conditions between species, and the integral role they play at many trophic levels in foodwebs, are ideal bioindicators for assessing environmental change (see McGeoch 2007 and references therein).

Invertebrates respond in different ways to the characteristics of urban environments, and as usual, there are winners and losers in response to various drivers (Jones and Leather 2012). The heterogeneity of habitats created by urbanisation can increase invertebrate abundance, and productive gardens associated with settlements in arid environments are a case in point (Gotlieb et al. 2011). Additionally, roadsides, parks, and "brown sites" (e.g. abandoned railway lines) can also provide forage and nesting sites (Hennig and Ghazoul 2012). Other factors, like pollution or an increase in impermeable surfaces are generally linked with declines. As a rule, extreme urbanisation is associated with a decline in invertebrate diversity and abundance, but moderate urbanisation, while also often associated with declines, may be associated with increased diversity in some cases (McKinney 2008). In addition, the shift in urban invertebrate communities is towards domination by generalist and opportunistic species (Jones and Leather 2012). Generalists tend to be of lower conservation concern and can be non-native species. Furthermore, although by definition, they are able to interact with many other species in ecological networks, the loss of specialists may have implications for species with closely evolved mutualisms with those specialists [e.g. loss of specialist pollinators of certain plant species (Pauw 2007)].

Habitat destruction and fragmentation tend to be associated with declines in invertebrate diversity, but again, most of this work has taken place in the Global North (e.g. Ahrné et al. 2009; Geslin et al. 2013). However, in a rare example from the Global South, Edge et al. (2008) were able to show that the critically endangered Brenton Blue Butterfly (*Orachrysops niobe*) was severely impacted by the destruction of natural thicket vegetation in urban environments. Furthermore, patterns are not always obvious if viewed merely in terms of abundance: for example, Cane et al. (2006) found that decreases in ground-nesting bees were offset by increases in cavity nesting bees in urban environments. Thus, it is crucial to assess not only changes in diversity and abundance, but also community composition and representation of life history traits to gain a clear understanding of how invertebrates respond to the various drivers associated with urbanisation. Increased fragmentation leads to smaller fragments and will host fewer species, with losses to key functional groups, such as pollinators, that will have implications for the plant species that depend on them (Harrison and Winfree 2015). For example, loss of a generalist oil-collecting bee species from the small natural habitat fragments in an urban matrix in and around Cape Town, South Africa, led to pollen limitation in six orchid species that depended on the bee, but there was no effect for a generalist plant [*Hemimeris racemosa* (Scrophulariaceae)], which also relied on other pollinators (Pauw 2007).

Areas that are highly urbanised, with high buildings, dominated by heat-absorbing and emitting surfaces, as well as additional heat production by traffic, have higher temperatures, referred to as the Urban Heat Island (UHI) effect. The increased temperatures and reduced humidity associated with UHIs have been found to influence the relative survival and phenotypic expression of butterfly species at rural sites in Belgium (Kaiser et al. 2016). How the UHI plays out for invertebrates in the Global South has yet to be investigated; however, cities and their heat islands may well present opportunities for understanding the effects of heating associated with future climate change on invertebrates (Harrison and Winfree 2015).

Urban areas are also associated with various kinds of pollution, not only of air and water, but crucially, also light, with marked effects for some invertebrates (see Box 1).

Air pollution can have surprising effects on invertebrates. Plants growing along roadsides have greater concentrations of nitrogen in their leaves than plants that do not, and this in turn is favourable to a variety of herbivorous invertebrates [e.g. aphids, (Spencer et al. 1988) or moths, (Port and Thompson 1980)]. Industrial melanism, with one of the most well-known examples being the case of the Pepper Moth (*Biston betularia*; Kettlewell 1955; Cook and Saccheri 2013), occurred when dark morphs of polymorphic species were selected because they blended into the dark soot-stained tree-trunks of industrial cities, and therefore escaped bird predation. Such pollution is now strongly legislated against in much of the Global North, and although cities in poor countries now have the worst levels of air pollution in the world (https://www.cbsnews.com/pictures/the-most-polluted-cities-in-the-world-ranked/51/), it is unclear whether this pollution produces conditions that might favour different morphs of any polymorphic species. Climate warming and urban heat islands may also select some morphs over others owing to the needs of thermoregulation, although this too does not seem to have been investigated.

Water pollution and changes to stream ecosystems also influence invertebrates in urban environments. The greater run-off associated with impervious surfaces, altered hydrology and geomorphology of streams, plus increased concentrations of nutrients, pesticides, metals, and other contaminants tend to reduce diversity and significantly change community composition of freshwater invertebrates (Paul and Meyer 2001). The combination of destruction of riparian vegetation, the "heat island" effect, and reduced groundwater recharge can see higher average water temperatures, and greater variation in water temperatures (LeBlanc et al. 1997). Freshwater taxa are particularly sensitive to water quality changes in urban environments, and unsurprisingly include the Ephemeroptera, Plecoptera, and Trichoptera (Hachmiiller and Matthews 1991). The news may not be all bad from a human point of view though, as pollution in water is thought to inhibit successful development of the *Anopheles* mosquitoes, effectively reducing malarial infection rates (Duchemin et al. 2003). Again, most work on how water pollution associated with urban environments impacts invertebrates has been conducted in the Global North. However, pollution in the Nile River, Cairo, was associated with declines in taxon richness of both insects and molluscs (Fishar et al. 2006).

As areas become more urbanised, the relative abundance of exotic to native plant species tends to increase (Gavier-Pizarro et al. 2010). This change is not necessarily associated with a change in invertebrate species richness overall, but again, certain species can be negatively affected (Samways et al. 1996). From a biomass point of view however, there are three good reasons to presume that invertebrate biomass should decline with greater abundance of exotic garden plants: (1) exotic plant species have escaped their invertebrate natural enemies, in their "new" environments (Kirichenko et al. 2013), (2) plant species selected for gardens are often selected for being relatively "pest free" in the first place e.g. *Lantana* species, and (3) indigenous invertebrates and introduced plant species have separate evolutionary histories (Tallamy 2004). This lack of shared evolutionary history is reflected in the fact that less than 10% of herbivorous insects feed on plants from more than three families (Bernays and Graham 1988).

Nevertheless, some invertebrate species can thrive in cities with exotic plants and offer the opportunity to explore not only urban entomology, but the socio-economic implications of invertebrate population demographics in the Global South. For example, the introduction of the invasive Polyphagous Shot Hole Borer beetle (PSHB; *Euwallacea whitfordiodendrus*) and its associated tree-killing fungus (*Fusarium euwallacea*) from southeast Asia into urban Africa, has important socio-ecological consequences for urban trees (Paap et al. 2018). The beetle and fungus are known to attack more than 200 tree species globally, causing damage and/or death to novel host tree species that have not evolved resistance to either of the pests (Eskalen et al. 2013). This is significant, as the beetle and fungus are now well established in more than 20 major city centres around South Africa and represents a mammoth management challenge to many already overstretched municipalities (Paap et al. 2018). This includes the metropolitan City of Johannesburg, which is considered to have the largest human-made forest in the southern hemisphere, with more than 10 million trees lining the streets (Fig. 2e) (Langer and McNamara 2011). Interestingly, the distribution of the host tree species in this city, which are mostly alien trees for landscaping purposes, is largely driven by people's income and social status (Shackleton et al. 2015). For example, the generally low-income settlements of Johannesburg have only recently been the target of urban greening projects (Lembani 2016), while the more affluent northern suburbs are lined by 100-year-old European oaks (*Quercus robur*) and London plane trees (*Platanus* spp.) (Schäffler and Swilling 2013). The societal consequences of this disparity are already apparent from the volume of social media devoted to the pest and measures for its control in the north of Johannesburg, while Soweto, in the south has been largely silent. Both locations nevertheless house people with a broad range of incomes, and possibly different attitudes to urban trees and urban pests (Soga and Gaston 2016), and which lends itself to comparative studies between these neighbouring locations, and perhaps further afield with towns in different climatic zones of the country.

7.1 Case Study: Global South Perspective on Urban Predators

Predators in the city are still rarely studied but they add substantially to understanding of functional diversity. Within the predator guild, raptors are among the better-studied groups in urban ecology (Magle et al. 2012), and many species manage to persist in cities and adapt to anthropogenic influences despite their sensitivity to global change (e.g. Love and Bird 2000; Boal and Dykstra 2018). To date, the majority of urban raptor research occurs in North America and Europe (e.g. Boal and Dykstra 2018; Kettel et al. 2018). However, findings in one region may not necessarily be representative of populations of the same species, or functional group, from other regions (Mannan and Steidl 2018). In fact, of the 389 studies published between 1990 and 2018 [ISI web of knowledge search 20 May 2018, term "urban raptor"], 82%

were conducted in the Global North (Europe, North America, most part of Asia and Oceania) and only 10% were conducted in South America, 5% in Africa, and 3% in India. Even if the proportion of land mass, as a proxy for the number of cities per region, is taken into account, the contributions only shift marginally: 77% from the Global North, and 23% from the Global South, whereby Europe contributes the most to urban raptor studies from a relatively small land area, and Africa contributes the least from a relatively large land area. Two recent reviews in ornithological research that specifically focused on urban raptors (Donázar et al. 2016; Kettel et al. 2018) underline this research gap further. In fact, Kettel et al. (2018) does not include a single study from the African continent. Donázar et al. (2016) focus on ecosystem services and African examples are limited to the vulture decline (Bamford et al. 2009; Ogada et al. 2016). This lack of research is despite Africa hosting its own unique biodiversity of raptor species (Amar et al. 2018), that over the last decades, numbers and diversity of raptors have declined dramatically (Garbett et al. 2018; McClure et al. 2018), and that urban areas might create valuable opportunities for public engagement and urban conservation of raptors.

This in itself might create a severe bias towards Northern systems, which may lead to quite narrow conclusions regarding urban raptor behaviour, ecology, and conservation. For example, one aspect that is largely missing from the current literature, is ecosystem service provision by raptors in cities. While in the Global North, these services are only highlighted regarding the predation of "pest" species (for instance pigeons and rodents; Sekercioglu 2006), we encounter a much broader variety in Global South systems, ranging from the accumulation of scavengers in cities due to limited waste management (Chamberlain et al. 2017), to high raptor densities associated with ritualised animal feeding, specifically in Muslim communities (Kumar et al. 2019). The extent to which such a high abundance of scavengers is due to climatic or biogeographic influences, socio-economic factors, or cultural perceptions is yet to be assessed. Regarding ecosystem services, the value of raptors has only been explored to some extent in more natural environments, for example, vultures scavenging in India, which has an estimated annual monetary value of $2 billion in terms of healthcare savings (Markandya et al. 2008). Quantitative data from the urban environment are still absent.

With large carnivores, human–wildlife conflicts arise from perceived threats to the safety of humans and domestic animals, which generate the need for management and removal strategies (Bjerke et al. 2003; Kleiven et al. 2004; Gavashelishvili and Lukarevskiy 2008; Redpath et al. 2013). Usually, only smaller carnivores are able to obtain their entire needs from urban resources, while larger species only incorporate urban areas in their extended home ranges and are typically transient visitors to scavenge human refuse (Beckmann and Lackey 2008; Gehrt 2010; Abay et al. 2011). This knowledge is, however, largely based on Northern systems, and patterns could be very different in tropical cities, where more powerful carnivores might become resident. An example of a large avian predator is the Crowned Eagle (*Stephanoaetus coronatus*) in Durban and Pietermaritzburg, South Africa, with 30 breeding pairs alone in the Metropolitan green-space system (McPherson et al. 2016a). Human–wildlife conflicts can occur because juvenile and immature eagles occasionally predate on pets, such as cats and small dogs, and domestic stock comprised approximately 6% of prey items provisioned to nests (McPherson et al. 2016b). There are, importantly very rare, reports on predatory attacks on children (Steyn 1983; Thomsett 2011)—although both records occurred outside urban environments—adding to the negative perception of larger carnivores living near humans.

Despite these possible threats to human safety, there is growing recognition of the important roles played by predators in sustaining biodiversity (Ritchie and Johnson 2009), because of their positive ecological influence through top-down regulation of trophic systems (Faeth et al. 2005). When populations of apex predators are reduced or go extinct, previously suppressed mesopredator populations may erupt, a phenomenon known as mesopredator release (Soulé et al. 1988; Courchamp et al. 1999; Crooks and Soulé 1999). This can lead to trophic cascades and the local extinction of shared prey (Courchamp et al. 1999). For instance, Vervet Monkeys (*Chlorocebus pygerythrus*), here considered mesopredators, are particularly successful urban exploiters in cities, and can predict human behaviours and respond accordingly (Ramkissoon 2006; Fuentes et al. 2007). This enables them to access urban areas that their natural predators, Leopard (*Panthera pardus*), African Rock Python (*Python natalensis*), and African Crowned Eagle (*Stephanoaetus coronatus*), cannot always access (Hart 2007); example from (McPherson 2015). These case studies underline that urban ecosystems can host species over many trophic levels and can potentially create refugia for predators.

8 Conclusions

Our synthesis has shown clear parallels between the Global North and South as to how animal diversity responds to the patterns and processes of urbanisation. This is especially clear when considering that there is a marked trend of decline in species diversity with increasing levels of urbanisation and a shift towards homogenous animal communities comprised of small-bodied generalist species. However, there are also some startling differences, most of which emerge due to vastly different human socio-cultural and socio-economic factors between cities of the Global North and South. The novel socio-ecological template of Global South cities represents an overlooked force in driving patterns of animal diversity in urban ecology and acknowledgement of these factors in urban ecological research will greatly improve understanding of the effects of urban development on biodiversity. Quantifying these factors might be difficult and represents a challenge to urban ecology of the Global South moving forward. This is incredibly relevant in the cities of the Global South, where environmental injustice persists, and certain sectors of society experience limited access to biodiversity due to their socio-economic status.

Finally, the synthesis highlights that the vast majority of studies on animal communities in the Global South have considered urbanisation only in spatial terms (i.e. the gradient approach) rather than in terms of analysing impacts of urbanisation over time. This illustrates a general problem in biodiversity monitoring as there are few rigorous data from long-term programmes in high biodiversity areas in the Global South. Nevertheless, some localised studies have shown some interesting possible indicators of urbanisation (e.g. declines in insectivorous species in Kampala, Uganda; Chamberlain et al. 2018). Furthermore, in many areas, the future is looking brighter, with monitoring programmes having been established in several African countries over the past decade (Wotton et al. 2020) and includes the use of citizen science projects such as the Southern African Bird Atlas Project (www.SABAP2.org.za) and iNaturalist (www.inaturalist.org/). Indeed, urban locations provide excellent opportunities for implementing citizen science projects that achieve the dual goal of monitoring biodiversity and helping link urban inhabitants with nature.

References

Abay GY, Bauer H, Gebrihiwot K, Deckers J (2011) Peri-urban spotted hyena (*Crocuta crocuta*) in Northern Ethiopia: diet, economic impact, and abundance. Eur J Wildl Res 57:759–765

Ackley JW, Wu J, Angilletta MJ, Myint SW, Sullivan B (2015) Rich lizards: how affluence and land cover influence the diversity and abundance of desert reptiles persisting in an urban landscape. Biol Conserv 182:87–92

Ahrné K, Bengtsson J, Elmqvist T (2009) Bumble bees (*Bombus* spp) along a gradient of increasing urbanisation. PLoS ONE 4:e5574

Alberti M, Marzluff JM, Shulenberger E, Bradley G, Ryan C, Zumbrunnen C (2003) Integrating humans into ecology: opportunities and challenges for studying urban ecosystems. Biosci 53:1169–1179

Alvaredo F, Chancel L, Piketty T, Saez E, Zucman G (2018) World inequality report 2018. Harvard University Press, London

Amar A, Buij R, Suri J, Sumasgutner P, Virani MZ (2018) Conservation and ecology of African raptors. In: Sarasola J, Grande J, Negro J (eds) Birds of prey. Springer, Cham, pp 419–455

Arnfield AJ (2003) Two decades of urban climate research: a review of turbulence, exchanges of energy and water, and the urban heat island. Int J Climatol 23:1–26

Ashley EP, Kosloski A, Petrie SA (2007) Incidence of intentional vehicle–reptile collisions. Hum Dimen Wildl 12:137–143

Babbit K, Tarr T (2002) Effects of habitat complexity and predator identity on predation of *Rana clamitans* larvae. Amphibia-Reptilia 23:13–20

Baldock KC, Goddard MA, Hicks DM, Kunin WE, Mitschunas N, Morse H, Osgathorpe LM, Potts SG, Robertson KM, Scott AV, Staniczenko PP, Stone GN, Vaughan IP, Memmott J (2019) A systems approach reveals urban pollinator hotspots and conservation opportunities. Nat Ecol Evol 3:363–373

Ballouard J-M, Ajtic R, Balint H, Brito JC, Crnobrnja-Isailovic J, Desmonts D, ElMouden EH, Erdogan M, Feriche M, Pleguezuelos JM, Prokop P, Sánchez A, Santos X, Slimani T, Tomovic L, Uşak M, Zuffi M, Bonnet X (2013) School children and one of the most unpopular animals: are they ready to protect snakes? Anthrozoös 26:93–109

Bamford AJ, Monadjem A, Anderson MD, Anthony A, Borello WD, Bridgeford M, Bridgeford P, Hancock P, Howells B, Wakelin J, Hardy IC (2009) Trade-offs between specificity and regional generality in habitat association models: a case study of two species of African vulture. J Appl Ecol 46:852–860

Batáry P, Kurucz K, Suarez-Rubio M, Chamberlain DE (2018) Non-linearities in bird responses across urbanisation gradients: a meta-analysis. Glob Change Biol 24:1046–1054

Beckmann JP, Lackey CW (2008) Carnivores, urban landscapes, and longitudinal studies: a case history of black bears. Hum Wildl Confl 2:168–174

Belaire JA, Westphal LM, Whelan CJ, Minor ES (2015) Urban residents' perceptions of birds in the neighborhood: biodiversity, cultural ecosystem services, and disservices. The Condor 117:192–202

Beninde J, Veith M, Hochkirch A (2015) Biodiversity in cities needs space: a meta-analysis of factors determining intra-urban biodiversity variation. Ecol Lett 18:581–592

Bennie J, Davies TW, Cruse D, Inger R, Gaston KJ (2018) Artificial light at night causes top-down and bottom-up trophic effects on invertebrate populations. J Appl Ecol 55:2698–2706

Bennie J, Duffy J, Davies T, Correa-Cano M, Gaston K, Bennie J, Duffy JP, Davies TW, Correa-Cano ME, Gaston KJ (2015) Global trends in exposure to light pollution in natural terrestrial ecosystems. Remote Sens 7:2715–2730

Bernays E, Graham M (1988) On the evolution of host specificity in phytophagous arthropods. Ecol 69:886–892

Bhatt D, Joshi KK (2011) Bird assemblages in natural and urbanised habitats along elevational gradient in Nainital district (western Himalaya) of Uttarakhand state, India. Curr Zool 57:318–329

Bickford D, Ng TH, Qie L, Kudavidanage EP, Bradshaw CJ (2010) Forest fragment and breeding habitat characteristics explain frog diversity and abundance in Singapore. Biotrop 42:119–125

Bjerke T, Østdahl T, Kleiven J (2003) Attitudes and activities related to urban wildlife: pet owners and non-owners. Anthrozoös 16:252–262

Blaustein AR, Wake DB, Sousa WP (1994) Amphibian declines: judging stability, persistence, and susceptibility of populations to local and global extinctions. Conserv Biol 8:60–71

Boal CW, Dykstra CR (2018) Urban raptors. Island Press, Washington, DC

Breuste J, Niemelä J, Snep RP (2008) Applying landscape ecological principles in urban environments. Landsc Ecol 23:1139–1142

Burkhardt K, Loxton H, Muris P (2003) Fears and fearfulness in South-African children. Behav Change 20:94–102

Cane JH, Minckley RL, Kervin LJ, Roulston TH, Williams NM (2006) Complex responses within a desert bee guild (Hymenoptera: Apiformes) to urban habitat fragmentation. Ecol Appl 16:632–644

Casey RE, Shaw AN, Massal LR, Snodgrass JW (2005) Multimedia evaluation of trace metal distribution within stormwater retention ponds in suburban Maryland, USA. Bull Environ Contam Toxicol 74:273–280

Chace JF, Walsh JJ (2006) Urban effects on native avifauna: a review. Landsc Urb Plan 74:46–69

Chamberlain DE, Henry DA, Reynolds C, Caprio E, Amar A (2019) The relationship between wealth and biodiversity: a test of the Luxury Effect on bird species richness in the developing world. Glob Change Biol 25:3045–3055

Chamberlain D, Kibuule M, Skeen R, Pomeroy D (2017) Trends in bird species richness, abundance and biomass along a tropical urbanisation gradient. Urban Ecosyst 20:629–638

Chamberlain D, Kibuule M, Skeen RQ, Pomeroy D (2018) Urban bird trends in a rapidly growing tropical city. Ostrich 89:275–280

Chamberlain DE, Reynolds C, Amar A, Henry DA, Caprio E, Batáry P (2020) Wealth, water and wildlife: landscape aridity intensifies the urban luxury effect. Glob Ecol Biogeogr 29:1595–1605

Colvin BA, Jackson WB (1999) Urban rodent control programmes for the 21st century. In: Singleton GR, Hinds LA, Leirs H, Zhang Z (eds) Ecologically-based management of rodent pests. Australian Centre for International Agricultural Research, Canberra, pp 243–258

Cook LM, Saccheri IJ (2013) The peppered moth and industrial melanism: evolution of a natural selection case study. Hered 110:207–212

Courchamp F, Langlais M, Sugihara G (1999) Cats protecting birds: modelling the mesopredator release effect. J Anim Ecol 68:282–292

Cregg BM, Dix ME (2001) Tree moisture stress and insect damage in urban areas in relation to heat island effects. J Arboric 27:8–17

Crooks KR, Soulé ME (1999) Mesopredator release and avifaunal extinctions in a fragmented system. Nat 400:563–566

Dacke M, Baird E, Byrne M, Scholtz CH, Warrant EJ (2013) Dung Beetles use the Milky Way for orientation. Curr Biol 23:298–300
Dacke M, Byrne MJ, Scholtz CH, Warrant EJ (2004) Lunar orientation in a beetle. Proc Royal Soc B 271:361–365
Dacke M, Nilsson DE, Scholtz CH, Byrne M, Warrant EJ (2003) Insect orientation to polarized moonlight. Nat 424:33
Davids R, Rouget M, Boon R, Roberts D (2016) Identifying ecosystem service hotspots for environmental management in Durban, South Africa. Bothalia 46:1–18
Dearborn DC, Kark S (2010) Motivations for conserving urban biodiversity. Conserv Biol 24:432–440
Dickie IA, Bennett BM, Burrows LE, Nunez MA, Peltzer DA, Porté A, Richardson DM, Rejmánek M, Rundel PW, Van Wilgen BW (2014) Conflicting values: ecosystem services and invasive tree management. Biol Inv 16:705–719
Doherty TS, Glen AS, Nimmo DG, Ritchie EG, Dickman CR (2016) Invasive predators and global biodiversity loss. Proc Nat Acad Sci 113:11261–11265
Donázar JA, Cortés-Avizanda A, Fargallo JA, Margalida A, Moleón M, Morales-Reyes Z, Moreno-Opo R, Pérez-García JM, Sánchez-Zapata JA, Zuberogoitia I, Serrano D (2016) Roles of raptors in a changing world: from flagships to providers of key ecosystem services. Ardeola 63:181–234
Dreistadt S, Dahlsten DL, Frankie GW (1990) Urban forests and insect ecology. Biosci 40:192–198
Duchemin J-B, Macintyre K, Warren M, Keating J, Robert V, Beier JC, Trape J-F (2003) Malaria transmission in urban sub-Saharan Africa. Am J Trop Med Hyg 68:169–176
Dunn RR, Gavin MC, Sanchez MC, Solomon JN (2006) The Pigeon Paradox: dependence of global conservation on urban nature. Conserv Biol 20:1814–1816
Dures SG, Cumming GS (2010) The confounding influence of homogenising invasive species in a globally endangered and largely urban biome: does habitat quality dominate avian biodiversity? Biol Conserv 143:768–777
Early R, Bradley BA, Dukes JS, Lawler JJ, Olden JD, Blumenthal DM, Gonzalez P, Grosholz ED, Ibañez I, Miller LP, Sorte CJ, Tatem AJ (2016) Global threats from invasive alien species in the twenty-first century and national response capacities. Nat Comm 7:12485
Edge DA, Cilliers SS, Terblanche RF (2008) Vegetation associated with the occurrence of the Brenton blue butterfly. S Afr J Sci 104:505–510
Egan RS, Paton PW (2004) Within-pond parameters affecting oviposition by wood frogs and spotted salamanders. Wetlands 24:1–13
el Jundi B, Warrant EJ, Byrne MJ, Khaldy L, Baird E, Smolka J, Dacke M (2015) Neural coding underlying the cue preference for celestial orientation. PNAS112:11395–11400
Entiauspe-Neto O, Perleberg T, de Freitas MA (2016) Herpetofauna from an urban Pampa fragment in southern Brazil: composition, structure and conservation. Check List 12:1–15
Eskalen A, Stouthamer R, Lynch SC, Rugman-Jones PF, Twizeyimana M, Gonzalez A, Thibault T (2013) Host range of fusarium dieback and its ambrosia beetle (Coleoptera: Scolytinae) vector in Southern California. Plant Dis 97:938–951
Evans KL, Chamberlain DE, Hatchwell BJ, Gregory RD, Gaston KJ (2011) What makes an urban bird? Glob Change Biol 17:32–44
Faeth SH, Bang C, Saari S (2011) Urban biodiversity: patterns and mechanisms. Ann N Y Acad Sci 1223:69–81
Faeth SH, Warren PS, Shochat E, Marussich WA (2005) Trophic dynamics in urban communities. Biosci 55:399–407
Fahrig L (2013) Rethinking patch size and isolation effects: the habitat amount hypothesis. J Biogeogr 40:1649–1663
Farrell BD (1998) "Inordinate fondness" explained: why are there so many beetles? Sci 281:555–559
Feldhamer GA, Farris-Renner KC, Barker CM (1988) *Dama dama*. Mamm Spec 317:1–8
Filloy J, Grosso S, Bellocq MI (2015) Urbanisation altered latitudinal patterns of bird diversity-environment relationships in the southern Neotropics. Urban Ecosyst 18:777–791

Fishar MR, Thorne R, Williams WP (2006) Physico-chemical conditions and macroinvertebrate fauna in the River Nile from Aswan to Cairo. Afr J Aquat Sci 31:247–259

Foord SH, Swanepoel LH, Evans SW, Schoeman CS, Erasmus BF, Schoeman MC, Keith M, Smith A, Mauda EV, Maree N, Nembudani N, Dippenaar-Schoeman AS, Munyai TC, Taylor PJ (2018) Animal taxa contrast in their scale-dependent responses to land use change in rural Africa. PLoS One 13: e0194336

Forbes AA, Bagley RK, Beer MA, Hippee AC, Widmayer HA (2018) Quantifying the unquantifiable: why Hymenoptera, not Coleoptera, is the most speciose animal order. BMC Ecol 18:21

Foster JJ, el Jundi B, Smolka J, Khaldy L, Nilsson DE, Byrne MJ, Dacke M (2017) Stellar performance: mechanisms underlying Milky Way orientation in dung beetles. Philos Trans 372:20160079

Frick WF, Puechmaille SJ, Willis CK (2016) White-Nose syndrome in Bats. In: Voigt CC, Kingston T (eds) Bats in the Anthropocene: conservation of bats in a changing world. Springer, Berlin, pp 245–262

Fuentes A, O'Neill N, Shaw E, Cortés JE (2007) Humans, monkeys, and the rock: the anthropogenic ecology of the Barbary macaques in the Upper Rock Nature Reserve, Gibraltar. Almoraima: revista de estudios Campo Gibraltareños 35:87–97

Fuller RA, Irvine KN, Devine-Wright P, Warren PH, Gaston KJ (2007) Psychological benefits of greenspace increase with biodiversity. Biol Lett 3:390–394

Gaertner M, Wilson JR, Cadotte MW, MacIvor JS, Zenni RD, Richardson DM (2017) Non-native species in urban environments: patterns, processes, impacts and challenges. Biol Inv 19:3461–3469

Galbraith JA, Beggs JR, Jones DN, Stanley MC (2015) Supplementary feeding restructures urban bird communities. Proc Nat Acad Sci 112:2648–2657

Garbett R, Herremans M, Maude G, Reading RP, Amar A (2018) Raptor population trends in northern Botswana: A re-survey of road transects after 20 years. Biol Conserv 224:87–99

Garden J, Mcalpine C, Peterson A, Jones D, Possingham H (2006) Review of the ecology of Australian urban fauna: a focus on spatially explicit processes. Aust Ecol 31:126–148

Garden JG, McAlpine CA, Possingham HP (2010) Multi-scaled habitat considerations for conserving urban biodiversity: native reptiles and small mammals in Brisbane, Australia. Landsc Ecol 25:1013–1028

Gaston KJ, Bennie J, Davies TW, Hopkins J (2013) The ecological impacts of nighttime light pollution: a mechanistic appraisal. Biol Rev 88:912–927

Gavashelishvili A, Lukarevskiy V (2008) Modelling the habitat requirements of leopard *Panthera pardus* in west and central Asia. J Appl Ecol 45:579–588

Gavier-Pizarro GI, Radeloff VC, Stewart SI, Huebner CD, Keuler NS (2010) Housing is positively associated with invasive exotic plant species richness in New England, USA. Ecol Appl 20:1913–1925

Gehrt SD (2010) The urban ecosystem. In: Gehrt SD, Riley SP, Cypher BL (eds) Urban carnivores: ecology, conflict, and conservation. John Hopkins University Press, Baltimore, MD, pp 3–11

Gehrt SD, Anchor C, White LA (2009) Home range and landscape use of coyotes in a metropolitan landscape: conflict or coexistence? J Mammal 90:1045–1057

Gehrt SD, Chelsvig JE (2003) Bat activity in an urban landscape: patterns at the landscape and microhabitat scale. Ecol Appl 13:939–950

Gehrt SD, Riley SP, Cypher BL (2010) Urban carnivores: ecology, conflict, and conservation. John Hopkins University Press, Baltimore, MD

Germaine SS, Wakeling BF (2001) Lizard species distributions and habitat occupation along an urban gradient in Tucson, Arizona, USA. Biol Conserv 97:229–237

Geschke A, James S, Bennett AF, Nimmo DG (2018) Compact cities or sprawling suburbs? optimising the distribution of people in cities to maximise species diversity. J Appl Ecol 55:2320–2331

Geslin B, Gauzens B, Thébault E, Dajoz I (2013) Plant pollinator networks along a gradient of urbanisation. PLoS ONE 8:e63421
Gilbert M, Sokha C, Joyner PH, Thomson RL, Poole C (2012) Characterising the trade of wild birds for merit release in Phnom Penh, Cambodia and associated risks to health and ecology. Biol Conserv 153:10–16
Godefroid S, Ricotta C (2018) Alien plant species do have a clear preference for different land uses within urban environments. Urban Ecosyst 21:1189–1198
Goodman SM (2011) Les chauvres-souris de Madagascar. Association Vahatra, Antananarivo
Gotlieb A, Hollender Y, Mandelik Y (2011) Gardening in the desert changes bee communities and pollination network characteristics. Basic Appl Ecol 12:310–320
Green RE, Cornell SJ, Scharlemann JP, Balmford A (2005) Farming and the fate of wild nature. Sci 307:550–555
Grimm NB, Faeth SH, Golubiewski NE, Redman CL, Wu J, Bai X, Briggs JM (2008) Global change and the ecology of cities. Sci 319:756–760
Gupta A (2002) Geoindicators for tropical urbanisation. Environ Geol 42:736–742
Gurnell J, Wauters LA, Lurz PW, Tosi G (2004) Alien species and interspecific competition: effects of introduced eastern grey squirrels on red squirrel population dynamics. J Anim Ecol 73:26–35
Haase D, Larondelle N, Andersson E, Artmann M, Borgström S, Breuste J, Gomez-Baggethun E, Gren Å, Hamstead Z, Hansen R, Kabisch N, Kremer P, Langemeyer J, Rall EL, McPhearson T, Pauleit S, Qureshi S, Schwarz N, Voigt A, Wurster D, Elmqvist T (2014) A quantitative review of urban ecosystem service assessments: concepts, models, and implementation. Ambio 43:413–433
Hachmiiller B, Matthews RA (1991) Effects of riparian community structure, sediment size, and water quality on the macroinvertebrate communities in a small, suburban stream. Northwest Sci 65:125–132
Hallam TG, McCracken GF (2011) Management of the Panzootic White-Nose Syndrome through culling of bats. Conserv Biol 25:189–194
Hamer AJ, McDonnell MJ (2008) Amphibian ecology and conservation in the urbanising world: a review. Biol Conserv 141:2432–2449
Hamer AJ, McDonnell MJ (2009) The response of herpetofauna to urbanisation: inferring patterns of persistence from wildlife databases. Aust Ecol 35:568–580
Happold M, Happold D (2013) Mammals of Africa Vol IV: hedgehogs, shrews and bats. Bloomsbury, London
Harrison T, Winfree R (2015) Urban drivers of plant-pollinator interactions. Funct Ecol 29:879–888
Hart D (2007) Predation on primates: a biogeographical analysis. In: Gursky SL, Nekaris KA (eds) Primate anti-predator strategies: developments in primatology—progress and prospects. Springer, Boston, MA, pp 27–59
Hau M (2001) Timing of breeding in variable environments: tropical birds as model systems. Horm Behav 40:281–290
Hennig EI, Ghazoul J (2012) Pollinating animals in the urban environment. Urban Ecosyst 15:149–166
Hill CJ (1996) Habitat specificity and food preferences of an assemblage of tropical Australian dung beetles. J Trop Ecol 12:449–460
Hoffman TS, O'Riain JM (2012) Troop size and human-modified habitat affect the ranging patterns of a chacma baboon population in the Cape Peninsula, South Africa. Am J Primatol 74:853–863
Hölker F, Moss T, Griefahn B, Kloas W, Voigt CC, Henckel D, Hänel A, Kappeler PM, Völker S, Schwope A, Franke S, Uhrlandt D, Fischer J, Klenke R, Wolter C, Tockner K (2010) The dark side of light: a transdisciplinary research agenda for light pollution policy. Ecol Soc 15:13
Hope D, Gries C, Zhu W, Fagan WF, Redman CL, Grimm NB, Nelson AL, Martin C, Kinzig A (2003) Socioeconomics drive urban plant diversity. Proc Nat Acad Sci 100:8788–8792
Houston WWK, Feehan JE, Runko S (1982) Methods for harvesting large numbers of dung beetles (Coleoptera: Scarabaeidae). Aust J Entomol 21:217–219
Howden HF, Howden AT, Storey RI (1991) Nocturnal Perching of Scarabaeine Dung Beetles (Coleoptera, Scarabaeidae) in an Australian Tropical Rain Forest. Biotrop 23:51–57

Hulme MF, Vickery JA, Green RE, Phalan B, Chamberlain DE, Pomeroy DE, Nalwanga D, Mushabe D, Katebaka R, Bolwig S, Atkinson PW (2013) Conserving the birds of Uganda's Banana-Coffee Arc: land sparing and land sharing compared. PLoS ONE 8:e54597

Ives CD, Lentini PE, Threlfall CG, Ikin K, Shanahan DF, Garrard GE, Bekessy SA, Fuller RA, Mumaw L, Rayner L, Rowe R, Valentine LE, Kendal D (2016) Cities are hotspots for threatened species. Glob Ecol Biogeog 25:117–126

Johnson MT, Munshi-South J (2017) Evolution of life in urban environments. Sci 358:eaam8327

Jones EL, Leather SR (2012) Invertebrates in urban areas: a review. Eur J Entomol 109:463–478

Jung K, Threlfall CG (2018) Trait-dependent tolerance of bats to urbanisation: a global meta-analysis. Proc Royal Soc B 285:20181222

Kaiser A, Merckx T, Van Dyck H (2016) The urban heat island and its spatial scale dependent impact on survival and development in butterflies of different thermal sensitivity. Ecol Evol 6:4129–4140

Kaltenborn BP, Bjerke T, Nyahongo J (2006) Living with problem animals: self-reported fear of potentially dangerous species in the Serengeti Region, Tanzania. Hum Dimen Wildl 11:397–409

Kamp J, Urazaliev R, Balmford A, Donald PF, Green RE, Lamb AJ, Phalan B (2015) Agricultural development and the conservation of avian biodiversity on the Eurasian steppes: a comparison of land-sparing and land-sharing approaches. J Appl Ecol 52:1578–1587

Kark S, Iwaniuk A, Schalimtzek A, Banker E (2007) Living in the city: can anyone become an "urban exploiter"? J Biogeogr 34:638–651

Karr J, Roth RR (1971) Vegetation structure and avian diversity in several new world areas. Am Nat 105:423–435

Kettel EF, Gentle LK, Quinn JL, Yarnell RW (2018) The breeding performance of raptors in urban landscapes: a review and meta-analysis. J Ornith 159:1–18

Kettlewell HB (1955) Selection experiments on industrial melanism in the Lepidoptera. Hered 9:323–342

Khera N, Mehta V, Sabata BC (2009) Interrelationship of birds and habitat features in urban greenspaces in Delhi, India. Urban For Urban Green 8:187–196

Kinzig AP, Warren P, Martin C, Hope D, Katti M (2005) The effects of human socioeconomic status and cultural characteristics on urban patterns of biodiversity. Ecol Soc 10:23

Kirichenko N, Péré C, Baranchikov Y, Schaffner U, Kenis M (2013) Do alien plants escape from natural enemies of congeneric residents? Yes but not from all. Biol Inv 15:2105–2113

Kleiven J, Bjerke T, Kaltenborn BP (2004) Factors influencing the social acceptability of large carnivore behaviours. Biodiv Conserv 13:1647–1658

Kruger DJ, Hamer AJ, Du Preez LH (2015) Urbanisation affects frog communities at multiple scales in a rapidly developing African city. Urban Ecosyst 18:1333–1352

Kumar N, Gupta U, Malhotra H, Jhala YV, Qureshi Q, Gosler AG, Sergio F (2019) The population density of an urban raptor is inextricably tied to human cultural practices. Phil Trans Royal Society B 286:20182932

Kumar N, Qureshi Q, Jhala YV, Gosler AG, Sergio F (2018) Offspring defense by an urban raptor responds to human subsidies and ritual animal-feeding practices. PLoS ONE 13:e0204549

Langer H, McNamara J (2011) African green city index. Economist Intelligence Unit, Munich

LeBlanc RT, Brown RD, FitzGibbon JE (1997) Modeling the effects of land use change on the water temperature in unregulated urban streams. J Environ Manage 49:445–469

Lembani RL (2016) Greening Soweto: calculating above-ground tree biomass, stored carbon and net economic value. MSc Thesis. University of the Witwatersrand, Johannesburg

Leong M, Dunn RR, Trautwein MD (2018) Biodiversity and socioeconomics in the city: a review of the luxury effect. Biol Lett 14:20180082

Leroy EM, Kumulungui B, Pourrut X, Rouquet P, Hassanin A, Yaba P, Délicat A, Paweska JT, Gonzalez J-P, Swanepoel R (2005) Fruit bats as reservoirs of Ebola virus. Nat 438:575–576

Leveau CM, Leveau LM (2005) Avian community response to urbanisation in the Pampean region, Argentina. Ornit Neotrop 16:503–510

Lewanzik D, Voigt CC (2014) Artificial light puts ecosystem services of frugivorous bats at risk. J Appl Ecol 51:388–394

Lin BB, Fuller RA (2013) Sharing or sparing? How should we grow the world's cities? J Appl Ecol 50:1161–1168

Longcore T, Rich C, Sullivan LM (2009) Critical assessment of claims regarding management of feral cats by trap-neuter-return. Conserv Biol 23:887–894

Love OP, Bird DM (2000) Raptors in urban landscapes: a review and future concerns. In: Fifth World Conference on Birds of Prey and Owls, British Columbia, pp 425–434

Lowe S, Browne M, Boudjelas S, de Poorter M (2000) 100 of the world's worst invasive alien species: a selection from the global invasive species database. The Invasive Species Specialist Group (ISSG), The World Conservation Union (IUCN), Auckland

Lowry H, Lill A, Wong BB (2013) Behavioural responses of wildlife to urban environments. Biol Rev 88:537–549

Loyd KA, Hernandez SM, Carroll JP, Abernathy KJ, Marshall GJ (2013) Quantifying free-roaming domestic cat predation using animal-borne video cameras. Biol Conserv 160:183–189

MacArthur RH, Wilson EO (1967) The theory of island biogeography. Princeton University Press, Princeton

MacGregor-Fors I, Morales-Perez L, Schondube JE (2012) From forests to cities: effects of urbanisation on tropical birds. In: Lepczyk CA, Warren PS (eds) Urban bird ecology and conservation. University of California Press, Berkeley, CA, pp 33–48

MacGregor-Fors I, Schondube JE (2011) Gray vs. green urbanisation: relative importance of urban features for urban bird communities. Basic Appl Ecol 12:372–381

Magle SB, Hunt VM, Vernon M, Crooks KR (2012) Urban wildlife research: past, present, and future. Biol Conserv 155:23–32

Malan G, Robinson ER (2001) Nest-site selection by Black Sparrowhawks *Accipiter melanoleucus*: implications for managing exotic pulpwood and sawlog forests in South Africa. Environ Manage 28:195–205

Malan G, Shultz S (2002) Nest-site selection of the crowned hawk-eagle in the forests of KwaZulu-Natal, South Africa, and Tai, Ivory Coast. J Raptor Res 36:300–308

Manfrin A, Singer G, Larsen S, Weiß N, van Grunsven RHA, Weiß NS, Wohlfahrt S, Monaghan MT, Hölker F (2017) Artificial light at night affects organism flux across ecosystem boundaries and drives community structure in the recipient ecosystem. Front Environ Sci 5:61

Mannan RW, Steidl RJ (2018) Demography of raptor populations in urban environments. In: Boal CW, Dykstra CR (eds) Urban raptors. Island Press, Washington, DC, pp 51–63

Markandya A, Taylor T, Longo A, Murty MN, Murty S, Dhavala K (2008) Counting the cost of vulture decline: an appraisal of the human health and other benefits of vultures in India. Ecol Econ 67:194–204

Markotter W, Randles J, Rupprecht CE, Sabeta CT, Taylor PJ, Wandeler AI, Nel LH (2006) Lagos bat virus, South Africa. Emerg Infect Dis 12:504–506

Marsh DM, Trenham PC (2001) Metapopulation dynamics and amphibian conservation. Conserv Biol 15:40–49

Martin RO, Sebele L, Koeslag A, Curtis O, Abadi F, Amar A (2014) Phenological shifts assist colonisation of a novel environment in a range-expanding raptor. Oikos 123:1457–1468

Marzluff JM (2017) A decadal review of urban ornithology and a prospectus for the future. Ibis 159:1–13

Mason CF (2000) Thrushes now largely restricted to the built environment in eastern England. Div Distrib 6:189–194

McClure CJ, Westrip JR, Johnson JA, Schulwitz SE, Virani MZ, Davies R, Symes A, Wheatley H, Thorstrom R, Amar A, Buij R, Jones VR, Williams NP, Buechley ER, Butchart SH (2018) State of the world's raptors: distributions, threats, and conservation recommendations. Biol Conserv 227:390–402

McDonald RI, Kareiva P, Forman RT (2008) The implications of current and future urbanisation for global protected areas and biodiversity conservation. Biol Conserv 141:1695–1703

McDonnell MJ, MacGregor-Fors I (2016) The ecological future of cities. Sci 352:936–938

McGarigal K, Cushman SA, Ene E (2012) FRAGSTATS V4: spatial pattern analysis program for categorical and continuous maps. University of Massachusetts, Amherst
McGeoch MA (2007) Insects and bioindication: theory and progress. In: Stewart AJ, New TR, Lewis O (eds) Insect conservation biology: Proceedings of the Royal Entomological Society's 23rd Symposium. CAB International, pp 144–174
McHale MR, Bunn DN, Pickett ST, Twine W (2013) Urban ecology in a developing world: why advanced socioecological theory needs Africa. Front Ecol Environ 11:556–564
McIntyre NE (2000) Ecology of urban arthropods: a review and a call to action. Ann Entomol Soc Am 93:825–835
McKinney ML (2006) Urbanisation as a major cause of biotic homogenisation. Biol Conserv 127:247–260
McKinney ML (2008) Effects of urbanisation on species richness: a review of plants and animals. Urban Ecosyst 11:161–176
McPhearson T, Pickett ST, Grimm NB, Niemelä J, Alberti M, Elmqvist T, Weber C, Haase D, Breuste J, Qureshi S (2016) Advancing urban ecology toward a science of cities. Biosci 66:198–212
McPherson SC (2015) Urban ecology of the crowned eagle *Stephanoaetus coronatus* in KwaZulu-Natal, South Africa. PhD Thesis. University of KwaZulu-Natal, Pietermaritzburg
McPherson SC, Brown M, Downs CT (2016a) Crowned eagle nest sites in an urban landscape: requirements of a large eagle in the Durban Metropolitan Open Space System. Landsc Urb Plan 146:43–50
McPherson SC, Brown M, Downs CT (2016b) Diet of the crowned eagle (*Stephanoaetus coronatus*) in an urban landscape: potential for human-wildlife conflict? Urban Ecosyst 19:383–396
Measey GJ, Tolley KA (2011) Investigating the cause of the disjunct distribution of *Amietophrynus pantherinus*, the endangered South African western leopard toad. Conserv Gen 12:61–70
Meerburg BG, Singleton GR, Kijlstra A (2009a). Rodent-borne diseases and their risks for public health. Crit Rev Microbiol 35:221–270
Meerburg BG, Singleton GR, Leirs H (2009b) The Year of the Rat ends-time to fight hunger! Pest Manage Sci 65:351–352
Møller AP (2009) Successful city dwellers: a comparative study of the ecological characteristics of urban birds in the Western Palearctic. Oecol 159:849–858
Montgomery MR (2008) The urban transformation of the developing world. Sci 319:761–764
Morling F (2014) Cape Town's cats: reassessing predation through kitty-cams. MSc Thesis. University of Cape Town, Cape Town
Müller N, Ignatieva M, Nilon CH, Werner P, Zipperer WC (2013) Patterns and trends in urban biodiversity and landscape design. In: Elmqvist T, Fragkias M, Goodness J, Güneralp B, Marcotullio PJ, McDonald RI, Parnell S, Schewenius M, Sendstad M, Seto KC, Wilkinson C (eds) Urbanisation, biodiversity and ecosystem services: challenges and opportunities. Springer, Dordrecht, pp 123–174
Myers G (2021) Urbanisation in the Global South. In: Shackleton CM, Cilliers SS, Davoren E, du Toit MJ (eds) Urban ecology in the Global South. Springer, Cham, pp 27–49
Nowak K, Wimberger K, Richards SA, Hill RA, le Roux A (2017) Samango Monkeys (*Cercopithecus albogularis labiatus*) manage risk in a highly seasonal, human-modified landscape in Amathole Mountains, South Africa. Int J Primat 38:194–206
Ogada D, Shaw P, Beyers RL, Buij R, Murn C, Thiollay JM, Beale CM, Holdo RM, Pomeroy D, Baker N, Krüger SC, Botha A, Virani MZ, Monadjem A, Sinclair AR (2016) Another continental vulture crisis: Africa's vultures collapsing toward extinction. Conserv Lett 9:89–97
Oke TR (1982) The energetic basis of the urban heat island. Quart J Royal Meteorol Soc 108:1–24
Oke TR (1995) The heat island of the urban boundary layer: characteristics, causes and effects. In: Cermak JE, Davenport AG, Plate EJ, Viegas DX (eds) Wind climate in cities. Springer, Dordrecht, pp 81–107
Özel M, Prokop P, Uşak M (2009) Cross-cultural comparison of student attitudes toward snakes. Soc Anim 17:224–240

Paap T, de Beer ZW, Migliorini D, Nel WJ, Wingfield MJ (2018) The polyphagous shot hole borer (PSHB) and its fungal symbiont *Fusarium euwallaceae*: a new invasion in South Africa. Aust Plant Path 47:231–237
Paul MJ, Meyer JL (2001) Streams in the urban landscape. Ann Rev Ecol Evol System 32:333–365
Pauw A (2007) Collapse of a pollination web in small conservation areas. Ecol 88:1759–1769
Perillo A, Mazzoni LG, Passos LF, Goulart VD, Duca C, Young RJ (2017) Anthropogenic noise reduces bird species richness and diversity in urban parks. Ibis 159:638–646
Perry G, Buchanan BW, Fisher R, Salmon M, Wise SE (2008) Effects of artificial night lighting on amphibians and reptiles in urban environments. In: Mitchell JC, Jung Brown RE, Bartholomew B (eds) Urban herpetology. Society for the Study of Amphibians and Reptiles, Salt Lake City, pp 239–256
Phalan B, Onial M, Balmford A, Green RE (2011) Reconciling food production and biodiversity conservation: land sharing and land sparing compared. Sci 333:1289–1291
Phillips K (1990) Where have all the frogs and toads gone? Biosci 40:422–425
Picker M, Griffiths C (2011) Alien and invasive animals. Struik Nature, Cape Town
Pillsbury FC, Miller JR (2008) Habitat and landscape characteristics underlying anuran community structure along an urban–rural gradient. Ecol Appl 18:1107–1118
Pomeroy D, Kibuule M (2017) Increasingly urban Marabou Storks start breeding four months early in Kampala, Uganda. Ostrich 88:261–266
Pope SE, Fahrig L, Merriam HG (2000) Landscape complementation and metapopulation effects on leopard frog populations. Ecol 81:2498–2508
Port GR, Thompson JR (1980) Outbreaks of insect herbivores on plants along motorways in the United Kingdom. J Appl Ecol 17:649–656
Potgieter LJ, Gaertner M, O'Farrell PJ (2019) Perceptions of impact: invasive alien plants in the urban environment. J Environ Manage 229:76–87
Purkayastha J, Das MD, Sengupta SS (2011) Urban herpetofauna: a case study in Guwahati City of Assam, India. Herpetol Notes 4:195–202
Qadeer MA (2000) Ruralopolises: the spatial organisation and residential land economy of high-density rural regions in South Asia. Urban Stud 37:1583–1603
Ramalho CE, Hobbs RJ (2012) Time for a change: dynamic urban ecology. Trends Ecol Evol 27:179–188
Ramírez-Restrepo L, MacGregor-Fors I (2017) Butterflies in the city: a review of urban diurnal Lepidoptera. Urban Ecosyst 20:171–182
Ramkissoon Y (2006) An assessment of the problem of vervet monkeys in the former Westville Borough: management implications. PhD Thesis. University of Kwa-Zulu Natal, Pietermaritzburg
Redman CL, Jones NS (2005) The environmental, social, and health dimensions of urban expansion. Pop Environ 26:505–520
Redpath SM, Young J, Evely A, Adams WM, Sutherland WJ, Whitehouse A, Amar A, Lambert RA, Linnell JD, Watt A, Gutiérrez RJ (2013) Understanding and managing conservation conflicts. Trends Ecol Evol 28:100–109
Ritchie EG, Johnson CN (2009) Predator interactions, mesopredator release and biodiversity conservation. Ecol Lett 12:982–998
Robb GN, McDonald RA, Chamberlain DE, Bearhop S (2008) Food for thought: supplementary feeding as a driver of ecological change in avian populations. Front Ecol Environ 6:476–484
Rust MK, Su N-Y (2012) Managing social insects of urban importance. Ann Rev Entomol 57:355–375
Samways MJ, Caldwell PM, Osborn R (1996) Ground-living invertebrate assemblages in native, planted and invasive vegetation in South Africa. Agric Ecosyst Environ 59:19–32
Sánchez-Bayo F, Wyckhuys KAG (2019) Worldwide decline of the entomofauna: a review of its drivers. Biol Cons 232:8–27
Sanz V, Caula S (2015) Assessing bird assemblages along an urban gradient in a Caribbean island (Margarita, Venezuela). Urban Ecosyst 18:729–746

Sazima I, D'Angelo GB (2013) Range of animal food types recorded for the Tegu Lizard (*Salvator merianae*) at an urban park in south-eastern Brazil. Herpetol Notes 6:427–430

Schäffler A, Swilling M (2013) Valuing green infrastructure in an urban environment under pressure: the Johannesburg case. Ecol Econ 86:246–257

Schneeberger K, Voigt C (2016) Zoonotic viruses and conservation of bats. In: Voigt CC, Kingston T (eds) Bats in the Anthropocene: conservation of bats in a changing world. Springer, Berlin, pp 263–292

Schoeman MC (2016) Light pollution at stadiums favors urban exploiter bats. Anim Conserv 19:120–130

Scott DM, Berg MJ, Tolhurst BA, Chauvenet AL, Smith GC, Neaves K, Lochhead J, Baker PJ (2014) Changes in the distribution of red foxes (*Vulpes vulpes*) in urban areas in Great Britain: findings and limitations of a media-driven nationwide survey. PLoS ONE 9:e99059

Sekercioglu CH (2006) Increasing awareness of avian ecological function. Trends Ecol Evol 21:464–471

Seto KC, Güneralp B, Hutyra LR (2012) Global forecasts of urban expansion to 2030 and direct impacts on biodiversity and carbon pools. Proc Nat Acad Sci 109:16083–16088

Seto KC, Sánchez-Rodríguez R, Fragkias M (2010) the new geography of contemporary urbanisation and the environment. Ann Rev Environ Res 35:167–194

Shackleton CM (2016) Do indigenous street trees promote more biodiversity than alien ones? Evidence using mistletoes and birds in South Africa. Forests 7:134

Shackleton CM, Cilliers SS, du Toit MJ, Davoren E (2021) The need for an urban ecology of the Global South. In: Shackleton CM, Cilliers SS, Davoren E, du Toit MJ (eds) Urban ecology in the Global South. Springer, Cham, pp 1–26

Shackleton S, Chinyimba A, Hebinck P, Shackleton C, Kaoma H (2015) Multiple benefits and values of trees in urban landscapes in two towns in northern South Africa. Landsc Urb Plan 136:76–86

Silva CP, García CE, Estay SA, Barbosa O (2015) Bird richness and abundance in response to urban form in a Latin American city: Valdivia Chile as a case study. PLoS ONE 10:e0138120

Silva-Rodríguez EA, Sieving KE (2011) Influence of care of domestic carnivores on their predation on vertebrates. Conserv Biol 25:808–815

Smart R, Whiting MJ, Twine W (2005) Lizards and landscapes: integrating field surveys and interviews to assess the impact of human disturbance on lizard assemblages and selected reptiles in a savanna in South Africa. Biol Conserv 122:23–31

Smolka J, Baird E, el Jundi B, Reber T, Byrne MJ, Dacke M (2016) Night sky orientation with diurnal and nocturnal eyes: dim-light adaptations are critical when the moon is out of sight. Ani Behav 111:127–146

Snodgrass JW, Casey R, Massal L (2007) Nitrogen pollution of stormwater ponds: potential for toxic effects on amphibian embryos and larvae. Appl Herpetol 4:19–29

Snodgrass JW, Casey RE, Joseph D, Simon JA (2008) Microcosm investigations of stormwater pond sediment toxicity to embryonic and larval amphibians: variation in sensitivity among species. Environ Pollut 154:291–297

Soga M, Gaston KJ (2016) Extinction of experience: the loss of human-nature interactions. Front Ecol Environ 14:94–101

Soga M, Yamaura Y, Koike S, Gaston KJ (2014) Land sharing vs. land sparing: does the compact city reconcile urban development and biodiversity conservation? J Appl Ecol 51:1378–1386

Soulé ME, Bolger DT, Alberts AC, Wrights J, Sorice M, Hill S (1988) Reconstructed dynamics of rapid extinctions of chaparral-requiring birds in urban habitat islands. Conserv Biol 2:75–92

Spencer HJ, Scott NE, Port GR, Davison AW (1988) Effects of roadside conditions on plants and insects I. atmospheric conditions. J Appl Ecol 25:699–707

Spoelstra K, van Grunsven RHA, Donners M, Gienapp P, Huigens ME, Slaterus R, Berendse F, Visser ME, Veenendaal E (2015) Experimental illumination of natural habitat–an experimental set-up to assess the direct and indirect ecological consequences of artificial light of different spectral composition. Philos Trans Royal Soc B 370:20140129

Steyn P (1983) Birds of prey of Southern Africa: their identification and life histories. Crown Helm, Birmingham

Stofberg M, Cunningham SJ, Sumasgutner P, Amar A (2019) Juggling a "junk-food" diet: responses of an urban bird to fluctuating anthropogenic-food availability. Urban Ecosyst 22:1019–1026

Stone EL, Jones G, Harris S (2009) Street lighting disturbs commuting bats. Curr Biol 19:1123–1127

Stone EL, Jones G, Harris S (2012) Conserving energy at a cost to biodiversity? Impacts of LED lighting on bats. Glob Change Biol 18:2458–2465

Streutker DR (2003) Satellite-measured growth of the urban heat island of Houston, Texas. Remote Sens Environ 85:282–289

Su S, Cassey P, Dyer EE, Blackburn TM (2017) Geographical range expansion of alien birds and environmental matching. Ibis 159:193–203

Tallamy DW (2004) Do alien plants reduce insect biomass? Conserv Biol 18:1689–1692

Tarboton WR (2001) Guide to nests and eggs of southern African birds. Randomhouse-Struik, Cape Town

Taylor PJ, Arntzen L, Hayter M, Iles M, Frean J, Belmain S (2008) Understanding and managing sanitary risks due to rodent zoonoses in an African city: beyond the Boston Model. Integ Zool 3:38–50

Taylor PJ, Cheney C, Sapsford C (1999) Roost habitat evaluation and distribution of bats (Order Chiroptera) in the Durban Metropolitan Region. Durban Mus Novit 24:62–71

Thomsett S (2011) On the African crowned eagle. African Raptors: The online home of African Raptor Interests, Part 3. http://www.africanraptors.org/simon-thomsett-on-the-african-crowned-eagle-part-3/

van Rensburg BJ, Peacock DS, Robertson MP (2009) Biotic homogenisation and alien bird species along an urban gradient in South Africa. Landscape Urban Plan 92:233–241

Vörösmarty CJ, McIntyre PB, Gessner MO, Dudgeon D, Prusevich A, Green P, Glidden S, Bunn SE, Sullivan CA, Liermann CR, Davies PM (2010) Global threats to human water security and river biodiversity. Nat 467:555–561

Walsh CJ (2000) Urban impacts on the ecology of receiving waters: a framework for assessment, conservation and restoration. Hydrobiol 431:107–114

Wells KD (2007) The ecology and behavior of amphibians. University of Chicago Press, Chicago, IL

Welz A, Jenkins A (2005) How green is the valley? Monitoring fish eagles along the Breede River. Afr Birds Birding 10:30–37

Wilby R (2007) A review of climate change impacts on the built environment. Built Environ 33:31–45

Williams ST, Maree N, Belmain SR, Keith M, Swanepoel LH (2018) Predation by small mammalian carnivores in rural agro-ecosystems: an undervalued ecosystem service? Ecosyst Serv 30:362–371

Woods M, McDonald RA, Harris S (2003) Predation of wildlife by domestic cats *Felis catus* in Great Britain. Mamm Rev 33:174–188

Wotton SR, Eaton MA, Sheehan D, Munyekenye FB, Burfield IJ, Butchart SH, Moleofi K, Nalwanga-Wabwire D, Ndang'ang'a PK, Pomeroy D, Senyatso KJ, Gregory RD (2020) Developing biodiversity indicators for African birds. Oryx 54:62–73

Wu J (2014) Urban ecology and sustainability: the state-of-the-science and future directions. Landsc Urb Plan 125:209–221

Zhang W, Li B, Shu X, Pei E, Yuan X, Sun Y, Wang T, Wang Z (2016) Responses of anuran communities to rapid urban growth in Shanghai, China. Urban For Urban Green 20:365–374

Ecosystem Provisioning Services in Global South Cities

Charlie M. Shackleton

Abstract All of humankind, urban and rural, rich and poor and whether living in the Global South or the Global North rely to some degree on provisioning ecosystem services such as food, water, timber, fibre and medicinal products. However, in contrast to the Global North the contextual forces shaping many urban livelihoods in the Global South necessitate that many people access the needed provisioning services themselves by directly harvesting or collecting from the immediate, local urban and peri-urban environments, or purchasing them from those who do. Thus, the need for and local availability of provisioning services has profound implications for not only the quality of urban livelihoods, but also the urban ecology in relation to extent and use of urban green infrastructure from where the provisioning services are sourced and the species within them. This chapter summarises compelling evidence from throughout the Global South regarding the extent of use of locally sourced provisioning services by urban communities. It shows that Global South contexts prompt the use of a diverse array of resource types and hundreds of different species in specific settings, much greater than found in Global North cities. The chapter explores the types, extent and nature of locally sourced provisioning services and contributions that they make to urban livelihoods via (1) direct household provisioning, (2) cash savings, (3) safety nets during adverse times, (4) income generation and (5) culture and traditions. It then explores the local-scale factors that influence the nature and extent of use, including proximity to harvest or collection site, access, local ecological knowledge, migrant status, culture and affluence. Thereafter, the social and ecological effects of local harvesting are considered, which potentially effect the sustainability of sought-after species and populations. The chapter ends with pointers on how the quality and quantity of locally produced provisioning services can be secured and enhanced and thereby better support urban livelihoods, especially of the urban poor, in the Global South.

Keywords Access · Collection · Impacts · Livelihoods · Local · Wealth

---

C. M. Shackleton (✉)
Department of Environmental Science, Rhodes University, Makhanda 6140, South Africa
e-mail: c.shackleton@ru.ac.za

C. M. Shackleton et al. (eds.), *Urban Ecology in the Global South*, Cities and Nature,
https://doi.org/10.1007/978-3-030-67650-6_8

1 Introduction

It is well established that the governance and socio-economic structures and dynamics of Global South cities are, on the one hand, diverse and constantly changing and on the other, typically distinct from their Global North counterparts in many ways (Myers 2021; Shackleton et al. 2021). Whilst each city is unique in one or more ways, and various zones with a single city may also be quite dissimilar, Global South cities on the whole differ from Global North ones in terms of higher rates of in-migration, high rates of land use transformation, lower levels of relative affluence and lower capacities to plan for and manage burgeoning demands for basic needs and services (such as housing, water, energy, greenspace, health, security and education) (Shackleton et al. 2021). This different socio-economic context sets the scene that contributes to the shaping of local needs for, and values attached to, biodiversity and nature. It is these large differences in context that undermine any search for global constants within the discipline of urban ecology. It is also these large differences in context that mean transferring knowledge and patterns developed from examination of Global North cities to Global South settings should be done with caution and critical reflection. Rather, there should be sharing of knowledge and experiences between Global South contexts, but this is hampered by a preponderance of literature from Global North settings (Roy et al. 2012; Shackleton 2012; Haase et al. 2014), with scant (albeit increasing) contributions from the Global South.

There is a growing multiplicity of frameworks for depicting and analysing the relationship between humans and nature, including ecosystem services, nature's contributions to people, biocultural diversity and social ecology, to name a few. Each emphasises particular aspects of the relationship and draws on particular terminology. This chapter does not explore the merits and demerits of each, nor does it wish to elevate one framework over another, even though it is framed around the term 'provisioning services'. The adoption of this term in the chapter is to use one that has been around for almost two decades, is common across the most widely used frameworks, i.e. the ecosystem services (MA 2003) and the nature's contributions to people framings (Diaz et al. 2018), and to provide compatibility with the structure and terminology of the entire book of which it is a part. Based on these framings, for the purposes of this chapter, provisioning services (PS), using the Common International Classification of Ecosystem Services (CICES) classification, are deemed to be 'all nutritional, non-nutritional material and energetic outputs from living systems as well as abiotic outputs (including water)' (Haines-Young and Potschin 2017). This definition accords well with the original definition of the Millennium Ecosystem Assessment (2005) and the subsequent one under The Economics of Ecosystems and Biodiversity synthesis (TEEB 2010). Within an urban context this includes a range of PS harvested from formal and informal, intact and transformed, private and public greenspaces within the urban matrix, as well as those farmed or produced by urban dwellers. However, this chapter will focus on the former, i.e. harvesting or foraging from the urban matrix, because urban agriculture is covered elsewhere in the book (Drescher et al. 2021).

In exploring the supply and use of PS in Global South cities this chapter has four objectives. The first is to demonstrate the variety of PS used within and across cities in the Global South and the spaces and places from where they are procured. The second is to show that many of the different PS are widely used by urban dwellers in such cities. The third is to consider the significance of such use for urban citizens generally as well as for specific groups, and for urban planning and management. The chapter will conclude with some thoughts of what it all means for understanding the urban ecology of Global South cities. Each of these will be addressed by drawing from relevant literature from across the Global South regions and cities and with, where possible, regular comparison to those in the Global North. The term 'cities' will be used throughout the chapter, even though the chapter includes examples and literature from larger and smaller urban complexes, such as megacities and towns, respectively.

2 Types and Sources of Provisioning Services Used

A diverse range of PS and species are used for a number of different purposes (Table 8.1). Use of some PS is common across the majority of Global South settings (e.g. water, firewood, wild foods), although the extent of use varies across locations. Others are city or region specific, such as shellfish and fish in coastal cities, widespread consumption of insects in Oaxaca (Mexico) (Hurd et al. 2019) or Zimbabwe (Manditsera et al. 2018) and bushmeat in central Africa (Fa et al. 2019) and parts of Amazonia (El Bizri et al. 2019). There have been no surveys in urban settings to assess how many different PS households or communities use relative to the full suite locally available. Most surveys are sectoral, i.e. assessing urban use of one or two specific PS such as firewood (Brouwer and Falcão 2004), bushmeat (Mbete et al. 2011) or medicinal plants (Messias et al. 2015). The broad-scale survey of Schlesinger et al. (2015) considered use of nine different biological PS (i.e. excluding water) in six cities in five sub-Saharan African countries. They reported that 74% of urban households used at least one locally procured PS, with an average of 3.2 per household out of the nine included in the survey. In northern Botswana, Joos-Vandewalle et al. (2018) reported that urban households used a mean of 9.2 wild resources out of a list of 18 inventoried (rural households used a mean of 11.2 of the 18). They reported that 80% or more of the urban households used firewood, wild fruits, vegetables, fish, reeds for mats, grass for brushes and waterlilies for food.

Typically water, wild foods, firewood and construction materials are amongst the most used PS by urban dwellers, but there is large variation within, and between, cities and countries. Work in Global North cities also reveals a diversity of species and products harvested from within public and private greenspaces (Hurley and Emery 2018; Tiwary et al. 2020), but perhaps with greater emphasis on wild foods and craft/decorative items. However, there is widespread use in the Global North of wild plants as natural home remedies to alleviate common ailments, such as coughs, fevers and indigestion (Becerra and Ingelhart 2008; Pieroni et al. 2013). Generally,

Table 8.1 The diversity of different provisioning services used by urban dwellers (excluding from urban agriculture [see Drescher et al. 2021])

Provisioning service	Type	Provisioning service	Type
Water	Fresh water	Construction	Timber
Wild foods	Vegetables		Bamboos and rattans
	Fruits and nuts		Thatching fibres
	Vertebrates		Clay and sand
	Fish, shellfish and amphibians	Medicines	Plant parts (roots, bark, leaves, bulbs, etc.)
	Insects and invertebrates		Chewing sticks
	Honey		Animals parts (fats, bones, organs, skins)
	Saps and resins	Tools/utensils	Wood for tool handles, spoons, etc.
	Eggs		Plants fibres for brooms, baskets, ropes, etc.
	Fungi	Craft/decoration	Plant fibres, seeds, flowers, fruits, nuts
Energy	Firewood and charcoal		Shells
	Organic wastes		Feathers
	Animal dung		

in the Global South home medication using wild plants and animals spans a greater range of species and ailments, and in many regions includes charms for protection against bad luck or malevolent forces and spirits (Bussmann et al. 2010; Williams and Whiting 2016).

The proportion of species available across different spaces within a city appears to be high, although to date determined from only a few sites. Poe et al. (2013) recorded that in Seattle (USA) 486 different plant species were foraged to some degree or another. Hurley and Emery (2018) considered only woody species (trees and shrubs) in New York (USA) and revealed that 83% of the 304 woody species in the city had one or more uses. Overall, they documented 581 different uses as many species had more than one use. In South Africa, Garekae (2020) used a plot approach to sample the vegetation in different urban landscapes in two medium-sized towns, identifying 163 species, of which 54% had one or more uses.

This rich diversity of PS is harvested from an array of different sites within the urban matrix, including formal urban parks, protected areas, informal greenspaces, urban forests and woodlands, abandoned lots, remnant vegetation, servitudes, private gardens, institutional grounds, cemeteries, riparian zones, roadsides and the peri-urban fringe (Slater and Twyman 2003). Which one or mix of these sites are most used is generally a function of their proximity to a user's place of residence, accessibility in terms of informal or formal rules, the availability and abundance of the PS or particular species sought after, and perceptions of personal safety within specific spaces (Mollee et al. 2017; Garekae and Shackleton 2020). For instance, Kaoma and Shackleton (2014) reported that in three towns in northern South Africa most firewood was harvested on the urban fringe, where it was most abundant, whilst wild fruits were mostly harvested from homestead plots, although not exclusively so. The suite of spaces from which PS are harvested is probably no different to those in Global North cities, but the application of municipal by-laws restricting harvesting in certain types of spaces, such as public parks, cemeteries and protected areas, is more likely to be enforced in Global North settings than in Global South ones, due to greater resources and number of personnel.

Whilst there is extensive self-collection of multiple PS in Global South cities, some urban consumers rather purchase them from local markets, for example firewood in Peru (e.g. Bennett et al. 2018), wild fruits in Democratic Republic of Congo (Termote et al. 2012) and medicinal plants in Madagascar (Randriamiharisoa et al. 2015). This may be driven by one or a combination of low availability of the PS in the immediate surrounds, insufficient household labour to collect the PS, limited knowledge of the sites and species, or sufficient household income to allow the option of purchasing rather than collection. Household surveys of use therefore cannot differentiate if the purchased PS were originally produced within the city, or were imported from neighbouring, or even distant, rural areas. It would require a market chain analysis for each purchased PS, such as for honey in the Philippines (Matias et al. 2018), *Ricinodendron huedelotii* products in Cameroon (Ndumbe et al. 2019) or firewood in Nicaragua (Baker et al. 2014) and South Africa (Guild and Shackleton 2018).

A major difference between the Global South and the Global North is the provision of fresh and safe water. Twenty-seven percent of the world's urban population does not have piped water to their house, nearly all of whom are in the Global South (UN Water 2010). There is almost universal coverage in the supply of piped, treated water in cities throughout the Global North (UN Water 2010), most often supplied from catchments and aquifers some distance away from the city. This is not the case for many urban dwellers in the Global South, with many having to collect water from wells, streams and rivers within their immediate environment, or water may be delivered by water-tanker according to an advertised scheduled, or purchased from private water vendors or shops. Those living in slums (approximately 828 million worldwide, of which over 95% are in the Global South) are particularly underserviced with respect to improved water and sanitation services. Dependency on alternative, often unreliable or unsanitary supplies underpins the markedly higher incidence of waterborne diseases, such as cholera, typhoid, dysentery and leptospirosis in the Global South (e.g. Ram et al. 2008; Torgerson et al. 2015). For example, over 99% of *Shigella* infections occur in the GS (Ram et al. 2008). The high rates of urban population growth and in-migration mean that increasing numbers are reliant on insecure water sources (UN Water 2010).

3 Extent of Use

Although the places where PS are accessed and the types of uses they are put to may be roughly similar between Global North and Global South cities, the extent of use is not. Multiple studies show high proportions of urban households in Global South cities making use of one or more locally accessed PS, such as firewood, wild foods, construction materials, medicinal plants and the like (Slater and Twyman 2003; Schlesinger et al. 2015; Joos-Vandewalle et al. 2018). For example, firewood or charcoal is the primary energy for hundreds of millions of GS urban households and small businesses (Arnold et al. 2006). And it is not necessarily restricted to the poor (Castán Broto et al. 2020), as illustrated in Maputo (Mozambique) where more than 50% of households in affluent neighbourhoods use charcoal as a domestic energy source (Brouwer and Falcão 2004). However, not all is harvested from within the urban matrix as some is imported from peri-urban and rural surrounds. Markets for wild foods (including fruits, vegetables, insects, fish, mushrooms and honey) abound in Global South cities indicating high demand and use; well illustrated by over 66 different wild foods available in markets in three cities in Cameroon, and with urban residents estimating that approximately 25% of their total food purchases are on wild foods (Sneyd 2013). Traditional medicines are also widely used; for example the review by de Medeiros et al. (2013) found that urban communities in Brazil used over 400 different plant species for medicinal purposes, with corresponding figures from a single urban market in Johannesburg in South Africa reporting 511 plant species traded (Williams et al. 2000). There is also extensive urban trade in animal products as medicines across the Global South (Ferreira et al. 2013; Kim and Song 2013; Nieman et al. 2019), which is rare in the Global North except for some immigrant communities. The participation rates in collection of plant and animal PS by households in the Global North is nowhere near such levels, which is one reason why it has been less studied in the Global North contexts (Mclain et al. 2014). Since the proportion of urban households using locally produced PS is much higher in the Global South cities, it is likely the quantities used are also greater.

The reasons underlying the high use in the Global South have not be systematically examined. However, across different studies (de Medeiros et al. 2013; Schlesinger et al. 2015; Garekae and Shackleton 2020), high use as been variously attributed to (i) cultural norms, especially amongst recent rural migrants to the city, (ii) sufficient levels of local ecological knowledge about useful species, (iii) ample, undeveloped spaces supporting populations of the desired resources and species and (iv) high levels of income poverty so that users save scarce cash sources by harvesting rather than purchasing the same PS or substitutes.

4 Nature of Use of Provisioning Services for Human Wellbeing

The high demand for PS in terms of both numbers of people and quantities consumed, denotes the importance of local urban and peri-urban environments in supporting human wellbeing in cities of the Global South. Citizens of the Global North use similar PS, but they are usually spatially and economically divorced from the sources of supply. Thus, for most Global North urban citizens the bulk, if not all, of their food, medicines, building materials, water and the like are produced elsewhere, maybe even in other countries, and transported to the city where they simply buy it from retail outlets. Thus, there is a disjuncture between their lifestyles and wellbeing, and the local environment within and adjacent to the city. In contrast, a significant proportion of Global South citizens rely to some extent on the local environment for one or more PS. Consequently, any decline in the quantity or quality of local environments can have direct effects on their wellbeing. Precisely how that manifests depends on the nature of the use of the PS across five different categories, namely (1) direct household provisioning, (2) cash savings, (3) cash income from sale, (4) safety-nets during household shocks and adversity and (5) cultural needs.

4.1 Direct Household Provisioning

Direct household provisioning relates to the daily or regular use of PS to meet household needs. Once considered primarily a component of rural livelihoods, it is increasingly being revealed as integral to many urban ones too (Schlesinger et al. 2015; Short Gianotti and Hurley 2016; Mollee et al. 2017; Shackleton et al. 2017; Charnley et al. 2018; Liao et al. 2018; Goswami and Nautiyal 2020; Shackleton et al. 2020; Tiwary et al. 2020). A good example is that of firewood or charcoal for cooking and heating. In 2010 the IEA reported that 50% of domestic urban energy use in sub-Saharan Africa was from firewood and charcoal (IEA 2010). Without sufficient locally accessible supplies of timber for firewood, households will have to change cooking and eating patterns by one or more of (1) purchasing alternative energy forms (e.g. gas, kerosene, electricity) for cooking and heating (thereby reducing the amount of cash available for other household needs), (2) making alterations to the number of meals they cook per day or (3) change the type of food to ones that require less cooking. Any of these alternatives could result in increased food insecurity and ill-health. For example, shortages of firewood can lead to fewer cooked meals, or the cooking of lower quality foods that can be cooked faster (Brouwer et al. 1989). Similar negative consequences for human wellbeing can be made around inadequate local supplies of potable water, or wild foods. Inadequate water supplies or quality compromise what food can be cooked, along with general hygiene and sanitation, and the increased risk of water borne diseases such as typhoid, cholera, dysentery and hepatitis. Such diseases are relatively rare within Global North cities, but are a constant threat in many Global South regions.

4.2 Cash Savings Through the Use of Locally Obtained PS

The value of locally sourced PS goes beyond just their role in providing consumptive resources. Being able to access a given PS locally at no financial cost (although there is the opportunity cost of labour), means that user households do not have to purchase the same or a substitute PS from a commercial outlet. In other words, it also saves them money; meaning that cash resources can be saved, invested or used for household needs that cannot be accessed directly from the local environment, such as school fees and books, or transport costs to work. This cash-saving function is most significant for poor urban households. For example, in a survey of 200 households in each of three towns in South Africa Kaoma and Shackleton (2015) found that the non-cash income provided through the collection of provisioning services from urban trees represented approximately one-fifth of total household cash and non-cash incomes. As might be expected, it was a lot higher (>30%) amongst households in poorer, informal settlements than formal ones.

4.3 Income from Sale of PS

Locally sourced PS allow some urban households to generate a cash income through selling them within the city limits. There are vibrant markets for firewood, wild foods, traditional medicines and crafts in many cities throughout the Global South. For example, Joos-Vandewalle et al. (2018) reported that 40% of the urban households in Kasane (Botswana) sold one or more wild-harvested provisioning products, whilst Nadkarni (2017) explores the role of the processing of wild products as catalysts for employment, poverty alleviation and development in small urban centres in India. Traders may rove the streets or residential neighbourhoods hawking their goods, or they may operate from fixed, well-known marketplaces or street corners. The number of urban households selling locally sourced PS is unknown, but is likely to run into millions across the Global South. Moreover, a direct count would be insufficient because it would need to differentiate between those traders who sourced their goods in rural areas beyond the city and transported them in, and those who harvested them within the city.

It is an equally difficult, and perhaps meaningless, task to determine the mean income from trade in urban PS. This is because of the wide range in dependency on such trade. For some vendors it is their only source of cash income. For others, it is a supplementary activity engaged in for a few hours per month, or once or twice a season to provide some 'top-up' cash income. On the other hand, attempting such quantification, as done in many rural contexts (e.g. Angelsen et al. 2014; Mugido and

Shackleton 2019), would provide better insights into the economic and non-economic importance of locally available urban PS, which might be useful in advocating for sufficient areas of urban green infrastructure to supply the required PS.

4.4 Safety-Nets

The importance of locally available PS comes to the fore during periods of household adversity. At a household scale (termed an idiosyncratic shock), this might be during times of heightened need resulting from retrenchment or death of a breadwinner, debilitating disease or financial losses. At a wider city-scale or larger (termed a covariate shock), it may be due to the effects of events such as drought, war or sudden economic downturns. During such adverse times, the option and ability to access and harvest local supplies of water, food and energy can literally mean the difference between life and death for some individuals and households. For others, it provides a coping strategy until such time they are able to recover from the negative effects of the shock. This may be manifest through one of several strategies (Shackleton and Shackleton 2004), including (1) increased use of a PS already used, (2) substitution of a purchased good with a local PS (e.g. use of locally harvested firewood instead of buying kerosene) or (3) temporary sale of a locally sourced PS to generate cash income to aid recovery from the shock. For example, Dalu and Shackleton (2018) investigated the coping responses of residents of informal housing zones to a major flooding event induced by unusually heavy rains in south-eastern South Africa. Using a questionnaire survey they interviewed residents who had suffered damage to their houses, as to how they recovered, either through collecting local materials (timber, thatching) to repair or rebuild their houses, or to sell local materials to earn cash to then purchase building materials. Overall, locally collected PS provided the equivalent of 57% of the costs of repairing or rebuilding the damaged structures, ranging from a mean of 46 to 70% in the three towns. At the opposite extreme, informal water sources and supplies become key during times of water supply interruptions or longer-term failure, such as in parts of Accra, Ghana (Tutu and Stoler 2016).

4.5 Cultural Needs

The use and consumption of some locally collected materials also fulfils certain cultural needs for some (e.g. Wehi and Wehi 2010; Grabbatin et al. 2011). Thus, it is both a PS and a cultural ecosystem service. Examples might include the culturally 'driven' consumption of wild insects in parts of Mexico (Hurd et al. 2019), bushmeat in urban areas of central Africa (Mbete et al. 2011) or use of traditional medicines associated with cultural belief systems in Peru (Ceuterick et al. 2011). Where cultural uses of PS are strong the demand does not wane due to the effects of urbanisation or mass education.

5 What Shapes Use of Local Urban Provisioning Services?

The currently relatively low number of studies and the diversity of species, spaces, end uses and contexts constrains the ability to make generalisations regarding the factors that shape the local reliance on and use of urban PS in either Global South or Global North settings. However, some that have been highlighted by more than one study are considered below. Although they have been presented individually, it is usually a complex mix of several that apply at any one time to a specific harvester, household or group of harvesters.

- Proximity

Local availability makes it easier and less time consuming to access the required PS such as wild plants in South Africa (Garekae and Shackleton 2020) or water in Madagascar (Boone et al. 2011). Harvesting or collection of PS from locations some distance from the home requires advance planning, more time and perhaps costs of transport to travel there and back. Additionally, knowledge of the species, abundance and season of particular species or products in local, proximal sites will be greater than for distant sites because local sites are observed more regularly during the course of other activities in the immediate neighbourhood, such as commuting to work, taking children to school or travelling to retail outlets. The likely greater use of more proximal sites requires that city planners seek to provide a good distribution of formal and informal green spaces throughout the urban matrix. Loss of sites and species, mainly through urban transformation, may result in declining use and ultimately knowledge of useful species, as reported to be the case in Florianópolis (Brazil) (Leal et al. 2018).

- Ease of access

Irrespective of whether PS are available relatively close by or further afield, physical access to the site is required to be able to harvest or collect the required products. However, not all sites are physically accessible because entry or use may be constrained by formal or informal tenure and usufruct regulations (Ginger et al. 2012). Access regulations may range from total prohibition of entry or harvesting activities within specific spaces (even if they are close to by), to no regulation at all. In between, there may be various formal or informal regulations and behaviours which limit or constrain access by particular groups of people, or at specific times or seasons, or for only designated activities or species. Thus, there may be trade-offs between proximity and access, necessitating that sometimes harvesters must travel to more distant, but more accessible, sites. Under any constrained access, harvesters may negotiate on the conditions of access with the land owner, management authority, or other user groups, or ignore the conditions and risk possible eviction or sanction. This becomes more challenging when the conditions of access shift as land use, land ownership or other users change (Hurley et al. 2008; Garekae 2020).

Direct access to a particular site may also be characterised in terms of any physical attributes which potentially impede ease of access. For example, major roads, railway lines or water bodies that have to be traversed may hamper access to particular green spaces and PS for some potential users, or necessitate that they must travel a more circuitous route that is more time consuming or costly. Some attributes of the site may also deter certain users, such as steep or uneven terrain, dense or spinescent vegetation, and the presence of potentially dangerous animals or other user groups. Degradation of sites due to pollution or dumping can also force users to seek alternative sources for vital PS. This is particularly so with respect to local water sources, but it may also apply to green spaces too, especially in contexts of weak environmental governance, characteristic of many Global South settings (Simon et al. 2021).

- Local ecological knowledge

Having sufficient knowledge of the local sites and access norms, as well as the species and how to harvest and process them, is a precondition for benefiting from many of the PS, such as wild foods and medicines. For example, Joos-Vandewalle et al. (2018) comment that the lack of skills regarding processing was the most likely reason why none of the households in their survey in northern Botswana were involved in mat weaving, beer brewing or wood carving. Various studies indicate that the primary avenues for acquiring such knowledge is by either transfer from elders to younger members within a family or by participating in harvesting groups and learning from experienced harvesters (e.g. Landor-Yamagata et al. 2018; Garekae and Shackleton 2020). However, in some cities, more in the Global North, there are foraging tours and classes (e.g. New York https://www.wildmanstevebrill.com/; London https://www.foragelondon.co.uk/, Melbourne https://www.eatthatweed.com/) or one can access relevant information from educational sources on the internet, in public media (such as magazines, newspapers) or pamphlets and books (e.g. Craft 2010; Rensten 2016). There is some commentary that local ecological knowledge about useful practices, species and the benefits they offer are declining in both rural and urban contexts internationally (e.g. Tsuchiya et al. 2014; Leal et al. 2018). This is often attributed to a mix of changing diets, increasing access to commercially produced substitute products, mass education systems, the aging of the knowledge holders and western cultural influences.

- Culture

Many urban residents who use locally available PS themselves do so because they view it as part of their culture (Wehi and Wehi 2010; Joos-Vandewalle et al. 2018). Precisely how culture is defined or viewed in such instances is rarely reported. But it seems to be at least partially reflective that the gathering or collecting practice has been done for a long time, with the respondent having learnt as a child about the species, uses and preparation from the elders in their own family or community. Thus, it is something they have practiced for a long time, as did the generation/s before them. Alternatively, it is part of the identity of relatively recent migrants to the city, either from rural areas or from different regions or countries. It may also be because the use of certain foods, or the manner in which they are prepared, is reserved for specific cultural ceremonies or celebrations. Additionally, there is a strong cultural dimension to the use of certain traditional medicines, especially those related to psychosomatic charms and beliefs and divination (e.g. Williams and Whiting 2016). Cultural beliefs, stories or practices may also underpin or contribute to the design and manufacture of utilitarian or craft materials for household use or sale (Cocks and Dold 2004; Grabbatin et al. 2011).

- Migrant status

One of the defining characteristics of most GS cities is the rapid rate of population growth (Myers 2021), partially driven by in-migration from rural areas. There is also some transnational migration in response to one or more of armed conflicts, political strife and uncertainty and weak economic conditions and prospects. Migrants to urban centres bring with them an enriching array of different worldviews, knowledge systems and cultures. For many migrants from deep rural areas, the harvesting of resources from local environments is a daily reality that they grew up with. Consequently, it is a familiar activity that they can employ whilst seeking alternative livelihood options in their new urban setting (Schlesinger et al. 2015). The high use of locally sourced PS by many migrants might not just be because of familiarity and need (e.g. 42% of the medicinal species used by Bolivian migrants in London (UK) were in common with those from Bolivia [Ceuterick et al. 2011]), but also due to deep cultural connections to nature and the use of certain species as medicines, food and the like (Ladio and Acosta 2019). In the latter case, the use of culturally prescribed species or resources might not diminish with time even if an urban livelihood is secured (Cocks and Dold 2004). Some migrants may also bring familiar species with them to their new settings and cultivate them in containers or small plots of land close to or further afield from the home or regularly import stocks (Fonesca and Balick 2018; Lagunes and Merçon 2021).

- Household affluence

There is some literature from the Global South that indicates that poorer urban dwellers are more likely to collect locally available PS, especially for basic household needs (such as firewood, water and common wild foods), than more affluent urban dwellers (Mollee et al. 2017; Garekae and Shackleton 2020). This may also be correlated with lower levels of education, and hence job and income security. However, this finding is based largely on interviews with harvesters and therefore perhaps overlooks that some affluent households may also use locally sourced PS, but do so through purchase rather than self-collection (Shackleton et al. 2020). This requires further examination. This association between household income and local harvesting is not replicated to the same extent in Global North settings, where more harvesters engage in foraging as a cultural or recreational activity which transcends economic classes (Grabbatin et al. 2011; Arrington et al. 2017), without ignoring that it can be a subsistence activity too for some (Arrington et al. 2017).

6 Effects of Local Procurement and Use

The collection or harvesting of locally available PS has ecological and social implications for the species, landscapes and people involved. These need to be considered by harvesters and management authorities if sustainable use is to be fostered.

6.1 *Ecological Effects*

Because of the widespread use of many of the PS for various purposes, the act of harvesting or collecting adds to the various other determinants and drivers of local species composition, diversity, functional diversity and distribution in public spaces throughout the city. This is a major difference to the Global North contexts, where harvesting intensity is much lower and for a narrower range of species and products.

The ecological effects of harvesting plant and animal species can be examined at various scales, ranging from genes to individuals, populations, species and landscapes (Ticktin 2004). However, there is a dearth of work on the effects and sustainability of plant and animal harvesting in urban areas compared to rural or 'natural' ones. It is conceivable that lessons and guidelines from rural contexts could be transferred to urban ones, certainly in the first instance. However, harvesting in both rural and urban contexts is not the only driver affecting plant, population and species vigour (however measured). Plants that are harvested are usually also subjected to other simultaneous stresses or drivers, such as competition, herbivory, fire, pests, water availability or invasive species. The relative intensities of some of these factors is likely to be very different in urban areas than in rural ones; for example fire and herbivory are likely to be lower. Moreover, the higher dependency on locally harvested resources for some peoples' livelihoods is also likely to increase the probability of overuse, or even local extinction, of some populations, such as through accumulative ringbarking of trees, but as yet this needs to be determined. On the other hand, there might be land management actions that could promote plant growth, such as irrigation, fertilisation and removal of competitive invasive species. Additionally, extended growing seasons and warmer ambient temperatures from the urban heat island effect may also benefit the growth dynamics or resilience of some harvestable species (Waffle et al. 2017; Li et al. 2019). Consequently, there is a dire need to investigate and understand the ecology of plant species providing key PS in different urban settings. The same applies to harvested animal species, some of which are favoured by urban environments and others less so (Reynolds et al. 2021). The effects of harvesting or collection on other species that may be dependent on the harvested species for food, habitat or other needs has not been examined.

The individual effects of water extraction are unlikely to have negative effects. However, if a specific water source is to serve a large number of people, the cumulative effects of extraction could result in a lowering of the water table. There is also the possibility of water quality impacts if residents use inappropriate chemicals when washing clothes in the water source or disposing of household or other waste too close to the water source.

6.2 Social Effects

Echoing the situation for ecological effects, there is a very little work on the social effects of collection of plant and animal PS specifically in urban areas. There is a rich and growing literature on the physical and psychological benefits of regular visits to and experiences of urban green spaces (van den Berg et al. 2015; Shanahan et al. 2015), which is where most of the PS are located. Additionally, since harvesting is frequently undertaken in the company of either family members or other harvesters (Hurley et al. 2015; Garekae and Shackleton 2020), there are benefits associated with the development of social networks and cohesion, as well as sharing of local ecological knowledge (Poe et al. 2013). The process of harvesting also promotes the maintenance of relevant local ecological knowledge of what species to use, where they are located, and how they should be harvested and prepared. Depending on how this is done, it may also contribute to the maintenance of particular cultural facets, such as the preparation of traditional foods or medicines, use of traditional methods or styles, or manufacture of cultural artefacts for household use or for sale (Grabbatin et al. 2011). Regular harvesting trips to predetermined localities are also likely to contribute to a sense of place and perhaps even a sense of identity (Poe et al. 2014).

On the downside, the necessity of having to collect PS from local greenspaces may expose the harvesters to some risks. These include physical harm from dangerous animals (such as snakes, scorpions, spiders, feral dogs), spinescent or poisonous plants, uneven and rocky terrain and thieves or other persons who may attempt to harm them (Garekae and Shackleton 2020). Females and elderly harvesters are more at risk in this respect, which is one of the motivations for some to conduct harvesting as a group activity. For wild foods and water, there is a potential risk of contamination by pollutants, heavy metals or chemicals (such as insecticides) (Sevik et al. 2020). There may also be some threat of arrest or other sanctions if harvesting is practised in spaces where it is prohibited by city authorities or private landowners. Ironically, in some settings undergoing rapid change there may be some negative stigma attached to harvesting wild resources for livelihoods. This is most frequently associated with a generational gap where some of the youth see it is an unsophisticated or traditional practice or a visualisation of poverty, as they aspire to lifestyles more aligned to the market economy and western ideals; for example the youth being less inclined to consume mopane caterpillars (*Gonimbrasia belina*) in Windhoek, Namibia (Nantanga and Amakali 2020).

7 Securing the Local Supply of Urban Provisioning Services

The preceding sections of this chapter have demonstrated that there is widespread collection and use of locally generated PS in cities throughout the Global South, and that the diversity of species and extent of use is markedly greater than in the Global North. Yet, in many settings, the supply or access is not secure, which undermines the livelihood security of millions of urban citizens in the Global South (Escobedo et al. 2015). For example, D'Souza and Nagendra (2011) describe how several lakes in Bengaluru (India) used by peri-urban communities for a range of purposes and PS (greens, fish, water for livestock), lost their livelihoods as the city expanded and converted many of the lakes to recreational areas. The lack of security is a consequence of a complex suite of factors, including (1) formal or informal development on harvesting spaces or water sources, (2) dumping on or pollution of harvesting spaces and water sources, (3) changing access rules, (4) competition for limited supplies and (5) declines in the abundance of some useful species or resources due to one or more of several factors such as herbivory, invasive species, pests and diseases, overuse and adverse weather events. In many instances, local harvesters have limited agency to curb or reverse any of these causes of declining access or supply. On the one hand, this necessitates that they are flexible and adaptive with respect to where and what they harvest (Hurley et al. 2008; Garekae 2020). On the other hand, city planners and authorities can be more proactive in planning and managing for a more secure local supply of PS to those who need or desire them. Core strategies in this regard are outlined below.

- Sufficient area of urban blue-green infrastructure

A vital strategy to maximise the local supply of PS is to plan for and ensure that there is a sufficient area of blue-green infrastructure within a city. How much is 'sufficient' will vary according to local contexts such as rates of urban growth, human population density, the proportion of a city's population that is reliant on locally produced PS, the quality and productivity of the blue and green infrastructure, the availability and affordability of substitutes and so on. However, because the demand for local PS in Global South cities, or specific parts of them, is higher than that typically observed in cites of the Global North, the total area of UGI required will also be higher than the prescriptions and targets set in the Global North. The high rates of urban growth in many Global South cities offer both opportunities and constraints with respect to providing sufficient blue-green urban infrastructure. An opportunity is the possibility to identify and secure land for UGI on the urban periphery well in advance of the physical growth of the city. Infrastructural development would then be excluded from the designated areas. The constraint is that even when green spaces are identified and designated, in the absence of strong governance they may still be occupied by new urban residents seeking spaces to build their own homes or small-scale business outlets.

- Equitable distribution of urban blue-green infrastructure

It is not just the amount of UGI in the city, but also that it should be equitably distributed throughout the city. There is growing evidence that this is rarely the case (Gerrish and Watkins 2018), and perversely it is often the poorer neighbourhoods and cities, that have less public and private UGI (McConnachie and Shackleton 2010; Li et al. 2018; Nero 2017; Venter et al. 2020), which has also been observed in cities of the Global North (e.g. Astell-Burt et al. 2014; Rigolon et al. 2018; Nesbitt et al. 2019). Given that proximity is a primary determinant of where harvesters choose to collect the required PS or collect water, it is vital that planners and city authorities aim to provide sufficient, quality UGI within reasonable vicinity of neighbourhoods dominated by residents who are most dependent on locally supplied PS. However, they must also be aware that such needs and their spatial distribution are not static, but will change in time as a city grows and develops. Even if the demand for local PS declines in a specific neighbourhood in the future, the other multiple functions of UGI will still be needed and provide benefits to residents and the city as a whole (Pauleit et al. 2021).

- Maintenance of natural sites and preferred species

Keeping urban green spaces in a more natural condition will provide a wider diversity of species and therefore potential PS. Additionally, natural sites are likely to be more familiar to new immigrants from surrounding rural areas, which will ease the transition from rural or urban. Offering more natural sites also lowers the costs of maintenance, which is an important consideration when operating budgets are constrained.

- Enrichment plantings

Although natural sites are preferred, some key species may be absent or in low abundance. Stocks could be enriched through planting species in high demand in suitable places throughout the city. For example, planting of fruit or nut trees in public parks, or in the grounds of public institutions (such as government offices, hospitals, schools and libraries) or along the streets. This encompasses concepts and designs such as edible urban landscapes and urban food forests (Russo et al. 2017; Kowalski and Conway 2019). Plants that supply craft fibres or decorative items can also be planted in places appropriate for their stature, such as the planting of sweetgrass used in basket weaving along the roadsides in Mount Pleasant (USA) (Grabbatin et al. 2011). Public land can be made available to citizens to practice urban agriculture (see Drescher et al. 2021) or incentives for doing so in private or allotment gardens. A core watchword here is the notion of multi-functionality of both spaces and species (du Toit et al. 2018; Hurley and Emery 2018). Planning and adopting a food forest or edible urban landscape approach does not mean that other more traditional urban forestry and landscaping must be abandoned. Fruit trees also provide other ecosystem services besides fruits (shade, aesthetics, carbon sequestration, stormwater mitigation and infiltration, air pollution and heat island mitigation, to mention a few), as do trees and plants with medicinal properties, or those that are useful for firewood and so on.

It is a case of recognising a new function, until now relatively neglected, to the urban forest and green space design and management, rather than abandoning old prescripts and benefits.

- Clear and inclusive access

Alongside the provision of sufficient and equitably distributed public blue-green infrastructure is the requirement that the access 'rules' or norms are clear and based on participation of the harvesters. The agreed rules should be publically available to anyone wishing to know them. Within a pro-poor lens, they should also indicate if harvesting is permitted solely for personal use, or that harvesting for commercial reasons is also possible or not. Any changes to the access 'rules' should be determined in consultation with the harvesters.

- Integration into pro-poor policies, strategies and programmes

Since some harvest locally produced PS because they lack the financial resources to purchase them or substitutes, any investment in securing the supply of priority PS can be seen as an investment in poverty mitigation or alleviation. Consequently, the PS considerations should be embedded in city pro-poor policies, strategies and programmes. Additionally, the provision of blue-green infrastructure could then be eligible for funding from poverty alleviation budgets.

- Monitoring of PS abundance and quality

A key consideration for the conservation and use of local urban PS in Global South cities is the need for urban authorities to develop and regularly update an inventory of what is available, where it is located, and its condition. Whilst this would also be necessary in the Global North, the rapid rates of land use change in and around Global South cities requires that more attention is given to monitoring of specific resources or spaces. The financial and human resource burden could be eased if the monitoring programme was designed and executed using a participatory or citizen science approach.

- Governance of urban green infrastructure for local PS

Harvesting of animal and plant materials from formal public urban green spaces is currently frequently prohibited or frowned upon (Shackleton et al. 2017), whilst it might be tolerated or ignored in informal green spaces. Consequently, there is a need for more supportive and inclusive policies and governance models that accommodate a greater diversity of activities in formal public green spaces and that are more explicit in what ecosystem services can be used. In acknowledging the importance of provisioning services to local citizens, planners and management authorities will be better able to anticipate the demand and plan spaces, species and, where necessary, regulations, accordingly. This is likely to be more successful if there is local level community engagement and participation from the very outset, shaped around environmental stewardship approaches and principles (Sardeshpande and Shackleton 2020).

8 The Future Harvest

This chapter has provided an overview of the extent, nature and determinants of the harvest of locally produced PS in urban settings of the Global South. In doing so, I have highlighted areas where the demand, patterns or determinants differ, or are likely to, from those in the Global North, which provides the impetus for different policies and strategies to meet the needs of urban citizens in the Global South. If such policies and strategies are to be contextually relevant, appropriate and responsive to the needs of urban dwellers in the Global South, then they need to be informed by appropriate research and understandings of urban ecology in Global South cities. For example, the everyday practices of urban foraging and urban agriculture are far more widespread in the Global South than the Global North, yet conceptual framings and research around these practices are dominated by contributions and understandings from the Global North.

The very different socio-economic contexts of the Global South has widespread implications for how ecosystem services are generated, managed, used and perhaps even viewed. This includes PS covered in this chapter. The high reliance on PS generated locally within the urban matrix or at the peri-urban periphery is in stark contrast to the Global North, where the source of most widely used PS is spatially divorced from the immediate, urban environment. This realisation then changes the prescriptions on how much UGI might be sufficient for a specific city, the purposes to which it is put, the values local people assign or place on it and the management and governance arrangements.

Some might hold the view that the current widespread reliance on locally generated PS will dwindle as Global South cities become more developed, efficient and better able to supply the basic needs of shelter, nutritional security and health to all citizens. This is hard to predict, not least because it is unknown how long it will take cities of the Global South to become such. In a progressively globalised world, achievements of universal targets of health, sanitation, education, food security and the like are increasingly reliant on national and even global agreements, economies and programmes rather than the what happens at local-scale directed and shaped by individual city authorities. Additionally, there is some evidence to the contrary, i.e. use of local PS persists through time for some households that might have lived in the city for decades (e.g. Schlesinger et al. 2015). This is because procurement of local PS is not solely due to need, but also that there is a strong cultural element to the use of some PS by some citizens, not to mention the recreational benefits. The complex interplay of local and boarder-scale determinants with tradition and culture makes the future of urban harvesting or foraging uncertain. But the ubiquitous prevalence of the practice and deep reliance of some urban dwellers on it, means that it will not disappear in the short or medium-term. It will certainly change in various ways in various cities and neighbourhoods, but the demand for and use of local PS is likely to be a feature of Global South cities for decades to come (as it is in the Global North too). Consequently, researchers need to provide the necessary understandings and models to allow city authorities and national governments to make informed policy choices and guide responsive, appropriate and equitable strategies to meet the needs of their citizens in this regard.

References

Angelsen A, Jagger P, Babigumira R, Belcher B, Hogarth NJ, Bauch S, Börner J, Smith-Hall C, Wunder S (2014) Environmental income and rural livelihoods: a global-comparative analysis. World Dev 64:S12–S28

Arnold JE, Köhlin G, Persson R (2006) Woodfuels, livelihoods and policy interventions: changing perspectives. World Dev 34:596–611

Arrington AB, Diemont SA, Phillips CT, Welty EZ (2017) Demographic and landscape-level urban foraging trends in the USA derived from web and mobile app usage. J Urban Ecol 3(1):jux006

Astell-Burt T, Feng X, Mavoa S, Badland HM, Giles-Corti B (2014) Do low-income neighbourhoods have the least green space? A cross-sectional study of Australia's most populous cities. BMC Public Heal 14:292

Baker K, Bull GQ, LeMay VM (2014) The use of fuelwood market segmentation and product differentiation to assess opportunities and value: a Nicaraguan case study. Energ Sustain Dev 18:58–66

Becerra RM, Inglehart AP (2008) Folk medicine use: diverse populations in a metropolitan area. Soc Work Health Care 21:37–58

Bennett A, Cronkleton P, Menton M, Malhi Y (2018) Rethinking fuelwood: people, policy and the anatomy of a charcoal supply chain in a decentralising Peru. Forests 9:533

Boone C, Glick P, Sahn DE (2011) Household water supply choice and time allocated to water collection: evidence from Madagascar. J Dev Stud 47:1826–1850

Brouwer R, Falcão MP (2004) Wood fuel consumption in Maputo, Mozambique. Biomed Bioenerg 27:233–245

Brouwer ID, Nederveen LM, den Hartog AP, Vlasveld AH (1989) Nutritional impacts of an increasing fuelwood shortage in rural households in developing countries. Prog Food Nutr Sci 13:349–361

Bussmann RW, Glenn A, Sharon D (2010) Healing the body and soul: traditional remedies for "magical" ailments, nervous system and psychosomatic disorders in northern Peru. Afr J Pharm Pharmacol 4:580–629

Castán Broto V, Arthur MD, Guibrunet L (2020) Energy profiles among urban elite households in Mozambique: explaining the persistence of charcoal in urban areas. Energ Res Soc Sci 65:101478

Ceuterick M, Vandebroek I, Pieroni A (2011) Resilience of Andean urban ethnobotanies: a comparison of medicinal plant use among Bolivian and Peruvian migrants in the United Kingdom and in their countries of origin. J Ethnopharmacol 136:27–54

Charnley S, McLain RJ, Poe MR (2018) Natural resource access rights and wrongs: nontimber forest products gathering in urban environments. Soc Nat Res 31:734–750

Cocks ML, Dold AP (2004) A new brush sweeps clean: the economic and cultural values of grass brushes in the Eastern Cape Province, South Africa. For Trees Livelih 13:33–42

Craft D (2010) Urban foraging: finding and eating wild plants in the city. Serviceberry Press, Minneapolis

Dalu MT, Shackleton CM (2018) The potential use of natural resources in urban informal settlements as substitutes for financial capital during flooding emergencies. Phys Chem Earth 104:18–27

de Medeiros PM, Ladio AH, Albuquerque UP (2013) Patterns of medicinal plant use by inhabitants of Brazilian urban and rural areas: a macroscale investigation based on available literature. J Ethnopharmacol 150:729–746

Diaz S, Pacual U, Srenseke M, Martin-López B, Watson RT, Molnár Z, Hill R, Chan KM, Baste IA, Brauman KA, Polasky S, Church A, Lonsdale M, Larigauderie A, Leadley PW, van Oudenhoven

AP, van der Plaat F, Schröter M, Lavorel S, Aumeeruddy-Thomas Y, Bukvareva E, Davies K, Demissew S, Eroul G, Failler P, Guerra CA, Hewitt CL, Keune H, Lindley S, Shirayama Y (2018) Assessing nature's contributions to people. Sci 359:270–272
Drescher AW, Isendahl C, Caridad Cruz M, Karg H, Menakanit A (2021) Urban and peri-urban agriculture in the Global South. In: Shackleton CM, Cilliers SS, Davoren E, du Toit MJ (eds). Urban ecology in the Global South. Springer, Cham, pp 293–324
D'Souza R, Nagendra H (2011) Changes in public commons as a consequence of urbanisation: the Agara lake in Bangalore, India. Environ Manage 47:840–850
du Toit MJ, Cilliers SS, Dallimer M, Goddard M, Guenat S, Cornelius SF (2018) Urban green infrastructure and ecosystem services in sub-Saharan Africa. Landsc Urb Plan 180:240–261
Eat that weed. https://www.eatthatweed.com/. Accessed 20 June 2020
El Bizri HR, Morcatty TQ, Valsecchi J, Mayor P, Ribeiro JE, Neto CF, Oliveira JS, Furtado KM, Ferreira UC, Miranda CF, Silva CH, Lopes VL, Lopes GP, Florindo CC, Chagas RC, Nijman V, Fa JE (2019) Urban wild meat consumption and trade in central Amazonia. Conserv Biol 34:438–448
Escobedo FJ, Clerici N, Staudhammer CL, Corzo GT (2015) Socio-ecological dynamics and inequality in Bogotá, Colombia's public urban forests and their ecosystem services. Urban For Urban Green 14:1040–1053
Fa JE, Wright JH, Funk SM, Márquez AL, Olivero J, Farfán MÁ et al (2019) Mapping the availability of bushmeat for consumption in Central African cities. Environ Res Lett 14:094002
Ferreira FS, Fernandes-Ferreira H, Léo Neto NA, Brito SV, Alves RR (2013) The trade of medicinal animals in Brazil: status and perspectives. Biodiv Conserv 22:839–870
Fonseca FN, Balick M (2018) Plant-knowledge adaptation in an urban setting: Candomblé ethnobotany in New York City. Econ Bot 72:56
Forage London and beyond. https://www.foragelondon.co.uk/. Accessed 20 June 2020
Garekae H (2020) Urbanisation, foraging and household food security in urban South Africa. PhD thesis, Rhodes University, Makhanda
Garekae H, Shackleton CM (2020) Urban foraging of wild plants in two medium-sized South African towns: people, perceptions and practices. Urban For Urban Green 49:126581
Gerrish E, Watkins SL (2018) The relationship between urban forests and income: a meta-analysis. Landsc Urb Plan 170:293–308
Ginger C, Emery MR, Baumflek M, Punam D (2012) Access to natural resources on private property: factors beyond right of entry. Soc Nat Res 25:700–715
Goswami M, Nautiyal S (2020) Transitional peri-urban landscape and use of natural resource for livelihoods. In: Roy N, Roychoudhury S, Nautiyal S, Agarwal S, Baksi S (eds) Socio-economic and eco-biological dimensions in resource use and conservation. Springer, Cham, pp 435–457
Grabbatin B, Hurley PT, Halfacre A (2011) "I still have the old tradition". The co-production of sweetgrass basketry and coastal development. Geoforum 42:638–649
Guild J, Shackleton CM (2018) South African informal urban fuelwood markets are resilient in the context of large-scale socio-economic change. Energ Pol 117:136–141
Haase D, Larondelle N, Andersson E, Artmann M, Borgström S, Breuste J, Gomez-Baggethun E, Gren Å, Hamstead Z, Hansen R, Kabisch N (2014) A quantitative review of urban ecosystem service assessments: concepts, models, and implementation. Ambio 43:413–433
Haines-Young R, Potschin M (2017) Common international classification of ecosystem services (CICES) v5.1: guidance on the application of the revised structure. Fabis Consulting, Nottingham
Hurd KJ, Shertukde S, Toia T, Trujillo A, Pétez RL, Larom DL, Love JJ, Liu C (2019) The cultural importance of edible insects in Oaxaca, Mexico. Ann Entomol Soc Am 112:552–559
Hurley P, Halfacre A, Levine N, Burke M (2008) Finding a "disappearing" nontimber forest resource: using grounded visualisation to explore urbanisation impacts on sweetgrass basket making in Greater Mt. Pleasant, SC. Prof Geog 60:1–23
Hurley PT, Emery MR (2018) Locating provisioning ecosystem services in urban forests: forageable woody species in New York City, USA. Landsc Urb Plan 170:266–275

Hurley PT, Emery MR, Mclain R, Poe M, Grabbatin B, Goetcheus CL (2015) Whose urban forest? The political ecology of foraging urban non-timber forest products. In: Isenhour C, McDonagh G, Checker M (eds). Sustainability in the global city: myth and practice. Cambridge, New York, pp 187–212
IEA (International Energy Agency) (2010) World energy outlook. IEA, Paris
Joos-Vandewalle S, Wynberg R, Alexander KA (2018) Dependencies on natural resources in transitioning urban centers of northern Botswana. Ecosyst Serv 30:342–349
Kaoma H, Shackleton CM (2014) Collection and use of urban tree products by households in poorer residential areas of three South African towns. Urban For Urban Green 13:244–252
Kaoma H, Shackleton CM (2015) The direct use value of urban tree non-timber forest products to household income in poorer suburbs in South African towns. For Pol Econ 61:104–112
Kim H, Song M-J (2013) Ethnozoological study of medicinal animals in Jeju island, Korea. J Ethnopharmacol 146:75–82
Kowalski JM, Conway TM (2019) Branching out: the inclusion of urban food trees in Canadian urban forest management plans. Urban For Urban Green 45:126142
Ladio AH, Acosta M (2019) Urban medicinal plant use: do migrant and non-migrant populations have similar hybridisation processes? J Ethnopharmacol 234:290–305
Lagunes AL, Merçon J (2021) Nature on the border: the reconfiguration of biocultural experiences by migrant women on the southern border of Mexico. In: Cocks ML, Shackleton CM (eds) Diversities of nature in cities: enriching belonging, wellbeing and bioculture. Earthscan/Routledge, London, pp 126–143
Landor-Yamagata JL, Kowarik I, Fischer LK (2018) Urban foraging in Berlin: people, plants and practices within the metropolitan green infrastructure. Sustain 10:1873
Leal ML, Alves RP, Hanazaki N (2018) Knowledge, use, and disuse of unconventional food plants. J Ethnobiol Ethnomed 14:6
Li D, Stucky BJ, Deck J, Baiser B, Guralnick RP (2019) The effect of urbanisation on plant phenology depends on regional temperature. Nat Ecol Evol 3:1661–1667
Li F, Wang X, Liu H, Li X, Zhang X, Sun Y, Wang Y (2018) Does economic development improve urban greening? Evidence from 289 cities in China using spatial regression models. Environ Monit Assess 190:541
Liao J, Zhang L, Liu Y, Li Q, Chen D, Zhang Q, Li J (2018) Diversity and utilisation of edible plants and macro-fungi in subtropical Guangdong Province. South China Forest 9:666
MA (Millennium Ecosystem Assessment) (2003) Ecosystems and human wellbeing: a framework for assessment. Island Press, Washington, DC
Manditsera FA, Lakemond CM, Fogliano V, Zvidzai CJ, Luning PA (2018) Consumption patterns of edible insects in rural and urban areas of Zimbabwe: taste, nutritional value and availability are key elements for keeping the insect eating habit. Food Sec 10:561–570
Matias DM, Tambo JA, Stellmacher T, Borgemeister C, von Wehrden H (2018) Commercialising traditional non-timber forest products: an integrated value chain analysis of honey from giant honey bees in Palawan, Philippines. For Pol Econ 97:223–231
Mbete RA, Banga-Mboko H, Racey P, Mfoukou-Ntsakala A, Nganga I, Vermeulen C, Doucet J-L, Hornick J-L, Leroy P (2011) Household bushmeat consumption in Brazzaville, the Republic of the Congo. Trop Conserv Sci 4:187–202
McConnachie MM, Shackleton CM (2010) Public green space inequality in small towns in South Africa. Habitat Int 34:244–248
McLain RJ, Hurley PT, Emery MR, Poe MR (2014) Gathering “wild” food in the city: rethinking the role of foraging in urban ecosystem planning and management. Local Environ 19:220–240
Messias MC, Menegatto MF, Prado AC, Santos BR, Guimarães MF (2015) Popular use of medicinal plants and the socioeconomic profile of the users: a study in the urban area of Ouro Preto, Minas Gerais, Brazil. Revista Brasil Plantas Medic 17:76–104
Millennium Ecosystem Assessment (2005) Ecosystems and human well-being: synthesis. Island Press, Washington, DC

Mollee E, Pouliot M, McDonald MA (2017) Into the urban wild: collection of wild urban plants for food and medicine in Kampala, Uganda. Land Use Pol 63:67–77
Mugido W, Shackleton CM (2019) The contribution of NTFPs to rural livelihoods in different agro-ecological zones of South Africa. For Pol Econ 109:e101983
Myers G (2021) Urbanisation in the Global South. In: Shackleton CM, Cilliers SS, Davoren E, du Toit MJ (eds). Urban ecology in the Global South. Springer, Cham, pp 27–49
Nadkarni M (2017) Non-timber forest products and small town economies. In: Denis E, Zérah MH (eds) Subaltern urbanisation in India: exploring urban change in South Asia. Springer, New Delhi, pp 601–614
Nantanga KK, Amakali T (2020) Diversification of mopane caterpillars (*Gonimbrasia belina*) edible forms for improved livelihoods and food security. J Arid Environ 177:104148
Ndumbe LN, Ingram V, Tchamba M, Nya S (2019) From trees to money: the contribution of njansang (*Ricinodendron huedelotii*) products to value chain stakeholder's financial assets in the South West region of Cameroon. For Trees Livelih 28:52–67
Nero BF (2017) Urban green space dynamics and socio-environmental inequity: multi-resolution and spatiotemporal data analysis of Kumasi, Ghana. Int J Remote Sens 38:6993–7020
Nesbitt L, Meitner MJ, Girling C, Sheppard SR, Lu Y (2019) Who has access to urban vegetation? a spatial analysis of distributional green equity in 10 US cities. Landsc Urb Plan 181:51–79
Nieman WA, Leslie AJ, Wilkinson A (2019) Traditional medicinal animal use by Xhosa and Sotho communities in the Western Cape province South Africa. J Ethnobiol Ethnomed 15:34
Pauleit S, Vasquéz A, Maruthaveeran S, Liu L, Cilliers SS (2021) Urban green infrastructure in the Global South. In: Shackleton CM, Cilliers SS, Davoren E, du Toit MJ (eds). Urban ecology in the Global South. Springer, Cham, pp 107–143
Pieroni A, Pardo-De-Santayana M, Firenzuoli F, Quave CL (2013) The European heritage of folk medicines and medicinal foods: its contribution to the CAMs of tomorrow. Evidence-based Complem Altern Med 827521
Poe MR, LeCompte J, McLain R, Hurley P (2014) Urban foraging and the relational ecologies of belonging. Soc Cult Geog, April:1–19
Poe MR, Mclain RJ, Emery M, Hurley RT (2013) Urban forest justice and the rights to wild foods, medicines and materials in the city. Hum Ecol 41:409–422
Ram PK, Crump JA, Gupta SK, Miller MA, Mintz ED (2008) Part II: analysis of data gaps pertaining to *Shigella* infections in low and medium human development index countries, 1984–2005. Epidem Infect 136:577–603
Randriamiharisoa MN, Kuhlman AR, Jeannoda V, Rabarison H, Rakotoarivelo N, Randrianarivony T, Raktoarivony F, Randrianasolo A, Bussmann RW (2015) Medicinal plants sold in the markets of Antananarivo. Madagascar. J Ethnobiol Ethnomed 11:60
Rensten J (2016) The edible city: a year of wild food. Boxtree, London
Reynolds C, Byrne MJ, Chamberlain DE, Howes CG, Seymour CL, Sumasgutner P, Taylor PJ (2021) Urban animal diversity in the Global South. In: Shackleton CM, Cilliers SS, Davoren E, du Toit MJ (eds). Urban ecology in the Global South. Springer, Cham, pp 169–202
Rigolon A, Browning M, Jennings V (2018) Inequities in the quality of urban park systems: an environmental justice investigation of cities in the United States. Landsc Urb Plan 178:156–169
Roy S, Byrne J, Pickering C (2012) A systematic quantitative review of urban tree benefits, costs, and assessment methods across cities in different climatic zones. Urban For Urban Green 11:351–363
Russo A, Escobedo FJ, Cirella GT, Zerbe S (2017) Edible green infrastructure: an approach and review of provisioning ecosystem services and disservices in urban environments. Agric Ecosyst Environ 242:53–66
Sardeshpande M, Shackleton CM (2020) Urban foraging: land management policy, perspectives, and potential. PLoS ONE 15(4):e0230693
Schlesinger J, Drescher A, Shackleton CM (2015) Socio-spatial dynamics in the use of wild natural resources: evidence from six rapidly growing medium-sized cities in Africa. Appl Geog 56:107–115

Sevik H, Cetin M, Ozel HB, Ozel S, Zeren Cetin I (2020) Changes in heavy metal accumulation in some edible landscape plants depending on traffic density. Environ Monit Assess 192:78

Shackleton CM (2012) Is there no urban forestry or greening in the developing world? Sci Res Essays 7:3329–3335

Shackleton CM, Cilliers SS, du Toit MJ, Davoren E (2021) The need for an urban ecology of the Global South. In: Shackleton CM, Cilliers SS, Davoren E, du Toit MJ (eds). Urban ecology in the Global South. Springer, Cham, pp 1–26

Shackleton CM, Drescher A, Schlesinger J (2020) Urbanisation reshapes gendered engagement in land-based livelihood activities in mid-sized African towns. World Dev 103:104946

Shackleton CM, Hurley PT, Dahlberg A, Emery MR, Nagendra H (2017) Urban foraging: a ubiquitous human practice but overlooked by urban planners, policy and research. Sustain 9:1884

Shackleton CM, Shackleton SE (2004) The importance of non-timber forest products in rural livelihood security and as safety-nets: evidence from South Africa. S Afr J Sci 100:658–664

Shanahan DF, Fuller RA, Bush R, Lin BB, Gaston KJ (2015) The health benefits of urban nature: how much do we need? Biosci 65:476–485

Short Gianotti AG, Hurley PT (2016) Gathering plants and fungi along the urban-rural gradient: uncovering differences in the attitudes and practices among urban, suburban, and rural landowners. Land Use Pol 57:555–563

Simon D, Goodness J, Lwasa S, Puppim de Oliveira JA, Macedo LV, Kavonic J, Hermansson Török E, Elmqvist T (2021) Urban governance of and for urban green and blue infrastructure. In: Shackleton CM, Cilliers SS, Davoren E, du Toit MJ (eds). Urban ecology in the Global South. Springer, Cham, pp 403–428

Slater R, Twyman C (2003) Hidden livelihoods? Natural resource-dependent livelihoods and urban development policy. ODI, London

Sneyd LQ (2013) Wild food, prices, diets and development: sustainability and food security in urban Cameroon. Sustain 5:4728–4759

TEEB (2010) The economics of ecosystems and biodiversity: mainstreaming the economics of nature: a synthesis of the approach, conclusions and recommendations of TEEB. TEEB, Geneva

Termote C, Everaert G, Meyi MB, Djailo BD, van Damme P (2012) Wild edible plant markets in Kisangani, Democratic Republic of Congo. Hum Ecol 40:269–285

Ticktin T (2004) The ecological consequences of harvesting non-timber forest products. J Appl Ecol 41:11–21

Tiwary A, Vilhar U, Zhiyanski M, Stojanovski V, Dinca L (2020) Management of nature-based goods and services provisioning from the urban common: a pan-European perspective. Urban Ecosyst 23:645–657

Torgerson PR, Hagan JE, Costa F, Calcagno J, Kane M, Martinez-Silveira MS, Goris MG, Stein C, Ko AI, Abela-Ridder B (2015) Global burden of Leptospirosis: estimated in terms of disability adjusted life years. PLoS Negl Trop Dis 9(10):e0004122

Tsuchiya K, Aoyagi M, Okuro T, Takeuchi K (2014) The potential of, and threat to, the transfer of ecological knowledge in urban areas: The case of community-based woodland management in Tokyo, Japan. Ecol Soc 19:25

Tutu RA, Stoler J (2016) Urban but off the grid: the struggle for water in two urban slums in greater Accra, Ghana. Afr Geog Rev 35:212–226

UN Water (2010) Water and urbanisation. https://www.unwater.org/water-facts/urbanisation/. Accessed Nov 2019

van den Berg M, Wendel-Vos W, van Poppel M, Kemper H, van Mechelen W, Maas J (2015) Health benefits of green spaces in the living environment: a systematic review of epidemiological studies. Urban For Urban Green 14:806–816

Venter ZS, Shackleton CM, van Staden F, Selomane O, Masterson VA (2020) Green apartheid: urban green space remains unequally distributed across income and race geographies in South Africa. Landsc Urb Plan 203:103889

Waffle AD, Corry RC, Gillespie TJ, Brown RD (2017) Urban heat islands as agricultural opportunities: an innovative approach. Landsc Urb Plan 161:103–114

Wehi PM, Wehi WL (2010) Traditional plant harvesting in contemporary fragmented and urban landscapes. Conserv Biol 24:594–604

Wildman Steve Brill. Getting closer to nature one bite at a time: the best foraging tours in the NY Tri-state region and beyond. https://www.wildmanstevebrill.com. Accessed 18 June 2020

Williams VL, Balkwill K, Witkowski ET (2000) Unravelling the commercial market for medicinal plants and plant parts on the Witwatersrand, South Africa. Econ Bot 54:310–327

Williams VL, Whiting MJ (2016) A picture of health? Animal use and the Faraday traditional medicine market, South Africa. J Ethnopharm 179:265–273

Understanding Urban Regulating Ecosystem Services in the Global South

Francisco J. Escobedo

Abstract The context of the Global South's urban ecosystems is often poorly understood or accounted for when applying the Urban Regulating Ecosystem Services (URES) concept, primarily because most available information is from the Global North. Thus, there is a need to better understand the unique socio-economic, political and environmental context of urban areas in the Global South when addressing issues related to urban ecosystem services, environmental quality, climate change and environmental justice. This chapter provides a more nuanced perspective on how URES can address urban environmental and climate change problems, thereby improving the human condition and policy formulation. It first defines URES using selected literature and discusses differences in the environmental and socio-economic characteristics and trajectories of cities in the Global South relative to the Global North. Several global examples overview the effectiveness and relevance of URES in the context of mitigation and adaptation to environmental pollution, extreme climatic events and sustainable development. Future research needs and environmental policy implications are also discussed. It concludes by emphasising the need for more interconnected and equal cooperation and knowledge exchange between Global North and South countries, when applying and studying URES concepts and practices.

Keywords Climate change · Environmental justice · Environmental quality · Human well-being · Urban ecosystem functions · Vulnerability

1 Introduction

Several chapters in this book (see Shackleton et al. 2021; Myers 2021; du Toit et al. 2021) have listed and explained the unique dynamics of urban socio-ecosystems in the Global South, their unique trajectories, processes and sociopolitical, economic, and environmental attributes and issues. Although great strides have been made in recent decades addressing these issues and in improving the overall well-being of citizens via health, qualilty of life, and governance; there are still pressing problems,

F. J. Escobedo (✉)
Faculty of Natural Science, Biology Department, Universidad del Rosario, Bogotá, Colombia

C. M. Shackleton et al. (eds.), *Urban Ecology in the Global South*, Cities and Nature,
https://doi.org/10.1007/978-3-030-67650-6_9

particularly in terms of environmental quality and its effects on human well-being. In the next decades most of the growth in the world's population will occur in cities in lower income countries of the Global South and these cities are often characterised by rapid peri-urban land use changes, poorly planned urban development and resource limited governments (Shackleton et al. 2021), all leading to a complexity of emerging environmental and public health issues and opportunities (WHO 2019). Although there is a growing tendency to focus on climate change effects in cities of the Global South, land conversion for resource extraction activities and overexploitation, along with urban development, are in reality the greatest drivers of biodiversity change (Maxwell et al. 2016). This, compounded with climate change, will detrimentally affect urban ecosystem structure, processes and subsequently the provision of ecosystem services and improvements in quality of life in the Global South.

It has now been over 20 years since the advent of the Ecosystem Services (ES) concept and several frameworks (i.e. Millennium Ecosystem Assessment, The Economics of Ecosystem and Biodiversity, Intergovernmental Science-Policy Platform on Biodiversity and Ecosystem Services) and seminal literature have documented how ecosystems and the components of nature benefit society in terms of well-being and its different socio-cultural, economic and environmental components (Costanza et al. 2017). These same sources have also adapted such frameworks and research to better understand these dynamics in urban social-ecological systems and in particular the Urban Ecosystem Services (UES) they provide (Roy et al. 2012; Haase et al. 2014). Several chapters in this book for example, demonstrate some of these ES and their relevance in the mitigation and adaptions of problems related to cities of the Global South (Pauleit et al. 2021; Shackleton 2021; Dobbs et al. 2021). Other studies have also addressed how ESs address air pollution removal (Escobedo et al. 2008), storm water runoff and flooding (Haase et al. 2014), water quality regulation and carbon offsets (Roy et al. 2012). This knowledge has advanced the concept of urban regulating ES and tools, approaches, typologies and even the urban Ecosystem Disservices (EDs), or the negative environmental, social and financial functions of ecosystems and their costs to society (Escobedo et al. 2011; Shackleton et al. 2016; von Döhren and Haase 2015). See Davoren and Shackleton (2021) for further consideration of urban EDs in the Global South.

However, context or the biophysical and sociopolitical realities of any urban or peri-urban area will be a common term used in the following section. As such, context is a key consideration when applying Urban Regulating Ecosystem Services (URES) in that much of this information on ES and ED concerning the environmental benefits and costs are from the Global North and other high-income societies (du Toit et al. 2018; Dobbs et al. 2019). Using the Common International Classification for Ecosystem Services (CICES) hierarchical typology, Regulation and Maintenance

Biotic-abiotic services (i.e. URES) are defined as "All the ways in which living organisms can mediate or moderate the ambient environment that affects human health, safety or comfort, together with abiotic equivalents" (Haines-Young and Potschin 2013). Hence, although necessary, these studies from the Global North and their application and context might not be sufficiently relevant to account for the Global South's realities of unique urbanisation patterns, pollution dynamics, resource limitations, climates, structural inequities, wealth-poverty disparities and policy and legal environments. Recent regional perspectives regarding urban ES from the Global South, such as Dobbs et al. (2019), Escobedo et al. (2019), and Lindley et al. (2018) have called for a more nuanced perspective on the society-environmental quality nexus that is relevant to the Global South.

Accordingly, given this mismatch in conceptual application of the URES concept, the aim of this chapter is to provide a newer and more nuanced perspective on the role of regulating ES in and for the Global South. In particular, the focus will be on environmental quality and climate regulation and how these specific problems affect human well-being and as such how URES can be used to improve the human condition. Although I will use "URES", this same approach and information in this chapter can also be used to account for other regularly used metaphors such as Nature-Based Solutions (NBS), Green Infrastructure and Nature's Contributions to People and their relevance to the URES discourse.

Therefore, the objectives of this chapter are three-fold. First, I provide an overview of how the Global South's urban environment and climate in the Global South are contextually different from that of the Global North and I will make the case that as a result, the function and importance of URES in the Global South are in some cases very different than in the Global North. Second, I indicate some differences between the environmental and socio-economic characteristics and trajectories of cities in the Global South relative to the Global North. I use this information to then synthesise the state of the art of URES from the Global South using a literature review based on key words and concepts that are related to URES. I focus particularly on recent perspective pieces from Africa, Latin America, and elsewhere and glean findings and information that are of relevance to environmental quality and well-being in urban and peri-urban areas in the Global South. Third, I synthesise the application, effectiveness and relevance of URES in the context of mitigation and adaptation of environmental pollution and extreme climatic events in the urban Global South. The role of policies, governance and environmental justice in relation to URES will then be discussed and how these mechanisms are similar, or different, from those of the Global North. I conclude by presenting some examples and case studies of URES in the Global South and some guidelines to propose a way forward to begin to evolve the URES and its use for addressing environmental problems in the Global South.

2 The Urban Environment and Social-Ecosystems of the Global South Are Different from the Global North

Several chapters in this book detail the unique characteristics and dynamics of urban ecosystem dynamics in the Global South relative to those in the Global North (Shackleton et al. 2021; Myers 2021; du Toit et al. 2021). However, in this chapter I focus on the role of urban ecosystem functions and processes in addressing environmental quality problems and the adverse impacts of extreme climate events in the Global South and how they are often different from those in the Global North. I begin by laying out the overall socio-economic development trajectories in the Global South and how these have direct relevance on pollution and extreme events. I begin by highlighting a key difference between the Global South's urban ecosystems and those of the north, in particular disparities between the Global North and South in several regularly used socio-economic and political metrics such as the percentage of total urban population, the respective countries' Gross Domestic Products, Gini Coefficients, levels of and access to infrastructure, access to proper health care and education and the transparency and planning regimes and the effectiveness of institutions, among others (Dobbs et al. 2019; Lindley et al. 2018). These factors are important in the Global South as they have direct implications on several environmental issues such as: energy use, population densities, development of the industrial and service sectors, and cleaner technologies. The sum of these realities has led to not only socio-economic disparities and inequities but different types and levels of the quality of air, water, soil, biodiversity and vulnerabilities to extreme weather events and pollution exposure (WHO 2019). In general, these environmental impacts have mostly been to the detriment of human well-being in the urban and peri-urban areas of the Global South. This detriment is exacerbated in the Global South, as well as some communities in the Global North, because of increased vulnerability and an inability to socio-politically adapt to these negative impacts from pollution and extreme climates (Heltberg et al. 2009).

Although Davoren and Shackleton (2021) detail the vulnerability of urban and peri-urban populations and ecosystems in the Global South, this chapter makes the argument for the importance and relevance of URES as key elements in maintaining human health and well-being as well as acceptable levels of environmental quality and mitigation of extreme events. Accordingly, given the ecosystem functions related to URES, we point out some priority risks to environmental quality and health as indicated by the World Health Organisation's (WHO 2019) characterisation of urban environments of the Global South. An overview of such environmental problems is important in that URES by definition serves to mitigate some of the impacts.

Nagendra et al. (2018) succinctly summarises the multiple and "statistically significant differences" in most environmental and socio-economic indicators (e.g. environmental quality, income inequality, education, infrastructure and health) between cities of the Global North and South. But in general, environmental problems such as poor sanitation and hygiene related to water quality has been estimated to annually kill at least 1.7 million people and ecosystem and climate change impacts another 150,000 deaths annually from extreme events (WHO 2019). Vector borne diseases (e.g. malaria) often related to ecosystem conversion and indoor air pollution from solid fuel burning for cooking and heating another 1.2 and 1.6 million, respectively (WHO 2019). Additionally, urban air pollution is estimated to kill 800,000 people and exposure to toxic substances, such as lead, kills over 230,000 (WHO 2019). All these problems will be compounded since it is estimated that the majority of the future world's population will live in Global South cities of less than one million (Nagendra et al. 2018).

Although problems associated with environmental pollution and extreme weather events are not exclusive to the Global South, again the reality is that there are greater numbers of vulnerable populations and a lack of access to the necessary resources and infrastructure in the Global South relative to the Global North (Füssel 2007). Indeed the impacts of climate change will be disproportionally borne by the Global South, which up to now has contributed the least to causing the problem but has the least resources to deal with it (Mora et al. 2013). This is primarily due to its dependence on climate sensitive sectors such as agriculture and fishing, low incomes and inequality and weak adaptive capacity of its populace and institutions (Heltberg et al. 2009). Extreme events such as floods, droughts and heatwaves will also exacerbate mortality, morbidity as well as the spread of vector borne diseases while undermining food safety and nutrition (Haines et al. 2006).

Similarly, due to the geography of the Global South, many of its biomes such as tropical and arid ecosystems, are less resilient to even minor changes in temperature and precipitation than more northern temperate ones in the Global North (Di Leo et al. 2016; Mora et al. 2013). This "geographic exposure" of the populations and biodiversity is also another factor impacting human well-being (Heltberg et al. 2009). For example, in tropical urban areas of low and medium income countries, temperatures are relatively stable, and a substantial number of people have less resources and air conditioning, thus these people spend less time indoors and in cars and experience greater temperatures and ambient air pollution (Haines et al. 2006). These extreme climate events will noticeably impact coastal urban areas in the middle latitudes. For example Mora et al.'s (2017) models show how the number of deadly heat days (i.e., heatwaves) in 2050 under a limited climate change mitigation scenario (Representative Concertation Pathway 8.5) will have greater impacts on cities in tropical, often coastal, highly populated regions in the Global South (See https://maps.esri.com/globalriskofdeadlyheat/#).

As will be discussed in the final section of this chapter, given the biophysical and socio-economic context of urban and peri-urban areas of the Global South, URES, along with appropriate NBS and green infrastructure practices, can provide both effective and efficient mitigation and adaptation to the environmental and climatic problems.

3 Overview of Urban Regulating Ecosystem Services in the Global South

Due to the lack of data, information, and relevant literature regarding the URES of the Global South, in this following section I first provide a general, conceptual outline of URES and their relevance in the Global South. I then present some relevant typologies and provide some actual examples of how URES are especially suited for mitigating many of the Global South's environmental problems. Overall, a characteristic of economic development in societies, and thereby cities of the Global South relative to those of the Global North, is the transitioning from economies primarily based on an agricultural sector to a shift towards industrial and service-based economies (Laband and Escobedo 2012). Accordingly, there is also a shift in low to medium income countries and their dependence on provisioning ecosystem services related to agriculture and natural resource extraction (Russo et al. 2017), to a development process whereby there is an increased importance of cultural ES, such as that typified in high-income countries (Angeluokski and Martínez Alier 2014; Laband and Escobedo 2012).

Regardless, URES are equally relevant in cities of the Global South and North as clean air, water and climate regulation are vital for all societies (Grimm et al. 2008). Also, along with this development, in general, comes greater environmental regulation, transparency and equity. The resulting shift in these values can be due to increased purchasing power, technological advances and adoption, reduced poverty levels, and increased access to education, among other factors (Laband and Escobedo 2012). In high and medium income economies typical of the Global North, the efficiency and effectiveness of technological, legal and policy instruments can curtail the effects of pollution. But, in many cities in the Global South that are transitioning from agriculture to industrial economies, the dependence on ecosystem processes and natural resources is still key due to the increased use of natural resources and energy inputs that eventually become precursors for pollution and other environmental effects (Nagendra et al. 2018). Given the Global South's limited resources, inequity, vulnerable populations, this is where URES can be particularly important given these structural problems. Such discussions are the basis for sustainable development and metrics such as the Gini Coefficient of Environmental Kuznets curves, which is a regular measure of the relationships between wealth and environmental quality (Angeluokski and Martínez Alier 2014).

It is a common global phenomenon that many cities in the Global South are transitioning to, or have become, more urban and industrialised societies (Laband and Escobedo 2012; Nagendra et al. 2018). This means increased energy use, transportation and industrial activities, and greater conversion of natural and agricultural land uses to urban ones (Derkzen et al. 2017). All these changes will lead to increased pollution emissions, more extreme and severe climate events and dense living conditions (Mora et al. 2013; WHO 2019). Accordingly, many of the Global South's urban areas are characterised by several pressing environmental quality problems that affect human health, well-being and even lives (Derkzen et al. 2017). As discussed earlier, as well as by Simon et al. (2021), as these cities in the Global South begin to grow and transition, in many cases the governance, governability (the ability of government to govern), transparency, government performance, and the overall participatory processes and capacities of institutions and governments can improve or for many cities in the Global South, become weak and resource limited (Lindley et al. 2018; Dobbs et al. 2019). As a result, this process of development and urbanisation makes "urban issues in the Global South distinctly different from those in the Global North" (Nagendra et al. 2018). It is because of this sociopolitical and environmental context of cities in the Global South that URES have a particular relevance to cities and the well-being of people, ecosystems and biodiversity in the Global South (Derkzen et al. 2017). Interestingly, a recent study from Colombia found that urban residents place greater importance on URES than on cultural or provisioning ecosystem services; a finding that is very different from studies in the Global North that find that most actors prioritise cultural over regulating ecosystem services (Escobedo et al. 2020).

As a result, these URES are particularly relevant and can be just as, if not more, effective and efficient means of reducing pollution levels and severe climate events in cities of the Global South. These URES are proven measures and strategies that help to improve environmental quality and enhance the capacity of remnant natural and semi-natural ecosystems to maintain the essential ecological process and life support systems, and are the ecosystem functions most relevant to addressing these environmental quality and climate change issues that affect lives, health and well-being (Table 9.1). These URES as such were first mentioned in the Millennium Ecosystem Assessment (Costanza et al. 2017), but have also been described as environmental processes, biotechnologies, phytoremediation, among other terms. In the following section, I will first analyse the use of the URES concept in the global literature, how

Table 9.1 Example of Regulating Ecosystem Services based on the Common International Classification of Ecosystem Services Version 4.3 typology (Haines-Young and Potschin 2013). Those in bold are of particular relevance to human health, safety and overall well-being

Division	Group
Regulation, mediation, maintenance and transformation of waste, toxics and other nuisance	Mediation by biota
	Mediation by ecosystems
Regulation, mediation, maintenance and transformation of flows	Mass flows
	Liquid flows
	Gaseous/air flows
Regulation, mediation, maintenance and transformation of physical, chemical, biological conditions	Atmospheric composition and climate regulation
	Water conditions
	Pest and disease control
	Soil formation and composition
	Lifecycle maintenance, habitat and gene pool protection

they are used in the Global North versus the Global South, and then I delve into their relevance for the urban realities of the Global South. Although we use URES, the final benefits are similar as those related to other metaphors such as Nature-Based Solutions, Green Infrastructure and Nature's Benefits to Society (Escobedo et al. 2019).

To assess the current state of the art on the topic of URES in the Global South, a June 2019 search in Google Scholar for "urban ecosystem services" identified nearly 1,250,000 articles, while refining this search to "urban ecosystem services" AND "regulating" OR "regulation" identified 2,800 results. An overview of these journal articles and reviews, books, and other types of publications show that topics discussed in the publications range from urban biodiversity, governance, green and blue infrastructure, planning, valuation, sustainability and many other topics. Although highly relevant to the content of this book, there are clearly topics outside the scope of this chapter.

To better focus on the topic at hand, another June 2019 literature review in SCOPUS for articles on URES-related articles, reviews and books and their respective title, abstract and keywords returned 3,371 publications on "Ecosystem services" AND "Urban", only four Global South countries ranked in the top 20 in terms of number of publications: South Africa (12th), Brazil (13th), India (16th) and Mexico (18th) for a combined total of 12% of global output. A search for "regulating ecosystem services" and "Urban" found only 51 articles. Out of 203 review publications in SCOPUS for "Ecosystem Services" AND "Urban"; Brazil, India, Indonesia and Colombia combined account for only 7% of all publications. Similarly, when using this same search string, of the 10 most cited articles only Kowarik (2011), Roy et al. (2012), and Haase et al. (2014) mention the under-representation of studies outside the United States, Europe and China. Haase et al.'s (2014) review also documents that 50% of all urban ES publications are specifically related to URES. However, Grimm et al. (2008) question the relevance of studies from the Global North to low and medium income countries in the Global South. Only Escobedo et al. (2011) and Guerry et al. (2015) specifically discuss the application of URES concepts using examples from Asia, Latin America and Africa.

Other recently published literature reviews and perspectives on urban ES and EDs have thoroughly discussed the topic of UES and its relevance to different regions in the globe and specifically the Global South (Dobbs et al. 2019; du Toit et al. 2018; Lindley et al. 2018). There is also a notable rise in English language publications from China. Indeed a literature review on the use of ES, Green infrastructure and NBS in the urban forest literature found that the United States, China, and larger European Union countries led "by a notable margin" in the production of English language publications on these topics (Escobedo et al. 2019). Hence given China's increasing economic and scientific influence, I will not discuss the URES concept relative to China. However, see Haase et al. (2014) and Yang et al. (2015) and citations therein for more detailed discussions of URES in China.

However, there is little information on URES from other eastern, central and southern Asian countries (Roy et al. 2012). Mukherjee and Kaoru (2015) discuss the role of urban green spaces in mitigating risks from extreme climate events and how

Asian countries' needs are different from those of the Global North; hence different frameworks and perspectives are warranted. In terms of flooding, the authors point out the damaging effects of urban flooding and the disproportionate impacts to vulnerable populations and minimal flood control infrastructure—relative to cities in the Global North—and how green spaces can reduce hydrologic peak flows, duration and frequencies. In terms of extreme heat events, the Global South's lack of resources, infrastructure, and public services can go beyond the direct effects on human well-being (decreased thermal comfort and increased morbidity and mortality) but can also lead to increased water use, collapse of energy supply networks and even deterioration of the transportation networks [e.g. increased temperatures degrade paved surfaces (Mukherjee and Takara 2018)]. Thus, the authors point out how urban green space implementation, as a means to provide URES, is lacking in many Asian countries such as China, even Japan and Korea when compared to most Global North countries. Overall, other reviews on URES from Asia, excluding China, are rare and more information is needed. In fact, Keniger et al.'s (2013) review of 57 studies on the benefits of interacting with nature found a bias towards high latitude countries of North America and Europe, while there were very few studies from Africa or South America. Interestingly, they point out that this bias away from tropical, highly biodiverse regions might be resulting in a bias towards benefits and not the problems associated with EDs such as disease vectors and poisonous animals (Shackleton et al. 2016), which are more common in the Global South (Davoren and Shackleton 2021).

Much of the global literature on urban environmental quality and URES of the Global South has been around for more than 10 years. A decade ago, Grimm et al. (2008) discussed the importance of the globalisation processes and resulting impacts to environmental quality, and Kowarik (2011) also mentioned that knowledge on the effects of biotic responses to urbanisation is increasing. Both these pieces concluded that at the time of writing, most knowledge on urban ecosystems, their functions and services were based on single case studies from very specific contexts, and hence, care was warranted when generalising the results to other less studied regions. Escobedo et al. (2011) also emphasised this, since urban ecosystems and their UES and URES depend on the biophysical and sociopolitical contexts specific to a city.

More recently however, other regions such as Africa and Latin America have begun to discuss the state of the art in UES knowledge and their relevance to the realities of the Global South (Dobbs et al. 2019; du Toit et al. 2018). Du Toit et al. (2018) for example reviewed 68 research articles from 20 Sub-Saharan African countries and found that 62% of these countries had no research on UES. The countries that were most studied were South Arica and Ghana. However, URES and provisioning ES got the most attention, in particular shading and water regulation were the most studied. Lindley et al. (2018) also discussed the UES concept from the perspective of Sub-Saharan African cities. The authors pointed out the realities of decision-makers in these countries involving URES are often made in contexts that lack appropriate governance and scientific foundations. However, they emphasised the importance of UES to livelihoods as well as safety and well-being of cities of the Global South. They noted the distinctness of UES in Africa and the importance of not only URES but provisioning and socio-cultural ES. Wangai et al. (2016) and McHale et al. (2013) also emphasise the spatial, socio-economic and ecological uniqueness of African countries and how these contexts and even theories are very different from research carried out in the Global North.

Dobbs et al. (2019) discussed the relevance of UES to countries in the regions of Latin America and the Caribbean (LAC). The authors reviewed 107 publications and found that only five of the 22 LAC countries had more than 10 publications on UES. More populous Brazil and Mexico had the most publications, while Colombia and notably Chile with much smaller populations were the countries with the most publications. Perhaps given the region's varied biomes and high biodiversity, the authors found that biodiversity conservation and URES studies were the publications that were most represented in the urban ES literature, while pest regulation, pollination and provisioning ES were studied the least. Other studies indicate that the region has a history in the last 10–15 years of publications on URES, particularly on the topic of air pollution removal related URES in countries such as Chile and Mexico (Escobedo et al. 2008). Other UES have however been less studied and there is little science-based information on provisioning, water quality and cultural ES (Dobbs et al. 2019). The authors concur with the previous reviews and perspectives from Sub-Saharan African countries in many aspects. For example, these authors also pointed out the realities of the region's inequities, governance structures and vulnerability to extreme climates. However, unlike Africa and parts of Asia, the LAC region is and has been for the last decades, one of the most highly urbanised regions in the world.

The previous discussion on the state of the art regarding UES and URES in the Global South based on recent regional perspectives indicates:

1. Despite the many environmental and quality of life issues in the Global South, there is a *relative* lack of information on URES in the Global South.
2. The breadth of topics related to URES from literature in the Global North and South do not all strictly use the URES term.
3. Although recent studies have provided regional summaries on UES and URES, the number of relevant publications from the Global South are still far behind those from the Global North.
4. There is clearly a need to account for the Global South's unique socio-economic, political and environmental contexts. Accordingly, in the following section I will first define URES and then discuss related literature to URES and its application to the Global South.

Literature from China, Latin America and South Africa is now beginning to shed light on the realities of URES in the Global South. However, a Scopus search for "ecosystem services" AND "urban" AND "air" OR "water" OR "climate" found that outside of China, there were no Global South countries in the top 10 list of countries publishing on URES. In addition, a following topical search for "Ecosystem services" AND "Urban" AND "Floods", found Brazil and South Africa each with five out of 171 total publications, while "Ecosystem services" AND "Urban" AND "Temperature" identified only Brazil with eight out of 227 total publications. Only the search for "Ecosystem services" AND "urban" AND "Pollution" found that Mexico and Brazil were the only two Global South countries with 21 out of 386 total publications. But Finland with a population of 5.5 million has the same number of publications (11) as that of Mexico with a population of 129 million.

Overall, the above URES-related literature from the Global South, other than China, indicates the lack of relevant research on URES in the Global South. There is nothing new in this north-south disparity in terms of URES literature (Escobedo et al. 2019; Nagendra et al. 2018). However, we can draw some conclusions from this set of available studies on URES related to water quality, flooding and air quality in cities of the Global South and this chapter's first section regarding environmental quality problems and extreme climate events of the Global South. First, the literature from the Global South is generally focused on URES as related to landslides and flooding. This information has particular relevance regarding human health, livelihoods and prevention of possible loss of life, particularly in more mountainous areas of the Global South. For example, the loss of 500 lives in Freetown (Sierra Leone) after heavy rains triggered land- and mudslides, which were catalysed by the removal of much of the vegetation on steep slopes (Cui et al. 2019). The role of remnant natural areas and green infrastructure has been documented to be a NBS in terms of increasing slope stability and water storage and infiltration, with subsequent storm water reduction and flooding reduction URES (Sidle et al. 2013; Mukherjee and Takara 2018), as well as providing provisioning resources for citizens to cope and rebuild lost structures, as was recorded after flooding in three towns in South Africa (Dalu and Shackleton 2018). These URES or NBS have particular relevance for reducing landslides in peri-urban areas and flooding, thus reducing damage to infrastructure, drainage and sewer system collapse and effects on human health and safety (Dobbs et al. 2019).

Air pollution problems that have affected the Global North have been the basis for increased research and subsequent development of technologies and policies to curtail the negative effects on air quality. Accordingly the literature on air quality has recently seen increased information and knowledge on the role of air quality improvement as a result of urban and peri-urban vegetation (Escobedo et al. 2008; Haase et al. 2014). This is particularly pressing, since, as described in previous sections, the impacts to human health and well-being are impacting societies and economies. However, as cities and societies in the Global South transition to more industrialised economies, other pollutants such as ozone are affecting environmental quality in many cities as typified by, but not restricted to, the large mega-cities of China, India and Mexico (Bell et al. 2006; Chan and Yao 2008; Lelieveld et al. 2015). This has implications, with increased NOx emissions and VOC emission from industrial and transportation emissions, combined with biogenic emissions from urban and peri-urban vegetation, can lead to subsequent ozone formation (i.e. EDS; Escobedo et al. 2011; von Döhren and Haase 2015).

Third, the literature from the Global North related to urban heat islands documents the impacts of extreme temperature events on human health and hence the climate regulation benefits of urban vegetation are beginning to be studied in the Global South (Di Leo et al. 2016; Roy et al. 2012). This is well developed in China (e.g. Zhou et al. 2014; Song et al. 2020), but with increasing studies elsewhere in the Global South, including Africa (e.g. Ngulani and Shackleton 2019; Simwanda et al. 2019) and Latin

America (Cui and De Foy 2012; Peres et al. 2018). Such studies include knowledge on the direct temperature reduction benefits as well as the indirect carbon storage and sequestration and subsequent mitigation of climate change effects by urban and peri-urban vegetation and soils (Dobbs et al. 2019; Lindley et al. 2018). Although, this has not been studied, the increased humidity, stable wind and temperatures in lower latitude, tropical climates can reduce the evapotranspiration function of urban trees and hence the benefits on cooling and thermal comfort might be minimal in these contexts. However, given the generally lower average CO_2 per capita emissions in the Global South, relative to the North, the potential to offset local and regional scale CO_2 emission via URES are greater (Escobedo et al. 2011).

Finally, much less studied are the role of urban and peri-urban ecosystem structures and biodiversity in not only mitigating dust, diseases and pests that detrimentally affect human health but can conversely also affect human health in the form of ecosystem disservices (Davoren and Shackleton 2021), like dust storms, vectors and subsequent diseases such as respiratory diseases, malaria, chikungunya, west nile, among other diseases (Escobedo et al. 2011; Shackleton et al. 2016; von Döhren and Haase 2015). However, in tropical contexts these URES can be complex and contradictory. For example, well-managed, subtropical urban forests in the Caribbean region have been shown to reduce hurricane related wind damage to homes and infrastructure by mitigating wind speeds and blocking flying debris (Escobedo et al. 2009). However, tropical cyclones in the eastern and southeastern Asia have been found to increase soil productivity, and the generation of wind-borne biomass debris is often used as fuel for cooking (Vink and Ahsan 2018). Similarly tree biomass and debris from hurricanes or maintenance and removal practices can be used for bioenergy production or urban provisioning ES (Escobedo et al. 2009; Russo et al. 2017).

4 Polices, Environmental Justice and Efficacy of Urban Regulating Ecosystem Services

Having reviewed the socio-environmental context of Global South cities, the URES literature, and the relevance and nuances of the concept, I now transition and discuss how URES can be used to address many of the problems affecting the increasingly urbanising societies of the Global South. Chapter "Urban Governance of and for Urban Green and Blue Infrastructure" (Simon et al. 2021) discusses the role of valuation and governance. Therefore, I now discuss the applicability and effectiveness of URES. The socio-environmental issues facing urban and peri-urban areas in the Global South are "wicked problems" and most decisions will have trade-offs, thus requiring the need to prioritise the scarce resources and account for issues of equity among provisioning ES that address immediate human needs, such as food, fibre, employment (Russo et al. 2017) over URES that address human health and safety, and other important cultural ES that might not necessarily be urgent for protecting human lives and physical health. Accordingly the next section briefly considers the concepts of *environmentalism of the poor* and *environmental justice*; perspectives that can help us understand issues of equity and sustainability when applying URES in the context of the Global South.

Anguelovski and Martínez Alier (2014) summarise environmentalism of the poor and environmental justice, as two perspectives based on social justice and the premise that human rights, human well-being and environmental quality are inseparable. This is an important consideration in that many conservation efforts aimed at conserving remnant flora or fauna are usually based on aspects of deep ecology and the preservation of ecosystems and biota often without including any human components of these systems. These efforts often advocated "pristine" conditions and therefore that humans and markets should be excluded from these spaces (Anguelovski and Martínez Alier 2014). Conversely, these same authors synthesise another "eco-efficiency" perspective based on engineering, resource management, economic and financial approaches that often quantify, monetise and optimise biodiversity and ecosystem processes. This latter approach, in particular, has relevance to urban Global South regions in Latin America and Asia that are implementing market-based instruments such as Payments for Ecosystem Services and carbon accounting and offset mechanisms as well as "green" technologies, monetary valuation and smart cities approaches (Escobedo et al. 2019). Such approaches, particularly in the Global South, can be controversial and have been referred to as neo-liberal approaches for the capitalisation and commodification of nature (Anguelovski and Martínez Alier 2014).

On the other hand, such policy instruments, approaches and metaphors like ESs and EDs quantification and valuation, market-based instruments, green infrastructure and NBS for example can, in many contexts, improve and foster the communication process of conveying the importance of urban nature and subsequent provision of URES and other ES (Di Leo et al. 2016; Escobedo et al. 2019). Given the complexity of dealing with the multitude of actors in cities, a more focused approach for promoting and advocating, or the use of URES instead of engineered approaches is to focus on influential and more powerful groups such as policy-makers and land managers to communicate these approaches (Dobbs et al. 2019). By also accounting for issues of sustainability and equity, a greater number of more vulnerable actors in the Global South could benefit directly from URES in terms of health, safety and well-being than more privileged groups and societies of the Global North.

Lindley et al. (2018) succinctly discuss how UES related planning, management and governance frameworks in the Global South are often "lambasted" for their weak enforcement, corruption and inefficacies. In addition the limited resources confronting many Global South cities means that management and planning strategies and activities associated with conserving and maintaining the ecosystems providing URES are lacking, as are the necessary requisites for participatory governance mechanisms and the monitoring and evaluation of URES initiatives and policies (Simon et al. 2021). However, cities and societies in the Global South are highly adaptive, with capacity to innovate (McHale et al. 2013; Nagendra et al. 2018). Escobedo et al. (2011), Dobbs et al. (2019), du Toit et al. (2018), Derkzen et al.

Table 9.2 Examples of success stories using URES in a few cities in the Global South

	Example	Source
Costa Rica	Payment for Ecosystem services schemes from Costa Rica that have been around for a very long time and have been the basis for other ecosystem service schemes and instruments	FONAFIFO, artículo 46 de la Ley 7575
Chile	Legislation on the use of tree planting in Chile as compensation for industrial particulate matter emissions	Resolución exenta 1260 de 25 November 2015 pagina 64.
Morocco	Restoration examples from Africa for peri-urban desertification/wind/dust reduction benefits	https://news.mongabay.com/2016/11/morocco-plants-millions-of-trees-along-roads-to-fight-climate-change/
South Africa	Strategic mapping and management of ecosystem services hotspots with the city of Durban	Davids et al. (2016)
China	Green belts—carbon sequestration	https://www.weforum.org/agenda/2019/03/plant-more-trees-young-forests-use-carbon-most-effectively/
Caribbean cities	Tropical wind storm mitigation	Escobedo et al. (2009)

(2017), Mukherjee and Takara (2018), and Wangai et al. (2016) present example case studies on the success and effectiveness of URES in addressing environmental quality and extreme climatic events (Table 9.2).

5 Conclusion and Way Forward

This chapter has highlighted the overall biophysical and economic, and to a limited extent the environmental justice, issues and differences between cities in the Global North and those of the Global South in terms of URES. Although I did not lay out the guidelines and considerations for the ecological based management of green spaces in the Global South, I point the reader to Simon et al. (2021) that specifically focus on the political and governance issues for more specific information and details. Reviews such as those of Haase et al. (2014) and Roy et al. (2012) point out the benefits of URES and how these are frequently studied, and the findings applied, in cities of high-income countries in North America, Europe, Australia and more recently China. However, the question remains: how relevant is information from Berlin, Stockholm, Baltimore or Phoenix to medium and large urban and peri-urban social-ecosystems in the Global South? Although valuable information has been provided to advance and promote UES, the knowledge production system has up until now been shaped by the Global North, despite the Global South "capacity to innovate and experiment" in issues related to URES (Nagendra et al. 2018).

Overall, as outlined by Dobbs et al. (2018), du Toit et al. (2018), and Lindley et al. (2018), the planning, management, and the policy formulation and implementation processes associated with UES and their role and importance in addressing environmental problems in the Global South are different and require more context-specific approaches, methods and implementation (Shackleton et al. 2021). Although, on average, the Global South does have in general greater environmental quality problems and more vulnerable populations than the Global North, there are key lessons to consider when applying URES in the Global South. As the previously mentioned authors summarised, basic issues related to food, clothing, shelter, health and equity in the Global South often override the environmental or cultural benefits. Similarly, education and research evidence, and access to these, is also often limited when making decisions regarding the role of UES. However, the above material all point to the potential that URES have in mitigating many of the problems associated with urbanisation, changing climates and development trajectories in cities of the Global South.

There is increasing interest on the future effects of climate change on cities of the Global South. However, land conversion, resource extraction and overexploitation, along with urban development, will be the greatest drivers of ecosystem change both in the Global South and North. However, as shown in this chapter, there is increased documentation of how URES can effectively mitigate many of these adverse environmental impacts. As such, the purpose of this chapter was to conceptually familiarise the reader with URES and provide a guide to better understand how urban ecosystem functions and processes can be used to solve many of the Global South's pressing environmental problems. By explaining, rather than providing a detailed systematic literature review or site specific details and numbers, some of the environmental and socio-economic characteristics and trajectories of cities in the Global South relative to those from the Global North, this chapter can be used to better educate users, practitioners and policymakers on how to better understand and apply URES. Accordingly, as a way forward I conclude with these take-home messages from information that can be gleaned from the above literature and information:

- Caution is necessary when applying the URES as the concepts were developed and applied in the Global North without accounting for the Global South's context of pollution, inequities and other environmental injustices. To these ends, other more recent concepts such as Nature-Based Solutions and Nature's Contributions to People could provide alternative frameworks to better understand how to more effectively communicate the benefits of urban green infrastructure and ecology for solving practical problems in the Global South. Specific URES practices and approaches from the Global North should be used to complement, not substitute, context-specific Global South knowledge and approaches,
- Although problems related to poverty and economic development need to be prioritised, the potential of URES in mitigating extreme temperatures and pollution levels in the Global South needs to be promoted in terms of incorporating

URES into policies, increased funding in research and the incorporation of UES concepts in environmental and extension education programs,
- URES-related benefits, costs and trade-offs need to be connected to other factors associated with multi-dimensional poverty; the creation of jobs, creating a sense of community are just as important as reducing temperatures and improving local-scale air quality,
- Cities in the Global South represent alternative social-ecological states with unique dynamics and organisational structures that can be used to study and develop relevant theories, approaches, methods and solutions related to URES,
- More cooperation and knowledge exchange is needed among societies of the Global South in applying the URES concepts and practices, less dependence on the Global North for knowledge. Indeed, recent experiences and information from China might be more relevant than other European and North American examples.

References

Anguelovski I, Martínez Alier J (2014) The 'environmentalism of the poor' revisited: territory and place in disconnected glocal struggles. Ecol Econ 102:167–176

Bell ML, Davis DL, Gouveia N, Borja-Aburto VH, Cifuentes LA (2006) The avoidable health effects of air pollution in three Latin American cities: Santiago, São Paulo and Mexico City. Environ Res 100:431–440

Chan CK, Yao X (2008) Air pollution in mag cities in China. Atmos Enviro 42:1–42

Costanza R, de Groot R, Braat L, Kubiszewski I, Fioramonti L, Sutton P, Farber S, Grasso M (2017) Twenty years of ecosystem services: how far have we come and how far do we still need to go? Ecosys Serv 28:1–16

Cui Y, De Foy B (2012) Seasonal variations in the urban heat island at the surface and the near surface and reduction due to urban vegetation in Mexico City. J Appl Meteorol Clim 51:855–868

Cui Y, Cheng D, Choi CE, Jin W, Lei Y, Kargel JS (2019) The costs of rapid and haphazard urbanization: lesson learned from the Freetown landslide disaster. Landslides 16:1167–1176

Dalu MT, Shackleton CM (2018) The potential use of natural resources in urban informal settlements as substitutes for financial capital during flooding emergencies. Phys Chem Earth 104:18–27

Davids R, Rouget M, Boon R, Roberts D (2016) Identifying ecosystem service hotspots for environmental management in Durban, South Africa. Bothalia 46:2118

Davoren E, Shackleton CM (2021) Urban ecosystem disservices in the Global South. In: Shackleton CM, Cilliers SS, Davoren E, du Toit MJ (eds) Urban ecology in the Global South. Springer, Cham, pp 265–292

Derkzen ML, van Teeffelen AJ, Nagendra H, Verburg PH (2017) Shifting roles of urban green space in the context of urban development and global change. Curr Opin Environ Sustain 29:32–39

Di Leo N, Escobedo FJ, Dubbeling M (2016) The role of urban green infrastructure in mitigating land surface temperature in Bobo-Dioulasso, Burkina Faso. Environ Dev Sustain 18:373–392

Dobbs C, Escobedo FJ, Clerici N, de la Barrera F, Eleuterio AA, MacGregor-Fors I, Reyes-Paecke S, Vásquez A, Camaño JD, Hernández HJ (2019) Urban ecosystem services in Latin America: mismatch between global concepts and regional realities? Urban Ecosyst 22:173–187

Dobbs C, Vasquez A, Olave P, Olave M (2021) Cultural urban ecosystem services. In: Shackleton CM, Cilliers SS, Davoren E, du Toit MJ (eds) Urban ecology in the Global South. Springer, Cham, pp 245–264

du Toit MJ, Cilliers SS, Dallimer M, Goddard M, Guenat S, Cornelius SF (2018) Urban green infrastructure and ecosystem services in sub-Saharan Africa. Landsc Urb Plan 180:249–261
Escobedo FJ, Bottin M, Cala D, Montoya DL (2020) Spatial literacy influences stakeholder's recognition and mapping of peri-urban and urban ecosystem services. Urban Ecosyst 23:1039–1049
Escobedo FJ, Giannico V, Jim CY, Sanesi G, Lafortezza R (2019) Urban forests, ecosystem services, green infrastructure and nature-based solutions: nexus or evolving metaphors? Urban For Urban Green 37:3–12
Escobedo FJ, Kroeger T, Wagner JE (2011) Urban forests and pollution mitigation: analyzing ecosystem services and disservices. Environ Poll 159:2078–2087
Escobedo FJ, Luley CJ, Bond J, Staudhammer C, Bartel C (2009) Hurricane debris and damage assessment for Florida urban forests. J Arboricult 35:100
Escobedo FJ, Wagner JE, Nowak DJ, De la Maza CL, Rodriguez M, Crane DE (2008) Analyzing the cost effectiveness of Santiago, Chile's policy of using urban forests to improve air quality. J Environ Manage 86:148–157
Füssel HM (2007) Vulnerability: a generally applicable conceptual framework for climate change research. Glob Environ Change 17:155–167
Grimm NB, Foster D, Groffman P, Grove JM, Hopkinson CS, Nadelhoffer KJ, Pataki DE, Peters DP (2008) The changing landscape: ecosystem responses to urbanisation and pollution across climatic and societal gradients. Front Ecol Environ 6:264–272
Guerry AD, Polasky S, Lubchenco J, Chaplin-Kramer R, Daily GC, Griffin R, Ruckelshaus M, Bateman IJ, Duraiappah A, Elmqvist T, Feldman MW (2015) Natural capital and ecosystem services informing decisions: from promise to practice. Proc Nat Acad Sci 112:7348–7355
Haase D, Larondelle N, Andersson E, Artmann M, Borgström S, Breuste J, Gomez-Baggethun E, Gren Å, Hamstead Z, Hansen R, Kabisch N (2014) A quantitative review of urban ecosystem service assessments: concepts, models, and implementation. Ambio 43:413–433
Haines A, Kovats RS, Campbell-Lendrum D, Corvalán C (2006) Climate change and human health: impacts, vulnerability, and mitigation. The Lancet 367:2101–2109
Haines-Young R, Potschin M (2013) Common international classification of ecosystem services (CICES): consultation on version 4, August-December 2012. EEA Framework Contract No EEA/IEA/09/003. Online: www.cices.euo
Heltberg R, Siegel PB, Jorgensen SL (2009) Addressing human vulnerability to climate change: toward a 'no-regrets' approach. Glob Environ Change 19:89–99
Keniger L, Gaston K, Irvine K, Fuller R (2013) What are the benefits of interacting with nature? Int J Environ Res Publ Heal 10:913–935
Kowarik I (2011) Novel urban ecosystems, biodiversity, and conservation. Environ Poll 159:1974–1983
Laband DN, Escobedo FJ (2012) Economic aspects and issues along an urban–rural gradient. In: Laband DN, Lockaby BG, Zipperer WC (eds) Urban-rural interfaces: linking people and nature. Wily, New York, pp 165–183
Lelieveld J, Evans JS, Fnais M, Giannadaki D, Pozzer A (2015) The contribution of outdoor air pollution to premature mortality rates on a global scale. Nat 525(7569):367–371
Lindley S, Pauleit S, Yeshitela K, Cilliers SS, Shackleton C (2018) Rethinking urban green infrastructure and ecosystem services from the perspective of sub-Saharan African cities. Landsc Urb Plan 80:328–338
Maxwell SL, Fuller RA, Brooks TM, Watson JE (2016) Biodiversity: the ravages of guns, nets and bulldozers. Nat News 536(7615):143
McHale MR, Bunn DN, Pickett ST, Twine W (2013) Urban ecology in a developing world: why advanced socioecological theory needs Africa. Front Ecol Environ 11:556–564
Mora C, Dousset B, Caldwell IR, Powell FE, Geronimo RC, Bielecki CR, Counsell CW, Dietrich BS, Johnston ET, Louis LV, Lucas MP (2017) Global risk of deadly heat. Nat Clim Change 7:501
Mora C, Frazier AG, Longman RJ, Dacks RS, Walton MM, Tong EJ, Sanchez JJ, Kaiser LR, Stender YO, Anderson JM, Ambrosino CM (2013) The projected timing of climate departure from recent variability. Nat 502(7470):183

Mukherjee M, Takara K (2018) Urban green space as a countermeasure to increasing urban risk and the UGS-3CC resilience framework. Int J Disast Risk Red 28:854–861
Myers G (2021) Urbanisation in the Global South. In: Shackleton CM, Cilliers SS, Davoren E, du Toit MJ (eds) Urban ecology in the Global South. Springer, Cham, pp 27–49
Nagendra H, Bai X, Brondizio ES, Lwasa S (2018) The urban south and the predicament of global sustainability. Nat Sustain 1:341
Ngulani T, Shackleton CM (2019) The degree, extent and value of air temperature amelioration by urban green spaces in Bulawayo, Zimbabwe. S Afr Geogr J 102:344–355
Pauleit S, Vasquéz A, Maruthaveeran S, Liu L, Cilliers SS (2021) Urban green infrastructure in the Global South. In: Shackleton CM, Cilliers SS, Davoren E, du Toit MJ (eds) Urban ecology in the Global South. Springer, Cham, pp 107–143
Peres LD, Lucena AJ, Rotunno Filho OC, Franca JR (2018) The urban heat island in Rio de Janeiro, in the last 30 years using remote sensing data. Int J Appl Earth Obs Geoinfo 64:104–116
Roy S, Byrne J, Pickering C (2012) A systematic quantitative review of urban tree benefits, costs, and assessment methods across cities in different climatic zones. Urban For Urban Green 11:351–363
Russo A, Escobedo FJ, Cirella GT, Zerbe S (2017) Edible green infrastructure: an approach and review of provisioning ecosystem services and disservices in urban environments. Agric Ecosyst Environ 242:53–66
Shackleton CM (2021) Provisioning services in Global South cities. In: Shackleton CM, Cilliers SS, Davoren E, du Toit MJ (eds) Urban ecology in the Global South. Springer, Cham, pp 203–226
Shackleton CM, Cilliers SS, du Toit MJ, Davoren E (2021) The need for an urban ecology of the Global South. In: Shackleton CM, Cilliers SS, Davoren E, du Toit MJ (eds) Urban ecology in the Global South. Springer, Cham, pp 1–26
Shackleton CM, Ruwanza S, Sanni GS, Bennett S, De Lacy P, Modipa R, Mtati N, Sachikonye M, Thondhlana G (2016) Unpacking Pandora's box: understanding and categorising ecosystem disservices for environmental management and human wellbeing. Ecosys 19:587–600
Sidle RC, Benson WH, Carriger JF, Kamai T (2013) Broader perspective on ecosystem sustainability: consequences for decision making. Proc Nat Acad Sci 110:9201–9208
Simon D, Goodness J, Lwasa S, Puppim de Oliveira JA, Macedo LV, Kavonic J, Hermansson Török E, Elmqvist T (2021) Urban governance of and for urban green and blue infrastructure. In: Shackleton CM, Cilliers SS, Davoren E, du Toit MJ (eds) Urban ecology in the Global South. Springer, Cham, pp 403–428
Simwanda M, Ranagalage M, Estoque RC, Murayama Y (2019) Spatial analysis of surface urban heat islands in four rapidly growing African cities. Remote Sens 11:1645
Song Y, Song X, Shao G (2020) Effects of green space patterns on urban thermal environment at multiple spatial-temporal scales. Sustain 12:6850
Vink K, Ahsan MN (2018) The benefits of cyclones: a valuation approach considering ecosystem services. Ecol Indic 95:260–269
von Döhren P, Haase D (2015) Ecosystem disservices research: a review of the state of the art with a focus on cities. Ecol Indic 52:490–497
Wangai PW, Burkhard B, Müller F (2016) A review of studies on ecosystem services in Africa. Int J Sustain Built Environ 5:225–245
WHO (2019) The health and environment linkages initiative, WHO and UNEP, New York, online: https://www.who.int/heli/en/
Yang G, Ge Y, Xue H, Yang W, Shi Y, Peng C, Du Y, Fan X, Ren Y, Chang J (2015) Using ecosystem service bundles to detect trade-offs and synergies across urban-rural complexes. Landscape Urban Plan 136:110–121
Zhou D, Zhao S, Liu S, Zhang L, Zhu C (2014) Surface urban heat island in China's 32 major cites: spatial patterns and drivers. Remote Sens Environ 152:51–61

Cultural Urban Ecosystem Services

Cynnamon Dobbs, Alexis Vasquez, Pilar Olave, and Magdalena Olave

Abstract There is an increasing recognition of the contribution of cultural ecosystem services to human wellbeing, however, research is scarce on the topic in the Global South. Evidence on urban cultural ecosystem services tends to focus on recreation and cities from the Global North. Cultural services such as social cohesion and cultural heritage are often overlooked from research. Global South realities are different from the Global North, given the proportion of indigenous communities, urbanisation dynamics, and environmental inequities. This chapter will address these topics by synthesising research carried out in the Global South and contrasting results with general discourses derived from studies in the Global North. The chapter closes by considering the way forward towards the research of ecosystem services in the Global South.

Keywords Cultural heritage · Environmental inequities · Landscape aesthetics · Social perception · Social cohesion

1 Introduction

There is an increasing recognition of the role of urban ecosystem services to the health and wellbeing of people living in cities (Roy et al. 2012; Haase et al. 2014). Ecosystem services are increasingly being integrated into urban planning and management for promoting resilient and sustainable cities (Gómez-Baggethun and Barton 2013; Pauleit et al. 2021; Cilliers et al. 2021).

Despite the recognised importance of cultural ecosystem services, recent reviews have shown that they are among the least researched urban ecosystem services, other than for recreation. Moreover, despite being the ones that provide better opportunities for impacting decision-making and promotion of people's commitment towards nature conservation (Milcu et al. 2013; Hernández-Morcillo et al. 2013; La Rosa et al.

C. Dobbs (✉) · P. Olave · M. Olave
Center for Modeling and Monitoring Ecosystems, Universidad Mayor, Santiago, Chile

A. Vasquez
Department of Geography, Universidad de Chile, Santiago, Chile

C. M. Shackleton et al. (eds.), *Urban Ecology in the Global South*, Cities and Nature,
https://doi.org/10.1007/978-3-030-67650-6_10

2016), they remain the most challenging to quantify. The same applies with respect to urban ecological research, with the majority of literature related to ecosystem services focusing on regulating services, with only few examples focusing on cultural ecosystem services. Different authors have emphasised the importance of increasing research on urban cultural ecosystem services (Haase et al. 2014; La Rosa et al. 2016) and also the need for more research from the Global South, especially from Latin America and sub-Saharan Africa (Haase et al. 2014; Dobbs et al. 2019).

Assessing cultural ecosystem services is complex since it requires a multidisciplinary approach, they are a reflection of local and individual values, are difficult to establish in a spatial unit, and are dependent on personal cultures, experiences, and preferences (Gee and Burkhard 2010; La Rosa et al. 2016). Most studies on cultural ecosystem services focus on recreation, tourism and aesthetic values, and there are only few studies focusing on the spiritual value, physical and mental well-being, cultural identity or social cohesion (Chan et al. 2012; Shackleton and Blair 2013; Andersson et al. 2015; Hernández-Morcillo et al. 2013; La Rosa et al. 2016). The importance of further understanding the role of cultural ecosystem services in cities is that they play a crucial role for social identity, local heritage, and cohesion (La Rosa et al. 2016).

Given the cultural, social, and historical background and the differences in urbanisation processes between the Global South and North, it is necessary to understand and be aware that local context is relevant for determining values and trends related to ecosystem services. This can aid in developing complementary frameworks to apply the ecosystem service approach in the Global South (Dobbs et al. 2019).

The chapter will try to address that by reviewing and synthesising urban cultural ecosystem services findings in the Global South. The chapter is divided into three main sections: an introductory section, a section on the cultural ecosystem services from urban areas in the Global South, and a discussion section which will contrast trends from the Global North and South. The chapter will close by considering the way forward towards the research of ecosystem services in the Global South.

2 Urban Cultural Ecosystem Services

Ecosystem services are defined as the benefits that people obtain from ecological processes, which have a direct or indirect effect on human wellbeing (Haines-Young and Potschin 2010). The mainstreaming of the concept globally occurred after the publication of the Millennium Ecosystem Assessment in 2005, emphasising the direct connections between people and nature (Daniel et al. 2012). The concept permeated not only conservation and biodiversity policies but also urban planning and policy documents addressing climate change adaptation, human health, urban forestry, green infrastructure, sustainability, among others, as a way to promote and justify the importance of nature in cities (Kabisch 2015), not just for biodiversity, ethical or moral reasons, but for the very quality of life of urban citizens.

Ecosystem services are typically classified into three categories: provisioning, regulating and maintenance, and cultural services (CICES Classification, www.cices.eu). Cultural services are defined as all the non-material, usually non-rival and non-consumptive outputs of ecosystems that have an effect on the physical and mental health of humans (Haines-Young and Potschin 2017). They are seen as the natural or semi-natural locations where the relation with nature triggers an improvement in the physical and mental health of people (Shanahan et al. 2015). Cultural ecosystem services include recreation and tourism, cultural diversity, knowledge systems, psychological and physical health, spiritual and religious values, educational values, inspiration, aesthetic values, sense of place, social cohesion, and patrimonial values (MEA 2005). Therefore, they have the particularity of being intangible and subjective (MEA 2005; Daniel et al. 2012), and thus may change in relation to changing experiences, values, and perceptions of local communities and urban dwellers. The effects of cultural ecosystem services can be perceived at multiple scales: individual, community, and society (Chen et al. 2019), such as mental health, sense of place or cultural identity, respectively (Chen et al. 2019; Jennings et al. 2016). Therefore, understanding the values that people ascribe to cultural ecosystem services requires identifying the reasons behind their enjoyment of a particular space or experience in that space. This requires the application of surveys or interviews or observational techniques that are demanding in resources and are limited in scale (Plieninger et al. 2013).

Another particularity of cultural services is that one acquires or experiences the benefits mostly by direct experiences of and in nature (Chen et al. 2019). Therefore, individuals require exposure to nature for the wellbeing benefits to be most evident, given that they directly affect humans' physical and mental health (Shanahan et al. 2016; Williams 2017). They are also recognised as a starting point to enrich and deepen the relation that the people have with their natural environment (Fish et al. 2016). This is of importance for increasing the commitment of people towards the maintenance of natural capital, through the development and strengthening of public policies related to appropriate and sustainable land planning and management (Chan et al. 2012).

The relevance of cultural services to human wellbeing is widely recognised. However, there are two characteristics that must be acknowledged from a strategic and planning perspective: (1) in contrast to regulating and provisioning services, cultural services cannot be replaced through technical or socioeconomic improvement (Plieninger et al. 2013; Hernández-Morcillo et al. 2013), and (2) some suggest that with the economic development of societies, there is an increasing use of and value attached to recreational and ecotourism ecosystem services when the more basic needs derived from provisioning and regulating services are solved (Guo et al. 2010). However, other cultural ecosystem services such as heritage or spiritual enrichment are more valued in developing countries, even more than regulating services (Ichumbaki and Pollard 2020). In considering the nature and importance of urban cultural ecosystem service in the Global South, this chapter will focus on five commonly recognised cultural services: landscape aesthetics, cultural heritage, social cohesion and sense of place, recreation and tourism, and psychological and physical health.

Landscape aesthetics are defined in the MEA (2005) as the aesthetic value of different aspects of ecosystems that are expressed as preferences for parks, scenic drives, or preferences for specific locations to live as shaped by some feature of the surrounding landscape. Later de Groot et al. (2010) defined the service as the appreciation of the scenery, while Chen et al. (2019) included the configuration of open space in the landscape, linking it to the relation between land use and land cover. The aesthetic quality of a landscape can be quantified by using perceptual surveys, where landscapes are ranked or chosen by different stakeholder groups (Daniel et al. 2012). Therefore, there is a recognition that culture, education, and socioeconomics has an effect on aesthetic preferences, giving importance to the understanding of the local context on evaluating this service. Landscape aesthetics are hypothesised to be pervasive across cultures and are governed by humanity's biophilic relation with nature (Chang et al. 2020), although at times urban planning and lifestyles might hinder or sever such relationships (Cocks and Shackleton 2021). Natural landscapes have been shown to be highly valued across a whole range of settings and contexts in terms of aesthetic preferences, leading to the establishment of biophysical models to quantify this service (Silvennoinen et al. 2001). However, the choice of a natural landscape and specific elements within it depends on the type and intensity of the connection that people have with nature (Chang et al. 2020), ranging from quite limited to essential, varying across countries (and how it is measured) and thereby highlighting the importance of assessing among and within country variation on landscape aesthetics preferences. For example, Lindemann-Matthies et al. (2013) showed that Swiss citizens preferred forests high in biodiversity, while Chinese showed no particular preference. Daams et al. (2019) found that attractive green spaces can increase property values of neighbouring houses by 2–9%. Czembrowski et al. (2019) found that real estate buyers in Stockholm were willing to pay more for properties that benefitted from aesthetics and nature features in green spaces of nearby houses.

Cultural heritage derives from the natural or semi-natural elements of the landscape that are associated with the identity of an individual, community, or society (Daniel et al. 2012), and therefore it is context dependent. Different cultures may have different heritage associations with the same ecosystem features, requiring understanding of the ecological and cultural contexts (Verschuuren 2006; Cocks and Shackleton 2021). Holland et al. (2017) found that 70% of papers considering cultural heritage as a cultural ecosystem service included natural landscapes, ranging from agricultural to forest and coastal areas. Some other papers, especially from Europe conceptualised cultural heritage as human-made remnants, such as archaeological sites, highlighting the importance of defining the meaning of the cultural services (Holland et al. 2017) when comparing studies, and especially those from the Global North to those in the Global South. Many societies place a high value on certain landscapes or species with respect to their cultural significance (MEA 2005), since they are linked to the legacy of the biophysical features, physical artefacts, and intangible attributes that an individual, group, or society have valued through generations and that are willing to maintain it for future generations (Czepczynski 2008). For example, the importance of wild olive (*Olea europea* subsp. *africana*) to Xhosa people in South Africa, where it is essential to serve ceremonial meat on the branches or platters made from this species (Dold and Cocks 2012), or *Ficus religiosa* as providing worship sites in urban areas of southern India (Jaganmohan et al. 2018). Their significance goes beyond just aesthetics as they are derived from a long-term interaction between site conditions and human influences and uses (Daniel et al. 2012). Cultural heritage includes tangible and intangible elements that can range from visual representations of cultural heritage in a landscape, such as vineyards and coffee plantations, to individual species linked to myths, stories, legends, and religious practices related to certain locations and ecosystem features (Garibaldi and Turner 2004; Daniel et al. 2012). Natural parks in South Africa and marine sites in the United Kingdom provide cultural and social values by the sense of identity that those sites inspire, similarly this occurs for the Satoyama landscape in Japan, the Waikaraka estuary in New Zealand, and the Arafura-Timor seascape in Southeast Asia (Chang et al. 2020). Preserving cultural heritage can trigger synergies with several other services and can support biodiversity (Takeuchi 2010). Cultural heritage is hard to associate to market economic techniques, since they are generally common goods rather than individually owned (Daniel et al. 2012). It is important to identify specific ecological landscape features to link to particular cultural heritage values of different stakeholders (Chen et al. 2019; Daniel et al. 2012), requiring elicitation of those values (Gregory and Trousdale 2009). A review by Holland et al. (2017) showed that a variety of methods can be used to assess cultural heritage, from valuation methods to interviews and participatory GIS.

Social cohesion relates to common areas where people gather for leisure, social activities and recreation, which in urban areas frequently occurs in formal green spaces and natural areas (Peters et al. 2010; Francis et al. 2012). It is recognised that the increase of social contact is a major pathway through which nature supports health (Hartig et al. 2014). Several studies, especially from the Global North, have shown that proximity to green spaces creates a sense of community, promotes social interactions and recreational activities (Jennings and Bamkole 2019). The design of green spaces may promote or hinder social cohesion (Fan et al. 2011), open park design encourages active recreational activities, shaded areas promote relaxation (Peters et al. 2010), accessibility and sidewalks are catalysts for peoples' encounters (Holtan et al. 2014; Ward-Thompson et al. 2016). The specific design of the green space for promoting social cohesion will depend on the views and preferences of the local community and on the distribution and accessibility of the space and, therefore, understanding the social environment surrounding the green space is necessary for creating settings that support social cohesion (Jennings and Bamkole 2019). This includes, safety and comfortable walking settings (Seaman et al. 2010), accessibility should be promoted to decrease the sense of loneliness and social support (Maas et al. 2009), and will also deepen on creating the sense of belonging, place attachment, and empowerment (Jennings and Bamkole 2019), especially important for more marginalised communities (Shinew et al. 2004; Peters et al. 2010). Pathways for social cohesion are also influenced by demographic factors such as age, ethnicity, culture, and socioeconomic status (Jennings and Bamkole 2019), therefore it is expected that differences occur within and between countries and cities.

Recreation and tourism relate to an appreciation of and support for ecosystems and services that provide opportunities for leisure, wonder, and education through direct interaction with or immersion in a natural or semi-natural state. This becomes particularly important for people living in urban environments (Daniel et al. 2012). Recreation and tourism become a service through physical exercise, aesthetic experiences, intellectual stimulation, inspiration, and other contributions to physical and psychological wellbeing (Chan and Satterfield 2015). This service is the easiest cultural ecosystem service to assess given that people actively choose where they spend leisure time based on the desired characteristics of a natural, semi-natural, or human-made area (Hernandez-Morcillo et al. 2013). A global study assessing cultural services from social media photographs showed that the most satisfaction was associated with recreation in places with abundant flora (Chang et al. 2020).

Urban lifestyles are often associated with low physical activity and frequently with high mental stress, both contributing to an increasing burden of diseases (Chen et al. 2019). Furthermore, urban inequities occur, to different degrees, in cities from the Global North and South resulting in different access to environmental resources, exposure to environmental risks and hazards, and differences in health status correlated with level of income, education, and social capital (Chen et al. 2019). Therefore, exposure to nature in urban areas becomes particularly relevant for addressing those inequities and adding to the improvement of health conditions. Less affluent communities often have less access to quality urban green space (Shen et al. 2017; Gerrish and Watkins 2018; Venter et al. 2020), and less health benefits from contact with nature (Mitchell and Popham 2008). There is a large amount of evidence from epidemiological research showing that exposure to nature can improve physical and mental health (Hartig et al. 2014; Shanahan et al. 2016; Fong et al. 2018). Even though causalities have not been thoroughly established, some short-term effects have been recognized, including positive mood, reduced stress levels, and improved cognitive capacities (Hartig et al. 2014). Causalities are believed to be related to multiple factors, that are working together with exposure to nature, and that the mechanisms in which nature acts might be related to specific population groups, highlighting the importance of having a strong geographical distribution of studies to establish general patterns on the physical and psychological benefits of nature (Chen et al. 2019).

Physical inactivity is a major contributor to increased levels of obesity, and an established risk factor for diabetes and cardiovascular diseases (Chen et al. 2019). People living close to green spaces tend to engage more in physical activity (Nesbitt et al. 2017), therefore, improving their physical health. Exercising has positive effects on improving body mass index and reducing obesity problems within a community (Lovasi et al. 2011). Living in green neighbourhoods attracts different age groups to spend more time outdoors in passive or active exercise, thereby improving physical health and likely longevity (Takano et al. 2002; Dushkova and Ignatieva 2020). Also, green spaces enable relief under heat stress events reducing negative outcomes of very high temperatures (Lafortezza et al. 2009). Mental health is also typically improved by exposure to nature. Social isolation, associated with urban living, can be a risk factor for mental disorders such as depression (Hidaka 2012; Lambert et al. 2015). This becomes particularly important in current times, where lockdowns have shown the need for living close by to green spaces or at least to view nature from one's residence, which can reduce anxiety and stress related to Covid-19. Psychological health is impacted in a positive manner by the ecosystem services provided by natural spaces and components of the cities. Being in contact with nature, through contemplation, and overlooking green spaces can reduce stress, improve cognition, and have a restorative effect (Grahn and Stigsdotter 2010; Carrus et al. 2015). Similar to physical health benefits, this can have positive implications for expenditure in public health (Browning et al. 2012).

3 Cultural Urban Ecosystem Services in the Global South

This section presents the results from a comprehensive literature search to identify the published research on cultural ecosystem services in the Global South. To achieve this, in March 2019 we conducted a systematic literature review, limited to peer-review articles including research and review articles and book chapters published in academic journals in English, Spanish, and Portuguese. We used the search engines of ISI Web of Knowledge and Scopus. Firstly, we looked at the global literature and secondly, we focused on countries from the Global South. We used as search terms "Urban Ecosystem Services" AND "Cultural" or "Recreation", or "Spiritual", or "Aesthetics", or "Tourism", or "Heritage", or "Inspiration", or "Sense of Place", or "Identity", or "Social cohesion", or "Patrimonial", or "Physical Health", or "Mental Health", or "Health". The search was restricted to 2005 onwards. For separating Global North and South we filtered the results of each search term by country.

The search at global level showed an increase in articles on cultural ecosystem services published in the last five years (Fig. 10.1), with a total of 181 articles, highlighting the growing interest in these services. Most of these articles have their origin in the Global North, with 139 articles (77%) belonging to this group, while only 42 (23%) were from the Global South. The countries that have the greatest quantity of papers are the United Kingdom, United States, Germany, and Spain.

We found 42 publications from the Global South since 2005, with an increase in publications since 2016 (Fig. 10.1b). Most of the publications coming from the Global South were case studies. When summarised by region, 23 publications (55%) come from Asia, and seven (17%) each from Africa and Latin America. In terms of the cultural services, the studies referred the most to recreation as the subject of study (17; 40%), followed by four on aesthetics and three on tourism. There were seven publications that included the analysis of various ecosystem services and three publications were related to the conceptual development of cultural ecosystem services using examples from the Global South and developed by authors from the Global South. Only one publication each was found dealing with connection to nature, education, inspiration, and tradition. The most common journals publishing on cultural ecosystem services from the Global South were *Ecosystem Services* (5), *Landscape and Urban Planning* (4), *Urban Forestry and Urban Greening* (4), and *Acta Ecologica Sinica* (3).

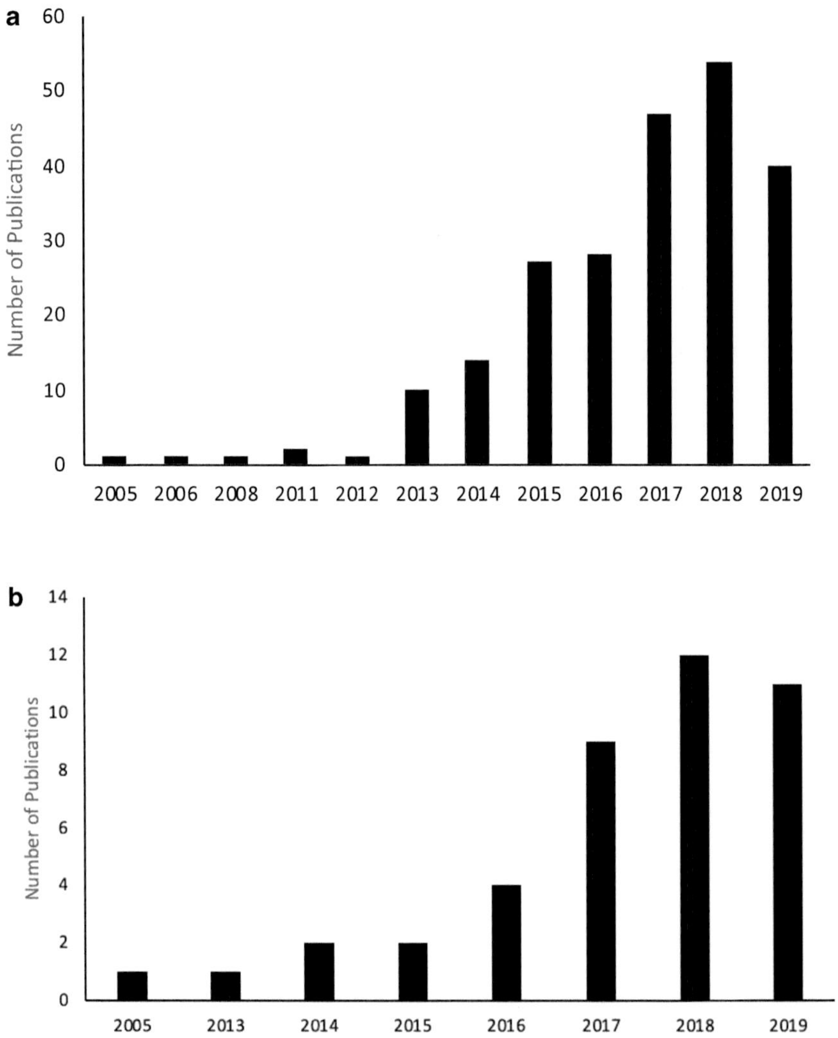

Fig. 10.1 (a) Number of publications on urban cultural ecosystem services globally since 2005 (obtained from a literature search in March 2019) (b) Publication on cultural ecosystem services from the Global South since 2005 (Fig. 10.1a shows an increase in publications since 2013. When looking at specific keywords the papers addressed, we found that the most common topics were health [41; 23%] and recreation [30; 17%], with only a few addressing sense of place, mental health, heritage, tourism, identity and social cohesion [only 2]. No papers were found related to urban cultural ecosystem services and inspiration or patrimonial related services.)

The scale of analysis and object of study varied widely in Global South studies, including parks, urban rivers, informal settlements, small, intermediate and large cities. We will use these papers to reflect on trends of urban cultural services occurring in the Global South.

4 Inequities in Cultural Ecosystem Services

One of the main problems of cities in Global South countries is the extreme inequity in the distribution of income, and of public and private investment. A pronounced inequitable distribution and access to urban green infrastructure has been widely reported in scientific publications in Latin American cities (Nero 2017; Vásquez 2017; Dobbs et al. 2019), Asia (Hussain et al. 2010; Maruthaveeran 2010; Lofti and Koohsari 2011; Ambrey and Shahni 2017; Dadashpour et al. 2017; Paul and Nagendra 2017) and Africa (McConnachie and Shackleton 2010; Willemse 2013; Donaldson et al. 2016; Nero 2017; Rigolon et al. 2018; Venter et al. 2020) (mostly South Africa), where high-income neighbourhoods have greater ratios of green infrastructure and the associated ecosystem services. Low-income populations tend to have lower proximity quality (Shackleton and Blair 2013; Vásquez and Dobbs 2020), leading to lower exposure to ecosystem services. These trends echo those of the Global North findings, where quantity, quality, and access to green spaces is better for higher income, more educated, and white citizens (Rigolon et al. 2018).

Although the scientific literature pays increasing attention to environmental justice issues, it seems that the social production of ecosystem services lens (Ernston 2013) has not been sufficiently used in urban cultural ecosystem services research in the Global South. Shackleton et al. (2018) studied green infrastructure and ecosystem services in small and medium-sized cities in South Africa, finding that about 70% of the low-income residents believed that public green space was insufficient, while only 48% of the high-income suburbs residents had that perception. In this case, the main use of green infrastructure was recreational, especially for family outings and sport. The authors did not report any differences according to socioeconomic level. Similarly, Dobbs et al. (2018) found that for Santiago (Chile), social inequities matched the inequalities presented in recreation, where there has been a decrease in the availability of green spaces in low-income areas in the last 10 years. Areas for recreation are generally needed by a larger number of low-income people given that most have little or no private green spaces at home, and hence public green space is an important space for recreational opportunities, often leading to those spaces being in high demand and hence overcrowded (Orta and Geneletti 2018). Therefore, there is also a need for improving not only the amount of public green space for recreational opportunities, but also their accessibility, distribution, and providing for multiple uses (Zhang and Zhou 2018). Income may not be the only factor that shapes access to public green spaces. For example, in South Africa, access to natural areas in peri-urban zones also relates to historical exclusion of non-white people from natural and touristic sites (Donaldson et al. 2016; Cocks et al. 2021). A gender perspective is also important to include when analysing access to green spaces or recreational services where some countries from the Global South, such as India, have women only green spaces, therefore their relation to nature is experienced in a different way than in Western culture (Iqbal 2018). This shows that the exploration of drivers to cultural ecosystem service provision should not be restricted to income-related variables and should definitely incorporate ethnicity and variables related to colonisation or migration to further understand trends in accessibility, distribution, and quality of cultural ecosystem services in urban areas.

An additional constraint is that large regions of the Global South are prone to periodic or frequent droughts leading to a decrease in green spaces or quality recreational areas. These trends have been discussed by Dobbs et al. (2018) in Santiago, Islam and Tuli (2017) in Bangladesh, and Li et al. (2018) by the physiography of the city and its socioeconomics and enlarging the inequity effect. According to Roy et al. (2018), people recognise the decline in the availability of trees and green infrastructure, and consequently, their experiences are affected by any decline in the quantity and quality of ecosystem service provision. Similar to Roy et al. (2018), Dobbs et al. (2018) also reported that the conditions of low-income neighbourhoods tend to get worse in terms of their environmental amenities over time. This becomes particularly relevant for poor communities that are more vulnerable to climate change impacts.

Measures to mitigate existing inequities in the distribution of green infrastructure and cultural ecosystem service provision should be taken while avoiding public interventions that produce or increase environmental injustice situations. With regard to the latter, Unnikrishnan and Nagendra (2015) pointed out that Private–Public Partnerships, common in Global South countries, can exacerbate inequities in access to urban cultural ecosystem services.

5 Perceptions on Urban Cultural Ecosystem Services

Roy et al. (2018) conducted surveys in Dar es Salaam, Tanzania, and they found that cultural ecosystem services are especially relevant to poor citizens. Using participatory methods, they assessed the four groups of ecosystem services proposed by MEA (2005) provided by green infrastructure and found that cultural ecosystem services were the most frequently identified group of services and that they are provided by most types of green infrastructure. The identified types of green infrastructure corresponded to the sea, public parks, gardens and playgrounds, house gardens, roadside plantations, and beaches. The only type of green infrastructure that, according to the community, does not provide cultural ecosystem services is natural vegetation, probably because it is relegated to less accessible and residual spaces. Preferences showed that low-income neighbourhoods do demand green infrastructure nearby for outdoor recreation, social encounters, and psychological health. A study from Delhi (India) showed that people valued green spaces and felt the need for more of them. They recognised that green spaces provide environmental, psychological, and mental health benefits (Paul and Nagendra 2017), while a study from Pakistan

(Hussain et al. 2010) showed that people recognised that green spaces can improve mental and physical health and that they improve the aesthetics of a city. Respondents also recognised feeling more relaxed and a reduced mental stress and tension, and a positive effect in their cognitive capacity, while also promoting social cohesion while spending more time in green spaces (Hussain et al. 2010). These examples allow us to recognise that findings from the Global North do apply to Global South, however, there is no specificity on the mechanisms in which benefits are delivered to the different communities living in these areas of the world, or they are not specific in addressing the type of green spaces that are perceived as providing the most benefits to specific communities. One benefit rarely mentioned in Global North studies, but recognised by poorer communities in the Global South, is the employment benefits for unskilled workers in creating and maintaining urban green infrastructure as shown by work in South Africa (Shackleton and Blair 2013; King and Shackleton 2020).

Another important issue in deprived neighbourhoods is that green infrastructure is often perceived as poorly maintained, which can lead to anxieties about safety (de la Barrera et al. 2016; Narh et al. 2020), which affects the provision of cultural ecosystem services and its valuation by the community. In extreme cases, negative behaviours in these spaces can lead to the production of cultural disservices which will have negative impacts on the population's wellbeing. Visitors to parks in Kuala Lumpur (Malaysia) recognise that to feel safe in a park it needs to have safety infrastructure, be clean and provide a pleasant and safe environment and have access to high-quality community and cultural facilities (Maruthaveeran 2010). The final benefits to wellbeing from green spaces were also related to neighbourhood crime rates, given that people avoid green spaces instead of visiting them (Ambrey and Shahni 2017). Visits to and experiences of nature in the Global South is highly determined by the possibility of being in a safe space, which is more prevalent than in the Global North given the existence of higher inequities in the Global South.

Perceptions on cultural services also influence aesthetics and spiritual connections with nature. Aesthetically desired plant species are defined by historical and cultural values that often respond to the heritage of colonial periods in the Global South (Abendroth et al. 2012). This may produce biocultural homogenisation which can impact the supply, value, and demand of cultural ecosystem services. For example, in Bangkok the trend to tree monocultures has led to a low species diversity, which impacts the diversity of cultural services that can be provided (Thaiutsa et al. 2008). In Bulawayo (Zimbabwe), some groups worship outdoors in nature, and require particular spaces, but these can be compromised by other users not respecting those spaces or littering them, or poor maintenance by the city authorities (Ngulani and Shackleton 2019). In Nigeria, Adedeji (2021) describes the close biocultural relationships of the Yoruba in Osogbo, which were suppressed under colonial rule and town planning that ignored sacred spaces and opportunities to interact with nature.

Perceptions on religious sites are rarely included in cultural ecosystem service research, even though its importance is commonly named in ecosystem service frameworks. Some studies from the Global South have shown the importance of green spaces for spiritual enrichment. De Lacy and Shackleton (2017) found that green spaces at sacred or religious sites tend to accentuate cultural ecosystem service provision and spiritual experience in medium-sized cities in South Africa. Thaiutsa et al. (2008) found that gardens within Buddhist temples in Bangkok (Thailand) were important for tree heritage conservation and for the provision of spiritual cultural ecosystem services. Sacred places in Bengaluru (India) are important habitats for plants of cultural significance and consequently, relevant in providing spiritual cultural ecosystem services, more than squares and parks (Jaganmohan et al. 2018). In certain cultural and religious contexts, specific green spaces, particular species, and even individuals have relevant meanings, and therefore are key providers of cultural ecosystem services, such is the case of *Ficus religiosa* associated with the enlightenment Gautama Buddha (Jaganmohan et al. 2018). Considering the importance of spiritual cultural ecosystem services, sacred and religious sites are relevant for synergies between conserving urban biodiversity, even endangered species (Adhikari et al. 2015), and the provision of spiritual cultural ecosystem services. Communities with strong place attachment and religious beliefs obtain greater cultural ecosystem services, which are normally rooted in traditional knowledge systems (Ballullaya et al. 2019). However, for these communities the concept of ecosystem services itself tends to be contradictory with communities' environmental ethics, their traditions, and belief systems.

6 Governance and Urban Cultural Ecosystem

Many Global South countries are characterised by poor urban governance, where weak public institutions are pressured by accelerated urbanisation processes led by private urban developers. Several countries in the Global South have struggled to incorporate green infrastructure into towns and cities in a systematic and planned way, frequently because they are more focused in infrastructure and service provision to rapidly increasing populations. Others have recently come to realise that the benefits of nature should be part of city planning and are actively incorporating them into their urban planning (i.e. Cape Town, South Africa; Bogota, Colombia).

Several publications have reported the scarcity of public green spaces in contrast with a large proportion of private green spaces, which by definition have limited access. For example, Shackleton et al. (2018) pointed out that 74% of urban green infrastructure in small and medium-sized towns in South Africa were private gardens rather than public spaces. Residential housing can also incorporate green spaces that could provide cultural ecosystem services, however, in the Global South we can commonly find high-density housing and slums which leaves very little space for private green space. This highlights the role of public spaces and street trees for providing cultural services such as aesthetics, and could eventually compensate for low per capita green space, which should also be incorporated in planning for enhancing ecosystem services (Arshad and Routray 2018).

Another aspect related to planning is the accessibility to green spaces. Paul and Nagendra (2017) evaluated the travelling distance to green spaces showing that most people traveled less than half a kilometre to access a park even though they were small, not necessarily well-maintained, highlighting the importance of better planning and designing public parks responding to the needs and preferences of the community.

It is common to find countries in the Global South that have adopted a neoliberal development model with strong privatisation and commodification processes of the urban space. This affects public green spaces such as parks, beaches, rivers, and forests. Unnikrishnan and Nagendra (2015) made a comparison of ecosystem services provided by public and privatised lakes in Bangalore (India) and concluded that public lakes provide a greater diversity of non-commercial uses and cultural services than private lakes. While it is true that both public and private lakes were relevant for recreational activities, such as walking and jogging, public lakes were more important as sacred elements. In some of the now privatised lakes, abandoned religious sites were found, indicating their past spiritual importance, and the now recent alienation of local communities from the water body.

Privatisation, and consequently the restricted access to ecosystem services, not only leads to lower cultural ecosystem service provision but also to lower cultural ecosystem service valuation. For example, Lagbas (2019), in a social valuation among college students of regulating and cultural ecosystem services provided by the Arroceros Forest Park (AFP) in Manila, Philippines, found that outdoor recreation and inspiration for culture were the least important ecosystem services. The authors explained the result from the restricted access character of the park, and therefore, students could not be users (Unnikrishnan and Nagendra 2015). This might be more important for low-income groups, as these communities benefit greatly from cultural ecosystem services provided by spaces that can be potentially privatised such as beaches, the shore, parks, and other public spaces (Roy et al. 2018).

7 The Way Forward

Urban communities the world over enjoy recreational services, from urban nature and green spaces. However, cities in the Global South more frequently have poor communities and have rural migrants on the outskirts of the city (Myers 2021), who also place considerable emphasis on provisioning ecosystem services (Shackleton 2021). This reinforces that for these communities urban green infrastructure is not important only for recreational services, but also for other ecosystem services that relate to their traditions: use of the land, economic, spiritual, aesthetics, and the like. Therefore, other cultural services might be of equal or higher importance than recreation to some communities.

Many poorer communities in and around Global South cities have some level of dependence on local natural resource extraction and subsistence, leading to a different relationship with nature than is commonly seen in cities of the Global North. In many Global South countries, indigenous people view and experience nature in ways very different to the Global North based on their own worldviews and following unique spiritual practices and uses which accord with their identity in relation to nature, and is reflected in their lifestyles and needs of urban nature (Shackleton and Cocks 2021). The experiences derived from these communities in terms of the valuation and preferences of cultural ecosystem services at different scales, from different ethnic, cultural and demographic groups can enrich the ecosystem service framework for urban areas. It is with cultural ecosystem services that it becomes more evident that local context is extremely relevant for planning with ecosystem services in urban areas.

For cultural ecosystem services, the relationship that people have with nature in urban areas is extremely relevant and that depends on the local context including social, economic, and cultural. Similar to other services, urban cultural services research mainly originated in the Global North. China has the greatest volume of publications from the Global South, especially in relation to recreation and aesthetics, with a particular cultural background determining the relationships with and needs of people from nature. We found a good breath of research on recreational services from Latin America, Asia, and Africa, but most other cultural ecosystem services are poorly represented. Further research needs to be developed in the Global South to better understand those relations in different socio-cultural contexts so as to improve recognition of the importance of urban ecosystem services in shaping livable cities. Especially lacking is research on the psychological and physical benefits derived from exposure to nature which can help to motivate greening policies in Global South cities.

Additional research is also needed for encouraging innovative and low-cost alternatives of collecting data. Collaborative research projects among Global South researchers and sharing existing data might offer new insights for urban ecosystem services. In contrast to other reviews on ecosystem services in urban areas, we found several publications addressing the demand or preference for cultural services, while there was less on assessing the supply of these services. Big data can be obtained from social media which is a low-cost means of obtaining information on the use and supply of cultural ecosystem services, such as through collecting geotagged pictures from a variety of cities. Sentiment analysis from twitter can disclose knowledge on recreation preferences, perceptions of nature, tourism, and aesthetics, among others. The geospatial tools can be used to assess mismatches between supply and demand for cultural ecosystem services to better understand environmental inequalities and trends of cultural ecosystem service provision in space and time.

There is some recognition of various inequities in the Global South in relation to urban ecosystem services in general, but also on cultural ecosystem services in particular. We found little mention on the governance or the importance of government for securing the equitable provision of cultural ecosystem services. Participatory processes including citizens, decision-makers, and politicians might facilitate data acquisition and aid in transferring knowledge and findings. Promoting translations to local languages can also aid in disseminating results and transferring knowledge to local actors. Eventually, this could underpin more robust planning and decision-making towards improving cultural ecosystem services in cities from the Global South. Some of the biggest challenges for cities in the Global South is finding context-specific information, instruments, or guidelines that can be integrated in the decision-making made by planners, managers, and even citizens.

References

Abendroth S, Kowarik I, Müller N, Von der Lippe M (2012) The green colonial heritage: woody plants in parks of Bandung, Indonesia. Landsc Urb Plan 106:12–22

Adedeji J (2021) Urban biocultural identity of Yorubas: intersection of philosophy and nature for wellbeing at Osun Sacred Grove UNESCO Site, Osogbo, Nigeria. In: Cocks ML, Shackleton CM (eds) Diversities of nature in cities: enriching belonging, wellbeing and bioculture. Earthscan/Routledge, London, p 16

Adhikari YP, Fischer A, Pauleit S (2015) Sustainable conservation perspectives for epiphytic orchids in the central Himalayas, Nepal. Appl Ecol Environ Res 13:753–767

Ambrey CL, Shahni TJ (2017) Greenspace and wellbeing in Tehran: a relationship conditional on a neighbourhood's crime rate? Urban For Urban Green 27:155–161

Andersson E, Tengö M, McPhearson T, Kremer P (2015) Cultural ecosystem services as a gateway for improving urban sustainability. Ecosys Serv 12:165–168

Arshad HS, Routray JK (2018) From socioeconomic disparity to environmental injustice: the relationship between housing unit density and community green space in a medium city in Pakistan. Local Environ 23:536–548

Ballullaya UP, Reshmi KS, Rajesh TP, Manoj K, Lowman M, Sinu PA (2019) Stakeholder motivation for the conservation of sacred groves in south India: an analysis of environmental perceptions of rural and urban neighbourhood communities. Land Use Pol 89:104213

Browning CR, Cagney KA, Iveniuk J (2012) Neighborhood stressors and cardiovascular health: crime and C-reactive protein in Dallas, USA. Soc Sci Med 75:1271–1279

Carrus G, Scopelliti M, Lafortezza R, Colangelo G, Ferrini F, Salbitano F, Agrimi M, Portoghesi L, Semenzato P, Sanesi G (2015) Go greener, feel better? The positive effects of biodiversity on the well-being of individuals visiting urban and peri-urban green areas. Landsc Urb Plan 134:221–228

Chan KMA, Satterfield T (2015) Managing cultural ecosystem services for sustainability. In: Potschin M, Haines-Young R, Fish R, Turner RK (eds) Routledge handbook of ecosystem services. Abingdon, Oxon, Routledge

Chan KM, Guerry AD, Balvanera P, Klain S, Satterfield T, Basurto X, Bostrom A, Chuenpagdee R, Gould R, Halpern BS, Hannahs N, Levine J, Norton B, Ruckelshaus M, Russel R, Tam J, Woodside U (2012) Where are cultural and social in ecosystem services? a framework for constructive engagement. Biosci 62:744–756

Chang CC, Cheng GJ, Le Nghiem TP, Song XP, Oh RR, Richards DR, Carrasco LR (2020) Social media, nature, and life satisfaction: global evidence of the biophilia hypothesis. Sci Rep 10:1–8

Chen X, de Vries S, Assmuth T, Dick J, Hermans T, Hertel O, Jense A, Jones L, Kabisch S, Lanki T, Lehmann I, Maskell L, Norton L, Reis S (2019) Research challenges for cultural ecosystem services and public health in (peri-)urban environments. Sci Total Environ 651:2118–2129

Cilliers SS, Breed CA, Cilliers EJ, Lategan LG (2021) Urban ecological planning and design in the Global South. In: Shackleton CM, Cilliers SS, Davoren E, du Toit MJ (eds) Urban ecology in the Global South. Springer, Cham, pp 365–401

Cocks ML, Shackleton CM (2021) Situating biocultural relations in city and townscapes: conclusion and recommendations. In: Cocks ML, Shackleton CM (eds) Diversities of nature in cities: enriching belonging, wellbeing and bioculture. Earthscan/Routledge, London, pp 241–268

Cocks M, Shackleton CM, Walsh L, Haynes D, Manyani A, Radebe D (2021) Decolonisation of nature in towns and cities of South Africa: incorporation of biocultural values. In: Cocks ML,

Shackleton CM (eds) Diversities of nature in cities: enriching belonging, wellbeing and bioculture. Earthscan/Routledge, London, pp 104–125
Czembrowski P, Laszkiewicz E, Kronenberg J, Engstrom G, Anderson E (2019) Valuing individual characteristics and the multifunctionality of urban green spaces: the integration of sociotope mapping and hedonic pricing. PLoS ONE 14(3):e0212277
Czepczyński M (2008) Cultural landscapes of post-socialist cities: representation of powers and needs. Ashgate Publishing, Farnham
Daams MN, Sijtsma FJ, Veneri P (2019) Mixed monetary and non-monetary valuation of attractive urban green space: a case study using Amsterdam house prices. Ecol Econ 166:106430
Dadashpour H, Azizi D, Asgharzade P (2017) Evaluating the livable capacity of urban neighborhoods in Tehran: a case study of Harandi, Takhti and Kosar neighborhoods. Geogr Urban Space Dev 3:53–68
Daniel TC, Muhar A, Arnberger A, Aznar O, Boyd JW, Chan KM, Costanze R, Elmqvist T, Flint CG, Gobster PH, Grêt-Regamey A, Lave R, Muhar S, Penker M, Ribe RG, Schauppenlehner T, Sikor T, Soloviy I, Spierenburg M, Taczanowska K, Tam J, von der Dunk A (2012) Contributions of cultural services to the ecosystem service agenda. Proc Nat Acad Sci 109:8812–8819
de Groot RS, Alkemade R, Braat L, Hein L, Willemen L (2010) Challenges in integrating the concept of ecosystem services and values in landscape planning, management and decision making. Ecol Complex 7:260–272
de la Barrera F, Reyes-Paecke S, Harris J, Bascuñán D, Farías JM (2016) People's perception influences on the use of green spaces in socio-economically differentiated neighborhoods. Urban For Urban Green 20:254–264
De Lacy P, Shackleton C (2017) Aesthetic and spiritual ecosystem services provided by urban sacred sites. Sustain 9(9):1628
Dobbs C, Escobedo FJ, Clerici N, de la Barrera F, Eleuterio AA, MacGregor-Fors I, Reyes-Paecke S, Vasquez A, Zea-Camaño JD, Hernandez HJ (2019) Urban ecosystem services in Latin America: mismatch between global concepts and regional realities. Urban Ecosys 22:173–187
Dobbs C, Hernández-Moreno Á, Reyes-Paecke S, Miranda MD (2018) Exploring temporal dynamics of urban ecosystem services in Latin America: the case of Bogota (Colombia) and Santiago (Chile). Ecol Indic 85:1068–1080
Dold T, Cocks M (2012) Voices from the forest: celebrating nature and culture in Xhosaland. Jacana Media, Johannesburg
Donaldson R, Ferreira S, Didier S, Rodary E, Swanepoel J (2016) Access to the urban national park in Cape Town: where urban and natural environment meet. Habitat Int 57:132–142
Dushkova D, Ignatieva M (2020) New trends in environmental health research: from geography of diseases to therapeutic landscapes and healing gardens. Geogr Environ Sustain 13:19–171
Ernstson H (2013) The social production of ecosystem services: a framework for studying environmental justice and ecological complexity in urbanised landscapes. Landsc Urb Plan 109:7–17
Fan Y, Das KV, Chen Q (2011) Neighborhood green, social support, physical activity, and stress: Assessing the cumulative impact. Heal Place 17:1202–1211
Fish R, Church A, Winter M (2016) Conceptualising cultural ecosystem services: a novel framework for research and critical engagement. Ecosys Serv 21:208–2017
Fong KC, Hart JE, Lames P (2018) A review of epidemiologic studies on greenness and health: updated literature through 2017. Curr Environ Heal Rep 5:77–87
Francis J, Giles-Corti B, Wood L, Knuiman M (2012) Creating a sense of community: the role of public space. J Environ Psychol 32:401–409
Garibaldi A, Turner N (2004) Cultural keystone species: implications for ecological conservation and restoration. Ecol Soc 9(3):1
Gee K, Burkhard B (2010) Cultural ecosystem services in the context of offshore wind farming: a case study from the west coast of Schleswig-Holstein. Ecol Complex 7:349–358
Gerrish E, Watkins SL (2018) The relationship between urban forests and income: a meta-analysis. Landsc Urb Plan 170:293–308

Gómez-Baggethun E, Barton DN (2013) Classifying and valuing ecosystems services for urban planning. Ecol Econ 86:235–245

Grahn P, Stigsdotter UK (2010) The relation between perceived sensory dimensions of urban green space and stress restoration. Landsc Urb Plan 94:264–275

Gregory R, Trousdale W (2009) Compensating aboriginal cultural losses: an alternative approach to assessing environmental damages. J Environ Manage 90:2469–2479

Guo Z, Zhang L, Yiming L (2010) Increased dependence of humans on ecosystem services and biodiversity. PLoS ONE 5:1–8

Haase D, Larondelle N, Andersson E, Artman M, Borgström S, Breuste J, Gomez-Baggethun E, Gren A, Hamstead Z, Hansen R, Kabisch N, Kremer P, Langermeyer J, Rall EL, McPhearson T, Pauleit S, Qureshi S, Schwarz N, Voigt A, Wurster D, Elmqvist T (2014) A quantitative review of urban ecosystem services assessments: concepts, models and implementation. Ambio 43:413–433

Haines-Young R, Potschin M (2010) The links between biodiversity, ecosystem services and human wellbeing. In: Raffaelli D, Frid C (eds) Ecosystem ecology: a new synthesis. Cambridge University Press, Cambridge, pp 110–139

Haines-Young R, Potschin MB (2017) Common International Classification of Ecosystem Services (CICES) V5.1 and guidance on the application of the revised structure. Available from www.cices.eu

Hartig T, Mitchell R, de Vries S, Frumkin H (2014) Nature and health. Ann Rev Publ Heal 35:207–228

Hernandez-Morcillo M, Plieninger T, Bieling C (2013) An empirical review of cultural ecosystem services indicators. Ecol Indic 29:434–444

Hidaka BH (2012) Depression as a disease of modernity: explanations for increasing prevalence. J Affect Disord 140:205–214

Holland H, Skrede J, Holmgaard SB (2017) Cultural heritage and ecosystem services: a literature review. Conserv Manage Archael Sites 19:210–237

Holtan MT, Dieterlen SL, Sullivan WC (2014) Social life under cover: tree canopy and social capital in Baltimore, Maryland. Environ Behav 47:502–525

Hussain G, Nadeem M, Younis A, Riaz A, Khan MA, Naveed S (2010) Impact of public parks on human life: a case study. Pak J Agric Sci 47:225–230

Ichumbaki EB, Pollard E (2020) Valuing the Swahili Cultural Heritage: a maritime cultural ecosystem services study from Kilwa, Tanzania. Conserv Manage Archaelogical sites. https://doi.org/10.1080/13505033.2020.1794344

Iqbal A (2018) How safe are women-only parks perceived to be? Secur J 31:859–881

Islam SN, Tuli SM (2017) Drought impacts on urbanization. In: Eslamian S, Elsamian F (eds) Handbook of drought and water scarcity: environmental impacts and analysis of drought and water scarcity. CRC Press, Boca Raton, pp 17–44

Jaganmohan M, Vailshery LS, Mundoli S, Nagendra H (2018) Biodiversity in sacred urban spaces of Bengaluru, India. Urban For Urban Green 32:64–70

Jennings V, Bamkole O (2019) The relationship between social cohesion and urban green space: an avenue for health promotion. Int Res Publ Heal 16:452

Jennings V, Larson L, Yun J (2016) Advancing sustainability through urban green space: cultural ecosystem services, equity, and social determinants of health. Int J Environ Res Publ Heal 13:196

Kabisch N (2015) Ecosystem service implementation and governance challenges in urban green space planning: the case of Berlin, Germany. Land Use Pol 42:557–567

King A, Shackleton CM (2020) Maintenance of private and public urban green space provides significant employment in Eastern Cape towns, South Africa. Urban For Urban Green 54:126740

La Rosa D, Spyra M, Inostroza L (2016) Indicators of cultural ecosystem services for urban planning: a review. Ecol Indic 61:74–89

Lafortezza R, Carrus G, Sanesi G, Davies C (2009) Benefits and well-being perceived by people visiting green spaces in periods of heat stress. Urban For Urban Green 8:97–108

Lagbas AJ (2019) Social valuation of regulating and cultural ecosystem services of Arroceros Forest Park: a man-made forest in the city of Manila, Philippines. J Urban Manage 8:159–177
Lambert KG, Nelson RJ, Jovanovic T, Cerda M (2015) Brains in the city: neurobiological effects of urbanisation. Neurosci Behav Rev S0149–7634(15):00103–00107
Li F, Sutton P, Nouri H (2018) Planning green space for climate change adaptation and mitigation: a review of green space in the central city of Beijing. Urban Reg Plan 3:55–63
Lindemann-Matthies P, Keller D, Li X, Schmid B (2013) Attitudes towards forest diversity and forest ecosystem services—a cross-cultural comparison between China and Switzerland. J Plant Ecol 7:1–9
Lotfi S, Koohsari MJ (2011) Proximity to neighborhood public open space across different socio-economic status areas in metropolitan Tehran. Environ Just 4:179–184
Lovasi GS, Grady S, Rundle A (2011) Steps forward: review and recommendations for research on walkability, physical activity and cardiovascular health. Public Health Rev 33:484–506
Maas J, van Dillen SM, Verheij RA, Groenewegen PP (2009) Social contacts as a possible mechanism behind the relation between green space and health. Heal Place 15:586–595
Maruthaveeran S (2010) Establishing performance indicators from the user perspective as tools to evaluate the safety aspects of urban parks in Kuala Lumpur. Pertanika J Soc Sci Humanit 18:199–207
McConnachie MM, Shackleton CM (2010) Public green space inequality in small towns in South Africa. Habitat Int 34:244–248
Milcu AI, Hanspach DA, Fisher J (2013) Cultural ecosystem services: a literature review and prospects for future research. Ecol Soc 18(3):44
Millennium Ecosystem Assessment (MEA) (2005) Ecosystem and human wellbeing: current state and trends. Island Press, Washington, DC
Mitchell R, Popham F (2008) Effect of exposure to natural environment on health inequalities: an observational population study. The Lancet 372:1655–1660
Myers G (2021) Urbanisation in the Global South. In: Shackleton CM, Cilliers SS, Davoren E, du Toit MJ (eds) Urban ecology in the Global South. Springer, Cham, pp 27–49
Narh SN, Takyi SA, Asibey MO, Amponsah O (2020) Garden city without parks: and assessment of the availability and conditions of parks in Kumasi. Urban for Urb Green 55:126819
Nero B (2017) Urban green space dynamics and socio-environmental inequity: multi-resolution and spatiotemporal data analysis of Kumasi, Ghana. Int J Remote Sens 38(23):6993–7020
Nesbitt L, Hotte N, Barron S, Cowan J, Sheppard SR (2017) The social and economic value of cultural ecosystem services provided by urban forests in North America: a review and suggestions for future research. Urban For Urban Green 25:103–111
Ngulani T, Shackleton CM (2019) Use of public urban green spaces for spiritual services in Bulawayo, Zimbabwe. Urban For Urban Green 38:97–104
Orta Ortiz MS, Geneletti D (2018) Assessing mismatches in the provision of urban ecosystem services to support spatial planning: a case study on recreation and food supply in Havana, Cuba. Sustain 10:2165
Paul S, Nagendra H (2017) Factors influencing perceptions and use of urban nature: surveys of park visitors in Delhi. Land 6:27
Pauleit S, Vasquéz A, Maruthaveeran S, Liu L, Cilliers SS (2021) Urban green infrastructure in the Global South. In: Shackleton CM, Cilliers SS, Davoren E, du Toit MJ (eds) Urban ecology in the Global South. Springer, Cham, pp 107–143
Peters K, Elands B, Buijs A (2010) Social interactions in urban parks: stimulating social cohesion? Urban For Urban Green 9:93–100
Plieninger T, Dijks S, Oteros-Rozas E, Bieling C (2013) Assessing, mapping and quantifying cultural ecosystem services at community level. Land Use Pol 33:118–129
Rigolon A, Browning MH, Lee K, Shin S (2018) Access to urban green space in cities of the Global South: a systematic literature review. Urban Sci 2(3):67
Roy S, Byrne J, Pickering C (2012) A systematic quantitative review of urban tree benefits, costs and assessment methods across cities in different climatic zones. Urban For Urban Green 11:351–363

Roy M, Shemdoe R, Hulme D, Mwageni N, Gough A (2018) Climate change and declining levels of green structures: Life in informal settlements of Dar es Salaam, Tanzania. Landsc Urb Plan 180:282–293

Seaman PJ, Jones R, Ellaway A (2010) It's not just about the park, it's about integration too: why people choose to use or not use urban greenspaces. Int J Behav Nutr Phys Act 7:78

Shackleton CM (2021) Ecosystem provisioning services in Global South cities. In: Shackleton CM, Cilliers SS, Davoren E, du Toit MJ (eds) Urban ecology in the Global South. Springer, Cham, pp 203–226

Shackleton CM, Blair A (2013) Perceptions and use of public green space is influenced by its relative abundance in two small towns in South Africa. Landsc Urb Plan 113:104–112

Shackleton CM, Blair A, De Lacy P, Kaoma H, Mugwagwa N, Dalu MT, Walton W (2018) How important is green infrastructure in small and medium-sized towns? Lessons from South Africa. Landsc Urb Plan 180:273–281

Shackleton CM, Cocks ML (2021) Urban nature and biocultural realities. In: Cocks ML, Shackleton CM (eds) Diversities of nature in cities: enriching belonging, wellbeing and bioculture. Earthscan/Routledge, London, pp 1–24

Shanahan DF, Bush R, Gaston KJ, Lin BB, Dean J, Barber E, Fuller RA (2016) Health benefits from nature experiences depend on dose. Sci Rep 6:28551

Shanahan DF, Fuller RA, Bush R, Lin BB, Gaston KJ (2015) The health benefits of urban nature: how much do we need? BioSci 65:476–485

Shen Y, Sun F, Che Y (2017) Public green spaces and human wellbeing: mapping the spatial inequity and mismatching status of public green space in the Central City of Shanghai. Urban For Urban Green 27:59–68

Shinew KJ, Glover TD, Parry DC (2004) Leisure spaces as potential sites for interracial interaction: community gardens in urban areas. J Leis Res 36:336–355

Silvennoinen H, Alho J, Kolehmainen O, Pukkala T (2001) Prediction models of landscape preferences at the forest stand level. Landsc Urb Plan 56:11–20

Takano T, Nakamura K, Watanabe M (2002) Urban residential environments and senior citizens' longevity in megacity areas: the importance of walkable green spaces. J Epidem Comm Heal 56:913–918

Takeuchi K (2010) Rebuilding the relationship between people and nature: the Satoyama Initiative. Ecol Res 25:891–897

Thaiutsa B, Puangchit L, Kjelgren R, Arunpraparut W (2008) Urban green space, street tree and heritage large tree assessment in Bangkok, Thailand. Urban For Urban Green 7:219–229

Unnikrishnan H, Nagendra H (2015) Privatising the commons: impact on ecosystem services in Bangalore's lakes. Urban Ecosys 18:613–632

Vásquez A (2017) Green infrastructure, ecosystem services and their contribution to address climate change in cities: the case of the coastal corridor of the river Mapocho in Santiago de Chile. Revista de Geografía Norte Grande 63:63–86

Vásquez A, Dobbs C (2020) Ensuring equitable green space to deprived social groups. In: Douglas I, Anderson P, Goode D, Houck M, Maddox D, Nagendra H, Tan PK (eds) The Routledge handbook of urban ecology, 2nd edn. Routledge, London p, pp 817–825

Venter ZS, Barton DN, Gundersen V, Figari H, Nowell M (2020) Urban nature in a time of crisis: recreational use of green space increases during the COVID-19 outbreak in Oslo, Norway. Environ Res Lett 15:104075

Verschuuren B (2006) An overview of cultural and spiritual values in ecosystem management and conservation strategies. In: Haverkort B, Rist S (eds) Endogenous development and bio-cultural diversity. Compas, Leusden, pp 299–325

Ward Thompson C, Aspinall P, Roe J, Robertson L, Miller D (2016) Mitigating stress and supporting health in deprived urban communities: the importance of green space and the social environment. Int J Environ Res Publ Heal 13:440

Willemse L (2013) A flowmap–geographic information systems approach to determine community neighbourhood park proximity in Cape Town. S Afr Geogr J 95:149–164

Williams F (2017) The nature fix: why nature makes us happier, healthier, and more creative. W.W. Norton, New York

Zhang S, Zhou W (2018) Recreational visits to urban parks and factors affecting park visits: evidence from geotagged social media data. Landsc Urb Plan 180:27–35

Urban Ecosystem Disservices in the Global South

Elandrie Davoren and Charlie M. Shackleton

Abstract As much as people benefit from the ecosystem services provided by nature, ecosystem disservices also impact daily lives. This is especially true for many urban communities in the Global South due to (1) the often greater diversity of ecosystem disservices and (2) higher vulnerabilities, which together can lead to greater impacts of ecosystem disservices in many Global South settings. This chapter provides an overview of the current understandings of urban ecosystem disservices and their assessment and management with an emphasis on the Global South, albeit with reference to the Global North where useful. This overview reveals a lack of detailed and systematic empirical research on urban ecosystem disservices generally, but even more so in the Global South, despite the greater diversity and vulnerability in Global South settings. This needs to be acknowledged in urban ecology framings of urban dynamics in the Global South and translated into better integration of both ecosystem services and disservices within common research, policy or management frameworks. It is only through such integration that appropriate context-relevant policy directions and management options can be identified, thereby promoting the wellbeing of urban citizens in the Global South.

Keywords Assessment · Ecosystem disservices · Impacts · Urban · Vulnerability

1 Introduction

There are two universal truths regarding the multiple relationships that humans have with nature. The first is that humans need nature. Nature provides humans with valuable, tangible and intangible ecosystem services and a lack of sufficient and high-quality ecosystem services greatly impacts both physical and mental well-being. These services include supporting or habitat services (Pauleit et al. 2021),

E. Davoren (✉) · C. M. Shackleton
Department of Environmental Science, Rhodes University, Makhanda 6140, South Africa
e-mail: e.davoren@ru.ac.za

C. M. Shackleton
e-mail: c.shackleton@ru.ac.za

C. M. Shackleton et al. (eds.), *Urban Ecology in the Global South*, Cities and Nature,
https://doi.org/10.1007/978-3-030-67650-6_11

provisioning (Shackleton 2021), regulating (Escobedo 2021) and cultural services (Dobbs et al. 2021). The importance of these benefits to humans and humanity has been emphasised by the Millennium Ecosystem Assessment (2005), The Economics of Ecosystems and Biodiversity (TEEB 2010), Mapping and Assessment of Ecosystems and their Services (Maes et al. 2013), the Common International Classification of Ecosystem Services (CICES, Haines-Young and Potschin 2013) and the Intergovernmental Platform on Biodiversity and Ecosystem Services (IPBES, Díaz et al. 2015).

The second undeniable truth is that nature isn't always kind to humans. Since the dawn of modern humans hundreds of thousands of years ago they have been attacked by wild animals and assailed by natural hazards, such as pests, diseases, droughts and wildfires, which have affected them directly or interrupted the flow or distribution of ecosystem services (Dunn 2010). Collectively these negative effects of nature on the physical and mental wellbeing of humans have come to be called ecosystem disservices, which have been defined as "ecosystem generated functions, processes and attributes that result in perceived or actual negative impacts on human wellbeing" (Shackleton et al. 2016). Both ecosystem services (ES) and disservices (ED) are created within and affect social-ecological systems (SES) throughout the world (Blanco et al. 2019).

Despite the ubiquity and significance of both ES and ED, considerably more research and policy attention has been given to understanding the former than the latter (Shackleton et al. 2016; Blanco et al. 2019). A systematic review by Blanco et al. (2019) found a total of 27,441 papers on ES published between 1976 and 2018. In contrast there were only 194 papers on ED, of which 46 were published in 2018. Roy et al. (2012) reviewed 115 research papers on urban trees, 74% of which originated from Global North (GN) countries, and of these only 18 papers considered problems and hazards (i.e. ED) associated with urban trees (Roy et al. 2012). Similarly, a review by Potgieter et al. (2017) on the role that alien plant species play in providing either ES or ED, found only one study on the ED provided by alien plants in a Global South (GS) country, compared to 95 studies conducted in Global North countries (Potgieter et al. 2017).

The above-mentioned results show a clear maldistribution of research on ED in general and specifically between GS and GN regions, both in number and in duration of research focus. This does not mean that ecosystem disservices were not researched in the past, but rather that they were simply not labelled as "ecosystem disservices" and some still aren't (Shackleton et al. 2016). However, if researchers, managers and policymakers are to truly understand the tangible and intangible values of nature and the relationships between humans and nature, a holistic appreciation is required of both the positives and negatives within the relationship, rather than just one or the other (Shackleton et al. 2016; Roman et al. 2021). Interestingly, the opposite applies with respect to newspaper media, where there is generally greater coverage of ED than ES (Lyytimäki 2014; McLellan and Shackleton 2019).

Although greater research attention to and understanding of ED are required, global ED assessments are limited by readily available data, a lack of resources (Lyytimäki 2017), a lack of will to research disservices (Shapiro and Báldi 2014;

Villa et al. 2014), a lack of application to multiple land uses and landscapes (with most having been done in agricultural (O'Farrell et al. 2007; Zhang et al. 2007) and urban landscapes (Lyytimäki et al. 2008; Lyytimäki and Sipilä 2009)) and a lack of research in Global South countries (Roy et al. 2012; von Döhren and Haase 2015; Potgieter et al. 2017). Developing a GS perspective of urban ecosystem disservices (UED) is necessary if they are likely to have greater or different effects in social-ecological systems, urban ecology and on human wellbeing in the GS than in the GN.

However, the current paucity of research on UED in both the GS and the GN makes it impossible to interpret if the patterns may indeed be different. But we would hypothesise that to be the case. For example, the diversity of ED is higher in regions of higher biodiversity (Dunn 2010; Herd-Hoare 2019), and since many GS countries are situated in biologically rich regions, it is likely that they will therefore experience a greater diversity of ED. Whether the hypothesised greater diversity of ED translates into a greater burden or frequency of negative impacts is unknown. However, the overall severity of many ED is higher in GS settings than experienced in the GN. For example, burdens of diseases such as malaria, cholera and bilharzia are far greater in the GS than the GN, as are snake bites, attacks by wild animals, landslides and floods. This is likely to be linked to the higher vulnerability of populations in the GS as a function of higher exposure, in conjunction to lower capacities (at state and household levels) to prevent, mitigate or adapt in the face of particular ED.

Within this framing, the aim of this chapter is to (1) provide an overview of the current knowledge and understanding of urban ecosystem disservices with an emphasis on the Global South, albeit with reference to the Global North where useful, (2) evaluate current UED assessment methods, (3) assess existing frameworks which incorporate both UES and UED and (4) provide insights into urban ecosystem disservices management in the Global South.

2 What Are Urban Ecosystem Disservices?

Simply put, urban ecosystem disservices (UED) are all the ecosystem disservices that are found in urban ecosystems (von Döhren and Haase 2015). According to Lyytimäki (2017) urban areas are characterised by different values, attitudes, lifestyles and varying degrees of tolerance towards any form of inconvenience related to urban green infrastructure (UGI). This makes the identification of UED difficult. What constitutes an UED and which are most relevant to a particular urban population, are determined by that community's context, heterogeneity, perceptions, prior experiences and scale (Escobedo et al. 2011). For example one individual might enjoy listening to birdsong, while another might consider the same birdsong noisy and disruptive (Cox et al. 2018). Urban ecosystem disservices can occur anywhere in a city or town, but marginalised communities or poor urban residents can be most affected, since they lack the financial resources and infrastructure to adequately manage or mitigate disservices (Dalu and Shackleton 2018), and are also most reliant on immediately available ES for many of their needs (Shackleton 2021).

3 Types of Urban Ecosystem Disservices

Blanco et al. (2019)'s systematic review of UED was repeated for the purpose of this chapter. The literature review was conducted using ISI Web of Science and Scopus, using the search terms "ecosystem" AND "disservice" OR "dis-service" OR "dys-service". ISI Web of Science found 284 records, of which 54 were published in 2019 and 38 in 2020 (as of August 2020). Scopus returned 368 records, of which 63 were published in 2019 and 56 in 2020 (as of August 2020). In both, ISI Web of Science and Scopus, the majority (79%) of the search articles originated from Global North countries, with only 21% of the records originating from Global South countries. Furthermore, most of these papers only mention disservices rather than reporting on empirical studies or frameworks in which ED are a major part. The papers found during this review were used in this chapter.

Based on reviews by Roy et al. (2012) and von Döhren and Haase (2015), UED were grouped into five categories, namely those that have a (1) visual/aesthetic, (2) ecological, (3) economic, (4) health or (5) psychological impact on human well-being. However, this categorisation ignores (6) cultural impacts of ED, which we argue should be added, as per Shackleton et al. (2016). We elaborate on all of these categories below. However, it must be appreciated that, conditional on the local context, any individual UED may have impacts on human wellbeing across more than one of these categories at the same time. For example, an invasive species may have ecological impacts by constraining the distribution or growth of a useful species, while also causing health impacts if it emits biogenic volatile organic compounds (BVOCs) or allergenic pollen and economic impacts if it damages infrastructure by cracking and uplifting paving.

3.1 Aesthetic Impacts

Aesthetic impacts consist of all UED that obstruct or negatively impact people's aesthetic value or appreciation of a specific ecosystem structure, process or service (Lyytimäki 2017). These types of UED are highly subjective, for example, some individuals may enjoy looking at the street trees outside their homes, while others perceive them as a nuisance blocking their view of the street or providing cover for would-be criminals (Delshammar et al. 2015; Gwedla and Shackleton 2019). In some instances, people perceive entire urban green infrastructures (UGI), such as brownfields/wastelands, forests, vacant land or wetlands, as unpleasant, unsightly or wild (Plieninger et al. 2013; Ostoić et al. 2017; Adegun 2018; Brun et al. 2018; Mathey et al. 2018; Riley et al. 2018). Many urban residents perceive the spontaneous wild vegetation, often referred to as "weeds", growing within the aforementioned UGIs as undesirable (Riley et al. 2018; Li et al. 2019).

The majority of research on UED with an aesthetic impact are related to people's perceptions of urban trees. Some perceive urban trees as UED when they obstruct views (Guo et al. 2019), decrease the aesthetic value of buildings or historic landmarks (Lyytimäki 2017), drip sap or sticky resin, fallen leaves, pollen, flowers or seeds litter the environment (Delshammar et al. 2015; Gwedla and Shackleton 2019; Suchocka et al. 2019) or are unmanaged and neglected if they contain dead branches (Lyytimäki 2017; Suchocka et al. 2019), or display pest invasions and diseases (Boyd et al. 2013; Urquhart et al. 2017). In their survey of 1100 people in 11 towns in Zimbabwe and South Africa, Shackleton and Mograbi (2020) reported that the four most commonly mentioned reasons for disliking specific trees were if they (1) were thorny, (2) were messy in dropping an abundance of flowers or fruits, (3) caused allergies and (4) were poisonous. In a Global North setting, Guo et al (2019) found that among 445 homeowners in Christchurch (NZ), 35% of respondents removed trees because they were damaged, diseased or dead, 9% because they dropped messy leaves, flowers, fruit or branches and 4% because they obstructed views.

Other UED in this impact category, include bird species that are considered noisy (Belaire et al. 2015; Cox et al. 2018; Echeverri et al. 2019), pets or other animals making loud or unpleasant noises (Lyytimäki 2014; Jégh-Czinege et al. 2020; McLellan and Shackleton 2019), the presence of bird and dog excrement (Belaire et al. 2015; Delshammar et al. 2015; Cox et al. 2018; Lyytimäki 2017) or the dung from cattle and other livestock that are perceived as unpleasant or unsightly (Shackleton et al. 2017). However, it is important to note that most studies that examined or mentioned UED impacting the aesthetic value of urban environments and UGI were concentrated in the Global North; only a few studies have been conducted in the Global South.

3.2 Ecological Impacts

Ecological impacts are UED that negatively affect ecosystem structure, processes and/or the services that they provide, such as temperature regulation, pollution or biological control (von Döhren and Haase 2015), as well as certain UED associated with birds, insects and animals. Numerous studies have highlighted the benefits of street trees for regulating temperature during summer months, however during winter months the opposite is true and shading can decrease the possibility for urban residents to access sunlight (Mesa et al. 2010; Roy et al. 2012; Lyytimäki 2014, 2017). For example, in Christchurch (NZ), 8–9% of the respondents stated that they removed trees from their property because they shaded their garden (9%) or home (8%) (Guo et al. (2019).

Urban trees can improve air quality by removing atmospheric pollutants and particulate matter, however, under high temperature conditions they act as emitting sources of biogenic volatile organic compounds (BVOCs) which may lead to the formation of ground-level ozone and secondary organic aerosols that affect air quality (Escobedo et al. 2011; Pataki et al. 2011; Baró et al. 2014; Lyytimäki 2017). Therefore, BVOC emissions negatively impact human health and wellbeing, as well as contribute to climate change (Pataki et al. 2011; Baró et al. 2014). BVOC emissions depend on both biotic (tree species and leaf biomass) and abiotic (daylight and air temperature) factors (Nowak et al. 2008). BVOC emissions are expected to increase due to climate change (Peñuelas and Llusià 2003).

Potgieter et al. (2017) conducted a literature review of 335 papers to identify UES and UED associated with alien plant species in urban areas. These studies recorded 337 alien plant species, which contributed to 27 different UED. Three of the species (*Ailanthus altissima* (19%), *Ambrosia artemisiifolia* (9%) and *Robinia pseudoacacia* (5%)) accounted for more than 30% of the recorded UED and 91 species (27% of all recorded species) provided both UES and UED (Potgieter et al. 2017). The most commonly studied UED were those that impacted human health, such as allergies or BVOC emissions. The most cited UED related to alien plants is their potential to become invasive (Escobedo et al. 2011; Wang et al. 2015) and the loss of endemic species (Escobedo et al. 2011; Roy et al. 2012, Lyytimäki 2017). Residents in Cape Town, South Africa, perceived the displacement of native plant species, increased water consumption and increased fire risk as the main UED associated with five invasive alien plant taxa (Potgieter et al. 2019b).

Globalisation, international plant trade, wood-based packaging and anthropogenic climate change have facilitated the introduction and spread of tree pests and diseases, resulting in more frequent epidemics and biodiversity loss (Boyd et al. 2013; Paap et al. 2017). As UED, pests and diseases decrease the functionality of trees, urban crops and other vegetation, affecting their ability to deliver valuable ecosystem services and potentially leading to plant mortality (Harwood et al. 2011; Boyd et al. 2013). The UK has experienced widespread tree loss due to diseases such as ash dieback, Dutch elm and canker stain disease (Harwood et al. 2011; Tsopelas et al. 2017; Hill et al. 2019). Urban forests and street trees in the United States have been impacted by sudden oak death, the emerald ash borer and Asian longhorn beetle (Cowett and Bassuk 2017; Paap et al. 2017; Alexander et al. 2020). The polyphagous shot hole borer, *Euwallacea whitfordiodendrus*, along with its fungal symbiont *Fusarium euwallaceae*, has emerged as an important urban invasive species in both southern California (US) and South Africa (Paap et al. 2018).

Birds provide a variety of ES, such as pollination, dispersal, pest control, beauty and inspiration and thus are vital components of urban ecosystems (Belaire et al. 2015). However, some species sometimes come into conflict with humans when they build their nests in or on people's homes (e.g. chimneys, gutters, roofs), damage ornamental trees and shrubs to build their nests (Belaire et al. 2015), or contribute to the spread of invasive alien species with generalised dispersal mechanisms (Whelan et al. 2015). According to Charles and Linklater (2013), human–avian conflicts tend to be species and situation specific. In South Africa, Zoeller et al. (2020) found that both the African Cuckoo and Cape Crow were perceived by respondents as ecosystem disservices because they reduce biodiversity (Zoeller et al. 2020). The

African Cuckoo practices brood parasitism which suppresses the nesting success of the host, while the Cape Crow reduces biodiversity through predation (Zoeller et al. 2020). Birds can also cause eutrophication of aquatic systems such as ponds, lakes and wetlands due to excessive inputs of nitrogen and phosphorus from their excrement (Whelan et al. 2015).

Ecological UEDs that are generally most prevalent in the Global South are those associated with the practice of keeping livestock in urban areas (Shackleton 2012; Shackleton et al. 2017; Drescher et al. 2021). In many Global South countries, livestock are held in urban areas for cultural, economic and food reasons (Lee-Smith 2010). Though highly valued by their owners, livestock damage street trees and contribute to tree mortality (Richardson and Shackleton 2014; Shackleton et al. 2017), increase soil compaction in urban green spaces and reduce plant litter (Shackleton et al. 2017), and their dung and the damage to vegetation can decrease aesthetic values (Shackleton and Njwaxu 2021). These disservices often lead to conflicts between livestock owners, other residents and municipalities, since owners require green spaces for their cattle to graze, while the municipality aims to provide adequate and accessible green spaces for all residents (Shackleton et al. 2017).

3.3 Economic Impacts

Economic impacts include all UED that negatively affect socio-economic structures and processes. These can incur both direct and indirect economic costs for maintenance and management of infrastructure (Roman et al. 2021). For example, the growth of tree roots can cause extensive damage to pavements by cracking and uplifting surfaces and kerbing (Blunt 2008; Lyytimäki 2017). According to Mullaney et al. (2015) pavement maintenance, repair or replacement, loss or replacement of street trees and compensation payments due to personal injury claims can have high-cost implications for local governments. McPherson (2000) found that fourteen Californian cities have reported a combined annual trip- and fall-payout of US$1.77 million due to pavement damage from tree roots and the highest single payment was US$120,000. In California and Australia, respectively, urban forestry managers and city residents both report infrastructure damage as one of the most common reasons for tree removal (Costello et al. 2000; Kirkpatrick et al. 2012).

Tree and shrub roots can also cause significant and costly problems by blocking and damaging sewer and water pipes (Östberg et al. 2012; Torres et al. 2017). The costs associated with the drainage problems include root removal, pipe replacement, as well as costs related to the damage and removal of trees (Randrup et al. 2001). According to McPherson et al. (2006) "repair costs typically range from US$100 for sewer roding (inserting a cleaning implement to temporarily remove roots) to US$1000 or more for sewer excavation and replacement". Ten per cent of the respondents in a survey by Guo et al. (2019) stated that they removed trees from their property because tree roots damaged drains, foundation or hard landscaping. Branches, especially falling ones, can cause damage to buildings, utilities and cars, or injure people, and fallen leaves can increase the braking distances of vehicles that can result in traffic accidents (Lyytimäki and Sipilä 2009; Delshammar et al. 2015). Microbial activity causes decomposition of construction wood (Lyytimäki and Sipilä 2009; Elam and Björdal 2020) and bird excrement can accelerate corrosion and damage buildings and historic landmarks (Lyytimäki and Sipilä 2009).

Indirect costs are more difficult and are less frequently assessed. One that has received more research attention than others is the decrease in property values due to people's negative perceptions of certain green spaces that are either poorly managed, harbour pests or wild animals or have high incidences of crime (Lyytimäki and Sipilä 2009). Cilliers and Cilliers (2015) found that the values of residential properties located adjacent to green spaces in Potchefstroom (South Africa) were lower than properties located further away due to perceptions of high crime rates.

The economic impacts of UED are perhaps most marked in terms of the costs associated with impairment of human health. These costs include the preventative measures to limit the impacts of widespread diseases such as malaria, reactive costs to contain an irregular health hazard, such as cholera, or losses in production when workers and employees are unable to work due to the effects of an ED, such as pollen allergies, snake and insect bites or bacterial diseases. For example, UNICEF (2004) states that the lost productivity due to malaria in sub-Saharan Africa is US$12 billion annually. Families living in malarial areas have to pay prevention costs, and when one or more members fall ill from malaria, they have less income, while incurring an average cost of one-quarter of their income (UNICEF 2004).

High economic costs are also a consequence of catastrophic ED such as floods, avalanches or non-seismic landslides, which result in extreme losses of public and private infrastructure and even human life. For example, over 4300 people are killed by landslides every year, with the highest number in Asia (Froude and Petley 2018). For many vulnerable households, hazards such as floods and landslides can mean the partial or complete loss of their homestead and belongings (not to mention loss of life) and access to basic services. Few have formal insurance strategies to cover the cost of such extreme losses. Many turn to locally sourced provisioning services to replace lost infrastructure, or become indebted to money-lenders, or have to relocate and depend on external aid (Dalu and Shackleton 2018).

3.4 Health Impacts

Health impacts include all UED that directly or indirectly affect human physical or mental health in an adverse way and can be divided into three groups (adapted from Cariñanos et al. 2017), namely (1) those derived from the inherent characteristics of urban vegetation or wildlife, (2) those in which an urban ecosystem supports the factor causing harm or discomfort and (3) those involving air pollutants (which we do not view as UED because pollution is generated by human actions and not ecosystem processes). However, this categorisation needs to be expanded to embrace those of non-biological origin, such as floods or wildfires. The first group includes the negative effects caused by BVOC emissions, pollen-related allergies and asthma (Arnold 2012; Cariñanos et al. 2017, 2019, 2020), vector-spread diseases such as Lyme disease, avian influenza and rabies (Lyytimäki et al. 2008; Lyytimäki and Sipilä 2009; Escobedo et al. 2011) and insect bites and animal attacks (Del Toro et al. 2012; Barua et al. 2013).

According to the World Allergy Organisation between 10 and 30% of the world population is affected by allergenic pollen and prevalence rates are increasing worldwide (Pawankar et al. 2013). Allergic reactions to tree pollen vary from irritation of the upper airways and eyes to severe asthmatic reactions requiring hospitalisation, and sometimes resulting in death (Bosch-Cano et al. 2011; D'amato 2000). The impact of allergens emitted by urban green spaces on human health has been extensively studied in the Global North (Cariñanos et al. 2019, 2020; Velasco-Jiménez et al. 2020), but there has been a shortfall of similar studies in the Global South. According to the European Academy of Allergy and Clinical Immunology more than 150 million Europeans (approx. 20% of the population) suffer from chronic allergic diseases (EAACI 2015), with an estimated cost of between €55–151 billion per year to the National Health Services (Zuberbier et al. 2014). However, according to Sousa-Silva et al. (2020) the "allergenic potency and cross-reactivity of most tree species are still largely unknown or based on undocumented methods".

A review by Moro et al. (2009) of more than two-hundred thousand cases handled by the Poison Control Centre of Milan over a four-year period revealed that over eight thousand of them were related to the ingestion of poisonous mushrooms or plants. The majority of these cases were children younger than fourteen, while some of the most serious cases involved adults who mistakenly identified a poisonous species as edible (Moro et al. 2009).

The practice of livestock farming in Global South countries is well documented (Lee-Smith 2010; Halloran and Magid 2013), however, the practice of backyard livestock and poultry farming in the United States has increased in recent years (McClintock et al. 2014; Tobin et al. 2015; Pires et al. 2019). According to McClintock et al. (2014) public interest in urban agriculture has rapidly increased in the United States, especially in terms of keeping urban livestock (pigs and goats), poultry (chickens, geese and ducks), rabbits and bees. This increase is attributed to a need to know where food comes from, production of healthier food, and in some cases cultural significance (McClintock et al. 2014). However, as previously mentioned, urban livestock are responsible for several UED, including diseases spread through livestock and poultry manure (Lupindu et al. 2014; Tobin et al. 2015; Ström et al. 2018; Pires et al. 2019).

At orders of magnitudes greater than any other UED that impacts human health, are the loss of life, wellbeing and productivity associated with a host of infectious diseases such as malaria, cholera, tuberculosis, Ebola and many more. For example, tuberculosis is an entirely treatable disease, yet more than one million people die from it every year, with similar numbers of deaths from malaria (Lozano et al. 2012). The considerable majority of deaths from communicable diseases are found in the GS, particularly more tropical countries, and are typically associated with and reinforce poverty (Bhutta et al. 2014). Non-infectious ED includes things like snake bites, which kill over 100,000 people a year, most in the GS (WHO 2010).

3.5 Psychological Impacts

Psychological impacts consist of all UED that cause negative feelings or emotions, such as discomfort, anxiety or fear due to, for example, animal excrement or plant litter causing disgust (Lyytimäki et al. 2008; Agbenyega et al. 2009; Limburg et al. 2010) or feelings of insecurity and fear of bodily harm when in urban green spaces (Tzoulas et al. 2007; Hofmann et al. 2012; Delshammar et al. 2015; Shackleton et al. 2015). For many urban residents, areas with uncontrolled or overgrown vegetation instil feelings of anxiety and fear, since these spaces can provide cover for dangerous or unattractive wildlife, such as snakes, spiders and insects, and facilitate crime or criminal activity by providing cover for criminals (Lyytimäki 2014; Shackleton et al. 2015; Lyytimäki 2017; Escobedo et al. 2018; Potgieter et al. 2019a). The expansion of urban areas into natural ones and the promotion of recreational outdoor activities can create an overlap of habitats which has led to increased human–wildlife conflicts (Bombieri et al. 2019).

There are two contradicting views on the relationship between vegetation and crime (Donovan and Prestemon 2012). Some studies have shown that vegetation, especially dense vegetation, can increase fear of crime (Nasar and Fisher 1993; Nasar et al. 1993; Donovan and Prestemon 2012; Adegun 2018), whereas others have shown that tall trees and well-maintained lawns can reduce the fear of crime (Kuo et al. 1998; Mouratidis 2019) or crime occurrence (Kuo and Sullivan 2001; Donovan and Prestemon 2012; Troy et al. 2012; Wolfe and Mennis 2012; Troy et al. 2016). In both Global North and South countries, studies have found that people experienced a fear of crime or being victims of crime in urban green spaces (Jansson et al. 2013; Sreetheran and Van Den Bosch 2014; Shackleton et al. 2015; Sreetheran and Van Den Bosch 2015; Lyytimäki 2017; Mouratidis 2019) and often these negative emotions lead to residents limiting their use of such spaces.

Potgieter et al. (2019a) found that criminal activity was unevenly distributed across the urban landscape of Cape Town and was either facilitated or limited by topographical features, such as hills, dunes or waterbodies, and physical features of the landscape, such as vegetation structure or type. Fear of crime is more prevalent among woman than men (Sreetheran and Van Den Bosch 2014). However, perceived personal safety is distinct from actual safety or security risks, since it is an experienced feeling and not an actual safety concern (Jansson et al. 2013). Escobedo et al. (2018) found that the occurrence of homicide in Bogota, Columbia, was not linked to the amount of urban green areas or tree density. Studies that have examined the relationship between crime and vegetation are predominantly conducted in Global North countries (especially North America and Europe); with very few studies in the Global South.

3.6 Cultural Impacts

Cultural ES are important to human wellbeing in the GS as well as the GN. Indeed, in the GN cultural UES such as recreational and educational services are among the most researched and directly managed UES (Roy et al. 2012). The same is likely to be true in the GS, but there still remains a paucity of research on cultural ES in the GS (Dobbs et al. 2021). However, the wider diversity of cultures, and hence views of and relationships with urban nature in the GS (Cocks and Shackleton 2021 and chapters therein), may well set the foundation for higher impacts of UED on cultural ES in the GS than the GN.

An example of a more GS perspective of cultural impacts of UED might be the loss of culturally important spaces or species from the urban ecosystem. Culturally sacred spaces are those that are used for particular ceremonies or rituals in accordance with particular cosmologies. Ngulani and Shackleton (2019) described the use of specific public green spaces in Bulawayo, Zimbabwe, for worship practices. In other cosmologies, access to particular forms of nature is vital in respecting and nurturing bonds with one's ancestors who "live in nature", as evidenced through Cocks et al. (2016) finding that 84% of urban Xhosa respondents in Makhanda (formerly Grahamstown), South Africa, said that it was necessary to have access to nature to communicate with their ancestors, and 92% affirmed that their household performs cultural rituals to communicate with their ancestors. Loss of quantity or quality of such spaces, due to UED such as invasive species, has marked implications for cultural identity and wellbeing.

The same applies to culturally important species that may be used in particular rituals, to treat physical or mental ailments that are culturally defined, to fashion culturally important symbols or artefacts, or to worship the spirits of individual trees. The daily street-side offerings and prayers under *Ficus religiosa* trees in towns and cities throughout southern India, is a classic example (Keswani 2017). These sorts of cultural ES are not as common in the GN where many dominant worldviews and cosmologies do not see humans as part of nature and vice versa (Shackleton and Cocks 2021). The loss of such culturally important species driven by UED such as pests and diseases, invasive species or overgrazing by livestock has not been studied, but because of the innateness of the human–nature relationship, it is likely to have been very keenly felt.

4 Vulnerability as the Mediator Between Ecosystem Disservices and Human Wellbeing

The human wellbeing impacts of UED at a specific time or place are dependent on the vulnerability of the affected community. An UED of similar nature, frequency and severity can have markedly different impacts in different countries, towns or neighbourhoods, because the resilience and adaptive capacity of the affected communities vary. Take a hypothetical example of the effects of floods in a small town in a relatively poor country in the GS compared to those of an affluent and well-resourced country in the GN. Flood damage, loss of life and immediate post-flood health concerns are far higher in the former than the latter. That is because communities are frequently more vulnerable in the former.

Vulnerability can be disaggregated into exposure risk (will a specific UED occur in a specific place), sensitivity (if it does occur, how many are exposed to the effects, and who) and response capacity (what is the capacity to cope, respond and learn for future, who has the capacity) (Adger 2006). All three dimensions can be analysed at different scales, including from individuals to households, civic groups, city authorities and state and national governments. In terms of risk, there is some commentary that more tropical regions have a greater incidence of ED. For example, Dunn (2010) speculated that regions with the highest biodiversity will also have the greatest diversity of harmful biota, which is borne out by several taxa, from the potentially wide-scale effects of dozens of infectious diseases (Bhutta et al. 2014), through to the dozens of species of venomous snakes in the tropics (compared to less than five in Europe and about a dozen in North America [WHO 2010]). Conservative estimates of global human fatalities from snake bites are approximately 100,000–125,000 per year, concentrated in SE Asia, South America and sub-Saharan Africa (WHO 2010; Geneviève et al. 2018).

Sensitivity is a complex variable to measure and model because it is dependent on so many factors, such as the magnitude or intensity of the ED experienced, the time span since a previous UED or other shock, through to population density and the economic and health status of the individual households or people affected. For instance, an infectious disease spreads more rapidly in areas with high population densities (especially if response capacities are low), or a young and healthy person is far less sensitive to an infectious disease or spider bite than the elderly or those with pre-existing co-morbidities. Given the wide range of UED, and the site specificity of sensitivity, it is difficult to conclude if sensitivity to UED is widely different in GS contexts compared to GN ones.

In contrast, there certainly are significant differences with respect to capacities to respond or cope with an UED, between GS and GN settings. The higher poverty levels in GS contexts, at both the household and state level, mean that there is less investment in pre-emptive strategies and limited budgets and skilled personnel to design and guide response strategies (Shackleton et al. 2021). Such stark inequities may also be apparent within towns and cities, where particular neighbourhoods may be more prone to certain UED as well as have less capacity to cope. This is particularly so with poorer neighbourhoods and informal or slum areas. For example, informal slums are often situated on urban lands that were deemed too hazardous for formal developments, such as on floodplains or on very steep slopes, making them more at risk to natural hazards (O'Hare and Rivas 2005; Dalu et al. 2018). Poorer neighbourhoods also typically have lower UGI and street trees to help mitigate some UED (Venter et al. 2020).

5 UED Assessment

UED are identified and defined based on an individual's or a community's context, heterogeneity and perception of their environment (Escobedo et al. 2011). Furthermore, new UED can be perceived as a result of changing ecosystems or changes in people's perceptions (Lyytimäki and Sipilä 2009; Roman et al. 2021). This makes assessment of UED complicated and requires the continued use of local knowledge (Lyytimäki and Sipilä 2009).

Over the last few decades, various studies have attempted to develop methods and assessment tools for quantifying both UES and UED (Pataki et al. 2011; Dobbs et al. 2011, 2014). However, these studies were predominantly conducted in the Global North and focused mainly on either street trees or urban forests (Dobbs et al. 2011; Escobedo et al. 2011; Dobbs et al. 2014; Speak et al. 2018). Several studies used surveys to determine peoples' perceptions of both UES and UED in urban areas (Plieninger et al. 2013; Teixeira et al. 2019; Larson et al. 2019). Delshammar et al. (2015) examined complaint records kept by municipal parks in Sweden to investigate public perceptions of UED, while Lyytimäki (2014) and McLellan and Shackleton (2019) conducted content analyses of newspapers in Finland and South Africa, respectively, to identify mentioned or extensively reported ecosystem disservices.

Pataki et al. (2011) proposed the identification and evaluation of biogeochemical cycles as a means of quantifying both UES and UED, for example, measuring gaseous emissions from trees and other vegetation to determine the extent of air pollution caused by urban vegetation. However, services and disservices that cannot be linked with ecosystem processes were excluded from Pataki et al.(2011)'s assessment. Wang et al. (2015) examined specific plant traits, including traits related to UES and UED, to determine how plant diversity varies in different residential areas in Beijing, China. Only two traits related to UED were assessed, namely, species invasive alien status and pollen allergenic potential (Wang et al. 2015). This Chinese study is the only one which provides a GS (Asian) perspective to simultaneously assess both UES and UED.

At a perceptions level, Shackleton and Mograbi (2020) reported the reasons that urban residents provided for favouring (the UES) or disliking (the UED) particular urban tree species in South Africa and Zimbabwe. Subsequently, Dobbs et al. (2011) used different indicators known to be related to UED, for example, the number of tree species susceptible to damage was used as an indicator of the disservice "Damage to infrastructure and risk to human safety". However, the study was limited to only a few disservices associated with urban trees and the methods used were site specific (both Global North and Northern hemisphere). Building on this, Dobbs et al. (2014) developed a meta-model for quantifying UES and UED under a social-ecological framework. Their study used aerial photography, thermal imagery, meta-data and forest inventories provided by the City of Melbourne to select spatially quantifiable UES and UED (Dobbs et al. 2014). This method can be applied to different urban areas where baseline information is available and can be a valuable tool for detecting UES and UED heterogeneity and identifying spatial patterns, for example identifying areas with high incidences of UED and areas with low incidences of provisioning or regulating services (Dobbs et al. 2014). However, only nine UES and two UED were assessed, namely trees with high allergenic potential and those with potential to damage infrastructure.

Some studies have assessed the net benefit of trees in urban environments by using selected UES and UED to develop performance or composite indicators (Tiwary et al. 2016; Speak et al. 2018). However, both studies only considered two UED, namely pollen allergenic potential and BVOC production (Tiwary et al. 2016; Speak et al. 2018). Other studies, like Escobedo et al. (2011) and Mullaney et al. (2015) conducted cost–benefit analysis of urban forests and street trees, respectively. The cost–benefit approach is appealing to policymakers and corporate entities, since both the benefits and associated costs are easily communicated and understood by all relevant stakeholders (Speak et al. 2018).

Most of the aforementioned approaches simply subtracted the harm of a few selected UED from the benefits of several UES (Dobbs et al. 2011; Escobedo et al. 2011; Dobbs et al. 2014; Mullaney et al. 2015; Tiwary et al. 2016; Speak et al. 2018), which is useful for a cost–benefit analysis, but fails to fully account for the full range of UES and UED and hence that the balance may vary (Roman et al. 2021). From a planning perspective, it is important to know how UES and UED are distributed across a city or town and from an analytical perspective, it is important to determine in which context biophysical ecosystem functions are experienced as either an UES or UED (Lyytimäki and Sipilä 2009; Escobedo et al. 2011; von Döhren and Haase 2015, 2019). Therefore, UES and UED assessments need to consider the distribution of (1) contextual benefits, (2) vulnerabilities of ecosystem functions and (3) vulnerabilities of local communities (von Döhren and Haase 2019).

Andersson et al. (2015) described an approach that considers the context along with the biophysical ecosystem functions for ES. According to von Döhren and Haase (2019) this approach, which builds on Service Providing Units, i.e. ecosystem elements and structures that produce ES, can be modified for UED by including concepts from natural hazards and risk assessments, such as the idea of vulnerabilities as contextual factors for UED. Von Döhren and Haase (2019) estimated the "risks associated with the UED of street trees, by incorporating the biophysical and social-technological dimensions that interact in forming these UED" in Berlin, Germany. Their study used "indicators that specifically reflect the most vulnerable parts of the urban system which correspond to each UED hazard" (von Döhren and Haase 2019). The term UED hazard refers to the UED potential, which is determined by the biophysical indicators, while the term UED risk refers to the UED hazard combined with the contextual vulnerability (von Döhren and Haase 2019), but has rarely included assessment of a community's vulnerability.

Though more comprehensive than previous assessments, von Döhren and Haase (2019) only measured ecological processes that contributed to the formation of four UED, namely pollen allergen potential, poison hazard of street trees, BVOC emissions and damage to infrastructure, which include damage to pavements by roots and falling trees or branches. Their approach is a good foundation for further development despite its limitations, namely; (1) the focus being solely on street trees, which means most urban green infrastructure, such as gardens, parks and urban forest, were excluded from the assessment, (2) only the most frequent street tree genera were sampled, which means not all street tree genera, or all streetscape vegetation, were accounted for, (3) unequal availability of data for all tree genera and vulnerabilities for the relevant UED categories and (4) lack of systematic detailed research on UED (von Döhren and Haase 2019).

Similarly, Potgieter et al. (2019a) developed a replicable approach to assess the role of urban vegetation in providing UES and UED at a local scale. They applied remote sensing techniques (integrating LiDAR data with high-resolution multispectral imagery) and supplementary spatial data to identify areas of high UES and UED provision, in Cape Town, South Africa (Potgieter et al. 2019a). Areas of high UES provision were characterised by the presence of large trees, which provide numerous, well-documented UES, while areas of low UES provision were characterised by little to no vegetation, dense informal housing, large buildings and bare ground (Potgieter et al. 2019a). Areas resulting in high UED coincided with areas densely invaded by alien plants, particularly along the urban edge, rivers and within wetlands, and areas associated with low UED occur outside the urban edge in uninvaded natural vegetation (Potgieter et al. 2019a). This assessment technique is a great starting point for assessing areas of either high or low UED, the next step would be to assess the situation on the ground by conducting surveys and then determining the actual number of UED present in an area.

The majority of the mentioned assessment techniques were site specific, meaning they were designed specifically for European and North American contexts and therefore, application in Global South cities would require extensive adaptation to correspond to their urban vegetation (i.e. grass, forb, shrub and tree species), climate and sociopolitical context. Furthermore, all of the aforementioned studies considered only a few UED, predominantly those associated with urban trees and forests, which means that many UED were not assessed. It is also noteworthy that none of the assessed UED included cultural or subjective UED, reflecting the difficulty in doing so. Nonetheless, this is an area ripe for further research and development of approaches and methods. Consequently, comprehensive evaluation methods, which account for multiple if not all UES and UED, still need to be developed for both the Global North and South. Future assessment methods should be context specific, include public sentiments and knowledge, take temporal and spatial scales into consideration and include a greater suite of UED.

6 UED Management

Managing ecosystems, including urban ones, to optimise the provision of ES while simultaneously minimising the occurrence or impacts of ED, is a major challenge (Shackleton et al. 2016; Potgieter et al. 2019b; Roman et al. 2021). This is further complicated by the possibility that management practices aimed at optimising specific UES, may in fact aggravate associated UED, while intervention strategies aimed at reducing UED may reduce UES (Shackleton et al. 2016). According to Potgieter et al. (2019b) "urban planners and managers are faced with many trade-offs in the decision-making process as each area (regional or local) is governed by different ecological, economic, and social variables". Stakeholders in urban areas often have opposing views regarding UES and UED, potentially leading to conflicts over particular management practices, especially with regard to invasive alien plants (Dickie et al. 2014; Shackleton and Shackleton 2016; Potgieter et al. 2019b).

Globally, countries have different approaches to managing invasive alien organisms in urban areas, which affect the number, distribution and intensity of invaders within cities and towns (Potgieter et al. 2017; Tsopelas et al. 2017). In terms of invasive alien plants, however, decisions around their management greatly depend on their capacity to either provide valuable ES or to create ED (Vaz et al. 2017; Potgieter et al. 2019b) and can often lead to conflicts of interest between managers and residents. For example, the removal of certain invasive tree species can be problematic, especially if those species provide provisioning and cultural services to the urban poor (Dickie et al. 2014; Potgieter et al. 2019b), such as invasive alien trees (*Acacia saligna* and *Eucalyptus* spp.) which are used as fuelwood by residents in Cape Town, South Africa (Potgieter et al. 2019b). Careful evaluation of the UES-UED dichotomy in urban ecosystems may allow conflicts to be mitigated and managed in more efficient ways (Dickie et al. 2014; Potgieter et al. 2017; Roman et al. 2021). However, management programmes are often closely linked to the availability of funding and city-planning priorities (Potgieter et al. 2017), and inadequately consider UES and UED within a common and integrated framework but rather seek to compartmentalise them and manage them independently (Shackleton et al. 2016; Roman et al. 2021).

Stricter border controls and screening is often the first step in preventing invasive pest and diseases from affecting local biodiversity or human health and well-being (Harwood et al. 2011), the lockdowns affected in most countries to control the spread of the recent covid-19 virus are a very real example (Atangana 2020). However, stricter border control requires sound knowledge of all real and potentially invasive organisms, which is hard to achieve. Therefore, government agencies and urban municipal and park managers also need to have strategies to manage or control new introductions of invasive species. These strategies can include developing reliable management programmes to prevent and stop the spread of invasive pests and diseases effectively, monitoring human health and trees to facilitate early detection of pathogen infection and creating resilient UGI (Harwood et al. 2011; Tsopelas et al. 2017; Urquhart et al. 2017). Using urban forestry as an example, municipal or park managers usually remove dead and infected trees to prevent pests and diseases from spreading to healthy trees and causing further tree loss (Boyd et al. 2013; Harwood et al. 2011). In the UK, park managers will also remove healthy trees if they are in close proximity to infected or dead trees (Harwood et al. 2011). The proper treatment and disposal of infected trees are key to managing disease outbreaks (Tsopelas et al. 2017).

Most Global North countries prioritise the development and management of public urban green spaces for human health and cultural ES such as recreation, exercise and education. In contrast many Global South countries often need to channel limited funding to other high-priority concerns, such as poverty alleviation and food insecurity (Shackleton et al. 2021), with lower budgets for UGI including public parks and street trees (Pataki et al. 2011; Chishaleshale et al. 2015). In Tanzania for example, rapid urbanisation has resulted in increased food insecurity (Wenban-Smith et al. 2016) and consequently a greater reliance on urban agriculture and the resulting susceptibility to diseases spread by livestock (Lupindu et al. 2014). This reliance on UA results in conflicting interests; on the one hand, livestock farming is a necessity for survival and on the other hand, public health is potentially compromised by the presence of livestock in urban areas. These conflicts are where local and regional governing bodies should get involved to help manage UED. UED management plans should be relevant to a city or towns specific environmental, economic and social contexts (Dobbs et al. 2011; Pataki et al. 2011).

7 Moving Towards a Policy-Relevant Framework

Assessing the presence and impact of UED in urban ecosystems is vital to the adequate management of both UES and UED. However, only a few attempts have been made to establish a comprehensive framework that takes both into account. Tiwary et al. (2016) developed a Performance Index framework for the purpose of facilitating decision-support to planners and practitioners. Their framework provides a repeatable metric for comparative evaluation of streetscaping design and development, for example planting a line of perennial shrubs versus seasonal woody trees. The framework was developed to account for both UES and UED, however, it only accounted for emissions of BVOC in street trees as an UED.

Alternatively, Bastian et al. (2012) proposed integrating the ecosystem services concept into spatial planning, and consequently developed an Ecosystem, Property, Potential and Services framework. This framework consists of three stages, (1) description and analysis of ecosystem and landscape properties (i.e. structure, processes and biophysical functioning), (2) identifying potential ES and (3) defining and valuing ES for human benefit (Bastian et al. 2012). Though this framework is comprehensive in its inclusion and assessment of ES, disservices are treated as an afterthought, but could be introduced without the need for significant restructuring. In the Nature's Contribution to People (NCP) framework, NCP are acknowledged to be either positive or negative according to the context, however, none of the 18 listed NCP correspond to a negative NCP (Díaz et al. 2018), which suggests that "ES and ED have yet to be integrated within a single and operational framework" (Blanco et al. 2019).

Understanding how people perceive urban ecosystem disservices is essential for environmental decision-making since it can indicate where management interventions are needed, by defining high-priority areas for control and determining the scale at which UED should be managed. By conducting more in-depth, systematic research on UED and developing a better understanding of the duality and integration of the UES-UED continuum, urban ecosystems could better be managed for human wellbeing without compromising ecosystem resilience (Lyytimäki 2015), achieve more balanced policies for sustainability (Schaubroeck 2017, Shackleton et al. 2016; Blanco et al. 2019) and develop an integrative and policy-relevant framework (Lyytimäki 2015) for urban ecosystem management.

To develop a policy-relevant framework, that minimises the effects of UED, while simultaneously maximising the benefits provided by UES, urban planners and policymakers need to be better informed of the ecosystem elements and processes that contribute to the production of both UES and UED (Pataki et al. 2011; von Döhren and Haase 2019; Roman et al. 2021), especially under different climate change scenarios. Policymakers should not assume that all UES and UED are equally important to all urban communities (Escobedo et al. 2011). For example, residents in one neighbourhood or town may regard damage caused by trees as more important than another, where fear of crime in urban green spaces is more of a concern. Simply put, municipalities will be more inclined to manage the UED that have the biggest impact on human wellbeing within their communities. There is no one-size-fits-all framework that can be used by both Global North and South countries. Policy-relevant frameworks need to be context, site and scale (i.e. specific UGS, city-wide or region) specific (Roman et al. 2021). Context refers to the environmental, economic, cultural, social and political realities of an urban community and the value and importance that they ascribe to both UES and UED (Dobbs et al. 2011; Escobedo et al. 2011).

8 Acknowledging Ecosystem Disservices in Ecosystem Thinking and Urban Planning

The ED concept has been the source of some controversy in the scientific community. Some argue that by acknowledging ED there is a risk of undermining biodiversity conservation (Villa et al. 2014), focusing too much attention on the negative impacts of nature (Shapiro and Báldi 2014) and ignoring the potential that every ecosystem has for creating either ES or ED, depending on an anthropocentric perspective (Saunders and Luck 2016). Others view the ED concept as a better framing to consider balancing the positive and negative effects of nature on human wellbeing, to assess its net contribution and to provide the basis for holistic and informed management of ecosystems (Dunn 2010; Lyytimäki 2015; Sandbrook and Burgess 2015; Shackleton et al. 2016; Schaubroeck 2017; Roman et al. 2021). However, developing holistic and well-informed management practices are complex, since a common ecosystem process or element can produce both ES and ED and the importance of both are dependent on peoples' attitudes and perceptions (Shackleton et al. 2016; Blanco et al. 2019). Currently, management is usually viewed as a choice between managing for UES or avoiding UED, rather than an integrated assessment of how specific management options will affect both (Shackleton et al. 2016; Roman et al. 2021).

Respondents in a study by Guo et al. (2019) cited several ED (12 disservices, ranging from 1 to 35% of respondents) as their reason for removing trees from their property, while simultaneously citing several ES as important reasons for both tree planting (15 services, ranging from 9 to 81% of respondents) and tree retention (15 services, ranging from 11 to 89% of respondents). Similarly, Plieninger et al. (2013) found that less than 30% of the respondents in their survey attributed disservices to sites within their communities, and Larson et al. (2019) found that the majority of respondents in their survey focused more on the services they receive from their environment than the disservices. Consequently, based on the results of these studies, acknowledging the presence of ED in urban ecosystems does not diminish the value of services provided to urban residents, if anything, creating public awareness of UED could contribute to conservation programmes.

Given the paucity of research on UED, in both the GS and the GN, a first step for urban planners should be to map the distribution and incidence of those UED with real impacts or the greatest potential negative impacts on human and ecological wellbeing, along with a vulnerability map of who is at risk of being most affected and where. Together, these two mapping exercises will provide the basis for urban design solutions (including nature-based solutions) to help minimise the negative effects of UED. Increasing efforts are being made to map ES production and consumption zones at various scales (e.g. Burkhard et al. 2012; Plieninger et al. 2013), thus, it is entirely possible to map equivalent zones of UED.

9 Moving Forward

Cities are dynamic social-ecological systems, providing both UES and UED to their inhabitants, and surrounding landscapes. As Lyytimäki and Sipilä (2009) stated "perceptions about ecosystem disservices have an increasing influence on how urban green areas are experienced, valued, used, managed and developed". But what is perceived as an UES or UED can vary greatly between countries, cities, towns, communities and individuals, and as shown in this chapter, there are some major differences in the nature and effects of UED between the Global South and the Global North. Therefore, local knowledge about UED is crucial in any management strategies and plans, and UED should be analysed within specific socioecological contexts and across appropriate spatial and temporal scales to better understand how they change and affect human wellbeing over time. Future research, in both the GS and the GN, should focus on:

1. Identifying, mapping and assessing both existing and potential UED, as well as their impact on human wellbeing, especially in the context of climate change. Climate change is impacting ecosystem functions, which in turn creates new kinds of services and disservices (Lyytimäki and Sipilä 2009). By continuously developing dynamic databases of perceived and actual ecosystem services and disservices, all cities and towns could benefit by expanding their knowledge and

developing context-specific management plans. This could be scaled-up through development of citizen science programmes and databases for reporting UED, especially in contexts where local research and monitoring capacity is limited.
2. Evaluating what attributes of urban trees and landscapes that urban citizens prefer or dislike (Shackleton and Mograbi 2020) and attempt to match species planting and landscape design to optimise the services and minimise the disservices (while acknowledging that there will all be potential conflicts between different groups). This would include developing a comprehensive ranking of all UED by tree species and size in different contexts, towards better matching of species to sites and contexts.
3. Assessing how the type, frequency and magnitude of UED for specific types of urban green infrastructure vary in relation to socio-economic and governance context, and how it changes through time (in relation to social and ecological dynamics).
4. Sharing of policy and management innovations and projects from Global South contexts that have (1) integrated ES and ED in a common management framework and (2) improved human wellbeing and urban sustainability through UED management.

This chapter has considered the types and effects of urban ecosystem disservices mostly within the Global South, but with relevant reference to the Global North. This analysis indicates that the diversity and impacts of UED are greater in the Global South. This has implications for urban planning and management, as well as basic models and framing of urban ecology as the interaction of humans and nature and natural systems in urban settings. If a substantial proportion of urban dwellers experience frequent or significant UED, their perceptions of nature and where it should be maintained will be very different to those living in more benign settings. Consequently, policies and programmes advocating for the benefits of urban nature and conservation of urban biodiversity will need to situate the policies and programmes within realistic and appropriate frameworks that resonate with lived realities of urban citizens in the Global South.

Acknowledgements This work was funded by the South African Research Chairs Initiative of the Department of Science and Innovation and the National Research Foundation of South Africa (grant no. 84379). Any opinion, finding, conclusion or recommendation expressed in this material is that of the authors and the NRF does not accept any liability in this regard.

References

Adegun OB (2018) When green is grievous: downsides in human-nature interactions in informal urban settlements. J Urban 11:347–361

Adger N (2006) Vulnerability. Glob Enviro Change 16:268–281

Agbenyega O, Burgess PJ, Cook M, Morris J (2009) Application of an ecosystem function framework to perceptions of community woodlands. Land Use Pol 26:551–557

Alexander K, Truslove M, Davis R, Stephens S, Zentz R (2020) A collaborative approach to preparing for and reacting to emerald ash borer: a case study from Colorado. For 93:239–253
Andersson E, McPhearson T, Kremer P, Gomez-Baggethun E, Haase D, Tuvendal M, Wurster D (2015) Scale and context dependence of ecosystem service providing units. Ecosys Serv 12:157–164
Arnold E (2012) Pollution-load zones of allergic tree pollen in Boston. Ecol Lett 8:1218–1234
Atangana A (2020) Modelling the spread of COVID-19 with new fractal-fractional operators: can the lockdown save mankind before vaccination? Chaos Solitons Fractals 136:109860
Baró F, Chaparro L, Gómez-Baggethun E, Langemeyer J, Nowak DJ, Terradas J (2014) Contribution of ecosystem services to air quality and climate change mitigation policies: the case of urban forests in Barcelona, Spain. Ambio 43:466–479
Barua M, Bhagwat SA, Jadhav S (2013) The hidden dimensions of human–wildlife conflict: health impacts, opportunity and transaction costs. Biol Conserv 157:309–316
Bastian O, Haase D, Grunewald K (2012) Ecosystem properties, potentials and services–The EPPS conceptual framework and an urban application example. Ecol Indic 21:7–16
Belaire JA, Westphal LM, Whelan CJ, Minor ES (2015) Urban residents' perceptions of birds in the neighbourhood: biodiversity, cultural ecosystem services, and disservices. Condor 117:192–202
Bhutta ZA, Sommerfeld J, Lassi ZS, Salam RA, Das JK (2014) Global burden, distribution and interventions for infectious diseases of poverty. Infect Dis Poverty 3:21
Blanco J, Dendoncker N, Barnaud C, Sirami C (2019) Ecosystem disservices matter: towards their systematic integration within ecosystem service research and policy. Ecosys Serv 36:100913
Blunt SM (2008) Trees and pavements—are they compatible? Arboric J 31:73–80
Bombieri G, Naves J, Penteriani V, Selva N, Fernández-Gil A, López-Bao JV, Ambarli H, Bautista C, Bespalova T, Bobrov V, Bolshakov V (2019) Brown bear attacks on humans: a worldwide perspective. Sci Rep 9:1–10
Bosch-Cano F, Bernard N, Sudre B, Gillet F, Thibaudon M, Richard H, Badot PM, Ruffaldi P (2011) Human exposure to allergenic pollens: a comparison between urban and rural areas. Enviro Res 111:619–625
Boyd IL, Freer-Smith PH, Gilligan CA, Godfray HCJ (2013) The consequence of tree pests and diseases for ecosystem services. Sci 342:6160
Brun M, Di Pietro F, Bonthoux S (2018) Residents' perceptions and valuations of urban wastelands are influenced by vegetation structure. Urban For Urban Green 29:393–403
Burkhard B, Kroll F, Nedkov S, Müller F (2012) Mapping ecosystem service supply, demand and budgets. Ecol Indic 21:17–29
Cariñanos P, Calaza-Martínez P, O'Brien L, Calfapietra C (2017) The cost of greening: disservices of urban trees. In: Pearlmutter D, Calfapietra C, Samson R, O'Brien L, Ostoic SK (eds) The urban forest. Springer, Cham, pp 79–87
Cariñanos P, Grilo F, Pinho P, Casares-Porcel M, Branquinho C, Acil N, Andreucci MB, Anjos A, Bianco PM, Brini S, Calaza-Martínez P (2019) Estimation of the allergenic potential of urban trees and urban parks: towards the healthy design of urban green spaces of the future. Int J Enviro Res Public Heal 16:1357
Cariñanos P, Ruiz-Peñuela S, Valle AM, De la Guardia CD (2020) Assessing pollination disservices of urban street-trees: the case of London-plane tree (*Platanus x hispanica* Mill. ex Münchh). Sci Total Enviro 737: 139722
Charles KE, Linklater WL (2013) Dietary breadth as a predictor of potential native avian–human conflict in urban landscapes. Wildl Res 40:482–489
Chishaleshale M, Shackleton CM, Gambiza J, Gumbo D (2015) The prevalence of planning and management frameworks for trees and green spaces in urban areas of South Africa. Urban For Urban Green 14:817–825
Cilliers J, Cilliers S (2015) From green to gold: a South African example of valuing urban green spaces in some residential areas in Potchefstroom. Town Reg Plan 67:1–12
Cocks M, Alexander J, Mogano L, Vetter S (2016) Ways of belonging: meanings of "nature" among Xhosa-speaking township residents in South Africa. J Ethnobiol 36:820–841

Cocks ML, Shackleton CM (2021) Diversities of nature in cities: enriching belonging, wellbeing and bioculture. Earthscan/Routledge, London
Costello LR, McPherson EG, Burger DW, Perry EJ, Kelley D (2000) Strategies to reduce infrastructure damage by tree roots: a symposium for researchers and practitioners. In: Proceedings of a Symposium. Western Chapter, International Society of Arboriculture, Cohasset, CA
Cowett FD, Bassuk N (2017) Street tree diversity in three northeastern US states. Arboric Urb for 43:1–14
Cox DT, Hudson HL, Plummer KE, Siriwardena GM, Anderson K, Hancock S, Devine-Wright P, Gaston KJ (2018) Covariation in urban birds providing cultural services or disservices and people. J Appl Ecol 55:2308–2319
D'amato G (2000) Urban air pollution and plant-derived respiratory allergy. Clin Exp Allergy 30:628–636
Dalu MT, Shackleton CM (2018) The potential use of natural resources in urban informal settlements as substitutes for financial capital during flooding emergencies. Phys Chem Earth 104:18–27
Dalu MT, Shackleton CM, Dalu T (2018) Influence of land cover, proximity to streams and household topographical location on flooding impact in the Eastern Cape, South Africa. Intn J Disast Risk Red 28:481–490
Del Toro I, Ribbons RR, Pelini SL (2012) The little things that run the world revisited: a review of ant-mediated ecosystem services and disservices (Hymenoptera: Formicidae). Myrmecol News 17:133–146
Delshammar T, Östberg J, Öxell C (2015) Urban Trees and ecosystem disservices—a pilot study using complaints records from three Swedish cities. Arboric Urb For 41:187–193
Díaz S, Demissew S, Carabias J, Joly C, Lonsdale M, Ash N, Larigauderie A, Adhikari JR, Arico S, Báldi A, Bartuska A (2015) The IPBES conceptual framework: connecting nature and people. Curr Opin Enviro Sustain 14:1–16
Díaz S, Pascual U, Stenseke M, Martín-López B, Watson RT, Molnár Z, Hill R, Chan KM, Baste IA, Brauman KA, Polasky S, Church A, Lonsdale M, Larigauderie A, Leadley PW, Van Oudenhoven APE, Van der Plaat F, Schröter M, Lavorel S, Aumeeruddy-Thomas Y, Bukvareva E, Davies K, Demissew S, Erpul G, Failler P, Guerra CA, Hewitt CL, Keune H, Lindley S, Shirayama Y (2018) Assessing nature's contributions to people. Sci 359(6373):270–272
Dickie IA, Bennett BM, Burrows LE, Nunez MA, Peltzer DA, Porté A, Richardson DM, Rejmánek M, Rundel PW, Van Wilgen BW (2014) Conflicting values: ecosystem services and invasive tree management. Biol Inv 16:705–719
Dobbs C, Escobedo FJ, Zipperer WC (2011) A framework for developing urban forest ecosystem services and goods indicators. Landsc Urb Plan 99:196–206
Dobbs C, Kendal D, Nitschke CR (2014) Multiple ecosystem services and disservices of the urban forest establishing their connections with landscape structure and sociodemographics. Ecol Indic 43:44–55
Dobbs C, Vasquez A, Olave P, Olave M (2021) Cultural urban ecosystem services. In: Shackleton CM, Cilliers, SS, Davoren E, du Toit MJ (eds) Urban ecology in the Global South. Springer, Cham, pp 245–264
Donovan GH, Prestemon JP (2012) The effect of trees on crime in Portland, Oregon. Enviro Behav 44:3–30
Drescher AW, Isendahl C, Caridad Cruz M, Karg H, Menakanit A (2021) Urban and peri-urban agriculture in the Global South. In: Shackleton CM, Cilliers, SS, Davoren E, du Toit MJ (eds) Urban ecology in the Global South. Springer, Cham, pp 293–324
Dunn RR (2010) Global mapping of ecosystem disservices: the unspoken reality that nature sometimes kills us. Biotrop 42:555–557
EAACI (European Academy of Allergy and Clinical Immunology) (2015) European academy of allergy and clinical immunology (eaaci): advocacy manifesto tackling the allergy crisis in Europe—concerted policy action needed. http://www.eaaci.org/outreach/public-declarations.html. Accessed on 18 Aug 2020

Echeverri A, Naidoo R, Karp DS, Chan KM, Zhao J (2019) Iconic manakins and despicable grackles: comparing cultural ecosystem services and disservices across stakeholders in Costa Rica. Ecol Indic 106:105454
Elam J, Björdal C (2020) A review and case studies of factors affecting the stability of wooden foundation piles in urban environments exposed to construction work. Int Biodeter Biodegrad 148:104913
Escobedo FJ (2021) Understanding urban regulating ecosystem services in the Global South. In: Shackleton CM, Cilliers, SS, Davoren E, du Toit MJ (eds) Urban ecology in the Global South. Springer, Cham, pp 227–244
Escobedo FJ, Clerici N, Staudhammer CL, Feged-Rivadeneira A, Bohorquez JC, Tovar G (2018) Trees and crime in Bogota, Colombia: is the link an ecosystem disservice or service? Land Use Pol 78:583–592
Escobedo FJ, Kroeger T, Wagner JE (2011) Urban forests and pollution mitigation: analysing ecosystem services and disservices. Enviro Pollut 159:2078–2087
Froude MJ, Petley DN (2018) Global fatal landslide occurrence from 2004 to 2016. Nat Hazards Earth Syst Sci 18:2161–2181
Geneviève LD, Ray N, Chappuis F, Alcoba G, Mondardini MR, Bolon I, Ruiz de Castañeda R (2018) Participatory approaches and open data on venomous snakes: a neglected opportunity in the global snakebite crisis. PLOS Negl Trop Dis 12:e0006162
Guo T, Morgenroth J, Conway T (2019) To plant, remove, or retain: understanding property owner decisions about trees during redevelopment. Landsc Urb Plan 190:103601
Gwedla N, Shackleton CM (2019) Perceptions and preferences for urban trees across multiple socio-economic contexts in the Eastern Cape, South Africa. Landsc Urb Plan 189:225–234
Haines-Young R, Potschin M (2013) Common international classification of ecosystem services (CICES): consultation on version 4, August-December 2012. EEA Framework contract No EEA/IEA/09/003. www.cices.eu
Halloran A, Magid J (2013) Planning the unplanned: incorporating agriculture as an urban land use into the Dar es Salaam master plan and beyond. Enviro 25:541–558
Harwood TD, Tomlinson I, Potter CA, Knight JD (2011) Dutch elm disease revisited: past, present and future management in Great Britain. Plant Pathol 60:545–555
Herd-Hoare S (2019) The relative role of ecosystem services and disservices in rural livelihoods in the Eastern Cape province, South Africa. Masters thesis, Rhodes University, Grahamstown
Hill L, Hemery G, Hector A, Brown N (2019) Maintaining ecosystem properties after loss of ash in Great Britain. J Appl Ecol 56:282–293
Hofmann M, Westermann JR, Kowarik I, Van der Meer E (2012) Perceptions of parks and urban derelict land by landscape planners and residents. Urban For Urban Green 11:303–312
Jansson M, Fors H, Lindgren T, Wiström B (2013) Perceived personal safety in relation to urban woodland vegetation: a review. Urban For Urban Green (12):127–133
Jégh-Czinege N, Faragó T, Pongrácz P (2020) A bark of its own kind–the acoustics of 'annoying' dog barks suggests a specific attention-evoking effect for humans. Bioacoustics 29:210–225
Keswani K (2017) The practice of tree worship and the territorial production of urban space in the Indian neighbourhood. J Urb Des 22:370–387
Kirkpatrick JB, Davison A, Daniels GD (2012) Resident attitudes towards trees influence the planting and removal of different types of trees in eastern Australian cities. Landsc Urb Plan 107:147–158
Kuo FE, Sullivan WC (2001) Environment and crime in the inner city: does vegetation reduce crime? Enviro Behav 33:343–367
Kuo FE, Bacaicoa M, Sullivan WC (1998) Transforming inner-city landscapes: trees, sense of safety, and preference. Enviro Behav 30:28–59
Larson KL, Corley EA, Andrade R, Hall SJ, York AM, Meerow S, Coseo P, Childers DL, Hondula DM (2019) Subjective evaluations of ecosystem services and disservices. Ecol Soc 24(2)
Lee-Smith D (2010) Cities feeding people: an update on urban agriculture in equatorial Africa. Enviro Urb 22:483–499

Li XP, Fan SX, Kühn N, Dong L, Hao PY (2019) Residents' ecological and aesthetical perceptions toward spontaneous vegetation in urban parks in China. Urban For Urban Green 44:126397

Limburg KE, Luzadis VA, Ramsey M, Schulz KL, Mayer CM (2010) The good, the bad, and the algae: Perceiving ecosystem services and disservices generated by zebra and quagga mussels. J Great Lakes Res 36:86–92

Lozano R, Naghavi M, Foreman K, Lim S, Shibuya K (+188 authors) (2012) Global and regional mortality from 235 causes of death for 20 age groups in 1990 and 2010: a systematic analysis for the Global Burden of Disease Study 2010. Lancet 380(9859): 2095–2128

Lupindu AM, Olsen JE, Ngowi HA, Msoffe PL, Mtambo MM, Scheutz F, Dalsgaard A (2014) Occurrence and characterisation of shiga toxin-producing *Escherichia coli* O157: H7 and other non-sorbitol–fermenting *E. coli* in cattle and humans in urban areas of Morogoro, Tanzania. Vector Borne Zoonotic Dis 14:503–510

Lyytimäki J (2014) Bad nature: Newspaper representations of ecosystem disservices. Urban For Urban Green 13:418–424

Lyytimäki J (2015) Ecosystem disservices: embrace the catchword. Ecosys Serv 12:136

Lyytimäki J (2017) Disservices of urban trees. In: Ferrini F, Konijnendijk van den Bosch CC, Fini A (eds) Routledge handbook of urban forestry. Routledge, New York, pp 164–176

Lyytimäki J, Sipilä M (2009) Hopping on one leg–the challenge of ecosystem disservices for urban green management. Urban For Urban Green 8:309–315

Lyytimäki J, Petersen LK, Normander B, Bezák P (2008) Nature as a nuisance? Ecosystem services and disservices to urban lifestyle. Enviro Sci 5:161–172

Maes J, Teller A, Erhard M, Liquete C, Braat L, Berry P, Egoh B, Puydarrieux P,Fiorina C, Santos F, Paracchini ML, Keune H, Wittmer H, Hauck J, Fiala I, Verburg PH, Condé S, Schägner JP, San Miguel J, Estreguil C, Ostermann O, Barredo JI, Pereira HM, Stott A, Laporte V, Meiner A, Olah B, Royo Gelabert E, Spyropoulou R, Petersen JE, Maguire C, Zal N, Achilleos E, Rubin A, Ledoux L, Brown C, Raes C, Jacobs S, Vandewalle M, Connor D, Bidoglio G (2013) Mapping and assessment of ecosystems and their services. An analytical framework for ecosystem assessments under action 5 of the EU biodiversity strategy to 2020. Publications Office of the European Union, Luxembourg

Mathey J, Arndt T, Banse J, Rink D (2018) Public perception of spontaneous vegetation on brownfields in urban areas: results from surveys in Dresden and Leipzig (Germany). Urban For Urban Green 29:384–392

McClintock N, Pallana E, Wooten H (2014) Urban livestock ownership, management, and regulation in the United States: An exploratory survey and research agenda. Land Use Pol 38:426–440

McLellan V, Shackleton CM (2019) The relative representation of ecosystem services and disservices in South African newspaper media. Ecosys People 15:247–256

McPherson EG (2000) Expenditures associated with conflicts between street tree root growth and hardscape in California, United States. J Arboric 26:289–297

McPherson EG, Simpson JR, Peper PJ, Maco SE, Gardner SL, Cozad SK, Xiao Q (2006) Midwest community tree guide: benefits, costs, and strategic planting. Gen. Tech. Rep. PSW-GTR-199. Department of Agriculture, Forest Service, Pacific Southwest Research Station, Albany, CA

Mesa A, Arboit M, De Rosa C (2010) Solar obstruction assessment model for densely forested urban environments. Archit Sci Rev 53:224–237

Millennium Ecosystem Assessment (2005) Ecosystems and human wellbeing: synthesis. Island Press, Washington, DC

Moro PA, Assisi F, Cassetti F, Bissoli M, Borghini R, Davanzo F, Della Puppa T, Dimasi V, Ferruzzi M, Giarratana T, Travaglia A (2009) Toxicological hazards of natural environments: clinical reports from Poison Control Centre of Milan. Urban For Urban Green 8:179–186

Mouratidis K (2019) The impact of urban tree cover on perceived safety. Urban For Urban Green 44:126434

Mullaney J, Lucke T, Trueman SJ (2015) A review of benefits and challenges in growing street trees in paved urban environments. Landsc Urb Plan 134:157–166

Nasar JL, Fisher B (1993) 'Hot spots' of fear and crime: a multi-method investigation. J Enviro Psychol 13:187–206
Nasar JL, Fisher B, Grannis M (1993) Proximate physical cues to fear of crime. Landsc Urb Plan 26:161–178
Nowak DJ, Crane DE, Stevens JC, Hoehn RE, Walton JT, Bond J (2008) A ground-based method of assessing urban forest structure and ecosystem services. Arboric Urb for 34:347–358
Ngulani T, Shackleton CM (2019) Use of public urban green spaces for spiritual services in Bulawayo, Zimbabwe. Urban For Urban Green 38:97–104
O'Farrell PJ, Donaldson JS, Hoffman MT (2007) The influence of ecosystem goods and services on livestock management practices on the Bokkeveld plateau, South Africa. Agric Ecosys Enviro 122:312–324
O'Hare G, Rivas S (2005) The landslide hazard and human vulnerability in La Paz, Bolivia. Geog J 171:239–258
Östberg J, Martinsson M, Stål Ö, Fransson AM (2012) Risk of root intrusion by tree and shrub species into sewer pipes in Swedish urban areas. Urban For Urban Green 11:65–71
Ostoić SK, Van den Bosch CCK, Vuletić D, Stevanov M, Živojinović I, Mutabdžija-Bećirović S, Lazarević J, Stojanova B, Blagojević D, Stojanovska M, Nevenić R (2017) Citizens' perception of and satisfaction with urban forests and green space: results from selected Southeast European cities. Urban For Urban Green 23:93–103
Paap T, Burgess TI, Wingfield MJ (2017) Urban trees: bridge-heads for forest pest invasions and sentinels for early detection. Biol Inv 19:3515–3526
Paap T, De Beer ZW, Migliorini D, Nel WJ, Wingfield MJ (2018) The polyphagous shot hole borer (PSHB) and its fungal symbiont Fusarium euwallaceae: a new invasion in South Africa. Austral Plant Pathol 47:231–237
Pataki DE, Carreiro MM, Cherrier J, Grulke NE, Jennings V, Pincetl S, Pouyat RV, Whitlow TH, Zipperer WC (2011) Coupling biogeochemical cycles in urban environments: ecosystem services, green solutions, and misconceptions. Front Ecol Enviro 9:27–36
Pauleit S, Vasquéz A, Maruthaveeran S, Liu L, Cilliers SS (2021) Urban green infrastructure in the Global South. In: Shackleton CM, Cilliers, SS, Davoren E, du Toit MJ (eds) Urban ecology in the Global South. Springer, Cham, pp 107–143
Pawankar R, Canonica GW, Holgate ST, Lockey RF, Blaiss M (eds) (2013)World Allergy Organisation (WAO) White book on allergy. World Allergy Organisation, Wisconsin
Peñuelas J, Llusià J (2003) BVOCs: plant defence against climate warming? Trends Plant Sci 8:105–109
Pires AF, Peterson A, Baron JN, Adams R, Martínez-López B, Moore D (2019) Small-scale and backyard livestock owners needs assessment in the western United States. PLoS ONE 14:0212372
Plieninger T, Dijks S, Oteros-Rozas E, Bieling C (2013) Assessing, mapping, and quantifying cultural ecosystem services at community level. Land Use Pol 33:118–129
Potgieter LJ, Gaertner M, Kueffer C, Larson BM, Livingstone SW, O'Farrell PJ, Richardson DM (2017) Alien plants as mediators of ecosystem services and disservices in urban systems: a global review. Biol Inv 19:3571–3588
Potgieter LJ, Gaertner M, O'Farrell PJ, Richardson DM (2019a) A fine-scale assessment of the ecosystem service-disservice dichotomy in the context of urban ecosystems affected by alien plant invasions. For Ecosys 6:46
Potgieter LJ, Gaertner M, O'Farrell PJ, Richardson DM (2019b) Perceptions of impact: invasive alien plants in the urban environment. J Enviro Mgmt 229:76–87
Randrup TB, McPherson EG, Costello LR (2001) Tree root intrusion in sewer systems: review of extent and costs. J Infrastruct Syst 7:26–31
Richardson E, Shackleton CM (2014) The extent and perceptions of vandalism as a cause of street tree damage in small towns in the Eastern Cape, South Africa. Urban For Urban Green 13:425–432
Riley CB, Perry KI, Ard K, Gardiner MM (2018) Asset or liability? Ecological and sociological tradeoffs of urban spontaneous vegetation on vacant land in shrinking cities. Sustain 10:2139

Roman LA, Conway TM, Eisenman TS, Koeser AK, Ordóñez Barona C, Locke DH, Jenerette GD, Östberg J, Vogt J (2021) Beyond "trees are good": disservices, management costs, and tradeoffs in urban forestry. Ambio 50:615–630
Roy S, Byrne J, Pickering C (2012) A systematic quantitative review of urban tree benefits, costs, and assessment methods across cities in different climatic zones. Urban For Urban Green 11:351–363
Sandbrook CG, Burgess ND (2015) Biodiversity and ecosystem services: not all positive. Ecosys Serv 12:29
Saunders ME, Luck GW (2016) Limitations of the ecosystem services versus disservices dichotomy. Conserv Biol 30:1363–1365
Schaubroeck T (2017) A need for equal consideration of ecosystem disservices and services when valuing nature; countering arguments against disservices. Ecosys Serv 26:95–97
Shackleton CM (2012) Is there no urban forestry in the developing world? Sci Res Essays 7:3329–3335
Shackleton CM (2021) Ecosystem provisioning services in Global South cities. In: Shackleton CM, Cilliers, SS, Davoren E, du Toit MJ (eds) Urban ecology in the Global South. Springer, Cham, pp 203–226
Shackleton CM, Cilliers SS, du Toit MJ, Davoren E (2021) The need for an urban ecology of the Global South. In: Shackleton CM, Cilliers, SS, Davoren E, du Toit MJ (eds). Urban ecology in the Global South. Springer, Cham, pp 1–26
Shackleton CM, Cocks ML (2021) Urban nature and biocultural realities. In: Cocks ML, Shackleton CM (eds) Diversities of nature in cities: enriching belonging, wellbeing and bioculture. Earthscan/Routledge, London, pp 1–24
Shackleton CM, Mograbi PJ (2020) Meeting a diversity of needs through a diversity of species: urban residents' favourite and disliked tree species across eleven towns in South Africa and Zimbabwe. Urban For Urban Green 48:126507
Shackleton CM, Njwaxu A (2021) Does the absence of community involvement underpin the demise of urban neighbourhood parks in the Eastern Cape, South Africa? Landsc Urb Plan 207: 104006
Shackleton CM, Shackleton RT (2016) Knowledge, perceptions and willingness to control designated invasive tree species in urban household gardens in South Africa. Biol Inv 18:1599–1609
Shackleton CM, Guild J, Bromham B, Impey S, Jarrett M, Ngubane S, Steijl K (2017) How compatible are urban livestock and urban green spaces and trees? An assessment in a medium-sized South African town. Int J Urb Sustain Dev 9:243–252
Shackleton CM, Ruwanza S, Sanni GS, Bennett S, De Lacy P, Modipa R, Mtati N, Sachikonye M, Thondhlana G (2016) Unpacking Pandora's box: understanding and categorising ecosystem disservices for environmental management and human wellbeing. Ecosys 19:587–600
Shackleton S, Chinyimba A, Hebinck P, Shackleton C, Kaoma H (2015) Multiple benefits and values of trees in urban landscapes in two towns in northern South Africa. Landsc Urb Plan 136:76–86
Shapiro J, Báldi A (2014) Accurate accounting: how to balance ecosystem services and disservices. Ecosys Serv 7:201–202
Sousa-Silva R, Smargiassi A, Paquette A, Kaiser D, Kneeshaw D (2020) Exactly what do we know about tree pollen allergenicity? Lancet Respir Med 8:10
Speak A, Escobedo FJ, Russo A, Zerbe S (2018) An ecosystem service-disservice ratio: using composite indicators to assess the net benefits of urban trees. Ecol Indic 95:544–553
Sreetheran M, Van Den Bosch CCK (2014) A socio-ecological exploration of fear of crime in urban green spaces–a systematic review. Urban For Urban Green 13:1–18
Sreetheran M, Van den Bosch CCK (2015) Fear of crime in urban green spaces–what the residents of Kuala Lumpur have to say? Urban For Urban Green 14:702–713
Ström G, Albihn A, Jinnerot T, Boqvist S, Andersson-Djurfeldt A, Sokerya S, Osbjer K, San S, Davun H, Magnusson U (2018) Manure management and public health: Sanitary and socio-economic aspects among urban livestock-keepers in Cambodia. Sci Total Enviro 621:193–200
Suchocka M, Jankowski P, Błaszczyk M (2019) Perception of urban trees by polish tree professionals vs. nonprofessionals. Sustain 11:211

TEEB (2010) The economics of ecosystems and biodiversity: ecological and economic foundation. Earthscan, London

Teixeira FZ, Bachi L, Blanco J, Zimmermann I, Welle I, Carvalho-Ribeiro SM (2019) Perceived ecosystem services (ES) and ecosystem disservices (EDS) from trees: insights from three case studies in Brazil and France. Landsc Ecol 34:1583–1600

Tiwary A, Williams ID, Heidrich O, Namdeo A, Bandaru V, Calfapietra C (2016) Development of multi-functional streetscape green infrastructure using a performance index approach. Enviro Pollut 208:209–220

Tobin MR, Goldshear JL, Price LB, Graham JP, Leibler JH (2015) A framework to reduce infectious disease risk from urban poultry in the United States. Public Heal Rep 130:380–391

Torres MN, Rodríguez JP, Leitão JP (2017) Geostatistical analysis to identify characteristics involved in sewer pipes and urban tree interactions. Urban For Urban Green 25:36–42

Troy A, Grove JM, O'Neil-Dunne J (2012) The relationship between tree canopy and crime rates across an urban–rural gradient in the greater Baltimore region. Landsc Urb Plan 106:262–270

Troy A, Nunery A, Grove JM (2016) The relationship between residential yard management and neighborhood crime: an analysis from Baltimore City and County. Landsc Urb Plan 147:78–87

Tsopelas P, Santini A, Wingfield MJ, Wilhelm de Beer Z (2017) Canker stain: a lethal disease destroying iconic plane trees. Plant Dis 101:645–658

Tzoulas K, Korpela K, Venn S, Yli-Pelkonen V, Kaźmierczak A, Niemela J, James P (2007) Promoting ecosystem and human health in urban areas using Green Infrastructure: a literature review. Landsc Urb Plan 81:167–178

UNICEF (2004) Fact sheet: Malaria—a global crisis. https://www.unicef.org/media/media_20475.html. Accessed Feb 2020

Urquhart J, Potter C, Barnett J, Fellenor J, Mumford J, Quine CP (2017) Expert risk perceptions and the social amplification of risk: a case study in invasive tree pests and diseases. Enviro Sci Pol 77:172–178

Vaz AS, Kueffer C, Kull CA, Richardson DM, Vicente JR, Kühn I, Schröter M, Hauck J, Bonn A, Honrado JP (2017) Integrating ecosystem services and disservices: insights from plant invasions. Ecosys Serv 23:94–107

Velasco-Jiménez MJ, Alcázar P, Cariñanos P, Galán C (2020) Allergenicity of the urban green areas in the city of Córdoba (Spain). Urban For Urban Green 49:126600

Venter ZS, Shackleton CM, Van Staden F, Selomane O, Masterson VA (2020) Green Apartheid: Urban green infrastructure remains unequally distributed across income and race geographies in South Africa. Landsc Urb Plan 203:103889

Villa F, Bagstad KJ, Voigt B, Johnson GW, Athanasiadis IN, Balbi S (2014) The misconception of ecosystem disservices: how a catchy term may yield the wrong messages for science and society. Ecosys Serv 10:52–53

von Döhren P, Haase D (2015) Ecosystem disservices research: a review of the state of the art with a focus on cities. Ecol Indic 52:490–497

von Döhren P, Haase D (2019) Risk assessment concerning urban ecosystem disservices: the example of street trees in Berlin. Germany. Ecosys Serv 40:101031

Wang H-F, Qureshi S, Knapp S, Friedman CR, Hubacek K (2015) A basic assessment of residential plant diversity and its ecosystem services and disservices in Beijing, China. Appl Geog 64:121–131

Wenban-Smith H, Fasse A, Grote U (2016) Food security in Tanzania: the challenge of rapid urbanisation. Food Sec 8:973–984

Whelan CJ, Şekercioğlu ÇH, Wenny DG (2015) Why birds matter: from economic ornithology to ecosystem services. J Ornithol 156:227–238

Wolfe MK, Mennis J (2012) Does vegetation encourage or suppress urban crime? evidence from Philadelphia, PA. Landsc Urb Plan 108:112–122

WHO (World Health Organisation) (2010) WHO guidelines for the production control and regulations of snake antivenom immunoglobulins. WHO, Geneva

Zhang W, Ricketts TH, Kremen C, Carney K, Swinton SM (2007) Ecosystem services and disservices to agriculture. Ecol Econ 64:253–260
Zoeller KC, Gurney GG, Heydinger J, Cumming GS (2020) Defining cultural functional groups based on perceived traits assigned to birds. Ecosys Serv 44:101138
Zuberbier T, Lötvall J, Simoens S, Subramanian SV, Church MK (2014) Economic burden of inadequate management of allergic diseases in the European Union: a GA2LEN review. Allergy 69:1275–1279

Urban and Peri-Urban Agriculture in the Global South

Axel W. Drescher, Christian Isendahl, María Caridad Cruz, Hanna Karg, and Alisara Menakanit

Abstract In the contemporary world, modernist Western thought supports stereotypical perceptions of cities (innovative, dynamic, progressive) as fundamentally different from the countryside (traditional, static, conservative). This leads to the effect of making near-residential agriculture obsolete; made redundant by transport technologies and an agricultural industry that rapidly moves large quantities of foodstuff from one region to another. However, the reality of urban food production in most countries of the Global South tells another story. This chapter uses recent and Pre-Columbian Maya case studies from four different regions of the world (Africa, South-East Asia, Middle America, and the Caribbean) to illustrate the importance of urban and peri-urban agriculture and food and nutrition security as fundamental aspects of the social-ecological resilience of cities. The chapter elaborates the differences of urban and peri-urban agriculture in the Global South and the Global North, including institutional support and regulations, gender roles, the importance for food and nutrition security (case studies on Cuba and Thailand), livelihoods, ecosystem services, urban ecology (case studies on Burkina Faso and Ghana), and the role of urban food commons (case study on Cuba). This demonstrates that food production is not "the antithesis of the city", but an urban activity that contributes to the resilience of cities.

A. W. Drescher (✉)
Department of Geography, University of Erlangen, Nuremberg, Germany
e-mail: axel.drescher@fau.de

C. Isendahl
Department of Historical Studies, University of Gothenburg, Gothenburg, Sweden
e-mail: christian.isendahl@gu.se

M. C. Cruz
Fundación Antonio Núñez Jiménez de la Naturaleza y el Hombre (FANJ), Havana, Cuba
e-mail: cary.cruz@fanj.cult.cu

H. Karg
Organic Plant Production and Agroecosystems Research in the Tropics and Subtropics, University of Kassel, Witzenhausen, Germany
e-mail: hanna.karg@gmx.de

A. Menakanit
Department of Horticulture, Kasetsart University, Bangkok, Thailand

C. M. Shackleton et al. (eds.), *Urban Ecology in the Global South*, Cities and Nature,
https://doi.org/10.1007/978-3-030-67650-6_12

Keywords Africa · South-East Asia · Latin America · Food and Nutrition Security · Urban and peri-urban agriculture · Urban resilience

1 Introduction

Food and nutrition security (FNS) are fundamental aspects of the social-ecological resilience of cities (Barthel and Isendahl 2013). Inclusively defined as the situation when "all people at all times have physical, social and economic access to food, which is safe and consumed in sufficient quantity and quality to meet their dietary needs and food preferences, and is supported by an environment of adequate sanitation, health services and care, allowing for a healthy and active life" (CFS 2012: 8), FNS are essentially timeless challenges for humanity (Isendahl and Barthel 2018; Vorster et al. 2011). Since the emergence of our species, human ingenuity, cooperation, and adaptation have generated a multitude of food systems in highly diverse sociol-ecological contexts, at different levels of complexity, and with varying degrees of efficacy to generate FNS. In general, contemporary urban food systems are the most complex, with the complexity of these systems absorbing unprecedented costs to maintain functions (Tainter and Allen 2019). These systems involve global networks of producers, distributors, and consumers; require vast land resources, high financial investment, high inputs, and advanced production and processing technologies for high output production; and depend on large-scale transport systems (Barthel et al. 2019). Early urbanisation is intimately linked to the emergence of agriculture (including both plant cultivation and rearing domesticated animals), and with strong spatial overlap: relatively high regional food production potential was a basic requirement for an evolving urban system. Indeed, the basic metabolism of cities and towns, past and present, is fundamentally sustained by agriculture (Graham and Isendahl 2018), and the global record of archaeological, historical, and observational data demonstrates considerable spatial diversity and temporal variation in how the basic urban metabolism of sustenance, i.e. feeding an urban population, is organised, and how these systems play out socially, economically, and environmentally.

In the contemporary world, modernist Western thought supports a stereotypical perception of cities as innovative, dynamic, and progressive and thus fundamentally different from the countryside which is depicted as traditional, static, and conservative (Isendahl 2012a; Isendahl and Barthel 2018). The conceptualisation of cities as bounded entities essentially different and detached from rural areas is particularly associated with the growth of the industrialised world; the technological innovations, socio-economic transformations, and emerging geopolitical asymmetry that embody "time-space compression" (Harvey 1990), particularly in the Global North (GN), and that increasingly contours a globalised world food order (Zeunert and Waterman 2018). The highly normative appreciation of cities as autonomous social systems that underpinned early twentieth-century urban planning in these regions (Eisenstadt 1966) simply found near-residential agriculture obsolete; purportedly made redundant by efficient transport technologies and an agricultural industry that rapidly moves large quantities of foodstuff from one world region to another. The progressive growth of the city outsourced agriculture, with the neoliberalisation of the food industry turning the place-based sociocultural practice of agriculture into faceless commodity production. Today, city models globally seek to erase the cultivation of food from the local and regional framework. In this "evolution", many urban populations have acquired a "citadine culture", a perception that increasingly distances them from the sources of their sustenance, from their innate capacity for creation, from local traditions, from their link with nature and green spaces, and from the very relations between human beings. This makes FNS of urban dwellers in the Global South (GS) particularly vulnerable to insufficient income, unhealthy food, and extreme events such as heavy storms, flooding, droughts, and pandemics. Although there are many benefits of global food trade, there is little doubt that there are severe trade-offs for livelihoods, long-term social-ecological sustainability, and equity (Barthel et al. 2019). Furthermore, from a long-term and comparative vantage point on global urban history (i.e. at time-scales up to several millennia and including the oft-neglected pre-modern cities of the GS), long-distance provisioning of bulk food goods is largely an anomaly (cf. Wallerstein 1974). Indeed, urban forms, ecologies, and foodways have interacted along a range of different developmental paths, generating considerable diversity.

The profit-making motivation of food production has turned agriculture into a practice of escalating habitat destruction and is directly responsible for global trends of deforestation, desertification, loss of biodiversity, soil degradation, and water pollution. Particularly since the green revolution of the 1960s onwards, the means and processes of the agro-industry is increasingly transforming agriculture towards monocultures, fossil-fuel dependence, mechanisation and use of sophisticated technologies, the application of synthetic fertilisers and insecticides, and a focus on food export and import rather than locally self-supporting systems (e.g. Barthel et al. 2019). Genetically modified organisms, the latest technological blanket solution to "end world hunger", have already demonstrated significant social and ecological trade-offs and costs (e.g. Brush 2001; Lappé and Collins 1979; Orzan et al. 2018). These trends notwithstanding, only 20% of the world's food is produced and marketed by agro-industry (FAO 2014a), and household-based smallholder farming remains indispensable for FNS and for lowering habitat destruction. Smallholder farming is particularly resourceful and valuable to support urban FNS and social-ecological sustainability when located in proximity to population centres.

Ultimately, FNS are inseparable from water and energy security, and the broad series of challenges and opportunities the food–water–energy nexus presents interact dynamically with urban form. How we recognise and address these are intimately linked to what kinds of cities we aspire to. How do we respond to calls for sustainable or resilient cities? Do ideas of the "intelligent city" and technological fetishing of digital "smart solutions" map out viable paths (for critiques see Colding et al. 2020; Hornborg 2001)? A basic counterargument is that it is the industrial–technological complex of, in particular, capitalism, that has led to the social and ecological degradation of the contemporary world in the first place.

This chapter discusses how different practices of urban and peri-urban agriculture (UPA) contribute to FNS in cities and, consequently, to urban resilience (Barthel and Isendahl 2013; Sitas et al. 2021). UPA in the GS is different from the GN in many respects, including institutional support and regulations, gender roles, the importance for FNS (see case studies on Cuba and Thailand below), livelihoods, ecosystem services, urban ecology (see case studies on Burkina Faso and Ghana below), and the role of urban food commons (see case study on Cuba below) (see also WinklerPrins 2017).

In the GN, UPA is generally well regulated, institutionalised, and recognised by the authorities (Drechsel and Dongus 2010). In the GS, however, legal frameworks for UPA are either non-existent or contradictory and overly complex (Baudoin and Drescher 2008). Although different forms of UPA have been a more common feature in the deeper history of cities globally than normally acknowledged, in the GS perhaps more so than in the GN, the historical roots of currently dominant forms in the GN (allotment and community gardens) are economic crisis (Binns and Nel 2014), military conflict, rapid urbanisation, and poverty during industrialisation. In more recent years, UPA in the GN is triggered by new movements like Alternative Food Networks, Guerrilla Gardening, Edible Cities, etc., that are closely related to the Green Cities Movement and a critique of the globalised food system, unsustainable farming practices, and socio-environmental injustices (e.g. Tornaghi 2014; WinklerPrins 2017). The Milan Urban Food Policy Pact (2015) and numerous Community Supported Agriculture (CSA) initiatives contribute to the popularisation of Alternative Food Networks activities. One goal is to achieve "short distance agriculture", which reduces food miles, is locally anchored, and socially and ecologically sound. These needs are as relevant in the GS. A major difference, however, is that informal food networks often already exist between cities and peri-urban peripheries in the latter, for instance in Sub-Saharan Africa (e.g. Haysom 2016; Skinner 2016). These networks embed traditional or indigenous agroecological knowledge, local food production, closely integrated producers and consumers, short, local value chains, minimal detours via middle-men distributors or outsourced food processing, high production and market transparency, and promote social interaction, reciprocity, social empowerment (Moustier and Renting 2015), participation, and civic engagement. Urban poverty groups and other vulnerable actors navigate through their "foodscapes" with creativity and a variety of adaptive solutions (Battersby 2012). Drechsel and Dongus (2010: 77) show that UPA is a "dynamic, viable and largely sustainable bright spot providing jobs and food for the cities" in Sub-Saharan Africa, as Smit et al. (1996) had suggested on a global level in their book "Urban Agriculture: Food, Jobs and Sustainable Cities". However, in the GS, Sub-Saharan Africa in particular, UPA is commonly perceived as a set of "temporary" and "transitional" poverty-related practices and remain understudied, insufficiently acknowledged in urban policy, and their capacity to strengthen urban resilience undervalued (e.g. Smart 2015).

Furthermore, there is a strong gender-perspective to UPA in the GS. It is estimated that 65% of urban farmers worldwide are women (van Veenhuizen 2006: 13). However, in contrast to the GN, the GS UPA is strongly triggered by the need to secure household food and income. Hence, in the GS, women dominate UPA small-scale subsistence production in most countries, and are key players in marketing fresh vegetables and other foods (de Neergaard et al. 2009; Hovorka and Lee-Smith 2006, see Thailand case study below).

Another significant difference between the GS and the GN is the extent of and reliance on livestock production and products within the urban limits (Shackleton et al. 2021), housed in backyards, and fed from undeveloped, open, and interstitial green spaces (Lee-Smith 2010). The keeping of livestock fulfils a variety of needs that vary between households and countries: food and nutrition, savings and investment, cultural and spiritual beliefs, draught and transport, and income generation. For example, Davenport et al. (2012) found that, on average, 28% of household cash and non-cash income was provided by livestock products to urban households in three towns in South Africa. The forms of livestock can range from poultry (e.g. pigeons, chickens, ducks, geese) to small stock (e.g. sheep, pigs) to large stock such as cattle, horses, or camels. The extent of livestock keeping (across all forms) in specific cities has rarely been enumerated, but some illustrative examples include Halloran and Magid (2013) who reported that over 70% of households in Dar es Salam (Tanzania) had livestock, Mundoli et al. (2015) who found that 30% of households in a part of Bangalore (India) did, and Shackleton et al. (2020) who reported that a mean of 25% of urban and 38% of peri-urban households across six secondary towns in five African countries kept one or more types of livestock. The keeping of livestock can have significant effects on urban ecology "in" the city through their need for food and water, and in turn the production of faeces that needs to be used or disposed. Food may be provided via imported feedstock, the cutting and carrying of fodder from urban green spaces, or letting the livestock range within the neighbourhood. The cutting and free-roaming strategies add a novel driver influencing plant species composition, diversity, and productivity in urban green spaces, as does the addition of animal dung, and the effects of trampling, none of which have been deeply considered in urban ecology studies in the GS (Shackleton et al. 2017). They also have a role in the urban ecology "of" the city through the interplay of the ecosystem services and disservices they provide, where and to who, how they are perceived by non-owners, and how they contribute to household food security, income, and culture, and hence resilience.

Drawing on a series of case studies, mostly concerning arable agriculture, from different spatio-temporal contexts, we argue that UPA is a salient feature of urban ecologies and metabolism when viewed from long-term and global vantage points, particularly in the GS. Hence, food production is not "the antithesis of the city", neither spatially nor systemically, but an integrated urban activity that contributes to the resilience of cities. Archaeological and historical data show how foodways influence urban form (e.g. Isendahl et al. 2020; Waterman 2018) and how the unintended consequences of urban planning and urban behaviours play out as causal relationships over the long term to increase or mitigate vulnerabilities to food and nutrition insecurity. This emphasises the important role that cross-cultural studies that put urban behaviours and strategies in the GS on equal footing with those in the GN can play to build a more complete frame of reference of cities. In particular, data from the GS strongly point to the idea that "rural food producers support urban food consumers" is far too simplistic to characterise urban sustenance patterns and metabolism, as well as how people and food are entangled in production, consumption, and distribution networks in complex urban systems. Failing to recognise this diversity and these complexities ultimately marginalises many alternative urban ecologies and strategic options to build FNS in cities that are surfacing from the GS and from archaeological and historical research (Barthel and Isendahl 2013; Isendahl and Barthel 2018; Isendahl and Smith 2013; Isendahl et al. 2020).

2 UPA as a Long-Term Practice: The Pre-Columbian Maya Lowlands

Surveying archaeological and historical records of urban systems in the GS, the key role that UPA has played for food nutrient and energy provisioning stands out. Among a plethora of pre-Industrial cases are the Aztecs and Maya of Central America, Aksum and Great Zimbabwe in Sub-Saharan Africa, and the Indus and Khmer cities of southern Asia (Isendahl and Smith 2013; Isendahl et al. 2020). These cases demonstrate, first, the wide range of pre-Industrial UPA practices in the GS and, second, the importance of local accessibility and diversity of food sources for urban resilience.

A good case in point of an integrated urban ecology that contests essentialist and binary conceptualisations of urban food consumption myopically set against rural food production are the agro-urban landscapes of the pre-Columbian Maya, in present-day southern Mexico, Guatemala, Belize, and Honduras (Graham and Isendahl 2018; Isendahl 2012b). Maya state-level polities and cities emerged during the early first millennium B.C.E., with a record of pre-Columbian urban history spanning over more than two millennia that suggests cycles of regional growth, decline, and re-organisation over the course of the Preclassic (2000 B.C.E.–250 C.E.), Classic (250–1000 C.E.), and Postclassic periods (1000–1550 C.E.) (Chase and Scarborough 2014). The complex regional settlement pattern included numerous cities and towns of varying sizes and longevity, some that functioned over more than a millennium. Among these, Caracol in western Belize was one of the largest. Today, the archaeological remains of monumental buildings, houses, agricultural terraces, and an extensive road network cover about 200 km^2 of tropically forested terrain. When abandoned in 900 C.E. it had been occupied for over 1500 years, with a maximum population of about 100,000 in 650 C.E. (Chase and Chase 2016).

Regional and interregional trade was an important component of the political and, to a lesser extent, social economy of the Maya, the latter including long-distance distribution networks of salt (McKillop 2002; Staller and Carrasco 2010). However, lacking beasts of burden and wheeled transport the overland movement of goods was time- and energy-consuming relative to pre-Industrial transportation technology in Eurasia and Africa (trade routes along the coast and rivers were probably used for moving goods on boats, at least seasonally, but many regions lacked permanent navigable streams) (Barthel and Isendahl 2013; Isendahl and Smith 2013). Transport energy deficiency constrained long-distance bulk trade in staples (e.g. maize) and other foodstuffs cultivable largely throughout the Maya region, and stimulated near-residential agriculture in urban population centres. In a "typical" Classic period Maya city, the main components were (1) sectors of domestic household residential units, dispersing from (2) urban centres of civic-ceremonial and elite residential building complexes that provided economic, political, social, and religious urban functions. The urban "sprawl" of residential units consisted of urban farmsteads with associated space for food production, including household gardens, infields at various stages of succession, orchards, and other productive green areas, thereby generating an integrated spatial patterning of land-use and ecology of a type that has been variously described as "agro-urban landscapes" (Graham and Isendahl 2018; Isendahl 2012b), "forest gardens" (e.g. Ford and Nigh 2015), "garden cities" (e.g. Dunning and Beach 2010), and "green cities" (e.g. Graham 1999).

The Maya diet was mainly vegetarian, with maize the most salient crop, but food diets varied temporally, spatially, and among social groups, and included a diverse range of cultivated and harvested plants. The vegetarian diet was supplemented by animal protein, including semi-domesticated (e.g. turkey) as well as non-domesticated (e.g. deer, armadillo, birds, and fish) species (White 1999). In the pre-Columbian western hemisphere (with the exception of the Andean llama), independence from the grazing-animal complex (Graham 1999) and the lack of beasts of burden have implications for the emergence of Maya agro-urban landscapes, relative to elsewhere in the GS (Isendahl et al. 2020). First, unless transported by vessels, people carried everything. Hence, the potential comparative advantage of particularly fertile soils and hydrological conditions levelled out fast with transport distance. Second, working the land simply with the aid of a digging stick, plough traction did not constrain the selection of field plots. Measures of intensification involved terracing and wetland farming. Third, with no grazing livestock, household FNS focused spatially on the residential dwelling and associated farming resources, and with less competition for land. Fourth, emphasis on hunting and fishing for animal foods provided an incentive to sustainably manage these resources. Fifth, without access to concentrated animal manure, organic waste generated from urban living (e.g. charcoal, harvesting and cooking waste, and human excrements) were potentially valued resources in urban agriculture and horticulture.

The data from pre-Columbian Latin America shows how large cities could sustain urban metabolism by closing the nutrient cycle. This brings us to another, a more recent example from the region: Cuba.

3 Contemporary UPA in Cuba: From Crisis Response to Sustainability Transition?

Today, UPA is practised in many cities of Latin America and the Caribbean, initiated, promoted, and managed by local individuals and communities, governments, universities, churches, non-governmental organisations (NGOs), and others. Practices of UPA are highly diverse, situated in very different contexts in terms of political and economic systems, land tenure, history, culture, and environment, as well as the interest and engagement among urban dwellers to participate. Contemporary UPA in the region needs to be framed within the region's critical political and economic situation, population growth, increasing urbanisation, and extreme events (Mougeot 2000). Brought together, these factors increase the risk of food and nutrition insecurity in cities, thus providing an incentive for UPA initiatives. In general, the challenges to expand UPA include availability of water, organic matter, seeds, tools, and other essential inputs, as well as access to space for cultivation in heavy competition over land for other urban uses. The growth of UPA in Cuba over the last three decades is a particularly interesting case, with Havana often cited as a contemporary sustainable urban food system and a generic best practice model with global applicability on how to increase FNS in cities, particularly in the GS (e.g. Altieri et al. 1999; Clouse 2014; Koont 2011; Schumacher Centre 2014; Viljoen and Howe 2005).

Following the revolution in 1959, Cuban politics and economy took shape in the shadow of a hostile and powerful neighbour-state to the north. In the early 1960s, the United States imposed a blockade (still in effect) that largely limited Cuba's options for international trade partners to the Soviet Union and its allies. Following a period of economic downturn in the 1980s, the Cuban economy worsened disastrously with the 1989–1991 disintegration of the Soviet Union and the fall of communist regimes in Eastern Europe (Koont 2011). The loss of former allies and major trade partners had a significant impact on Cuban FNS, reducing food imports and decreasing the availability of essential inputs, fossil fuels in particular, and thus the ability to produce and transport food within the territory. This led to a period of crisis policy (the so-called *período especial*) characterised by poor state finances, large-scale malnutrition, and food and nutrition insecurity. The emerging crisis was publically announced on 27 December 1987 "when the Central Committee of Cuba's Communist Party called for action nationwide to promote intensive horticulture, using a technology known as 'organoponics'" (Thomas 2014: 10). Organoponics (Fig. 1; Box 1) are managed by horticultural labour employed by a government agency within the Ministry of Agriculture. At each site of organoponics there is a small shop or vending point, offering fresh vegetables and herbs grown without agrochemicals (which are prohibited in urban areas) directly to the inhabitants of surrounding neighbourhoods. In addition, food and nutrition insecurity triggered the massive growth of urban dwellers cultivating any available space in the cities (Cruz and Medina 2003).

Box 1: Organoponics

"With the onset of the período especial, organoponic gardens proved ideal for growing crops on poor soils in small urban spaces. A typical organoponic

Fig. 1 Typical organoponic in Alamar (Havana) (photo by C. Cruz)

garden is started by making furrows in the soil, then lining the rows with protective barriers of wood, stone, bricks or concrete. The soil quality is gradually improved through the incorporation of organic matter; as organic content increases, so do the levels of soil nutrients and moisture (and the height of the bed).

Organopónicos—the term applies to both the technology and the garden – can be applied on building sites, vacant lots and roadsides, and arranged in terraces on sloping land. Soil can be tailored, using specific mixtures, to specific crops" (Thomas 2014: 10).

The popular growth of UPA was soon supported by the lower levels of government, and in 1994 the Cuban government acknowledged UPA as an important strategy to combat urban food and nutrition insecurity, supporting its growth throughout the country (Cruz and Medina 2003, Drescher et al. 2000; Koont 2011; Novo and Murphy 2000; Santandreu et al. 2009). For instance, in the early 1990s, the local government of Havana (with a population of c. 2 million home to about 20% of the Cuban population; in total, 75% of the national population live in urban areas), organised information campaigns and public meetings, as well as used the media (radio and TV) to actively encourage the capital's population to cultivate all unoccupied spaces in the urban fabric, most of which are state property, to produce food for consumption, and assured access to technical support, seeds, and agricultural equipment. A couple of years later, orchards operated by the general public occupied more than 1500 hectares, and from 1989 to 2000, Cuba increased its fresh vegetable production by nearly 140% (Pérez-López 2002).

The National Urban Agriculture Programme was created in 1997 as a section of the Ministry of Agriculture to support the establishment of UPA in all municipalities in the country. Furthermore, in 2000, UPA was recognised as a key component of urban morphology to be taken into consideration in the Territorial Planning of all Cuban cities, thus not only emphasising the practical role of UPA to build FNS, but also strengthening the status of farming as an urban function of rank in the urban planning sector. Since 2011, land for UPA is included in the Cuban government's Economic and Social Policy Guidelines (Forero-Niño 2011). Thus, Cuba builds on the experience of mitigating food and nutrition insecurity in cities by UPA in a time of crisis to build strategies for long-term FNS, acknowledging the diverse additional benefits of UPA: community inclusion, local engagement, high productivity, low dependency on inputs, and low food mileage.

Three decades since the economic crisis began, thousands of people in Havana are now producing food in residential patios and urban field plots and on rooftops and balconies. Estimates suggest that between 300,000 and 400,000 Cubans are presently engaged in UPA (Gonzales-Corzo 2016), managing 50,000 hectares of productive urban space to produce more than 70% of the Cuban population's consumption of fresh vegetables and herbs (Altieri and Funes-Monzote 2012). Nevertheless, a conservative estimate indicates that forms of commercial agriculture (e.g. organoponics) located in Havana occupy less than 10% of the area that is used to produce food; with most food production taking place in about 95,000 backyards and plots of less than 800 m^2 between individual houses and low buildings that characterise the city (FAO 2014b). In 2014, in Cuba as a whole, there were almost 90,000 gardens (*patios*), 6400 intensive gardens, and 4000 high-yielding organoponics (FAO 2014b).

In Cuban UPA, the relationship between consumers and producers is direct, without intermediators and without transportation costs. Practices, particularly the organoponics, generate formal employment, taking advantage of the potential workforce of the city. UPA also favours inclusion of the most vulnerable sectors of the society such as elderly people (Cruz and Medina 2003).

Operating both from public and official engagement and serving the community with government support, the growth of Cuban UPA suggests the transformation of crisis-generated survival production into a functioning infrastructure for household, neighbourhood, and city self-sufficiency generated from bottom-up initiatives and top-down capacity-building. In pre-1989 Havana, UPA was almost non-existent. Hence, the 1990s saw the development of a practical knowledge-culture of UPA unrelated to earlier long-term traditions in the Latin American and Caribbean region's deeper pre-Columbian past.

There are, however, several challenges to a transition towards greater reliance on UPA for FNS. First, there is an unmet demand for composted organic material. The waste the city generates contains c. 60% organic matter, but since Havana's municipal waste management system does not separate different kinds of waste and a centralised system for collecting, composting, and distributing organic waste is lacking (Lloréns et al. 2008), it forms a vast but underutilised potential source. A particular challenge is the uneven availability of fuel and spare parts for servicing garbage trucks, which contributes to leaving mixed waste dumps in the urban landscape. Second, rainwater harvesting needs to be developed to reduce pressure on public water distribution systems. Third, prices of products at selling points are relatively high in relation to a household's average income. Fourth, although space for UPA needs to be included in urban planning, competition for urban space (e.g. high demand to provide housing and to ensure the growth of the tourism industry important to the state economy) limits opportunities to expand, in particular, the state-owned organoponics. This indicates that it is necessary to first develop strategies to balance the allocation of urban space for "traditional" urban functions (infrastructure, housing, public buildings, businesses, etc.) with UPA, and second, to intensify the promotion of and support to small household-based or collective communitarian production units. Recognising the value of maintaining healthy soils in the urban and peri-urban landscape is an important supportive step to safeguard the permanent coexistence of UPA with other urban land-uses.

Since 1993, NGOs have become increasingly involved in the growth of UPA in Havana as well as in other Cuban cities. The Havana-based Fundación Antonio Núñez Jiménez de la Naturaleza y el Hombre (FANJ) is a key NGO that aims to contribute to a sustainability transition by tutoring urban and peri-urban farmers in permaculture. Developed by Australians Bill Mollison and David Holmgren in the mid-1970s, permaculture systems, whether in urban or rural environments, aim to mimic the diversity, stability, productivity, and resilience of natural ecosystems, and, by needing very low energy inputs, have a high degree of circularity and self-sufficiency (Mollison and Holmgren 1990). The ethical and design principles of permaculture involve holistic thinking and local action. At the same time, the permaculture approach is quite flexible as it permits each system to be designed in a unique way, drawing on local conditions and individual preference and knowledge (Fig. 2).

FANJ set up the permaculture training programme in Havana in 1994 with the purpose to support ecological and organic household- and community-based UPA. From 2000 onwards, capacity-building began to scale-up beyond the capital towards the national level, taking advantage of the competencies developed by previous participants in the training programme. FANJ is now assisted by a voluntary network of permaculture promoters, as well as of other NGOs and local or national government agencies. At present, there is a collaborative network of hundreds of people, organised in 28 community groups across the country (five in Havana), exchanging experiences, practices, and seeds, based on local resources and acquired and shared knowledge of practice. An infrastructure of support facilities includes training programmes, access to technical and horticultural expert advice, and seed-producing farms (Thomas 2014). An important part of permaculture training programmes is to raise awareness of the urgency to modify lifestyles that contribute to habitat exhaustion, to develop a sense of assuming responsibility and commitment to solve local problems, and the importance to gain the social, economic, and ecological capabilities that permaculture generates.

The growth of UPA in Cuba demonstrates its contribution to FNS over short- to mid-term crises (i.e. here at a time-scale of a few years to about a generation).

Fig. 2 Permaculture farm in Havana (photo by A. Drescher)

In addition to addressing issues of FNS, UPA production and marketing systems have the potential to contribute to poverty reduction, local economic and community development, social inclusion of marginalised groups, as well as to enhance urban environmental management by increasing biodiversity, soil health, and the productive reuse of organic waste. However, implementing and safeguarding the production of food in the cities of the GS faces many challenges. Ongoing efforts to support UPA as a practice that functions over the longer term (i.e. here, several generations), as it did in the pre-Columbian cities of the Maya lowlands, indicate the complexity of factors influencing urban food systems. Often, the complexities of UPA is not fully understood by regional and urban planners, city administrators, policymakers, researchers, and communities, hence constraining its potential in a sustainability transition of urban and peri-urban areas in the GS. In particular, competition and conflicting interests in the use of urban space is a major constraint that not only limits the potential growth of UPA but also threatens existing production sites. The nature and efficacy of governance to address such conflicts is of central concern. The mega-city of Thailand's capital Bangkok is an interesting example of governance conflicts as well as of the local opportunities that have been created by institutional confusions and disconnections, as discussed in the following section.

4 UPA in Bangkok: Managing an Urban Food System

Elucidating challenges of governance, we draw on recent research in Bangkok, Thailand, to detail the different components of the local urban food system (producers,

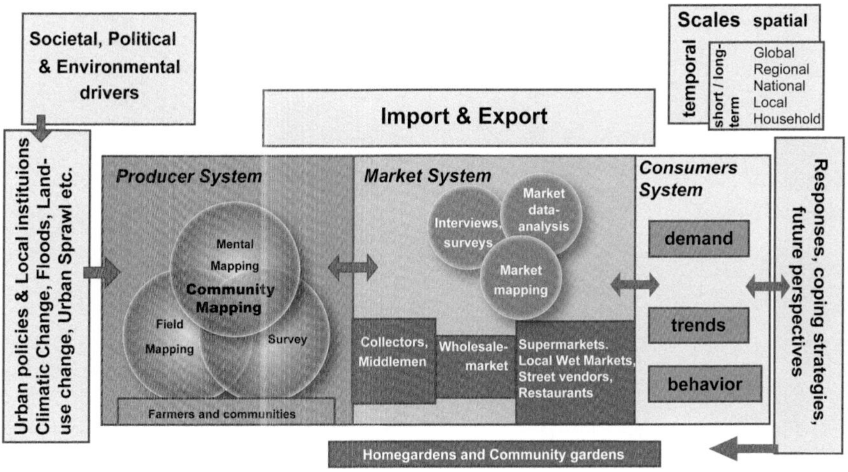

Fig. 3 Conceptual framework of the food system and research approach in Bangkok (*Source* Drescher et al. 2015, modified)

marketers and street vendors, consumers, governance, and the role of local institutions). The urban food system is not static but subject to change and fluctuation depending on external and internal societal, political, and environmental drivers, as well as on changing trends, demands, and behaviour of consumers (Fig. 3). A closer look at the complex marketing system depicts the importance of UPA for the urban food supply. The following analysis of urban policies and local institutions helps to explain why UPA has decreased despite the designation of urban space for agricultural land-use in the city's comprehensive urban development plans.

With a population of c. 10.2 million, Bangkok is one of the world's 33 megacities (UN 2019). Located on Thailand's central plain on the east side of the Chao Phraya River, Bangkok has developed in a region with a deep history of agricultural production. In the late nineteenth century, investments in roads and railways set Bangkok on a modernising trajectory, transforming it from a water-based city focused on the Chao Phraya River (with a network of canals facilitating not only the movement of goods and people, but also agricultural activities) to a land-based city with urban development expanding towards the east. Today, the pace of urban development is fast, with built areas expanding and threatening agricultural areas, despite recent protective legislation (Nagasawa and Patanakano 2013). By 1980, the Ministry of Interior had designated Bangkok's eastern districts as *agricultural areas* and the following year a cabinet resolution announced the western districts as *agricultural preservation areas* (Suttipong 1993). The agricultural areas in both city districts are included in Bangkok's four most recent Comprehensive Plans of 1992, 1999, 2006, and 2013 (Fig. 4), designated as one of the two agricultural land-use types: (1) rural and agriculture and (2) rural and agricultural conservation. In practice, there is little difference, with both types mainly used for different forms of farming (Fig. 4). However, some other activities are allowed under certain conditions.

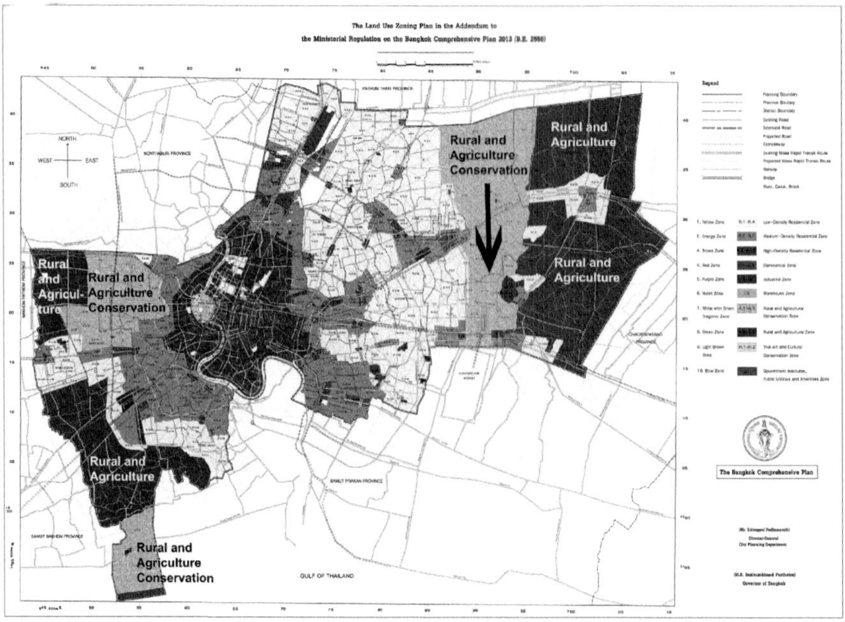

Fig. 4 The 2013 Bangkok Comprehensive Plan (B.E. 2556) with the two types of agricultural land-uses marked (*Source* adapted from the City Planning Department, BMA 2013 http://cpd.bangkok.go.th:90/web2/NEWCPD2556/open%20space%20plan%202556_eng.jpg)

For instance, markets covering less than 1000 m^2 are permitted if located at least 12 m from a road or within 500 m from a Sky Train station (since 1999 Bangkok's extensive public transport system). Small-scale industries no larger than 100 m^2 may be allowed under ministerial regulation, as are cemeteries, which are regulated by the Bangkok Metropolitan Administration (BMA; see below). However, several other land-uses, such as office space over 500 m^2, hotels, conference and exhibition centres, amusement parks, etc., are prohibited (City Planning Department, BMA 2013). Despite the inclusion of agricultural land-use in the city's Comprehensive Plans and these protective regulations, several studies demonstrate that the extent of agricultural areas is continuously decreasing (Jongkroy 2009; Menakanit et al. 2018; Pasama 1993; Policy and Planning Office, City Planning Department, BMA 2012). In the following, we address the questions why and how there is a diminishing trend, ultimately to examine factors that reduce the capacity of UPA to provide FNS and build urban resilience in Bangkok.

According to the Agricultural Extension Subdivision (2014), agriculture was practised in 26 of Bangkok's 50 districts, mainly in the western and eastern peri-urban zones (Fig. 5). But agriculture is not only practised in areas designated for this kind of land-use in Bangkok's Comprehensive Plan, as can be seen in Fig. 4. Unsurprisingly, farming outside of these land-use types are particularly susceptible to urban

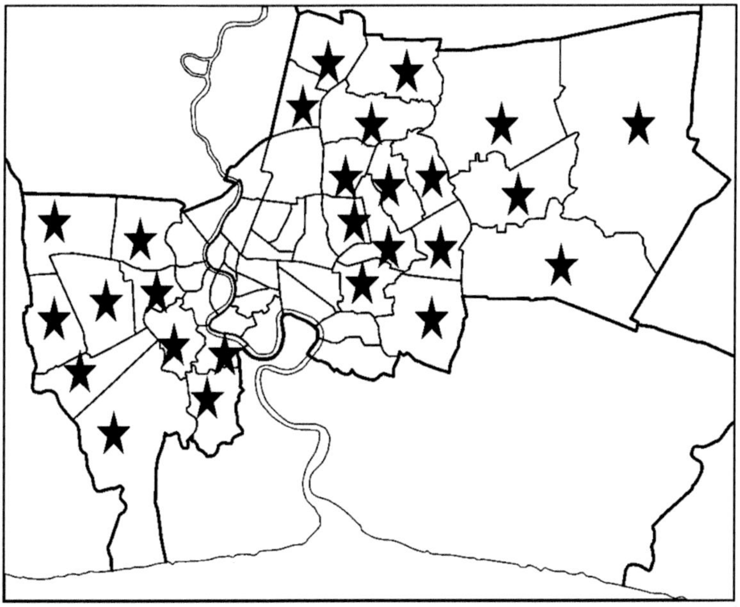

Fig. 5 Bangkok's 26 agricultural districts (*Source* Agricultural Extension Subdivision 2014—https://commons.wikimedia.org/wiki/File:Thailand_Bangkok_location_map.png)

development, but also rural and agricultural zones have decreased from 674 km^2 in 1992 to 617 km^2 in 2013 due to urbanisation (Posuk et al. 2018).

Recent data from the Bangkok Agricultural Extension Office (2017) show that rice production dominates Bangkok UPA, followed by ornamental plants, fruit trees, and vegetable production (Table 1). Rice is mainly produced in the eastern districts, with Nong Chok the largest production district. The other three crops are mostly cultivated in the western districts and closer to the city centre. Nong Kham is the most important production district for flowering plants and fruit trees, while Thawi Watthana is the largest vegetable production district (Fig. 6 and Table 2).

Menakanit (2019) found that two main government agencies are involved in the governance of Bangkok's UPA: BMA and the Ministry of Agriculture and Cooperatives. BMA, with an elected governor as the chief executive, has the main responsibility of the city's governance (BMA 2019). The administrative structure of BMA

Table 1 Bangkok production areas of 4 major crops in 2017

Crop	Production area (km^2)
Rice	132.30
Flowering plants	7.32
Fruit trees	5.37
Vegetables	3.19

Source Bangkok Agricultural Extension Office 2017

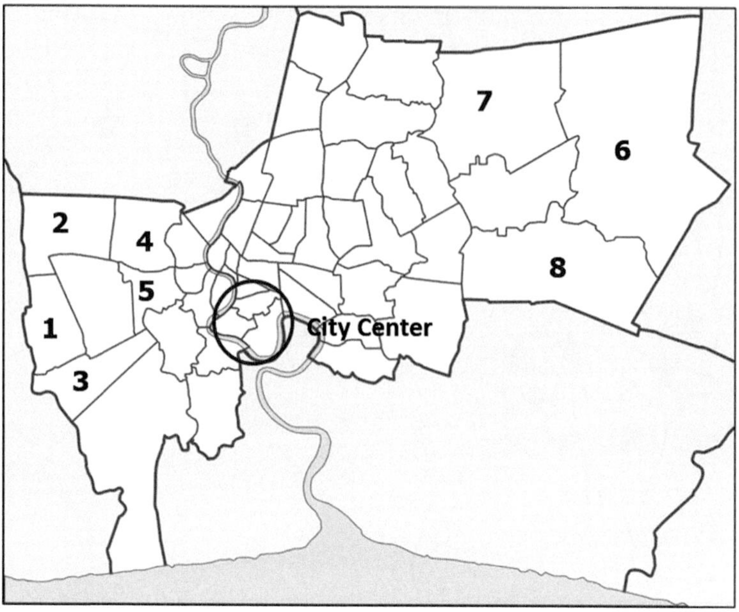

Fig. 6 The top three production districts in Bangkok of four major crops in 2017 (*Source* Bangkok Agricultural Extension Office 2017); for numbered districts see Table 2

Table 2 The top three production districts in Bangkok of four major crops in 2017

Crop	Map location	District	Production area (km^2)
Flowering plants	1	Nong Kham	3.10
	2	Thawi Watthana	1.46
	5	Bang khae	1.40
Fruit trees	1	Nong Kham	0.95
	2	Thawi Watthana	0.83
	3	Bang Bon	0.81
Vegetables	2	Thawi Watthana	1.62
	4	Taling Chan	0.42
	5	Bang Khae	0.71
Rice	6	Nong Chok	83.12
	7	Klong Samwa	23.60
	8	Lad Krabang	18.20

Source Bangkok Agricultural Extension Office 2017

is composed of three main offices (the Secretary to the Governor, the Bangkok Metropolitan Council, and the Civil Service Commission), 16 departments, and an office each for the city's 50 districts. The 16 departments cover different special functions of the BMA (e.g. medical service, health, education, public works, drainage and sewerage, environment, social development, and city planning). The 50 district offices provide services at the district level. In addition, a series of national government agencies are involved in particular aspects of city administration that affects UPA, for instance, road networks, construction, and maintenance (the Department of Highways and the Ministry of Transport) and the public water supply and infrastructure (the Metropolitan Waterworks Authority, a state enterprise under the Ministry of Interior). Furthermore, all major development programmes in and around the metropolitan area, such as the Comprehensive Plans for urban planning and development, are approved by the national cabinet (International Urban Development Association 2013). The complex structure of the city's governance makes the management of Bangkok's UPA quite complicated.

The BMA Act of 1985 does not include UPA as one of the BMA's main duties, instead of focusing on the governance of other urban services and functions. There are no policies, protective regulation, or supporting infrastructure for urban and peri-urban farmers at this level. The administrative vision of "green open space" usually simply emphasises "parks or tree-filled space" rather than productive agricultural green areas. This void generates incoherence in the management system of the city as a whole, because while the BMA does not recognise UPA, several sub-agencies do, namely:

1. The Social Development Department deals with career promotion, including agriculture. Their activities include training workshops and transfer of agricultural knowledge and technologies to farmers and the public, supporting 50 Agricultural Technology Transfer and Service Centres, and running five Agricultural Extension and Education Centres.
2. The City Planning Department's major task is, according to the Town Planning Act of 1975, to implement the Bangkok Comprehensive Plan and to monitor that built-up areas in each district agree with it. Their work, however, focuses on physical aspects only, which does not protect Bangkok's agricultural area. Their neglect to safeguard UPA areas in Bangkok are related to prioritising socio-economic growth.
3. The district offices of Bangkok's 26 districts with UPA have staff that supports agricultural training and capacity-building. They work together with local farmers and staff from the Department of Agricultural Extension, Ministry of Agriculture and Cooperatives through the Agricultural Technology Transfer and Service Centres. However, since agriculture is not one of BMA's main duties, they perform only routine tasks.

A key problem is budget allocation: BMA provides only 73,680 baht/year (c. 2385 USD) for the running of each Agricultural Technology Transfer and Service Centre (Agricultural Extension Subdivision 2014). This follows a pattern where urban farmers face an unfavourable situation in terms of financial support compared with rural farmers. Rural farmers are supported by different funding sources, such as the budget from provincial strategic plans or from government agencies dealing with agricultural, educational, economic, and social development. In terms of budget management and compensation, rural management is less complicated and more effective than BMA. The provincial governor has overall responsibility for all government agencies of the province, and takes the final decision with support from other government agencies. For example, if provincial farmers are hit by a natural disaster, the governor can proclaim a state of disaster to quickly allocate a budget for compensation. In Bangkok, however, the Bangkok Permanent Secretary (coordinating all City Departments) must make a notice to the Director General of the Department of Disaster Prevention and Mitigation at the Ministry of Interior to ask for such measures (Chiangchaisakulthai pers. com. 2018). This adds further steps to the action plan and delays the government's response. Moreover, the relief process for Bangkok UPA farmers is even more complicated, since only farmers that have registered with the Department of Agricultural Extension are eligible for government compensation. The registration requires formal land rental contracts, which most UPA farmers lack (Chiangchaisakulthai pers. com. 2018). Since more than 93% of Bangkok farmers rent the land without having formal contracts, the majority of UPA farmers are relatively vulnerable to extreme events, such as natural disasters (Charoensap pers. com. 2018).

The Ministry of Agriculture and Cooperatives is responsible for policy-making and implementation of Thailand's agriculture and cooperative plans. Since Bangkok is one of 77 provinces, Bangkok's UPA is under the Ministry's responsibility. The Department of Agricultural Extension is the agency mainly responsible for Bangkok's UPA and applies the Ministry of Agriculture and Cooperatives' policies in the city. However, since policies are mostly designed for "large plot" rural practices, there is a mismatch with the context and challenges of UPA.

4.1 *Urban Markets and Street Food*

The findings of a market survey in 2013–2014 show that wholesale and local (wet) markets serve as a hub for FNS in the Greater Bangkok Area (Drescher et al. 2015; Tsuchiy et al. 2015). Wetmarkets are a collection of individual stands where fruit, vegetables, meat, and other products are sold. They are traditional markets that, in contrast to modern retailers, mainly target lower income groups (Shipmann and Qaim 2011). The major wholesale markets, such as Talad Thai and Si Mum Muang, are central nodes in the complex food supply chain (Boossabong 2017). Producers deliver their products at these wholesale markets, where local markets and street vendors buy products to sell all over the city. Street vending forms a key element in the food chain, especially in the inner city of Bangkok, and is an important part of local food culture (Bhowmik 2005). In 2005, it is estimated that there were around 100,000 official and informal street vendors in a city (with a population of seven million (Bhowmik 2005; UN 2019). The Food and Agriculture Organisation (FAO 2020) states that street vendors "represent a significant part of urban food consumption for millions of low-and-middle-income consumers, in urban areas on a daily basis. Street foods may be the least expensive and most accessible means of obtaining a nutritionally balanced meal outside the home for many low-income people, provided that the consumer is informed and able to choose the proper combination of foods". Indeed, 70% of the Thai population belongs to the lower income class (Kawazu 2013). Nearly two out of five consumers buy street food for lunch and about one of three for supper. But most frequented by consumers, independent of age and gender, are local wetmarkets, though at some wetmarkets street food vendors are selling as well. Most relevant for consumers is proximity to food, which highlights the great local value of central Bangkok's street food vendors. The fact that more than two-thirds of the products sold by street vendors are non-processed vegetables shows the close relation between street vending and UPA.

Drescher et al. (2015) tracked, based on data from the wholesale market, the origin and destination of most vegetables traded in Bangkok, as well as their destination within the system. Nearly 92% of all vegetables traded originate in Thailand, followed by China (5%), and Laos and India. There are significant differences concerning the origin of the different vegetable categories in relation to the distance to Bangkok's wholesale markets. The main national production areas for the wholesale markets in Bangkok are the provinces in the north (Chiang Mai) and west (e.g. Suphanburi, Ratchaburi, Kanchanaburi, and Tak). Vegetables that wilt easily or are very sensitive to mechanical damage (such as leafy vegetables) come from regions around Bangkok as well as from UPA in the vicinity of both wholesale markets (intra-urban production). In contrast, root, bulb, and tuber vegetables have low susceptibility to wilting and mechanical damage and can therefore be transported from regions that are relatively far from markets (Hardeweg 2008). Hence, a considerable share of leafy vegetables at Bangkok's urban markets are grown in close distance to the city, e.g. in Pathum Thani (north), Nakhon Pathom (west), and Suphanburi (north-west) Provinces, demonstrating the importance of regional production for the FNS of a mega-city. It is remarkable that Pathum Thani, with a surface area of only 120 km^2, is the third most important region of origin of vegetables, although sub-urbanisation is currently taking place and many cropping areas are in transformation, thus reducing the cropping areas considerably. Nevertheless, this region has an advantage because of the proximity to both wholesale markets in Bangkok. 52% of products from Si Mum Muang Market stay in the Bangkok Metropolitan Area and 85% in the Greater Bangkok Area (including neighbouring provinces).

According to market survey responses, one of the most important quality criteria for vegetables is the level of pollution with pesticides (Drescher et al. 2015). A major problem is that non-toxic and less-toxic vegetables are exported to, for instance, the European Union, whereas highly polluted and toxic vegetables are consumed locally in Thailand. The Thailand Pesticide Alert Network (ThaiPAN) found high toxic contamination on 41% of sampled vegetables and fruits (Bangkok Post 2019). Health and freshness aspects are key criteria of vegetable buyers in Bangkok; however, these aspects are heavily dependent on education and income levels (Roitner-Schobesberger et al. 2008).

Indigenous vegetables such as cha-om (Acacia shoot, *Senegalia pennata*), bai ya nang (*Tiliacora triandra*), dok salid (cowslip creeper, *Telosma cordata*), paak khom/phakkhom-suan (*Amaranthus* spp.), and many others are very important components of the diet for people in Thailand and Bangkok. Furthermore, many of the indigenous species have a much better nutrient value than regular vegetables (Yang and Keding 2009). The main supplier of indigenous vegetables for the wholesale markets in Bangkok is the neighbouring Chachoengsao Province, about 80 km east of Bangkok's city centre. The data of the study underpin the importance of indigenous vegetables in local food habits, with more than 150 different vegetable species being traded in Bangkok markets. More than one-fourth of the respondents grow their own vegetables and fruits. In this way, they improve their FNS, even if most of them cultivate on a small scale.

While there are several documented benefits of UPA, particularly to the urban poor, Bangkok UPA is left with no direction because of different authorities' miscommunication, lack of planning, and appropriate policies. Thailand may have abundant food so the population of Bangkok does not realise the importance of UPA. However, if the goal is development of sustainable cities, it is crucial to educate both policymakers and the public on the importance of UPA and maintain urban FNS long-term (e.g. Isendahl and Barthel 2018).

The Bangkok case shows that complex systems of governance in mega-cities are poorly adapted to the particular opportunities, benefits, and challenges of UPA and FNS. Next, we present and discuss results on temporal and spatial patterns of urban crop production in Tamale, Ghana, and Ouagadougou, Burkina Faso to examine the spatial dynamics of UPA interacting with urban development in two smaller cities in sub-Saharan Africa.

5 UPA Spatial Dynamics in Tamale and Ouagadougou

Tamale is the third largest city in Ghana, with a population of about 313,000 (2015), and is the administrative centre of the Northern Region. Ouagadougou is Burkina Faso's capital city with a population of about 2.3 million (2015) (OECD/SWAC 2018). Both cities are rapidly growing, with annual growth rates of 4.5% in Tamale and 3.8% in Ouagadougou (Ghana Statistical Service 2013; Institut National de la Statistique et de la Démographie 2015).

Drechsel and Dongus (2010) classified urban crop production as (1) open-space production of high-value products on undeveloped urban land and (2) mostly subsistence gardening in backyards. Both forms of UPA are found in Tamale and Ouagadougou, albeit with major differences between the cities, as is the production of urban livestock (Box 2). As part of the UrbanFoodPlus project (2013–2018) a sample of urban agricultural areas were mapped in situ in Tamale and Ouagadougou during the rainy season (2014) and the dry season (2015) to obtain information on field sizes, crop diversity, and spatial patterns of urban crop production (Karg 2016). In Tamale, the main irrigated farming sites were located in the city centre and relied

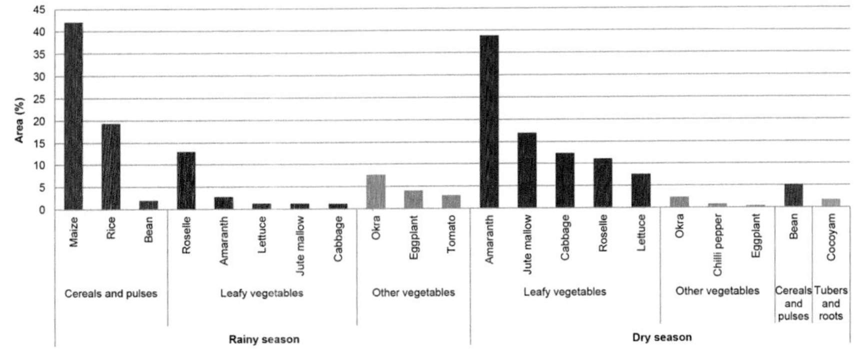

Fig. 7 Crops grown on urban open space agricultural sites in Tamale (*Source* Karg 2016)

on various irrigation water sources, including groundwater, piped, and gutter water, with varying degrees of contamination. During the dry season, around 25 ha were under cultivation within the urban area (Karg and Schlesinger 2018). During the rainy season, the area under cultivation increased considerably when people make use of the many small spaces that have emerged in the course of the unplanned expansion of the city. In Ouagadougou, most irrigated open space farming sites were located close to drinking water reservoirs (*barrages*) in the city centres with a high groundwater level and wells for irrigation. Other irrigation water sources included streams and industrial wastewater. During the dry season, vegetables and fruits were cultivated on around 527 ha within the urban area (Karg and Schlesinger 2018). Rain-fed cereal production was, unlike in Tamale, a minor activity in Ouagadougou.

In Tamale, crop composition changed noticeably across the two seasons (Fig. 7). Cereals such as maize and rice covered more than 60% of the cultivated area during the rainy season. During the dry season, leafy vegetables, in particular indigenous leafy vegetables such as amaranth and jute mallow, covered 90% of the area. The production of perishable crops susceptible to wilting is common in urban areas and has been reported for other cities in Africa and Asia where the lack of cool storage does not allow for long-distance transportation of fresh produce (Moustier 2007).

In Ouagadougou, a large number of different vegetables were produced in the urban area, covering 73% (rainy season) to 90% (dry season) of the mapped area under cultivation (Fig. 8). Cereal production played a minor role in the rainy season, with maize and rice covering 18% of the area, far less than the scale of cereal production in Tamale. One reason for this relates to legislation, according to which the cultivation of tall crops in urban Ouagadougou is prohibited (Bellwood-Howard et al. 2018). The most common crop was lettuce, an exotic leafy vegetable, reflecting the status of Ouagadougou as a Francophone capital city with an international dining scene, as compared with Tamale where indigenous leafy vegetables outweighed the exotic ones (Fig. 7). The diversity of vegetables was also significantly higher in Ouagadougou than in Tamale.

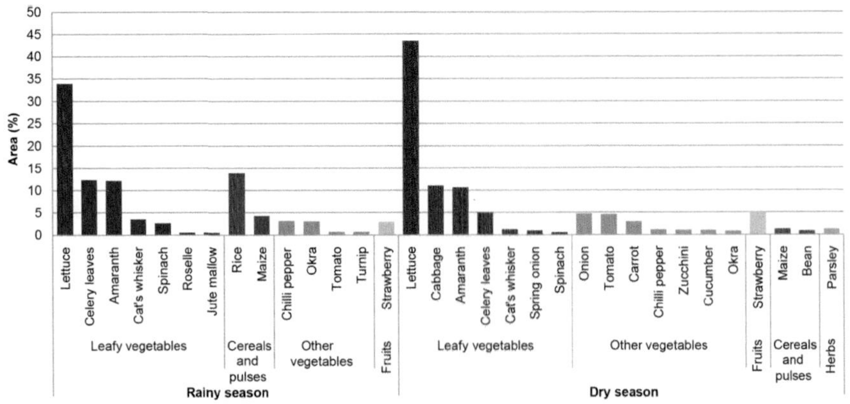

Fig. 8 Crops grown on urban open space agricultural sites in Ouagadougou (*Source* Karg 2016)

Box 2: Livestock in Tamale and Ouagadougou

Livestock farming is another important element of UPA in these cities. An exploratory survey on urban and peri-urban agriculture found that 93% (n = 251) and 63% (n = 154) of all interviewed crop farmers in Tamale and Ouagadougou, respectively, also owned some livestock and/or poultry (Bellwood-Howard et al. 2015).

In Tamale, 90% of urban farming households, comprising isolated and open-space farms, were engaged in livestock rearing as compared with 99% of peri-urban farming households (Table 3). The share of farm households that owned three or more species was higher in peri-urban villages (74%) than in urban areas, where more farms own only one livestock species, mostly poultry.

In Ouagadougou, the proportion of livestock-owning farms was lower than in Tamale, where, among urban farms, only 48% owned livestock as compared with 87% of peri-urban farm households (Table 3). Like in Tamale, most farm households in peri-urban villages around Ouagadougou owned more than one livestock species (86%) and only a minority of farmers in urban areas had several livestock species (17% on average).

Source Bellwood-Howard et al. (2015).

Not only did the composition of cultivated crops vary across seasons but also the farming location and area. In Ouagadougou, the farming site Boulmiougou illustrates how the lack or abundance of water influences spatial production patterns (Fig. 9). There, vegetable production was limited to the elevated areas in the rainy season, and the production of rice in the lower-lying land. As previously waterlogged areas dried out in the course of the dry season, the area under cultivation expanded. At the same time, cultivation in the centre of the more elevated area dropped. Similar patterns have been observed in Tamale.

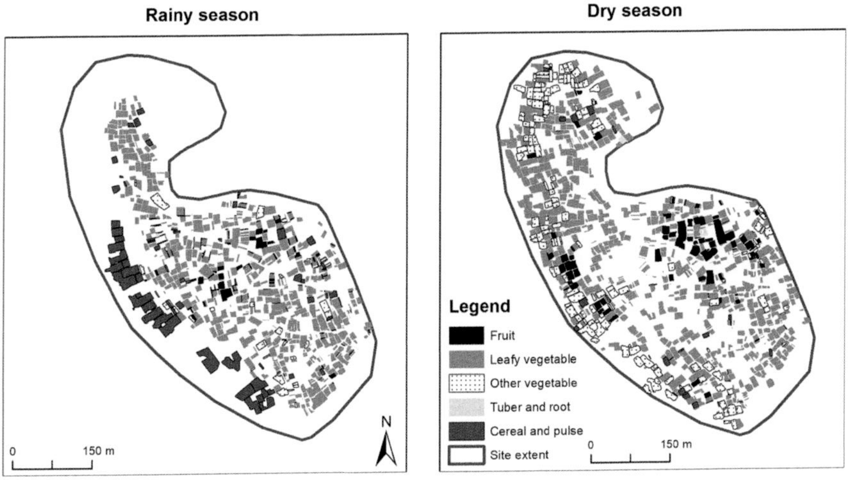

Fig. 9 Crop production in Boulmiougou, Ouagadougou (*Source* modified from Karg 2016)

Table 3 Livestock ownership of farmers in Ouagadougou and Tamale, by farm type

	Ouagadougou	Tamale
Urban isolated farms (%)	49	86
Urban open-space farms (%)	48	95
Peri-urban villages (%)	87	99
Total farms with livestock (%)	63	93

Source Bellwood-Howard et al. (2015: 13–14) (modified)

Thus, water availability, or the lack thereof, determines where particular types of crops are produced. Another factor for the choice of crop is economic considerations. For instance, in the dry season, the demand for vegetables exceeds the supply, when (leafy) vegetables cannot be cultivated elsewhere due to the lack of rain, and therefore it is highly profitable for farmers to grow vegetables in the dry season (Gerstl et al. 2002).

In Ouagadougou, open space sites played a greater role in production and marketing than in Tamale. In Tamale, on the other hand, isolated urban plots, including backyards, were more important for food production and a substantial proportion of crops produced in these sites were marketed (Bellwood-Howard et al. 2018). These differences are partly related to the history of urban development of the two cities. Ouagadougou's greater size and longer history as a capital city as well as formal land policies may have prevented the emergence of those interstitial spaces that have developed in Tamale in the course of rather unplanned urban development (Bellwood-Howard et al. 2018). In Ouagadougou, land is state-owned and the city's consolidated core was formally planned, with wide streets separating gridded neighbourhoods (Fig. 10a). Formalising the informal settlements in the urban fringe, however, has been lagging behind. In Tamale, traditional villages are absorbed into the urban structure as the city expands. Figure 10b shows a mix of traditional

a

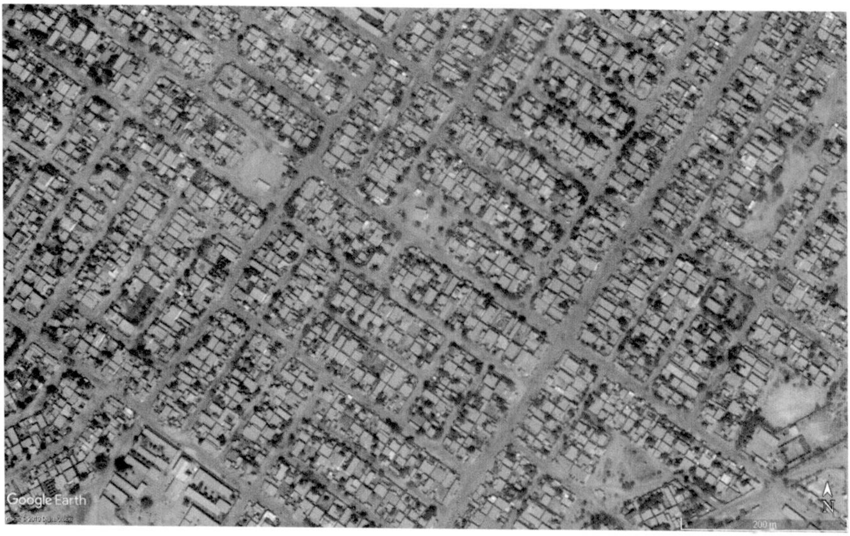

b

Fig. 10 a–b. Gridded layout in Ouagadougou (a; above) and absorbed village in Tamale (b; below) (*Sources* GoogleEarth, February 17, 2019 and December 21, 2018)

and modern housing, but still represents a quite "rural" structure of pathways, dirt roads, and backyards that are typically more prevalent in rural areas. Thus, traditional housing layouts remain, including spaces for farming between dwellings (Bellwood-Howard et al. 2018). Land-use planning has historically been weakly enforced in Tamale, which has to do with the prevailing traditional land ownership. Traditional chiefs own 90% of the land in the city, which they hold in trust for the community. However, chiefs are taking advantage of rising land values and lease lands to estate developers for personal gain (Nchanji et al. 2017), which may eventually also put pressure on farming sites.

In summary, different urban development trajectories have resulted in different urban agricultural landscapes. In Tamale, farming on interstitial spaces is more common due to weak implementation of urban plans, whereas in Ouagadougou, formal urban planning and legislation has prevented such sites from emerging. In the long term, however, farmers will probably lose the competition for space in the course of rapidly growing cities.

6 Concluding Discussion

Multiple crises (climate change, environmental degradation, social inequalities, pandemics) are already endemic to the global community, but we should not ignore that social, economic, and environmental sustainability demand restraint in consumption levels on the part of the GN and the affluent, in particular. Abundance, assumed in contemporary myths of wealth and growth (Worster 1993, 2016), is not compatible with sustainability (Daly 1996). We must also consider that, as an old Chinese proverb says, every crisis generates opportunities. However, at times there is a lack of capacity to find and take advantage of those opportunities, as well as to select the processes or actions that foster engagement, inclusion, and self-determination. The neoliberal structure of our global economy is not supporting these values, as illustrated in the context of the current Covid-19 pandemic. Ultimately, the leading role belongs to people, because deficiencies also generate wisdom, commitment, and creativity. If people take ownership of processes and actions, support may disappear, but seeds remain, which can then grow and multiply.

There are demonstrable results that UPA contributes to improving quality of life by the provision of healthy food, the economic gain of not buying food or marketing a part of the production, benefiting from better sanitary conditions, the health effects of an increase in urban green spaces, and the creation of social networks building a perception of a city closer to nature and urban dwellers to each other. UPA can facilitate place-making of joint action among populations, local governments, and civil society. It is a source of technical and productive employment and a component of communities' social economy and networks. Last but not least, UPA typically fosters social inclusion of vulnerable or marginalised sectors of the population, such as the elderly and the disabled, and contributes to the environmental education of children and young people in cities and reconnect them to food production and the environment. These are vital health, social, economic, and educational benefits.

UPA cannot be approached from a rural agrarian perspective because both the landscape and general context are very different. The support of technologies promoted by agricultural extension have to be adapted to the place and the requirements of the changing context of urban environments. It is important to ensure that a city's institutions of governance and services recognise UPA as part of the urban fabric, contribute to its design, and establish solid relationships with other urban functions. The visions of what people want their neighbourhood or city to be like, are still not understood sufficiently. There is clearly a need for greater direct citizen participation in urban planning and management.

Some local experiences refer to financial and input constraints as a problem for UPA. There is no doubt that these are important issues, especially to initiate processes. However, the sustainability of UPA depends very much on the priority that is given, in capacity-building, to the identification and forms of use of the urban dweller's own local resources and the waste that is produced. It is also a priority to establish personal, family, neighbourhood, and community spaces that, due to the capacity reached and the use of local potential, can be multiplied and scaled-up. Those place- and space-specific social relations motivate, multiply, and make experiences visible and tangible. In addition, we need to foster human relations on the basis of cooperation, not competition, and encourage alliances between people, organisations, local government entities, and other movements.

Considering the examples from the very different spatio-temporal contexts discussed in this chapter, we see that UPA has multiple expressions everywhere, whether it is reintroduced, controlled, or simply left alone by urban authorities. The examples from the pre-Columbian Maya and contemporary Cuba, Thailand, Ghana, and Burkina Faso briefly outlined here show that urban planning, or its absence and failures, create systems that may be resilient to a certain degree, but are not necessarily desired by the population and do not contribute to sustainable urban development. Backyard gardens are the most prominent private production sites worldwide, and their importance is still widely overlooked. However, the African case studies show that backyard gardening has been undermined by formal urban planning in Ouagadougou while it was "supported" by the lack of planning in Tamale. In most African countries, UPA is a visible reality and people are dependent on these activities to meet some or all of their food and income needs. In view of the rapidly growing cities in Africa, this will prevail for decades to come and requires a response by urban planners. While in the African case studies questions of seasonality of crop production, and conflicts with customary and modern land tenure prevails, the example from Thailand perhaps most obviously shows the institutional confusion that governs, or not, the productive use of public spaces. Nevertheless, it also reveals the great potential of single city districts to support urban agriculture. The Thailand case also shows how important UPA is for feeding cities. Particularly in Cuba and Thailand, the issue and awareness of product quality and the demand for organic production of vegetables without the use of pesticides and mineral fertilisers is growing. The studies also show how important it is to look at the entire food system and all of its components to clearly understand how food for cities is produced, transported, marketed, and how it finally reaches the consumer.

Cuba is probably still the best example of how this vital activity can be integrated into cities. However, one must consider how state-controlled systems collide with individually managed systems. Still, Cuba tries to meet an emerging goal that is growing in significance (not in the GS), namely, food sovereignty. The problem of transport, which is currently (2019/2020) worsening again in Cuba, is one of the main arguments for increasing local production. The answer is small-scale production systems independent from machinery and huge inputs, prevalent all over the world to feed the population and much more resilient to shocks than are high-input industrial agriculture. Reducing food mileage has entered international discourses and debates, leading to concepts of the "edible city", CSA, Alternative Food Networks, and other comparable ideas in many parts of the world, especially in the GN. Many of these elements are endemic to the GS. This is maybe an anchor to learn more from the GS food systems for the future sustainable design of urban food systems globally. The social practice of UPA has a history as ancient as the city itself. Rejuvenating these systems to feed an urban population may potentially have cascading positive impacts and prevent them from further marginalisation.

References

Agricultural Extension Subdivision (2014) Career Promotion Division, Social Development Department, BMA. Agricultural Technology Transfer and Service Center Handbook. Bangkok: Chulalongkorn University Printing House

Altieri MA, Companioni N, Cañizares K, Murphy C, Rosset P, Bourque M, Nicholls CI (1999) The greening of the "barrios": urban agriculture for food security in Cuba. Agric Hum Val 16:131–140

Altieri MA, Funes-Monzote FR (2012) The paradox of Cuban agriculture. Month Rev 63:23–33

Bangkok Agricultural Extension Office (2017) 2017 fundamental agricultural data. http://www.bangkok.doae.go.th/bankkok20.htm. Accessed 1 July 2019

Barthel S, Isendahl C (2013) Urban gardens, agriculture, and water management: sources of resilience for long-term food security in cities. Ecol Econ 86:224–234

Barthel S, Isendahl C, Vis BN, Drescher A, Evans DL, van Timmeren A (2019) Global urbanisation and food production in direct competition for land: leverage places to mitigate impacts on SDG2 and on the earth system. Anthrop Rev 6:71–97

Baudoin W, Drescher A (2008) Urban agriculture for sustainable poverty alleviation and food security. FAO, Rome

Bellwood-Howard I, Häring V, Karg H, Roessler R, Schlesinger J, Shakya M (2015) Characteristics of urban and peri-urban agriculture in West Africa: results of an exploratory survey conducted in Tamale (Ghana) and Ouagadougou (Burkina Faso). International Water Management Institute, Vol. 163, Colombo

Bellwood-Howard I, Shakya M, Korbeogo G, Schlesinger J (2018) The role of backyard farms in two West African urban landscapes. Landsc Urban Plan 170:34–47

Bhowmik SK (2005) Street vendors in Asia: a review. Econ Polit Weekly, XL(May 28–June 4):2256–2264

Binns JA, Nel E (2014) The significance of urban agriculture in food security and sustainable livelihoods in response to economic restructuring in Zambia's Copperbelt Province. Global Education Programme International Development Research Fund Final Report, 112. https://www.mfat.govt.nz/. Accessed 29 June 2020

BMA (2019) http://www.bangkok.go.th/main/page.php?and328andl=en. Accessed 2 Mar 2019

Bangkok Post (2019) 41% of sampled fruits, vegetables contaminated. 26 June. https://www.bangkokpost.com/thailand/general/1702276/41-of-sampled-fruits-vegetables-contaminated. Accessed 28 June 2020

Battersby J (2012) Beyond the food desert: finding ways to speak about urban food security in South Africa. Geografiska Annaler: Series B, Human Geography 94(2):141–159

Boossabong P (2017) Articulating public agencies, experts, corporations, civil society and informal sector in planning food systems in Bangkok. In: Cabannes Y, Marocchino C (eds) Integrating food into urban planning: a history and theory of pebbles in a pebble heathland landscape. UCL Press, Los Angeles, pp 184–201

Brush SB (2001) Genetically modified organisms in peasant farming: social impact and equity. Indiana J Glob Legal Stud 135–162

CFS (2012) Committee on world food security: coming to terms with terminology CFS 2012/39/4, Thirty-ninth Session, FAO Rome, Italy, October 15–20. http://www.fao.org/bodies/cfs/cfs39/en/. Accessed 29 June 2020

Chase AF, Chase DZ (2016) Urbanism and anthropogenic landscapes. Ann Rev Anthrop 45:361–376

Chase AF, Scarborough VL (eds) (2014) the resilience and vulnerability of ancient landscapes: transforming Maya archaeology through IHOPE. Wiley-Blackwell, Hoboken

City Planning Department, BMA (2013) Bangkok comprehensive plan 2013. http://cpd.bangkok.go.th:90/web2/NEWCPD2556/land%20use%20plan%202556_eng.jpg. Accessed 29 June 2020

Clouse C (2014) Cuba's urban farming revolution: how to create self-sufficient cities. The Architect Rev 17 March

Colding J, Colding M, Barthel S (2020) The smart city model: a new panacea for urban sustainability or unmanageable complexity? Enviro Plan B—Urb Analyt City Sci 47:179–187

Cruz MC, Medina RS (2003) Agriculture in the city: a key to sustainability in Havana. Ian Randle Publishers, Kingston, Jamaica

Daly HE (1996) Beyond growth: the economics of sustainable development. Beacon Press, Boston

Davenport NA, Shackleton CM, Gambiza (2012) The direct-use value of municipal commonage goods and services to urban households in the Eastern Cape, South Africa. Land Use Pol 29:548–557

de Neergaard A, Drescher AW, Kouamé C (2009) Urban and peri-urban agriculture in African cities. In: Shackleton CM, Pasquini MW, Drescher AW (eds) African indigenous vegetables in urban agriculture. Earthscan, London, pp 67–96

Drechsel P, Dongus S (2010) Dynamics and sustainability of urban agriculture: examples from sub-Saharan Africa. Sustain Sci 5:69–78

Drescher AW, Holmer RJ, Glaser R, Hoschek M, Pariyanuj C (2015) Mapping vegetables – understanding the food system of greater Bangkok, Thailand: a web-based collaborative research environment. In: Proceedings of the Regional Symposium on Sustaining Small-Scale Vegetable Production and Marketing Systems for Food and Nutrition Security (SEAVEG2014) AVRDC—The World Vegetable Center, Publication No. 15, 785:254–268

Drescher AW, Jacobi P, Amend J (2000) Urban agriculture—a response to crisis? Urban Agriculture Magazine No. 1, RUAF Foundation, The Netherlands

Dunning NP, Beach T (2010) Farms and forests: spatial and temporal perspectives on ancient Maya landscapes. In: Martini IP, Chesworth W (eds) Landscapes and societies. Springer, Berlin, pp 369–389

Eisenstadt SN (1966) The basic characteristics of modernisation. In: Eisenstadt SN (ed) Modernisation, protest and change. Prentice-Hall, Eaglewood Cliffs, pp 1–19

FAO (2014a) The state of food and agriculture—innovation in family farming. FAO, Rome

FAO (2014b) La agricultura urbana y periurbana en América Latina y el Caribe: compendio de estudios de casos. FAO, Rome

FAO (2020) Food for the cities. FAO, Rome

Ford A, Nigh R (2015) The Maya forest garden: eight millennia of sustainable cultivation of the tropical woodlands. Left Coast Press, Walnut Creek

Forero-Niño L (2011) The Cuban government approves guidelines to reform Cuba's economic model and develops an implementation strategy. Law Busin Rev Am 17:761
Gerstl S, Cissé G, Tanner M (2002) Economic impact of urban agriculture on home gardeners in Ouagadougou. Urb Agric Magz 7:12–15
Ghana Statistical Service (2013) 2010 population and housing census: National Analytical Report, Accra
Gonzales-Corzo M (2016) Urban agriculture in Cuba: an update. ASCE, https://www.ascecuba.org/asce_proceedings/urban-agriculture-in-cuba-an-update/. Accessed 28 June 2020
Graham E (1999) Stone cities, green cities. Archaeol Pap Am Anthrop Assoc 9:185–194
Graham E, Isendahl C (2018) Neotropical cities as agro-urban landscapes: revisiting low-density, agrarian-based urbanism. In: Ekblom A, Isendahl C, Lindholm K-J (eds) The resilience of heritage: cultivating a future of the past. Uppsala University, Uppsala, pp 165–180
Halloran A, Magid J (2013) Planning the unplanned: incorporating agriculture as an urban land use into the Dar es Salaam master plan and beyond. Environ Urban 25:541–558
Hardeweg B (2008) The spatial distribution and interregional dynamics of vegetable production in Thailand. Masters dissertation, University of Hannover, Hannover
Harvey D (1990) Between space and time: reflections on the geographical imagination. Ann Assoc Am Geog 80:418–434
Haysom G (2016) Alternative food networks and food insecurity in South Africa. Working Paper 33, PLAAS and Centre of Excellence on Food Security, University of Western Cape, Cape Town
Hornborg A (2001) The power of the machine: global inequalities of economy, technology, and environment. Altamira Press, Walnut Creek
Hovorka A, Lee-Smith D (2006) Gendering the urban agriculture agenda. In: Van Veenhuizen R (ed) Cities farming for the future: urban agriculture for green and productive cities. RUAF Foundation and IDRC, Montreal, pp 125–144
Institut National de la Statistique et de la Démographie (2015) Annuaire statistique 2014. Ministère de l'économie et des finances, Burkina Faso
International Urban Development Association (2013) Bangkok metropolitan area: Thailand. https://inta-aivn.org/en/481-inta/activitities/exchange/roundtables/20122013-inbetween/1681-bangkok-metropolitan-area. Accessed 9 Mar 2019
Isendahl C (2012a) Investigating urban experiences, deconstructing urban essentialism. Urban Glob Enviro Change Viewpoints 8:25–28
Isendahl C (2012b) Agro-urban landscapes: the example of Maya lowland cities. Antiq 86:1112–1125
Isendahl C, Barthel S (2018) Archaeology, history, and urban food security: integrating cross-cultural and long-term perspectives. In: Zeunert J, Waterman T (eds) Routledge handbook of landscape and food. Routledge, London, pp 61–72
Isendahl C, Smith ME (2013) Sustainable agrarian urbanism: the low-density cities of the Mayas and Aztecs. Cities 31:132–143
Isendahl C, Smith ML, Stark MT, Sulas F, Barthel S (2020) Urban ecology in the ancient tropics: foodways and urban forms. In: Douglas I, Anderson P, Goode DA, Houck ME, Maddox D, Nagendra N, Tan PY (eds) Routledge handbook of urban ecology (2nd ed). Routledge, London, pp 13–23
Jongkroy P (2009) Urbanisation and changing settlement patterns in peri-urban Bangkok. Kasetsart Journal (Soc. Sci) 30:303–312
Karg H (2016) The role of locality in feeding cities: spatial analysis of agriculture, markets and food flows in West African urban food systems. PhD thesis, University of Freiburg, Freiburg
Karg H, Schlesinger J (2018) Open space farming. In: Karg H, Drechsel P (eds) Atlas of West African urban food systems: examples from Ghana and Burkina Faso. IMWI, Colombo, pp 15–29
Kawazu N (2013) Consumer trends and expansion of retail markets in growing ASEAN economies. Nomura Research Institute, Tokyo, pp 6–13
Koont S (2011) Sustainable urban agriculture in Cuba. University Press of Florida, Gainsville

Lappé FM, Collins J (1979) World hunger: ten myths. Institute for Food and Development Policy, San Francisco

Lee-Smith D (2010) Cities feeding people: an update on urban agriculture in equatorial Africa. Enviro Urbanis 22:483–499

Lloréns MD, Torres ML, Alvarez H, Arrechea AP, García JA, Aguirre SD, Fernández A (2008) Characterisation of municipal solid waste from the main landfills of Havana city. Waste Mgmt 28:2013–2021

McKillop H (2002) Salt: white gold of the Ancient Maya. University Press of Florida, Gainsville

Menakanit A (2019) Urban agriculture: a void in Bangkok metropolitan management. Veridian E-Journal 12:1136–1154

Menakanit A, Vudipong D, Prapassara N, Nath P (2018) Causes of vegetable production area reduction in Thawi Watthana District, Bangkok Metropolitan Administration. Veridian E-Journal 11:2843–2858

Milan Urban Food Policy Pact (2015) An international pact signed by 210 cities from all over the world with more than 450 million inhabitants. http://www.milanurbanfoodpolicypact.org. Accessed 30 June 2020

Mollison B, Holmgren D (1990) Permaculture one: a perennial agriculture for human settlements, 6th edn. Tagari Publications, Tyalgum Australia

Mougeot LJ (2000) Urban agriculture: definition, presence, potentials and risks, and policy challenges. In: Bakker N, Dubbeling M, Guendel S, Sabel-Koschella U, Zeeuw HD (2000) Growing cities, growing food: urban agriculture on the policy agenda: a reader on urban agriculture. Deutsche Stiftung fur Internationale Entwicklung (DSE), Zentralstelle fur Ernahrung und Landwirtschaft pp 1–42

Moustier P (2007) Urban horticulture in Africa and Asia, an efficient corner food supplier. Acta Horticult 762:145–158

Moustier P, Renting H (2015) Urban agriculture and short chain food marketing in developing countries. In: de Zeeuw H, Drechsel P (eds) Cities and agriculture: developing resilient urban food systems. Earthscan/Routledge, London, pp 121–138

Mundoli S, Manjunath B, Nagendra H (2015) Effects of urbanisation on the use of lakes as commons in the peri-urban interface of Bengaluru, India. Intn J Urb Sustain Dev. 7:89–108

Nagasawa R, Patanakano B (2013) Urbanisation and its influences on the suburban landscape changes in Bangkok Metropolitan Region, Thailand. J Japan Agric Sys Soc 29:29–39

Nchanji EB, Bellwood-Howard I, Schareika N, Chagomoka T, Schlesinger J, Drescher A, Glaser R (2017) Assessing the sustainability of vegetable production practices in northern Ghana. Intn J Agric Sustain 15:321–337

Novo MG, Murphy C (2000) Urban agriculture in the city of Havana: a popular response to a crisis. In: Bakker N, Dubbeling M, Gündel S, Sabel-Koshella U, de Zeeuw H (eds) Growing cities, growing food: urban agriculture on the policy agenda. Zentralstelle für Ernährung und Landwirtschaft (ZEL), Feldafing, pp 329–346

OECD/SWAC (2018) Africapolis (database), www.africapolis.org. Accessed 5 Feb 2019

Orzan G, Pricop A, Mihalache MM (2018) GMO's: the trade-offs between economic benefits and perceived health risks. Annales Universitatis Apulensis Series Oeconomica, Alba Iulia 1(20):1–5

Pasama S (1993) A study to identify suitable patterns for development and conservation of suburban agricultural area: a case study of Western Bangkok greenbelt. Master Thesis in Urban and Regional Planning, Chulalongkorn University, Bangkok

Pérez-López JF (2002) The Cuban economy in an unending special period. Cuba in Transition 12:507–521

Policy and Planning Office, City Planning Department, BMA (2012) Bangkok agricultural report. http://cpd.bangkok.go.th:90/web2/strategy/DATA54_55/1AG.pdf. Accessed 29 Aug 2019

Posuk S, Kajita Y, Petchsasithon A (2018) Comparative analysis of city planning and land use change in Bangkok, Thailand, by using remote sensing and GIS. MATEC Web of Conferences 192, 02064 https://www.matecconferences.org/articles/matecconf/abs/2018/51/matecconf_iceast2018_02064/matecconf_iceast2018_02064.html. Accessed 30 June 2020

Roitner-Schobesberger B, Darnhofer I, Somsook S, Vogl CR (2008) Consumer perceptions of organic foods in Bangkok, Thailand. Food Pol 33:112–121

Santandreu A, Gómez Perazzoli A, Terrile R, Ponce M (2009) Urban agriculture in Montevideo and Rosario: a response to crisis or a stable component of the urban landscape? Urban Agriculture Magazine No. 22, RUAF Foundation, Ottawa

Schipmann C, Qaim M (2011) Modern food retailers and traditional markets in developing countries: comparing quality, prices, and competition strategies in Thailand. Appl Econ Persp Pol 33:345–362

Schumacher Centre (2014) Cuban sustainable food systems delegation October 27–November 2. Organised by the Schumacher Centre for a New Economics in coordination with the Christopher Reynolds Foundation. Report by Greg Watson. Massachusetts

Shackleton CM, Cilliers SS, du Toit MJ, Davoren E (2021) The need for an urban ecology of the Global South. In: Shackleton CM, Cilliers, SS, Davoren E, du Toit MJ (eds) Urban ecology in the Global South. Springer, Cham, pp 1–26

Shackleton CM, Drescher A, Schlesinger J (2020) Urbanisation reshapes gendered engagement in land-based livelihood activities in mid-sized African towns. World Dev 103:104946

Shackleton CM, Guild J, Bromham B, Impey S, Jarrett M, Ngubane M, Steijl K (2017) How compatible are urban livestock and urban green spaces and trees? An assessment in a medium-sized South African town. Intn J Sustain Urb Dev 9:243–252

Sitas N, Selomane O, Hamann M, Gajjar SP (2021) Towards equitable urban resilience in the Global South within a context of planning and management. In: Shackleton CM, Cilliers, SS, Davoren E, du Toit MJ (eds) Urban ecology in the Global South. Springer, Cham, pp 347–364

Skinner C (2016) Informal food retail in Africa: a review of evidence. Consuming Urban Poverty Project Working Paper No. 2, African Centre for Cities, University of Cape Town. https://www.gov.uk/dfid-research-outputs/informal-food-retail-in-africa-a-review-of-evidence. Accessed 30 June 2020

Smart J (2015) Urban agriculture and economic change in the Zambia Copperbelt: the cases of Ndola, Kitwe and Luanshya. PhD thesis, University of Otago

Smit J, Ratta A, Nasr J (1996) Urban agriculture: food, jobs and sustainable cities. Habitat II, UNDP, New York

Staller J, Carrasco M (eds) (2010) Pre-Columbian foodways: interdisciplinary approaches to food, culture, and markets in ancient Mesoamerica. Springer, New York

Suttipong P (1993) A Study to Identify Suitable Patterns for Development and Conservation of Suburban Agricultural Area: A Case Study of Western Bangkok Greenbelt. Master Thesis in Urban and Regional Planning. Graduate School, Chulalongkorn University

Tainter JA, Allen TF (2019) Energy gain and the evolution of organisation. In: Isendahl C, Stump D (eds) The Oxford handbook of historical ecology and applied archaeology. Oxford University Press, Oxford, pp 558–577

Thomas G (2014) Growing greener cities in Latin America and the Caribbean: an FAO report on urban and peri-urban horticulture in the region. FAO, Rome

Tornaghi C (2014) Critical geography of urban agriculture. Prog Hum Geog 38:551–567

Tsuchiy K, Hara Y, Thaitakoo D (2015) Linking food and land systems for sustainable peri-urban agriculture in Bangkok Metropolitan Region. Landsc Urb Plan 143:192–204

UN Department of Economic and Social affairs (2019) 2019 revision of world population prospects. United Nations, New York

van Veenhuizen R (ed) (2006) Cities farming for the future: urban agriculture for green and productive cities. RUAF Foundation and IIRR publishers, Ottawa

Viljoen A, Howe J (2005) Cuba: laboratory for urban agriculture. In: Viljoen A, Bohn K, Howe J (eds) Continuous productive urban landscapes: designing urban agriculture for sustainable cities. Elsevier, Amsterdam, pp 146–191

Vorster HH, Kruger A, Margetts BM (2011) The nutrition transition in Africa: can it be steered into a more positive direction? Nutr 3:429–441

Wallerstein I (1974) The modern world-system I: capitalist agriculture and the origins of the european world-economy in the sixteenth century. Academic Press, New York
Waterman T (2018) Taste, foodways, and everyday life. In: Zeunert J, Waterman T (eds) Routledge handbook of landscape and food. Routledge, London, pp 517–530
White CD (ed) (1999) Reconstructing ancient Maya diet. University of Utah Press, Salt Lake City
WinklerPrins AM (ed) (2017) Global urban agriculture: convergence of theory and practice between north and south. CAB International, Boston
Worster D (1993) The wealth of nature: environmental history and the ecological imagination. Oxford University Press, New York
Worster D (2016) Shrinking the earth: the rise and decline of natural abundance. Oxford University Press, New York
Yang RY, Keding GB (2009) Nutritional contributions of important African indigenous vegetables. In: Shackleton CM, Pasquini MW, Drescher AW (eds) African indigenous vegetables in urban agriculture. Earthscan, London, pp 105–143
Zeunert J, Waterman T (2018) Routledge handbook of landscape and food. Routledge, London

Towards Equitable Urban Resilience in the Global South Within a Context of Planning and Management

Nadia Sitas, Odirilwe Selomane, Maike Hamann, and Sumetee Pahwa (SP) Gajjar

Abstract Over the last decade, the need to advance urban resilience research and practice has been emphasised, especially for safeguarding important ecosystem services that are critical for human wellbeing in and around cities. The pace and scale of changes in the Anthropocene make this imperative even more pressing, especially within the context of rapid urbanisation, infrastructure deficits and rising inequalities in the Global South. In this chapter, we contribute to these advances by highlighting the value of taking a social-ecological resilience approach to understanding the contested nature of urban resilience from multiple perspectives. We discuss important principles for exploring and building resilience and how these principles can be used within the context of urban resilience using two examples from the Global South. This chapter also outlines how an intersectionality approach can advance conceptualisation and implementation of resilience by focusing explicitly on how power mediates certain relationships and interdependencies in social-ecological systems. Foregrounding the role of power can assist with better understanding the opportunities for adaptation and transformation and 'putting the last, first' in resilience research and practice. We conclude by exploring some key gaps, blindspots and opportunities for a research-action agenda for urban resilience going forward.

Keywords Adaptation · Cities · Equity · Nature based solutions · Social-ecological systems · Transformation

---

N. Sitas (✉)
Centre for Complex Systems in Transition, Department of Conservation Ecology and Entomology, Stellenbosch University, Stellenbosch, South Africa
e-mail: nadiasitas@sun.ac.za

O. Selomane · M. Hamann
Centre for Complex Systems in Transition, Stellenbosch University, Stellenbosch, South Africa
e-mail: odirilwes@sun.ac.za

M. Hamann
e-mail: maike@sun.ac.za

S. P. Gajjar
Fellow, PlanAdapt, Berlin, Germany
e-mail: spahwagajjar@plan-adapt.org

C. M. Shackleton et al. (eds.), *Urban Ecology in the Global South*, Cities and Nature,
https://doi.org/10.1007/978-3-030-67650-6_13

1 Introduction

The future of the world is an urban one. This will be especially evident in cities and urban spaces in the Global South which are predicted to account for more than half of the world's population by 2050 (UN DESA 2018). Even though urban spaces account for less than 3% of the surface of the earth, over 70% of the world's energy-related carbon emissions stem from urban areas (IPCC 2014; Mitchell et al. 2018), and they contribute to around 75% of global GDP (Elmqvist et al. 2019). This makes urban areas extremely important focal points for navigating global sustainability, equity and resilience. Cities and towns are where challenges linked to the Anthropocene (a new geological epoch dominated by humans) will play out (Bai et al. 2018). Around 20% of the nearly four billion urban dwellers live in informal settlements (~ 800 000 000 people), often disconnected from the main urban infrastructure. This means those that are the most socially, economically and environmentally vulnerable are set to be hit hardest by Anthropocene challenges such as water and food insecurity and increased exposure to extreme events (UN DESA 2018). People will increasingly need access to basic services, including safe shelter, food and water, but rapid urbanisation, especially into informal settlements, means that urban governance, planning and management systems often struggle to keep up with current in-migration rates (du Toit et al. 2018; Simon et al. 2021).

At the same time, the capacity of social-ecological systems to continue to contribute to human wellbeing through the provision of ecosystem services (see Chaps. 8–10) is being eroded. This reduces the capacities of these systems to absorb, adapt and transform in response to multiple pressures (McPhearson et al. 2015; Díaz et al. 2019). Urban ecosystem services include regulation of air and water quality, flood mitigation, temperature regulation during heatwaves, contribution to good physical and mental health and promotion of social and cultural wellbeing (Depietri et al. 2012; Jenerette et al. 2011; du Toit et al. 2018; Keeler et al. 2019). As urban spaces have higher concentrations of people and infrastructure, the exposure to natural hazards often results in greater numbers of fatalities and socio-economic losses in comparison to rural landscapes (Dickson et al. 2012). While all urban systems are vulnerable to the impacts of natural hazards and global change processes, urban dwellers, especially in the Global South, are more exposed due to a number of features which exacerbate their risk (Davoren and Shackleton 2021). In Africa for example, these features include poverty, high population growth rates, informality and persistent transhumance patterns that connect urban and rural landscapes, coupled with a heavy reliance on natural resources and environmental degradation due to the impacts of urban sprawl (O'Farrell et al. 2019; Shackleton et al. 2021). In addition, socio-economic inequalities are increasing within nations and can lead to negative feedbacks with the environment, further undermining the resilience of social-ecological systems (Hamann et al. 2018). Thus, healthy, functioning urban ecosystems have an especially critical role to play in enhancing the resilience of urban landscapes and communities into the future.

The Sustainable Development Goals (SDGs), New Urban Agenda and other global instruments have foregrounded the need to find and build socio-technical and ecological coalitions to achieve their foundational principle to 'leave no one behind' (Simon et al. 2021). It is unprecedented that so many states have committed to a more resilient and just future, but how these commitments are implemented remains a daunting task. Although many cities are rising to the challenge (e.g. C40 Cities initiative), research and policy still largely work within disciplinary or sectoral silos and limited resources are available to address research-policy-implementation divides, impeding further development progress. Furthermore, one of the biggest challenges for sustainable, resilient and equitable development is addressing the dire power asymmetries evident in urban areas across the globe and a need to 'put the last, first' (Ravera et al. 2016; Borie et al. 2019; Görgens and Ziervogel 2019).

Urban areas in the Global South face many uncertainties around future trajectories of change which pose challenges to current and future urban planning (e.g. sea-level rise, political stability and resource availability). While many cities have persisted over centuries, the pace and scale of the Anthropocene pose many new and coupled challenges for urban spaces. This necessitates a shift in focus towards understanding, planning and building 'resilient' cities and urban spaces to deal with constant change. In this context, the fact that the majority of infrastructure is yet to be constructed presents immense opportunities for (re)imaging equitable, resilient and multifunctional cityscapes (Nagendra et al. 2018). While the term 'resilience' has been used across science, policy and practice domains, definitions differ even within the urban discourse (Meerow et al. 2016; Meerow and Stults 2016). A criticism of resilience theory has been that, because it emerged mainly from the natural sciences, the inclusion of people as well as social and institutional dynamics, including an acknowledgement of power and politics, has been under-represented in resilience research (Cote and Nightingale 2012; Brown 2015; Ravera et al. 2016). However, in recent years advances have been made in resilience praxis through the concept of 'negotiated resilience' (Harris et al. 2018) which is grounded in political ecologies of the Global South. Accordingly, a focus on social-ecological resilience (SER), with equal emphasis on the 'social' and the 'ecological' and including a more nuanced understanding of different conceptualisations and contestations of coupled human–natures, offers fertile ground for co-developing responses that can lead to more equitable futures (Reyers et al. 2018). A social-ecological approach to resilience is especially important for urban areas in the Global South where there is a disproportionate reliance on local nature and ecosystem services for wellbeing and livelihoods (Kumar and Yashiro 2014; Chaplin-Kramer et al. 2019). Applying a SER approach to urban planning can help with urban adaptation strategies and identify opportunity contexts for transformation in the face of challenges, assisting cities to prepare for both expected and unexpected disturbances.

2 Urban Resilience in the Anthropocene

Cities will be increasingly exposed to severe climate change impacts such as extreme heat, droughts, flooding and coastal storm hazards (Rosenzweig et al. 2011; Hobbie and Grimm 2020). At the same time, cities are experiencing unprecedented growth and rising inequalities (Glaeser et al. 2009; Nagendra et al. 2018). Especially in the Global South, urban development is likely to be largely informal, hampered by a shortage of resources and governance capacity, as well as legacies of colonial spatial planning (Myers 2021). The risk is that ad hoc and business-as-usual approaches to urban development in a time of climate crisis, ecological tipping points and societal upheaval will not only have negative environmental impacts, but also severely limit options for a healthy, fair and desirable future for all urban residents (Leach et al. 2018). Many global initiatives that focus on urban resilience therefore place an emphasis on building stakeholder capacity and applying a holistic, or systems-based, approach to sustainable development, to address challenges at multiple scales and from multiple perspectives (Fernández et al. 2016; Elmqvist et al. 2019). Thus, the potential leadership role of local governments and urban communities in sustainable development is significant (Ziervogel et al. 2019b). Cities are economic and financial centres, cultural melting pots and innovation hubs. As such, they have the ability to catalyse change, experiment with novel solutions and share their knowledge and experiences across the world. Especially cities in the Global South have a strong imperative, as well as unique circumstances and creative capacities, to advance the global sustainability agenda (Nagendra et al. 2018; O'Farrell et al. 2019; Ziervogel 2019).

In response to many of these urban challenges, there has been an increase in the number of donor-led initiatives that are explicitly focusing on building resilience across scales through strengthening institutional capacities (e.g. C40 Cities Initiative,[1] 100 Resilient Cities,[2] FRACTAL,[3] City Resilience Program[4]), as well as research-action networks that have been established (e.g. Future Earth Urban Knowledge-Action Network,[5] Urban Resilience Hub[6]) with support from civil society organisations and Non-Governmental Organisations (e.g. Slum/Shack Dwellers International, The Nature Conservancy, World Wildlife Fund). Each of these initiatives has a slightly different definition of urban resilience, and while diversity is one of the principles of resilience, definitions are often built on a number of implicit assumptions that need to be surfaced to understand questions related to 'resilience of what' and 'resilience for whom?' (Carpenter et al. 2001; Brown 2015; Olsson et al. 2015; Meerow et al. 2016; Meerow and Newell 2019).

[1] https://www.c40.org/.

[2] https://www.100resilientcities.org/.

[3] http://www.fractal.org.za/.

[4] https://www.worldbank.org/en/topic/disasterriskmanagement/brief/city-resilience-program.

[5] https://futureearth.org/networks/knowledge-action-networks/urban/.

[6] http://urbanresiliencehub.org/.

3 Urban Resilience from Multiple Perspectives

Urban resilience is not a new term. It has been used widely in many fields when the future of cities and urban spaces is spoken about, and one of the Sustainable Development Goals (Goal 11) explicitly calls for 'inclusive, safe, resilient and sustainable' cities. However, definitions of what resilient cities look like and how they function differ depending on the application of the term. This has resulted in resilience being labelled a 'boundary concept', a concept that enables a common understanding to be built among diverse actors and user-groups by facilitating conceptual communication about the multidimensional nature of resilience (Star and Griesemer 1989; Mollinga 2008). While this malleability can allow for collective sense-making and co-production of knowledge (Reyers et al. 2015; Sitas et al. 2016), it can also result in difficulties when it comes to operationalising resilience-building activities, and can hamper monitoring resilience through the use of indicators and other metrics (Quinlan et al. 2016). Meerow et al. (2016) identify six conceptual tensions in urban resilience literature, the first five of which are also evident in broader resilience literature: (1) differences in those studies focusing on equilibrium compared to non-equilibrium resilience; (2) normative conceptualisations of resilience (positive, neutral or negative); (3) differences in the ways in which system change is perceived (i.e. whether systems persist, transition or transform); (4) adaptation as opposed to general adaptability and (5) timescale of action. The sixth tension according to Meerow et al. (2016) relates specifically to urban resilience and is linked to how 'urban' is defined as this is different based on the various disciplines or theoretical constructs through which the term is viewed. The second tension identified by Meerow et al. (2016) has particular resonance with those working with environmental and social justice approaches. The use of resilience, whether it is for advancing theory, designing or implementing policies, or undertaking resilience-building activities, is not neutral. Many questions relating to urban governance and resilience initiatives are linked to questions around justice and ethics (Allen et al. 2017; Ziervogel et al. 2017; Borie et al. 2019). The normative nature of resilience requires that narratives on urban resilience and the intended outcomes that are envisioned through them, need to be negotiated through deliberative platforms that are cognisant of power asymmetries. The concept of disruptive resilience looks at the potential of embracing informality, autonomous innovation and distributed data sources, while harnessing the unique properties of urban areas to devise new and experimental responses to the COVID-19 pandemic and extending this learning to climate resilience, over time (Bahadur and Dodman 2020). However, it needs to be considered alongside the hard-fought notions of justice, ethics and social, process-driven negotiation brought into the rubric of resilience.

4 Urban Resilience from a Social-Ecological Systems Perspective

Social-ecological systems are characterised by complex and adaptive interactions across scales in space, time and organisational hierarchies. These cross-scale dynamics can result in novel or nonlinear behaviour that may cause surprising shifts in the way the system is structured and functions (Levin et al. 2013). A social-ecological systems perspective therefore provides a lens to better understand how complex systems like cities deal with change, uncertainty and surprise (Reyers et al. 2018). In this context, a resilient city is one that navigates change and surprise (through persisting, adapting and/or transforming) in ways that enable a sustained supply of a desired set of ecosystem services that are important for human wellbeing. A set of key principles have been suggested for building resilience in social-ecological systems, three of which focus on system properties and four of which concern system governance (Fig. 1) (Biggs et al. 2012). We outline how these principles can be used within the context of urban resilience (Table 1) and provide examples from the Global South covered in more detail within this book to illustrate some of the dimensions.

5 Urban Social-Ecological Resilience Policies and Practices

Cities face complex sustainability challenges, and are increasingly looking for creative, cost-effective solutions that deliver multiple co-benefits to a range of urban residents (Keeler et al. 2019). While there is still a long way to go in terms of integrating a SER lens into urban planning and management practices, there are some examples of how this has been done in specific places and with diverse sectors and collaborations. Nagendra et al. (2018) identify unique southern opportunities for urban development that allow for innovations grounded in local realities, including: (1) leapfrogging over traditional development trajectories through experimenting with green infrastructure, producing and using lower emission-intensive materials and lowering the barriers of entrance by promoting the use of appropriate technologies; (2) using decentralised and bottom-up approaches in decision-making processes to address infrastructure deficits and minimise path dependencies and 'lock-in' effects linked to large infrastructure projects and (3) nurturing and innovating with informal governance processes that build on existing social capital. This last point has become increasingly important during response activities to global crises such as the COVID-19 pandemic (Alonge et al. 2019).

One strategy for incorporating a SER approach in urban planning and management is based on 'nature-based solutions' (or green infrastructure) which include street trees, parks and open spaces, rain gardens, green roofs, small-scale urban farming and living/natural shorelines (Keeler et al. 2019). Nature-based solutions aim to build resilience by promoting public health and safety, enhancing quality of life and reconnecting urban residents to natural systems. In addition, they restore

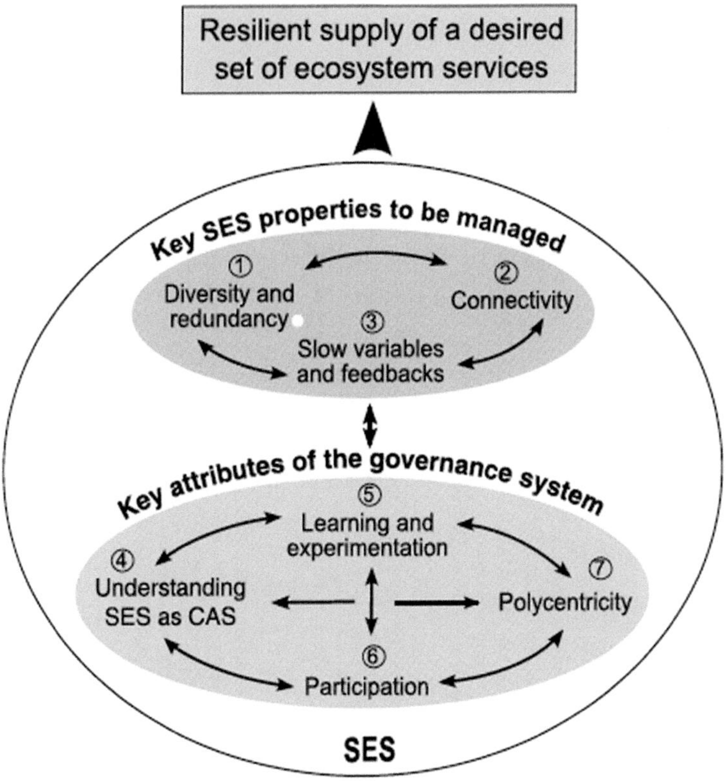

Fig. 1 Seven principles of resilience according to Biggs et al. (2015) grouped into those that relate to generic social-ecological system properties to be managed and those that relate to key properties of social-ecological system governance. Abbreviations: Complex adaptive system (CAS); Social-ecological system (SES). From Biggs et al. (2015) with permission

and improve natural hydrologic and ecological processes in urban environments, thereby contributing to climate change adaptation (Gajjar et al. 2016). Many cities are exploring such nature-based or ecosystem-based approaches to adaptation, as they often provide a suite of co-benefits that go beyond their primary adaptation goal (Kabisch et al. 2017). For example, a river revitalisation project in Lilongwe, Malawi, adopted an 'urban tinkering' approach to turn organic waste collected from the river into compost, which can be used on local farming lots and sold to private companies. As a result, both waste management and livelihoods in the area were improved (ICLEI CBC 2019). Urban tinkering may be a particularly useful way to implement nature-based adaptation in the Global South, as it provides an alternative to lengthy, capacity-intensive and top-down planning processes. Instead, urban tinkering is based on small-scale, 'safe-to-fail' experimentation, making use of existing structures and relationships and seeing urban informality as an opportunity for flexibility and innovation (Elmqvist et al. 2018; ICLEI CBC 2019). River restoration projects

Table 1 The principles of resilience identified by Biggs et al. (2015) as related to urban resilience, with examples from urban systems in the Global South

Principle of resilience	How they relate to urban resilience	Examples/link to chapters in book
1. Maintain diversity and redundancy In a SES, components such as species, landscape types, knowledge systems, actors, cultural groups or institutions all provide different options for responding to change and dealing with uncertainty and surprise	Cities rely acutely on a range of natural resources (usually outside of their core boundaries) to continue supplying water, food, clean air and many other services for their inhabitants. This high reliance means that they cannot usually afford to lose the supply of these services. Maintaining multiple lines of supply and storage facilities for these resources ensures that they are buffered from shocks and unexpected changes that inevitably occur. This same principle also applies to how institutions are structured, and how a city maintains its 'economic complexity', and diversity of skills	The City of Cape Town's water supply relied only on dams which were negatively affected by drought, having a more diverse water supply would have mitigated the impacts of the drought in 2018 (Rodina 2019; Ziervogel et al. 2019a) Chapter 15
2. Connectivity Connectivity can be both a good and a bad thing. High levels of connectivity can facilitate recovery after a disturbance but highly connected systems can also spread disturbances faster	The reliance of urban areas on the resources outside of their boundaries means that they have built complex connections to the 'outside world' through infrastructure, trade, policies and other mechanisms. This connectivity is a double-edged sword. Cities are able to import food from thousands of kilometres away which enables them to thrive without necessarily producing sufficient food for themselves. But this also makes cities vulnerable to changes in areas beyond their control (e.g. changes in oil prices affect food prices locally by increasing transport costs)	Exploring coupled water systems between urban and rural systems in Ethiopia (Chignell and Laituri 2016); Understanding the difference in ecosystem service use between urban and rural systems (Hamann et al. 2015) Chaps. 5 and 14

(continued)

Table 1 (continued)

Principle of resilience	How they relate to urban resilience	Examples/link to chapters in book
3. Manage slow variables and feedbacks SES can often be 'configured' in several different ways. In other words, there are many ways in which all the variables in a system can be connected and interact with one another. Often, slow-changing variables play an important role in determining what configuration a system is in. When slow variables reach a threshold, a system can 'flip' into a different state	As SES, cities in the Global South are made up of multiple interconnected and complex relationships. On some level, there are processes that are changing rapidly (e.g. in-migration). While these changes are fast, they are 'easy' to see and track (i.e. one can see the increases in informal settlements). On the other hand, there are less visible, slow-moving processes that are harder to track and attribute to single causes, because they usually start far earlier than their impacts manifest. For instance, the loss of highly skilled people ('brain drain') from cities is usually a slow-moving process whose impacts are felt when a critical threshold is reached	Peres and du Plessis (2013) look at the interconnections between slow (press) and faster (pulse) disturbances for community resilience in an informal settlement in Tshwane, South Africa
4. Learning and experimentation Knowledge of a SES is always partial and incomplete. Efforts to enhance the resilience of SES must therefore be supported by continuous learning and experimentation	Recognising that learning is continuous and knowledge will always be limited, there are emerging concepts which encourage constant learning in urban areas, rather than following blueprints	Urban tinkering builds on evolutionary theory but focuses on designing new and redesigning existing urban structures by modifying and molding 'existing traits and forms' (see Elmqvist et al. 2018). This is one of several examples of embracing continuous learning as a critical process of building resilient and sustainable cities

(continued)

Table 1 (continued)

Principle of resilience	How they relate to urban resilience	Examples/link to chapters in book
5. Broaden participation Participation through active engagement of all relevant stakeholders is considered fundamental to building social-ecological resilience. It helps build the trust and relationships needed to improve legitimacy of knowledge and authority during decision-making processes	An understanding of equity—which is essential to achieve sustainability outcomes (see Leach et al. 2018)—suggest a broadened conceptualisation of equity (McDermott et al. 2013) that includes not just redistribution of resources, but also elevation of different voices, and understanding the role of context and other structural factors in achieving resilience and sustainable outcomes. This is even more pertinent in cities, where people and resources are in constant flux. Designing systems that are adaptable to these constantly changing conditions while maintaining inclusivity is both a big challenge and a critical necessity	Current protests around the world reflect the neglect of this principle. People do not see themselves represented in the decisions taken by governments, municipalities or cities within which they live (Fernández et al. 2016)
6. Polycentricity Polycentricity, a governance system in which multiple governing bodies interact to make and enforce rules within a specific policy arena or location, is considered to be one of the best ways to achieve collective action in the face of disturbance or change	Urban spaces are complex adaptive systems, made up of interconnected relationships operating across time and space, thus governance structures need to be flexible to respond to changing contexts and couplings. Polycentric governance is a complex form of governance with multiple centres of decision-making, each of which operates with some degree of autonomy	The notion of 'incremental urbanism' or co-production platforms as practised by Justicia Hidrica (Water Justice movement) in Latin America, demonstrates a decentralised way of collectively managing urban water resources (see Allen et al. 2017)

(continued)

Table 1 (continued)

Principle of resilience	How they relate to urban resilience	Examples/link to chapters in book
7. Understanding SES as complex and adaptive For us to continue to benefit from a range of ecosystem services, we need to understand the complex interactions and dynamics that exist between actors and ecosystems in a SES	Cities are complex systems of interacting people and natures in densely built spaces, serviced by infrastructure and managed by social and political organisations (Bettencourt 2015). Attempts to better manage urban spaces along only a few of these features rarely succeed. A more holistic, complex adaptive systems understanding of urban spaces is needed, where all aspects of urban life come together	See Harrison et al. (2014) for a tool to promote urban resilience thinking in municipalities based on systems thinking. The idea is to link different disciplines, practitioner communities and knowledges for an improved understanding of urban SES

in eThekwini, South Africa, the Aller River Pilot Project and the Palmiet River Rehabilitation Project, were initiated as part of the city's resilience journey (Martel and Sutherland 2019). Both projects pursued intense community interaction to build trust between city officials and community members, employing the Eco-Champs—who were existing community members living and working in the catchment—as intermediaries. A form of community-based monitoring is conducted regularly, using a river walk methodology. Engaging with eco-clubs in local schools and organising regular community events and cleanliness drives, are additional practices for achieving sustained community interest.

Below we outline two case studies that highlight how resilience has been integrated into urban planning processes.

6 Learning from Drought in Cape Town, South Africa

According to the World Population Review, Cape Town is South Africa's second largest city after Johannesburg and the metropolitan area is home to over four million people. With the population growing by 7.1% from 2011 to 2016, and expected to grow in the future, the city will not cope with continued in-migration without concerted efforts to address housing, infrastructure and employment demands (Small 2017). A major shock to Cape Town's social-ecological system was the recent extreme drought which started in 2015 (Matikinca et al. 2020). This period saw the catchment areas that supply the city suffer the driest period since the 1930s. Because of the city's near-exclusive reliance on surface water, this drought made it very vulnerable. As a response, the city started a campaign to enlist households, businesses and citizens to help deal with the drought, using a suite of price and non-price mechanisms (Matikinca et al. 2020; Simpson et al. 2020). Some affluent households replaced lawns and water-sensitive plants with alternatives requiring less water, the public reduced their personal water use and, in some cases, people collected greywater for use in toilet flushing. Others invested in water-saving devices such as low-flow taps, water-efficient shower heads and smaller toilet cisterns. However, for many, dealing with limited water supply and access to basic water and sanitation services was nothing new. Many residents, especially those in informal settlements, have been coping with chronic water deficits and inequities in water infrastructure and access for decades (Smith and Hanson 2003; Enqvist and Ziervogel 2019). Nevertheless, the city's response to the drought crisis was swift. In 2017, a Chief Resilience Officer was appointed as the main champion of resilience in the city and has guided the development of Cape Town's Resilience Strategy, which was approved towards the end of 2019. In its resilience strategy, the City of Cape Town lists five pillars as important for its resilience, i.e. aiming to be a (1) compassionate, holistically healthy city, (2) connected, climate-adaptive city, (3) capable, job-creating city, (4) collectively, shock-ready city and (5) collaborative, forward-looking city. These pillars highlight the complexities of a fast-changing city with multiple stressors including mental health, poverty, climate change, unemployment and connectivity to the global landscape. Ecological infrastructure has become part of the solution toolbox, alongside human-made infrastructure. Cape Town was also part of the '100 Resilient Cities' network from 2016. This network comprised 100 cities from across the world (10 of which are in Africa) that committed to building urban resilience around the social, economic and physical challenges of the twenty-first century. The complexity of the city—with challenges ranging from poverty to drought—makes it difficult for one strategy to address everything. Emerging analysis of the drought responses in Cape Town, and the mental models underpinning them, suggests some shortcomings related to whether the solutions put forward are long term and stable over time (Simpson et al. 2020). A number of research projects have since been initiated to track how the collaborations formed in this time of crisis continue to provide solutions for a sustainable Cape Town in the future (Enqvist and Ziervogel 2019; Ziervogel et al. 2019a).

7 Building Back Better, but for Whom? Chennai, India

Chennai, a megacity on the eastern coast of India, is home to just over seven million people, 29% of whom live in informal structures, according to the 2011 census. Following the Asian tsunami in 2004 and alternating periods of droughts, cyclones and floods, Chennai embarked on several resilience-building initiatives and studies (Joerin et al. 2014; Greeshma and Kumar 2016; Manohar and Muthaiah 2016). One of these included joining the 100 Resilient Cities initiative in 2014 just before devastating floods in 2015 which resulted in over 400 deaths and displaced many vulnerable communities living in marginal landscapes, mostly in informal settlements. At the heart of these resilience-building initiatives was a drive to understand key drivers of change, how they were interlinked, and where important interventions could be made that could both enhance the resilience of Chennai to extreme weather events and enable it to adapt and transform to demographic and land-use changes (Padgham and Jabbour 2014; Arabindoo 2016; Idicheria et al. 2016; Yadav and Rahman 2016). Through broad stakeholder engagement with citizens and experts from academia, the private sector, civil society and government, Chennai co-developed a Resilience Strategy[7] founded on a vision of an enlightened, just and integrated Chennai. Through this engagement, the city also appointed a Chief Resilience Officer in 2018 to spearhead the initiative and engage with relevant authorities within city governance structures in an attempt to work beyond sectoral silos. The resilience strategy is based on five pillars, each with a number of goals associated with them: (1) Healthy and planned urbanisation ('urbanising responsibly'); (2) water systems ('carving a resilient future around our water resources'); (3) disaster preparedness ('making Chennai a prepared city'); (4) governance ecosystem ('together we lift Chennai') and (5) Chennai's vulnerable communities ('valuing the city's vulnerable'). As this

[7] Chennai Resilience Strategy 2019, Rockerfeller Foundation, found online at: https://100resilientcities.org/wp-content/uploads/2019/06/Resilience-Strategy-Chennai-English.pdf.

is a new assessment, and many of the impacts of the implementation activities are yet to be initiated or realised, concerns have been raised that responses to previous disasters have exacerbated inequalities. While disasters often provide opportunities for different actors to deepen their activities (e.g. NGOs assisting communities) and unlock resources for rebuilding infrastructure and forging new relationships, these opportunities can also facilitate certain agendas to be furthered, which often privilege the powerful elite at the expense of marginalised groups (Janakarajan et al. 2007; Jayarman 2016; Jain et al. 2017).

8 Urban Social-Ecological Resilience as Transformation

The principle of leaving no one behind necessitates that implementation activities aligned with the SDGs are aimed at improving the lives of the poorest and most vulnerable, irrespective of their gender, race, sexuality, age, ability, religion or location of residence (Görgens and Ziervogel 2019). In urban planning, many of these categories are either tackled in isolation (e.g. age or ability), or are addressed in an additive manner (e.g. focusing on poor, disabled women), when in reality these categories interact in complex and often surprising ways. Improving the understanding of these interconnections and how they are mediated by power at both local and broader scales is fundamental to identifying important leverage points in systems where transformative actions that benefit both people and the biosphere that supports them, can be implemented (Pelling and Manuel-Navarrete 2011; Abson et al. 2017). This is especially important as marginalised groups are the most vulnerable, but often also have unique capacities for building resilience and adapting to change.

Urban inequities are the result of complex, interconnected relationships which are experienced differently depending on social positions or categories, geographies and historical legacies. Access to essential ecosystem services within urban spaces, and exposure to risks, are affected by multiple social variables including race (especially in the case of countries with a history of racial segregation, like South Africa), ethnicity, gender, sexuality, class, age, nationality, geographic location and legal status, among other social categories (Ernstson 2013; Wolch et al. 2014; Cutter et al. 2006; Nagendra et al. 2018; Venter et al. 2020). The flow of ecosystem services, how people benefit from these services and the underlying resilience of the systems that co-produce the services are shaped by these intersecting variables, resulting in an inequitable distribution and access to these ecosystem services (Geneletti et al. 2020).

Text Box 1: Using an intersectionality lens to boost social-ecological resilience in eThekwini, South Africa

According to the City of eThekwini's Resilience Journey through the Rockefeller Foundation's 100 Resilient Cities initiative, there are six 'levers of change' that are essential in building eThekwini's resilience:

Lever 1: Strengthen local communities and build social cohesion;
Lever 2: Improve effectiveness of education and skills development;
Lever 3: Promote economic growth in line with twenty-first-century trends and opportunities;
Lever 4: Manage environmental assets effectively;
Lever 5: Create a more inclusive and integrated spatial plan and
Lever 6: Improve municipal effectiveness

What we know from resilience and complexity thinking is that all of these levers are interconnected, some in reinforcing ways while others will result in trade-offs where some citizens benefit at the expense of others. To address issues related to inequality and implementing the city vision of making eThekwini Africa's most liveable city, an intersectionality lens can help identify implementation activities linked to each of the levers of change with a focus on balancing power asymmetries. Some activities might include identifying and restoring strategic ecosystem service areas that are critical for mitigating against disasters (through local employment, education and small business support). By coupling this with a better understanding of the lived realities, vulnerabilities and experiences of residents living in these areas, opportunity contexts for interventions can be surfaced that can bolster social-ecological resilience of people and communities. Here, questions such as 'who benefits, who dominates and who gets left behind?' are good starting points. While it is important to determine who has power *over* what or whom, it is also critical to stimulate power *with* whom to identify entry points for mobilising or nurturing collective action and facilitating new alliances for change.

Using an intersectional approach therefore enables a deeper reflection on power, and how it mediates the interlinkages between intersecting social variables. Who benefits how from the biosphere? An intersectional approach maintains that human wellbeing, equity and biosphere protection is most effective when pursued collectively and enables the broadening of multiple possible transformative pathways. It also seeks to explore options beyond just reducing environmental burdens, to understanding the distribution of the burdens and benefits. Applying an intersectional lens to urban planning also helps to consider how the scale, connectivity and speed of resource flows affect the wellbeing of people differently (Maciejewski et al. 2021). Where resources are flowing (and their associated feedbacks), how they flow (which might be mediated by infrastructure or institutional arrangements) and how fast they flow (rapidly urbanising spaces in the Anthropocene) are all mediated by certain informal and formal governance mechanisms (see Text Box 1). Due to their complex nature, cities force us to think beyond the local to better understand urban footprints, cross-scale connections and telecoupling with far-away places and rural areas, and compel us to think beyond persistence (Boillat et al. 2018). Cityscapes are changing along with development priorities, and transformations are necessary. To understand where to intervene therefore requires a systems-based approach grounded within a framework that considers critical intersections and dimensions of equity.

9 Gaps, Blindspots and Opportunities for a Research-Action Agenda for Urban Resilience

Urban resilience research and practice have increased rapidly in the last decade, with a strong focus on understanding how urban social-ecological systems can be more resilient in light of the rapid pace and scale of change in the Anthropocene. A number of scholars working on urban resilience have outlined key gaps and opportunities for advancing SER praxis. Elmqvist et al. (2019) identify four challenges for future research relating to (1) whether urbanisation results in diversification or simplification of coupled human-nature systems, (2) whether rising connectivity between cities is becoming a force of its own in shaping governance and the biosphere, (3) how to better understand the role of connectivity and whether being 'too' connected can result in new vulnerabilities and (4) the role of cities for nudging global development towards more positive futures. Meerow et al. (2016) and Meerow and Newell (2019) suggest that resilience researchers, practitioners and policymakers should ask themselves five key questions to get a more nuanced understanding of competing interests, burdens and benefits. These include asking questions in relation to 'who?', 'what?', 'when?', 'where?', and 'why?'. For example, who determines what is desirable for an urban system, whose resilience is prioritised? And who is included (and excluded) from the system? What shocks and stresses should the urban system be resilient to? What networks and sectors are included or excluded in the urban system? Is the focus of the work on generic or specific resilience? Is the focus of the work on rapid or slow changes or disturbances, short-term or long-term resilience, and concerned with resilience of present or future generations? Where are the spatial boundaries of the urban system? Are some areas more prioritised over others? And does building resilience in some areas affect resilience in other areas? What is the goal of building resilience, what are the underlying motivations for building resilience? Is the focus of the work on process or outcomes?

For the African context, O'Farrell et al. (2019) suggest and expand on three areas of intervention for advancing urban resilience: (1) co-developing knowledge with diverse actor groups and facilitating and allowing for experimentation; (2) mobilising knowledge and ensuring that it is accessible for diverse end-users (from researchers, practitioners, policymakers to citizens), and that communication on sustainability and resilience should be engaging and user-useful and (3) developing or subverting institutions to allow for effective management and governance. Similarly, Fernández et al. (2016) suggest that to break resilient patterns of inequality, powerful sectors that have benefited from neoliberal policies need to be overcome by strengthening social movements and collective action.

As resilience is becoming increasingly adopted in cities to develop policy and inform implementation activities, decision makers are becoming more aware about the important system components that constitute key relationships and feedbacks in a city. While this adoption is happening both in the Global North and South, research on and from cities in the Global South, as well as southern concepts, theories and practices are not prominent or visible in urban sustainability publications (Nagendra et al. 2018). In addition, many urban 'solutions' developed within the context of the Global North are not applicable to southern cities, for example new waste technologies that do not take into account manual and informal waste picker networks, expensive subway systems or exclusively market-driven urban renewal processes. There is a need for a new type of urban research premised on the conditions, constraints and dynamics of the Global South (Pieterse and Simone 2013; Groffman et al. 2017; Simone and Pieterse 2018; Shackleton et al. 2021).

In summary, the evolution of the resilience framework, with a focus on social-ecological resilience, emphasises the need to define 'resilience of what' and 'for what', which indirectly also highlights what is being left out. This approach is not a panacea, but it helps to guide thinking about key features, relationships and feedbacks within systems. Drawing on an intersectionality approach can further advance how resilience can be conceptualised in a more nuanced way to account explicitly for how power mediates certain relationships, in an attempt to better understand opportunities for transformation. This highlights that while most of the research on resilience, especially within an urban context, focuses on building or boosting resilience, there are many instances where resilient feedbacks or traps have locked the system in states which exacerbate inequitable benefit flows. A social-ecological resilience lens can be used to understand where to intervene in a system to undermine these feedbacks and create more just and sustainable urban futures.

References

Abson DJ, Fischer J, Leventon J, Newig J, Schomerus T, Vilsmaier U, von Wehrden H, Abernethy P, Ives CD, Jager NW, Lang DJ (2017) Leverage points for sustainability transformation. Ambio 46:30–39

Allen A, Walnycki A, von Bertrab É (2017) The co-production of water justice in Latin American cities. In: Allen A, Griffin L, Johnson C (eds) Environmental justice and urban resilience in the Global South. Palgrave Macmillan, New York, pp 175–193

Alonge O, Sonkarlay S, Gwaikolo W, Fahim C, Cooper JL, Peters D (2019) Understanding the role of community resilience in addressing the Ebola virus disease epidemic in Liberia: a qualitative study (community resilience in Liberia). Glob Heal Action 12:1662682

Arabindoo P (2016) Unprecedented natures? An anatomy of the Chennai floods. City 20:800–821

Bai X, Dawson RJ, Ürge-Vorsatz D, Delgado GC, Barau AS, Dhakal S, Dodman D, Leonardsen L, Masson-Delmotte V, Roberts DC, Schultz S (2018) Six research priorities for cities and climate change. Nat 555:23–25

Bahadur AV, Dodman D (2020) Disruptive resilience: an agenda for the new normal in cities of the global South. IIED Brief: 1–4

Bettencourt LM (2015) Cities as complex systems. In: Furtado BA, Sakowski PA, Tovolli MH (eds) Modeling complex systems for public policies, IPEA, Brasilia, pp 217–236

Biggs R, Schlüter M, Biggs D, Bohensky EL, BurnSilver S, Cundill G, Dakos V, Daw TM, Evans LS, Kotschy K, Leitch AM (2012) Toward principles for enhancing the resilience of ecosystem services. Ann Rev Enviro Res 37:421–448

Biggs R, Schlüter M, Schoon ML (eds) (2015) Principles for building resilience: sustaining ecosystem services in social-ecological systems. Cambridge University Press, Cambridge
Borie M, Ziervogel G, Taylor FE, Millington JD, Sitas R, Pelling M (2019) Mapping (for) resilience across city scales: An opportunity to open-up conversations for more inclusive resilience policy? Enviro Sci Pol 99:1–9
Boillat S, Gerber JD, Oberlack C, Zaehringer JG, Ifejika Speranza C, Rist S (2018) Distant interactions, power, and environmental justice in protected area governance: a telecoupling perspective. Sustain 10(11):3954
Brown K (2015) Resilience, development and global change. Routledge, London
Carpenter S, Walker B, Anderies JM, Abel N (2001) From metaphor to measurement: resilience of what to what? Ecosys 4:765–781
Chaplin-Kramer R, Sharp RP, Weil C, Bennett EM, Pascual U, Arkema KK et al (2019) Global modeling of nature's contributions to people. Sci 366(6462):255–258
Chignell SM, Laituri MJ (2016) Telecoupling, urbanisation, and the unintended consequences of water development aid in Ethiopia. In: Wessel GR, Greenberg JK (eds) Geoscience for the public good and global development: toward a sustainable future. Geological Society of America, pp 25–135
Cote M, Nightingale AJ (2012) Resilience thinking meets social theory: situating social change in socio-ecological systems (SES) research. Prog Hum Geog 36:475–489
Cutter SL, Emrich CT, Mitchell JT, Boruff BJ, Gall M, Schmidtlein MC, Burton CG, Melton G (2006) The long road home: race, class, and recovery from Hurricane Katrina. Enviro Sci Pol Sustain Dev 48:8–20
Davoren E, Shackleton CM (2021) Urban ecosystem disservices in the Global South. In: Shackleton CM, Cilliers SS, Davoren E, du Toit MJ (eds) Urban ecology in the Global South. Springer, Cham, pp 265–292
Depietri Y, Renaud FG Kallis G (2012) Heat waves and floods in urban areas: a policy-oriented review of ecosystem services. Sustain Sci 7: 95–107
Díaz S, Settele J, Brondízio ES, Ngo HT, Guèze M, Agard J, Arneth A, Balvanera P, Brauman KA, Butchart HM, Chan KA, Garibaldi LA, Ichii K, Liu J, Subramanian SM, Midgley GF, MiloslavichP Molnár Z, Obura D, Pfaff A, Polasky S, Purvis A, Razzaque J, Reyers B, Chowdhury R, Shin YJ, Visseren-Hamakers IJ, Willis KJ, Zayas CN (eds) (2019) Summary for policymakers of the global assessment report on biodiversity and ecosystem services of the Intergovernmental Science-Policy Platform on Biodiversity and Ecosystem Services. IPBES secretariat, Bonn
Dickson E, Baker JL, Hoornweg D, Asmita T (2012) Urban risk assessments: an approach for understanding disaster and climate risk in cities. World Bank, New York
du Toit MJ, Cilliers SS, Dallimer M, Goddard M, Guenat S, Cornelius SF (2018) Urban green infrastructure and ecosystem services in sub-Saharan Africa. Landsc Urb Plan 180:249–261
Elmqvist T, Siri J, Andersson E, Anderson P, Bai X, Das PK, Gatere T, Gonzalez A, Goodness J, Handel SN, Török EH (2018) Urban tinkering. Sustain Sci 13:1549–1564
Elmqvist T, Andersson E, Frantzeskaki N, McPhearson T, Olsson P, Gaffney O, Takeuchi K, Folke C (2019) Sustainability and resilience for transformation in the urban century. Nat Sustain 2:267–273
Enqvist JP, Ziervogel G (2019) Water governance and justice in Cape Town: an overview. Wiley Interdisc Rev Water 6(4):e1354
Ernstson H (2013) The social production of ecosystem services: a framework for studying environmental justice and ecological complexity in urbanised landscapes. Landsc Urb Plan 109:7–17
Fernández I, Manuel-Navarrete D, Torres-Salinas R (2016) Breaking resilient patterns of inequality in Santiago de Chile: challenges to navigate towards a more sustainable city. Sustain 8(8):820
Gajjar SP, Revi A, Jain G, Bazaz A (2016) Bangalore, India. In: Bartlett S, Satterthwaite D (eds) Cities on a finite planet: towards transformative responses to climate change. Routledge, London, pp 41–62

Geneletti D, Cortinovis C, Zardo L, Esmail BA (2020) Planning for ecosystem services in cities. Springer, Cham

Glaeser EL, Resseger M, Tobio K (2009) Inequality in cities. J Reg Sci 49:617–646

Görgens T, Ziervogel G (2019) From "No One Left Behind" to "Putting the Last First": centring the voices of disabled people in resilience work. The Palgrave handbook of disability and citizenship in the Global South. Palgrave Macmillan, Cham, pp 85–102

Greeshma P, Kumar KM (2016) Disaster Resilience in vulnerable cities through neighbourhood development: a case of Chennai. Procedia Techn 24:1827–1834

Groffman PM, Cadenasso ML, Cavender-Bares J, Childers DL, Grimm NB, Grove JM, Hobbie SE, Hutyra LR, Jenerette GD, McPhearson T, Pataki DE (2017) Moving towards a new urban systems science. Ecosys 20:38–43

Hamann M, Berry K, Chaigneau T, Curry T, Heilmayr R, Henriksson PJ, Hentati-Sundberg J, Jina A, Lindkvist E, Lopez-Maldonado Y, Nieminen E (2018) Inequality and the biosphere. Ann Rev Enviro Res 43:61–83

Hamann M, Biggs R, Reyers B (2015) Mapping social-ecological systems: identifying 'green-loop' and 'red-loop' dynamics based on characteristic bundles of ecosystem service use. Glob Enviro Change 34:218–226

Harris LM, Chu EK, Ziervogel G (2018) Negotiated resilience. Resil 6:196–214

Harrison P, Bobbins K, Culwick C, Humby TL, La Mantia C, Todes A, Weakley D (2014) Urban resilience thinking for municipalities. Wits University Press, Johannesburg

Hobbie SE, Grimm NB (2020) Nature-based approaches to managing climate change impacts in cities. Phil Trans Royal Soc B 375(1794):20190124

ICLEI CBC (2019) Handbook 3: Innovative approaches to planning in an African context: Urban tinkering in Malawi. In: The value of urban natural assets when planning for resilient Africa cities: consideration and decision-making processes. ICLEI CBC, Cape Town

Idicheria C, Neelakantan A, Graft A, Banerjee A, Kumar K (2016) Transforming Chennai: a research report in building micro, small and medium enterprise resilience to water related environmental change. Okapi and Mercy Corps, New Delhi

Intergovernmental Panel on Climate Change (IPCC) (2014) Mitigation of climate change. Contribution of Working Group III to the Fifth Assessment Report of the Intergovernmental Panel on Climate Change, Cambridge University Press, Cambridge

Jain G, Singh C, Coelho K, Malladi T (2017) Long-term implications of humanitarian responses: the case of Chennai. IIED Working Paper, IIED, London

Janakarajan S, Butterworth J, Moriarty P, Batchelor C (2007) Strengthened city, marginalised peri-urban villages: Stakeholder dialogues for inclusive urbanisation in Chennai, India. In Peri-Urban Waterconflicts-Supporting Dialogue and Negotiations; Technical Paper Series 50; In: Butterworth J, Ducrot R, Faysse N, Janakarajan S (eds.), International Water Sanitation Centre, Delft, The Netherlands, pp 51–75

Jayarman N (2016) Chennai's seashore beautification and fisherfolk resistance, India. Environmental Justice Atlas. Retrieved from, https://ejatlas.org/conflict/the-beautification-of-chennais-seashore-and-fisherfolk-resistance-india. Accessed on March 2020

Jenerette GD, Harlan SL, Stefanov WL, Martin CA (2011) Ecosystem services and urban heat riskscape moderation: water, green spaces, and social inequality in Phoenix, USA. Ecol Appl 21:2637–2651

Joerin J, Shaw R, Takeuchi Y, Krishnamurthy R (2014) The adoption of a climate disaster resilience index in Chennai, India. Disasters 38:540–561

Kabisch N, Korn H, Stadler J, Bonn A (eds) (2017) Nature-based solutions to climate change adaptation in urban areas. Linkages between science policy and practice. Springer, Cham

Keeler BL, Hamel P, McPhearson T, Hamann MH, Donahue ML, Prado KA, Arkema KK, Bratman GN, Brauman KA, Finlay JC, Guerry AD (2019) Social-ecological and technological factors moderate the value of urban nature. Nat Sustain 2:29–38

Kumar P, Yashiro M (2014) The marginal poor and their dependence on ecosystem services: evidence from South Asia and Sub-Saharan Africa. In: von Braun J, Gatzweiler F (eds) Marginality. Springer, Dordrecht, pp 169–180

Leach M, Reyers B, Bai X, Brondizio ES, Cook C, Díaz S, Espindola G, Scobie M, Stafford-Smith M, Subramanian SM (2018) Equity and sustainability in the Anthropocene: a social–ecological systems perspective on their intertwined futures. Glob Sustain 1(e13):1–13

Levin SA, Xepapadeas T, Crepin AS, Norberg J, de Zeeuw A, Folke C, Hughes T, Arrow K, Barrett S, Daily G, Ehrlich P, Kautsky N, Mäler KG, Polasky S, Troell M, Vincent JR, Walker B (2013) Social-ecological systems as complex adaptive systems: modeling and policy implications? Env Dev Eco 18:111–132

Maciejewski K, Currie P, O'Farrell P (2021) Social-ecological connectivity in Global South cities. In: Shackleton CM, Cilliers SS, Davoren E, du Toit MJ (eds) Urban ecology in the Global South. Springer, Cham, pp 347–364

Manohar L, Muthaiah KT (2016) Towards resilience in Chennai. Intn Plan Hist Soc Proc 17:252–260

Matikinca P, Ziervogel G, Enqvist JP (2020) Drought response impacts on household water use practices in Cape Town, South Africa. Water Pol 22:483–500

Martel P, Sutherland C (2019) governing river rehabilitation for climate adaptation and water security in Durban, South Africa. In: Cobbinah PB, Addaney M (eds) The geography of climate change adaptation in urban Africa. Palgrave Macmillan, Cham, pp 355–387

McDermott M, Mahanty S, Schreckenberg K (2013) Examining equity: a multidimensional framework for assessing equity in payments for ecosystem services. Environ Sci Pol 33:416–427

McPhearson T, Andersson E, Elmqvist T, Frantzeskaki N (2015) Resilience of and through urban ecosystem services. Ecosys Serv 12:152–156

Meerow S, Newell JP (2019) Urban resilience for whom, what, when, where, and why? Urban Geogr 40:309–329

Meerow S, Newell JP, Stults M (2016) Defining urban resilience: a review. Landsc Urb Plan 147:38–49

Meerow S, Stults M (2016) Comparing conceptualisations of urban climate resilience in theory and practice. Sustain 8(7):701

Mitchell LE, Lin JC, Bowling DR, Pataki DE, Strong C, Schauer AJ, Bares R, Bush SE, Stephens BB, Mendoza D, Mallia D (2018) Long-term urban carbon dioxide observations reveal spatial and temporal dynamics related to urban characteristics and growth. Proc Nat Acad Sci 115:2912–2917

Mollinga PP (2008) The rational organisation of dissent: Boundary concepts, boundary objects and boundary settings in the interdisciplinary study of natural resources management. ZEF working paper series #33, Bonn

Myers G (2021) Urbanisation in the Global South. In: Shackleton CM, Cilliers SS, Davoren E, du Toit MJ (eds) Urban ecology in the Global South. Springer, Cham, pp 27-49

Nagendra H, Bai X, Brondizio ES, Lwasa S (2018) The urban south and the predicament of global sustainability. Nat Sustain 1:341–349

Padgham J, Jabbour J (eds) (2014) Building urban resilience: assessing urban and peri-urban agriculture in Chennai, India. United Nations Environment Programme (UNEP), Nairobi

O'Farrell P, Anderson P, Culwick C, Currie P, Kavonic J, McClure A, Ngenda G, Sinnott E, Sitas N, Washbourne CL, Audouin M (2019) Towards resilient African cities: shared challenges and opportunities towards the retention and maintenance of ecological infrastructure. Glob Sustain 2(e19):1–6

Olsson L, Jerneck A, Thoren H, Persson J, O'Byrne D (2015) Why resilience is unappealing to social science: theoretical and empirical investigations of the scientific use of resilience. Sci Adv 1(4):e1400217

Pelling M, Manuel-Navarrete D (2011) From resilience to transformation: the adaptive cycle in two Mexican urban centers. Ecol Soc 16(2):11

Peres E, Du Plessis C (2013) The threat of slow changing disturbances to the resilience of African cities. Think tank on Resilient Urban Systems in Transition (TRUST), Pretoria

Pieterse EA, Simone AM (2013) Rogue urbanism: emergent African cities. Jacana, Johannesburg

Quinlan AE, Berbés-Blázquez M, Haider LJ, Peterson GD (2016) Measuring and assessing resilience: broadening understanding through multiple disciplinary perspectives. J Appl Ecol 53(3):677–687

Ravera F, Iniesta-Arandia I, Martín-López B, Pascual U, Bose P (2016) Gender perspectives in resilience, vulnerability and adaptation to global environmental change. Ambio 45:235–247
Reyers B, Nel JL, O'Farrell PJ, Sitas N, Nel DC (2015) Navigating complexity through knowledge coproduction: Mainstreaming ecosystem services into disaster risk reduction. Proc Nat Acad Sci 112:7362–7368
Reyers B, Folke C, Moore ML, Biggs R, Galaz V (2018) Social-ecological systems insights for navigating the dynamics of the Anthropocene. Ann Rev Enviro Res 43:267–289
Rosenzweig C, Solecki WD, Hammer SA, Mehrotra S (eds) (2011) Climate change and cities: first assessment report of the urban climate change research network. Cambridge University Press, Cambridge
Rodina L (2019) Water resilience lessons from Cape Town's water crisis. Wiley Interdisc Rev Water 6:e1376
Shackleton CM, Cilliers SS, du Toit M, Davoren E (2021) The need for an urban ecology of the Global South. In: Shackleton CM, Cilliers SS, Davoren E, du Toit MJ (eds) Urban ecology in the Global South. Springer, Cham, pp 1–26
Simon D, Goodness J, Lwasa S, Puppim de Oliveira JA, Macedo LV, Kavonic J, Hermansson Török E, Elmqvist T (2021) Urban governance of and for urban green and blue infrastructure. In: Shackleton CM, Cilliers SS, Davoren E, du Toit MJ (eds) Urban ecology in the Global South. Springer, Cham, pp 403–428
Simone A, Pieterse E (2018) New urban worlds: inhabiting dissonant times. Wiley, Hoboken
Simpson NP, Shearing CD, Dupont B (2020) Gated adaptation during the cape town drought: mentalities, transitions and pathways to partial nodes of water security. Soc Nat Res 33:1041–1049
Sitas N, Reyers B, Cundill G, Prozesky HE, Nel JL, Esler KJ (2016) Fostering collaboration for knowledge and action in disaster management in South Africa. Curr Opin Enviro Sustain 19:94–102
Small K (2017) 2016 community survey Cape Town trends 1996 to 2016. Organisational Policy and Planning Department, Cape Town
Smith L, Hanson S (2003) Access to water for the urban poor in Cape Town: where equity meets cost recovery. Urb Stud 40:1517–1548
Star SL, Griesemer JR (1989) Institutional ecology 'translations' and boundary objects: amateurs and professionals in Berkeley's Museum of vertebrate zoology, 1907–39. Soc Stud Sci 19:387–420
United Nations, Department of Economic and Social Affairs, Population Division (2018) The world's cities in 2018: data booklet (ST/ESA/ SER.A/417). United Nations, New York
Venter ZS, Shackleton CM, Van Staden F, Selomane O, Masterson VA (2020) Green apartheid: urban green infrastructure remains unequally distributed across income and race geographies in South Africa. Landsc Urb Plan 203:103889
Wolch JR, Byrne J, Newell JP (2014) Urban green space, public health, and environmental justice: the challenge of making cities 'just green enough'. Landsc Urb Plan 125:234–244
Yadav M, Rahman Z (2016) The social role of social media: the case of Chennai rains—2015. Soc Netw Anal Mining 6(1):101
Ziervogel G (2019) Building transformative capacity for adaptation planning and implementation that works for the urban poor: Insights from South Africa. Ambio 48:494–506
Ziervogel G, New M, Liu W (2019a) Making cities water-wise and climate-resilient–lessons and experience from the cape town drought. Landsc Archit Front 7:94–100
Ziervogel G, Pelling M, Cartwright A, Chu E, Deshpande T, Harris L, Hyams K, Kaunda J, Klaus B, Michael K, Pasquini L (2017) Inserting rights and justice into urban resilience: a focus on everyday risk. Environ Urban 29:123–138
Ziervogel G, Satyal P, Basu R, Mensah A, Singh C, Hegga S, Abu TZ (2019b) Vertical integration for climate change adaptation in the water sector: lessons from decentralisation in Africa and India. Reg Enviro Change 19:2729–2743

Social-Ecological Connectivity in Global South Cities

Kristine Maciejewski, Paul Currie, and Patrick O'Farrell

Abstract Cities are designed and built to be inhabited by humans, and therefore rely on connectivity to ensure the flow of services, movement, communication and agency for people as well as the flow of benefits or services from nature to people. In this chapter, we focus on the connectivity between the social and ecological components that urban systems are comprised of. We describe the social fabric as the social connection, how people connect with people, which is particularly relevant to the Global South, given the high density of people and rapid population growth in urban centres. People are also impacted by their surrounding environment which forms the focus of the second component where we look at the relationship between *people and nature.* This social-ecological connection represents natures' contribution to people, or the provision of ecosystem services. The long-term sustainable use of these ecosystem benefits can only be ensured through the protection of nature, which relies on establishing ecological connectivity, connecting *nature with nature.* The relationships between these elements result in an emergent outcome which we understand as the city. Influencing the city therefore requires understanding these social and ecological components and how to ensure connectivity.

Keywords Corridors · Ecological infrastructure · Ecosystem services · Green spaces · Social-Ecological systems

K. Maciejewski (✉)
Centre for Complex Systems in Transition, School of Public Leadership, Stellenbosch University, Stellenbosch, South Africa

P. Currie
ICLEI Africa, Cape Town, South Africa

P. O'Farrell
FitzPatrick Institute of African Ornithology, DST-NRF Centre of Excellence, University of Cape Town, Cape Town, South Africa

C. M. Shackleton et al. (eds.), *Urban Ecology in the Global South*, Cities and Nature,
https://doi.org/10.1007/978-3-030-67650-6_14

1 Introduction

Connectivity can be defined as the quality, state or the capacity to connect. Frequently it references human's ability to reach or communicate with others. In urban ecology the concept of connectivity speaks to the linkages within and between social-ecological systems and the porosity of these in enabling and facilitating the movement and spread of species within cities and between cities and adjacent areas (Muratet et al. 2013).

Urban areas contain more than half of the world's human population but are also home to some of the world's most fertile and biodiverse landscapes. Because of this, they have the potential to support biodiversity, and for many, provide the only opportunity for connecting citizens with nature. Finding new ways to incorporate and integrate nature and ecosystem services into the planning and management of urban landscapes is important. This is especially so in the Global South due to their rapid expansion to accommodate ever more people, which results in rapid and large-scale land transformation and fragmentation. This frame of reference and mindset of cognitive reality informs the types of decisions that people take into the future, particularly those affecting nature, which in turn affect the flows of benefits or services from nature to people. Connectivity is therefore a vital component of urban systems.

In 2011 the global population reached the seven billion mark and is expected to increase to nine billion by 2050. Urbanisation, driven by natural births in cities, urban incorporation of peripheral areas and communities and the influx of people from rural to urban areas is a global phenomenon (Myers 2021). In the nineteenth century urbanisation was seen as a predominantly Global North process. However, this process has now shifted and is seen to be a core issue of the Global South (United Nations 2011). The rate of urbanisation is currently most rapid in global cities of Africa, Latin-America, Asia, the Pacific and the Caribbean (Onodugo and Ezeadichie 2019). Projections by the United Nations (UN) indicate that more than four billion or 77% of urban residents of the world will be in developing regions by 2025 (Onodugo and Ezeadichie 2019). This translates to a total increase of 186% or an equivalent of 2.61 billion new urban citizens in Latin America, Asia and Africa starting from 1990. This rapid burgeoning of cities in the Global South requires a purposeful, informed and forward-looking response by city governments and developers alike to ensure that cities develop in ways that promote urban connectivity towards improved resilience to shocks.

In this chapter we discuss urban connectivity according to three specific components: connectivity of people with people, of nature with people, and of nature with nature. The functioning of the urban system is based largely on the presence of complex natural systems therein: each of these systems cannot function effectively without the other. Ensuring ecological connectivity between natural areas, through corridors or the embedding of nature in cities is vital for its renewal and for it to perpetuate itself (Beninde et al. 2015), and thereby support human wellbeing. The concept of social-ecological connectivity highlights the interconnectedness between the social and ecological components in urban landscapes. These interactions are often captured in a rather anthropocentric framing of ecosystem goods and services, by which nature is understood as providing, regulating, and supporting human survival, comfort and activity.

Taking only a socio-economic perspective on connectivity, the urban landscape can be understood as enabling services, movement, communication and agency for its citizens through networked infrastructures, such as water pipes, energy reticulation, communication towers, transportation systems and the flows which move through them, such as water, information or vehicles. Much focus on socio-economic connectivity centres on these built features, particularly in the Global South, where much of the future infrastructure has yet to be built. However, connectivity is further supported by the people and the social ties they produce. In this way, personal or social infrastructures must be considered as part of wider urban connectivity considerations.

In this chapter we provide a brief introduction to connectivity and how this differs across disciplines (social and ecological), and then address ecological, socio-ecological and socio-economic connectivity, and its importance, in urban landscapes of the Global South.

2 Understanding Connectivity

All cities depend on the flow of energy and other natural resources to satisfy human needs through consumption as well as local production, trade and services. Cities are designed and built to be inhabited by humans and therefore draw on infrastructure which can be defined as the physical components of interrelated systems that provide commodities and services (Childers et al. 2019). The provision of fresh drinking water to houses is typically facilitated by a network of pipes that connects to large water reservoirs within or external to the city. Similarly, an electrical main and associated circuit breaker panel connects a household's electrical devices to the grid to provide electricity. In many urban settings pipes and infrastructure transport wastewater and sewage from urban dwellings to wastewater treatment plants where contaminants are removed and converted into an effluent. However, in many Global South settings such infrastructural connectivity is absent, underdeveloped or overburdened with resulting negative effects on human wellbeing (Grimm et al. 2008a).

During the twentieth century, ecological considerations have inspired broader thinking about urban environments. Modern planning approaches such as landscape urbanism, ecological urbanism and ecological landscape urbanism highlight the importance of integrating nature and natural processes into city planning (Mostafavi and Doherty 2016; Cilliers et al. 2021). In addition to these broad-scale theoretical approaches, multiple local initiatives and interventions that seek to connect urban developments with nature have emerged. The concept of Ecological Infrastructure first appeared in 1984 which was developed by the United Nations Educational, Scientific and Cultural Organisation's (UNESCO) Man and Biosphere Programme. Several decades later the term Urban Ecological Infrastructure was developed at the 2013 International Ecopolis Forum for New Urbanisation. This concept comprises all parts of a city that support ecological structures and functions, including ecosystem services that directly affect human wellbeing (Childers et al. 2019). Examples include Green Infrastructure (GI), Green-Blue Infrastructure (GBI), Urban Green Space (UGS) and Nature-Based Solutions (NBS) (Pauleit et al. 2021). The term Nature-Based Solutions has gained considerable traction, particularly in Europe, although this concept is focussed more on goal-orientated engineering.

3 Understanding Connectivity Within Social-Ecological Systems

Urban landscapes are complex and comprised of social and ecological components that are inextricably linked, forming a social-ecological system (Grimm et al. 2008b). The interactions between the social and ecological elements in cities help regulate ecosystem functions which provide essential ecosystem services to urban populations (Egerer et al. 2020). These services enhance the wellbeing of urban residents by improving physical and mental health and providing the basic materials for a good life (Díaz et al. 2018). Yet the maintenance and management of ecosystem services across urban landscapes is challenging because of high spatial heterogeneity in land use composition and structure and often rapid rates of change. Urban ecosystem services that support human wellbeing are derived from multiple and diverse biophysical elements coupled with social, technological and economic features that vary across the landscape (Wiggering et al. 2003).

Biophysical and social heterogeneity of landscapes influences connectivity of ecosystem services through the facilitation of ecosystem service flows (Mitchell et al. 2013). Landscape facilitation of flows occurs when, for example, organisms that provide ecosystem services (e.g. pollination, dispersal) can easily move across habitats, meaning the habitats are more connected through these service providers. Landscape resistance occurs when barriers (such as roads or inhospitable land covers) exist to movement. Ecosystem services are thus better facilitated when there is enhanced landscape connectivity that facilitates species movement, ecosystem processes and ecological flows (Egerer et al. 2020).

As social-ecological systems are composed of social and ecological components, we focus here on the connectivity between these components and then address them in a combined or integrated way. The social fabric that makes up an urban landscape is described as the social connection, how *people connect with people*. In the Global South this requires particular attention given (1) the high density of people that live in urban environments and (2) rapid population growth in urban centres. People are also impacted by their surrounding environment which forms the focus of the second component where we look at the relationship between *people and nature*. This social-ecological connection represents natures' contribution to people, or the provision of ecosystem services. The long-term sustainable use of these ecosystem benefits can only be ensured through the protection of nature, which relies on establishing ecological connectivity, connecting *nature with nature*.

3.1 People with People

A city emerges as a result of large numbers of people coming together and investing into the area according to diverse interests and activities. This form of urban agglomeration is based on the interconnection of work and industrial opportunities, and amenities, such as housing, transportation, water and sanitation, food and energy access and leisure (Turok 2013). The relationships between these elements result in an emergent outcome which is understood as the city: in this way, influencing the city requires understanding how different elements within the city interact (Grimm et al. 2008b). The city can also be understood by tracking the metabolism of a city, i.e. how various infrastructures convey flows of resources and information (e.g. Kennedy et al. 2007; Bai 2007; Dijst et al. 2018). To improve the connectivity in a city, one must interrogate, not just how nature supports itself and urban citizens, but how urban citizens interact with built environments and with each other. The layout of the street network, the design of streets, buildings and public spaces, and the manner in which people access their goods and services influence the way people connect with one another. Socio-technical transition and political ecology discourses need to be brought into the conversation when designing cities, particularly in the Global South (Newell and Cousins 2015; Currie and Musango 2016). These socio-technical systems co-create social configurations which shape how people participate in the city and how people relate to each other therein (Swilling and Annecke 2012). Smart city discourses, which seek to automate the person-infrastructure interface, offer great possibilities for connecting people, but they have also been critiqued as entrenching new forms of social exclusion through personal convenience (e.g. van Rensburg et al. 2019; Agbali et al. 2019).

The most visible interactions in cities are typically those compelled by built systems. In this way, the shape and form of a city is often most noted by the ways in which people move through the city. Here, the universal public transport systems navigating dense buildings, or the sprawling neighbourhoods connected by wide roadways and the clearest juxtaposition. The city skyline is often a notable feature that expresses one of the key characteristics of a city: density. This density is praised for the ability to provide services to greater numbers of people using fewer resources (Barles 2009; Fernández 2014) but is also lamented for restricting access to nature (Grimm et al. 2008a), or for contributing to negative outcomes such as spreading contagions or limiting access by emergency services (Betterncourt et al. 2007). Missing in these images of cities are often the people that demand and convey urban services, and who are perhaps more visible and influential in cities of the Global South (Parnell and Pieterse 2014; Pieterse and Simone 2017).

Cities of the Global South are generally characterised by large informal, economies that have grown loosely around a small planned settlement (Onodugo and Ezeadichie 2019). Global South urbanisation is fraught with many challenges due to rapid urbanisation and many governments' lack of preparedness to meet the associated challenges. Such urban diseconomies include unemployment, crime, traffic congestion, paving or degradation of natural areas, poor waste treatment and disposal, and inconsistent distribution of vital services. However, where challenges exist, many informal enterprises may emerge to fill gaps in services (Turok 2013).

One such emergence is that of waste pickers, individuals who gather recyclable materials and convey them to buy-back centres or recycling businesses. These individuals make money off these collected materials and offer a unique contribution to the urban economy. While some may think that they are in competition with municipal or private collection of refuse, they provide a bridging service or connection between two value chains. A municipality's responsibility is to service residents and businesses by collecting (treating) and disposing of wastes and, depending on the sophistication of their waste management system, this may entail transfer to a landfill or any manner of composting, waste to energy processes, landfilling and alternative waste treatment. If a recycling industry is established in the city, there will be material value chains to process used materials for reuse in some form. Where limited separation of wastes takes place, waste pickers fulfil the role of connecting the municipality's service value chain to the material value chain of recycling companies (Godfrey and Oelofse 2017).

A typical response to the emergence of informal settlements in many cities is to demolish the settlement and evict people (Pieterse 2009; Huchzermeyer 2010; Muchadenyika 2015). This approach has the simple result of ensuring a new settlement emerges elsewhere. Where informal settlements are upgraded in situ, or the residents are offered alternative settlement, there is often a missing conceptualisation of what these settlements entail. They are not simply houses, but additionally are functioning economies, providing both monetary and non-monetary services to the community (Smit et al. 2017). Establishing new settlements often overlooks the need for space to undertake economic activities, and relocation often overlooks the connections to how people will make a living, and access to transport routes and means, that link them to key areas within the city. The multi-faceted nature of informal neighbourhoods means any interventions to upgrade or develop them, must acknowledge the existing relationships which exist therein, along with the broad spatiality of the city. Methodologies such as those followed by Slum Dwellers International (enumeration and community savings) (SDI 2016) provide the basis for communities to understand their assets, negotiate with municipal officials and offer their own funds as leverage for community development. They further provide a basis for community members to check in with each other and identify if any support is needed (SDI 2016).

The advent of mobile information communication technologies (ICT) has provided the opportunity to leapfrog over laid communication networks and has given citizens the possibility to hold the world's knowledge in their pockets. The potentials held by this have yet to be fully realised across sectors, as many societies in the Global South are still caught in a developmental path dependency set by the Global North. Educational systems, health systems, office work and finance systems have yet to adapt fully to the potentials offered by decentralised communication. This has also raised questions about whether density will remain the prerequisite for community living, or if the ICT revolution will foster de-urbanisation. The authors believe this is an unlikely trend, given the natural tendency for people to gather, the resource savings of servicing dense neighbourhoods, as well as the value offered by tacit interactions, which so underpin human spiritual needs. Nevertheless, ICT poses numerous opportunities for personal and community growth. ICT has also provided the opportunity to visualise many of the hidden flows in cities. These include tracking peri-urban food production using drones, visualising informal transportation routes (Digital Matatu 2015) or mapping enterprises in informal settlements (SLF 2016). The value of social connectivity for the achievement of wider urban wellbeing is immeasurable, particularly in concert with natural amenities that enable or support social interaction. Investing in green public space in a city can facilitate community interaction, by providing spaces for work, recreation or trade. For example, attempts to improve food environments in Dar es Salaam have been connecting informal vendors and green space (FAO 2020). Municipal Demonstration Gardens can also bring communities together around food and the environment (Drescher et al. 2021), such as in Antananarivo (Madagascar), where citizens can learn how to make their own vegetable gardens from available materials. The need and value of human-human interaction is not innately different between the Global North and South, but it manifests slightly differently based on form and function of the cities therein. For example, the prevalence of decentralised informal economies creates different forms of interpersonal interaction from formalised centres of commerce: a bustling food market with competing vendors has a different interpersonal requirement from a supermarket queue, particularly one with self-service checkout. This section has aimed to suggest that the value of these interpersonal requirements should not be overlooked and lost as cities are guided to 'formalise' their systems.

As cities of the Global South grow and develop the necessary built environments, attention must be kept on social infrastructures. These infrastructures have the benefit, not only of providing a single service, but multiple tacit benefits in addition. The value of decentralised ICT systems must not be understood in isolation from the value offered from interpersonal interactions. These include opportunities for learning, improvements to mental health, regulators of wealth and social inequity and the creation of infrastructures of solidarity. Infrastructures of solidarity are perhaps the best contributers to resilience, as they provide the basis for responding collectively to shocks and enabling the sharing of resources.

3.2 *People with Nature*

Expanding from the connections between people, there are clear relationships that exist between people and nature, even in urban settings where nature is often limited or removed. Humans are intimately and inextricably connected to nature and their environments. These connections that contribute to personal wellbeing have been much researched over the last 20–25 years. This section links to chapters of this book that consider the ecosystem services in urban landscapes, including provisioning (Shackleton 2021), regulating (Escobedo 2021) and cultural (Dobbs et al. 2021).

Understanding of connectivity between people and nature has been recently advanced by the work of the Intergovernmental Platform on Biodiversity and Ecosystem Services (IPBES), which was established in 2012. The focus of IPBES is to strengthen the science-policy interface for biodiversity and ecosystem services, through better understanding of the requirements for conservation and sustainable use and management of biodiversity for long-term human wellbeing and survival (Díaz et al. 2015). The IPBES developed a conceptual framework that highlights the connections between nature (i.e. biodiversity, ecosystem structure and functioning), and its contributions to people (for example food, flood water regulation, crop pollination and recreational opportunities) and how these connections are important in providing a good quality of life for people (Díaz et al. 2015). It also makes explicit the role of both direct and indirect drivers, institutions and anthropogenic assets in influencing connections and ensuring benefits from, nature.

The IPBES conceptual framework focusses on six interlinked components that represent the natural and social systems that operate at various scales in time and space (Díaz et al. 2015). These components, which are described below, highlight how they connect people with nature, with a focus on Global South cities.

i. *Nature*—refers to the diversity of living organisms and their interactions among themselves and their environment. In urban landscapes nature covers an enormous range of different green infrastructure, including green spaces, parks and gardens, urban river systems, and road verges. All of these nature areas have intrinsic value and potentially thereby contribute to the wellbeing of urban residents (Swilling and Annecke 2012). Within Global South cities the importance of nature and its intrinsic value often tends to be overlooked as socio-economic pressures, and the immediacy of poverty, result in low investment, sprawling developments (often informal) impacting nature (O'Farrell et al. 2019).
ii. *Nature's contributions to people* or the ecosystem goods and services describe the multiple connections between people and nature. All urban natural areas provide a range of varying benefits to people. For example, the value of small-scale beekeeping in a city can be quantified in terms of the monetary value it generates and also the quantified pollinating services that are provided by the bees (Edwards and Dixon 2016). A green roof that includes soil and plants can diminish stormwater, absorb heat and potentially enhance pollination services. Despite ecosystem services being impacted by urban developments in Global South cities, the benefits that nature provides are extremely important. Natural areas act as wellbeing safety nets, providing critical resources to people, especially the poorest urban residents. Provisioning services such as firewood, water, building materials and foods are harvested where possible. In addition to these

material benefits, the areas provide non-material benefits linked to cultural practices, such as places to perform certain cultural rites and initiations. Another critical point of connectivity is that between cities and the surrounding rural landscapes and the role they play in suppling cities with critical ecosystem services, such as water and food. In fact, many of the critical resources that cities depend on have been outsourced to the surrounding rural landscapes (O'Farrell et al. 2012), more so in the Global North than the Global South. It must be noted that nature also provides multiple ecosystem disservices (Shackleton et al. 2016; Davoren and Shackleton 2021), which impact negatively on people (such as malaria and other diseases, the threat of wild animals, tree branches damaging infrastructure, or allergens and poisons causing ill-health). In Global South cities people do not have the same level of infrastructure development to distance themselves from the negative impacts of nature. In many instances the high degree of connectivity and immediacy of nature places them at greater risks in contrast with cities in the Global North (Davoren and Shackleton 2021).

iii. *Anthropogenic assets*—refer to the built infrastructure, health facilities, technology as well as knowledge systems found in a city, and urban form is constituted by the infrastructure required for water, shelter, energy supply and transportation (Grimm and Schindler 2018). This is the environment directly surrounding urban societies, that they connect with on a daily basis. In the Global South, anthropogenic assets are limited and therefore require specific focus. For example, drinking water supply may be reliant on functional water filtration plants, and waste management which is fundamentally linked to human health (Grimm and Schindler 2018). Ecosystems can play a vital role in supporting anthropogenic assets, such as in purifying water so that treatment requirements and their associated costs are significantly reduced. Finding nature-based solutions to assist in alleviating some of the development challenges in Global South cities is appropriate given budget constraints and the advantages offered through self-sustaining natural interventions.

iv. *Institutions and governance systems*—affect the ways that people organise themselves and their interactions with nature at different scales (Simon et al. 2021). For example, this may represent local government, or judicial systems that determine the access to, or control the allocation and distribution of nature and access to people. These are referred to as indirect drivers because they do not directly affect nature but do so through their effects on anthropogenic drivers. Within some Global South cities institutional processes tend to be overwhelmed by multiple competing needs, however there are clear opportunities to work to enhance nature and benefits to people through informal networks thereby enabling people to diversify opportunities through the maintenance of multiple ecosystem services.

v. *Forces or drivers acting on nature and natural areas*—there are multiple different drivers, both natural and anthropogenic, that affect urban nature. Natural direct drivers include climate and weather patterns, as well as extreme events such as droughts or floods. Anthropogenic drivers include the harvesting of resources and the transformation of natural systems and areas, leading to land degradation, the emission of carbon that drives climate change, the pollution of soils, water and air, and the introduction of non-indigenous species. In emerging and rapidly developing cities of the Global South these direct impacts on nature, particularly anthropogenic drivers, are highly visible. Transformation of natural areas through sprawling informal development characterises urbanisation processes here (O'Farrell et al. 2019). Furthermore, natural drivers have the potential to be highly impactful, due to degraded ecosystem services and limited investment in grey infrastructural alternatives.

vi. *Good quality of life*—is the achievement of a fulfilled life. Within an urban context this could be represented by well-structured food systems, access to shelter, health, education, security, equity, freedom of choice. Human wellbeing is a concept often used to define the state of physical and mental health of individuals. Thus in the cities in the Global South, much developmental emphasis is placed on provisioning services and basic survival—foundational in terms of life. This is often at the cost of nature and the natural components of ecosystems that make up these cities. To work towards a good quality of life the critical elements underpin peoples' relational values with nature and the intrinsic value nature has (Himes and Muraca 2018), will need to be considered, managed and in some instances restored.

3.3 Nature with Nature

The provision of ecosystem services that benefit people in a city is maintained and regulated by nature. Green spaces such as parks and gardens, provide important ecosystem services that help regulate temperature, sequester carbon and bring a sense of peace through the aesthetic character of nature. Exposure to nature can also help citizens improve focus and concentration, which is known as Attention Restoration Theory, or ART (Ohly et al. 2016). Attention Restoration Theory has demonstrated how exposure to natural environments help people recover from physiological stress and mental fatigue (Berto 2014). Nature provides calming responses that mediates the negative effects of stress and enhances positive emotions. These instrumental values therefore play an important role in maintaining human wellbeing. However, nature also has intrinsic values, i.e. the perspective that nature has value in its own right, independent of human use.

In 1985 Michael Soule, an American biologist, best known for his work in promoting the idea of conservation biology, introduced the idea that 'species have value in themselves, a value neither conferred nor revocable but springing from a species' long evolutionary heritage and potential' (Soulé 1985). Intrinsic value is the value that an entity has in itself, for what it is, or as an end. This idea is in contrast with instrumental value, which is the value that something has as a means to a desired or valued end. Instrumental value is always derivative from the value of something else, whereas intrinsic value or non-instrumental value, has value as an end or value in themselves.

The green spaces in cities hold both intrinsic and instrumental values (Klain et al. 2017). They are however embedded in, and surrounded by non-natural infrastructure; by concrete, roads, buildings, parking lots, highways, residential houses, shopping malls, shanty-towns. Urban sprawl may encroach into green spaces, further breaking up, dividing and fragmenting remaining natural areas (Angel et al. 2012). Consequently, some urban green spaces often become isolated or placed under selective pressures through the development of human-made infrastructure that can lead to various environmental and socio-economic problems. This fragmentation of natural landscapes results in habitat loss, with increasing distances between natural vegetation remnants. This further decreases biotic populations and results in the loss of species and the disruption of provisioning services.

Increasing connectivity between green spaces can provide an effective strategy to address biodiversity decline and associated ecosystem services reductions (Egerer et al. 2020; Pauleit et al. 2021). This is particularly important in an urban environment where the natural spaces are encroached upon and fragmented as populations increase (Nor et al. 2017). Ecological connectivity contributes to both the maintenance of recreation spaces and associated opportunities, but also plays an important role in maintaining ecological functioning. This can facilitate the persistence of ecological processes, such as pollination, nutrient cycling and photosynthesis in fragmented urban landscapes. Connecting natural ecosystems in urban environments can also have important environmental management implications, such as in the control, diversion and management of flood waters (Egerer et al. 2020).

Ecological corridors, defined as 'linear strips of land which differs from the surrounded matrix' (Forman 1995), exhibit heterogeneity characteristics when compared to the surrounding landscape. Ecological corridors were originally proposed to connect isolated habitats of wild animals by establishing migration and movement corridors, thereby enhancing access and connectivity between populations (Shi et al. 2018). In the 1980s the International Union for the Conservation of Nature (IUCN) applied the concept of ecological corridors to global conservation strategies, which shifted the sole function of natural habitat towards embracing the notion of multi-functionality. In recent years the concept of ecological corridors has been extended and applied to the management of landscape aesthetics, recreation, as well as historical and cultural protection (Peng et al. 2017).

Similar to ecological corridors, the concept of 'urban ecological corridors' refers to linear, or ribbon ecological landscapes, which have the functions of natural habitat, green open space in urban areas (Peng et al. 2017). These corridors provide a 'conduit function' that enables the flow of ecosystem services and ecological processes across and within urban landscapes. Connecting nature with nature also facilitates the movement of organisms across an urban landscape. Small mammals, such as rodents or shrews are often found in city gardens and play an important role in many ecosystems. They are a prey source for many urban predators, including owls and cats and they are the predators of soil invertebrates (Vergnes et al. 2013). Shrews are particularly sensitive to habitat fragmentation and therefore rely on corridors to reach domestic gardens. Without ecological corridors, the functioning of shrews as metapopulations could be compromised (Vergnes et al. 2013). The movement of propagules and plants are also maintained by the establishment of ecological corridors across an urban landscape. Such movement is important for the maintenance of genetic diversity and the long-term persistence of populations and ecological communities. The wider the ecological corridors, the better it is for habitat quality. However, in an urban landscape, land resources and space is often limited. The construction of urban ecological corridors not only needs to consider the ecological benefits, but also take into account the spatial demands from the natural environment. It is therefore particularly important to set a reasonable width when connecting 'nature' with 'nature'. Different studies have proposed different widths; 3–12 m, 12–30 m, 60–100 m and 100–200 m as the suitable width of different types of biodiversity conservation corridors (Peng et al. 2017). However, these studies have failed to understand the complex nature of urban landscapes and the different variables that need to be taken into consideration. Suitable corridor widths need to be defined according to the types of species that may be utilising these green networks. The surrounding infrastructure also plays an important role; the size of the city, the density of people, the cultural identity and value of urban dwellers. At a broader scale, the spatial location also plays an important role, as well as the types of disturbances that may occur. For example, in Southeast Asia, where almost 50% of the population live in urban areas (UN Population Division 2011), the conservation of nature and wellbeing of the urban environment relies on the richness and abundance of bird species. The Eurasian tree sparrow (*Passer montanus*) is a dominant species found in Southeast Asia that has adapted to urban areas and can survive in highly built-up areas but requires green cover for breeding success (Nor et al. 2017). Bird dispersal is a major driver of plant propagules. Specifically, this granivorous bird eats small grains or seeds which it carries from one habitat patch to another, therefore acting as mobile links connecting isolated habitats in fragmented landscapes (Nor et al. 2017). The size of the habitat patch and vegetation density plays an important role in facilitating the movement of this bird, an important indicator and driver of urban ecosystem health (Nor et al. 2017).

In another example, a study in Melbourne, Australia, demonstrated that the urban matrix can provide suitable habitat for gliding mammals to move and forage (Caryl et al. 2013). The sugar glider (*Petaurus breviceps*) relies on high-contrast land-cover edges between residential properties, specifically tall trees a relatively short distance apart. As residential housing is usually the dominant land use within urban areas in Melbourne, connectivity plans need to account for the behavioural needs of the target species, as structural connectivity may not be enough to maintain landscape connectivity for this nocturnal species in urban areas (Caryl et al. 2013).

In Johannesburg, the largest city in South Africa, the establishment of green spaces and green infrastructure results from well within the policies of City of Johannesburg (CoJ) and Johannesburg City Parks (Burton and Rogerson 2017). However, the establishment of ecological corridors between urban green spaces is a lot more challenging and requires careful planning. The Kliprivier Urban Biodiversity Corridor (KUBiC) project was established as an initiative to develop ecology through the promotion, protection and enhancement of biodiversity and urban upliftment (Burton and Rogerson 2017). This green corridor stretches 13 km and links the centre of

Table 1 The three dimensions that the Kliprivier Urban Biodiversity Corridor Project (KUBiC) aimed to achieve

Elements	Description
Environmental	– Establish self-sustaining attractive habitats adding to the biodiversity – Improve the linkages between fragmented formal, informal and natural areas of open space areas – Promote biodiversity in the mixed use developed urban areas – Improve aesthetics along the corridor and into surrounding urban areas – Mitigate carbon emissions and air pollution
Economic	– Promote tourism and investment – Improve the visual experience and impression of the south and along Kliprivier Drive – Improve property value and increase investment opportunities –Reduced costs of maintenance and urban management along the roadside – Improve accessibility experience into and out of the city from the south –Improve the environment for business to do its business
Social	– Improve living conditions of those living next to Kliprivier Drive – Improve working conditions of those working along Kliprivier Drive – Provide job creation and employment opportunities – Reduce social problems and costs – Improve open space areas for active and passive recreation – Strengthen civic identify and pride of communities

Source KUBiC Launch Presentation 2012

Johannesburg to the southern periphery of the city. The idea behind the establishment of this ecological corridor was to beautify the areas situated along a major arterial roadway, creating a unique experience when entering and leaving Johannesburg (Burton and Rogerson 2017). This is achieved through the development of three major elements, namely: environmental elements, economic elements and social elements (Table 1). These three elements suggested that the project would serve a sustainable approach towards developing and maintaining the corridor within the south of Johannesburg.

The Kliprivier Sustainability Association, a conservation association, was a key partner on the KUBiC. This organisation focussed on the Klipriviersberg Nature Reserve (KNR) along the green corridor which serves as a major focus for the biodiversity aspect of the project (Burton and Rogerson 2017). This reserve is seen as the 'Jewel of the South' and local residents and communities understood the need to preserve its integrity through sustainable development. The focus of the Kliprivier Sustainability Association was therefore to find suitable and appropriate social and economic developments that will protect, promote and enhance the value of this reserve as well as surrounding natural assets, including the ridges, grasslands and wetlands (Burton and Rogerson 2017). The establishment of the corridor encouraged animal migration and encouraged the development of biodiversity zones within highly developed areas. These biodiversity zones promote energy flows and provides a bundle of ecosystem services to the surrounding areas.

In Singapore an ecological corridor, known as the Eco-Link, was constructed over the Bukit Timah expressway to re-establish a connection that was severed in 1985 between the city's Bukit Timah and Central Catchment nature reserves (Chong et al. 2010). This corridor is planted to represent a forest to facilitate the movement of organisms between the two reserves. The long-term view is to restore the ecological balance in these fragmented habitats and rectify the loss of biodiversity (Chong et al. 2010).

Green roofs and walls also play an important role in conserving biodiversity in urban landscapes. Even though these vegetated patches may be small and poor in species diversity and abundance compared to gardens and parks, they may act as stepping stones facilitating the movement of organisms between patches. These patches are particularly important and often visited by birds and insects that also rely on corridors, or smaller habitat patches scattered across an urban landscape. Green rooftop systems provide a private, safe and secure natural environment as opposed to green parks, which are often avoided in certain areas due to crime, particularly in the Global South (Labuschagne and Zulch 2016). Green rooftop systems act as insulation as the green plantings of the roof shade the building from solar radiation, and in winter the growing mechanism provides an additional layer of insulation (Carpenter 2004). There are many additional benefits to green roofs, which include slowing the spread of fire between buildings, controlling runoff rainwater in urban environments, assisting in stormwater management, and mitigating the urban heat island. Studies have also found that green rooftop systems help reduce stress, reduce patient recovery time, improve worker productivity and even increase property values (Labuschagne and Zulch 2016). In the Global South green rooftops may even help alleviate food security issues as they may be used as urban agriculture space. Green rooftops enhance community participation in the food system and opens up new job opportunities for local people.

4 Conclusions

This chapter has provided an overview of the role that connectivity plays in urban settings, with an emphasis on the Global South. We have presented connectivity as a socio-ecological issue concerned with the way and degree to which resources, species or social actors disperse, migrate or interact across ecological and social landscapes. We have demonstrated the various aspects of connectivity by describing the connection between people in cities; how people interact with one another and with the built environment. Cities are, in part, composed of large numbers of people living together in close proximity. This form of urban agglomeration is based on the interconnection of homes, industrial and work opportunities, and amenities, such as transportation, water and sanitation, food and energy access, and leisure. The relationships between these elements result in an emergent outcome which we understand as the city: in this way, influencing the city requires understanding how different elements within the city interact. People also need to interact with their surrounding natural landscape—the green spaces found within an urban setting.

The relationship between people and nature and how humans are inextricably connected to their environment plays a fundamental role in maintaining personal wellbeing. Nature provides various ecosystem services in an urban setting, including the provisioning, regulating and cultural ecosystem services. The values attributed to these services are closely coupled to, and dependent on, maintaining the connection between people and nature, particularly in an urban setting by bringing people into contact with nature, thereby ensuring nature's contributions are appreciated and persevered for future generations. The retention of nature in cities or green spaces, can only be ensured by maintaining their ecological function. Enabling ecological processes, such as pollination, nutrient cycling, photosynthesis to continue is a challenge given the fragmented nature of urban landscapes.

In the Global South, the interactions between people and nature requires careful consideration given the high levels of dependence on ecosystem services and potential vulnerabilities imposed by poverty and issues associated with global change.

References

Agbali M, Trillo C, Fernando T, Oyedele L, Ibrahim IA, Olatunji VO (2019) Towards a refined conceptual framework model for a smart and sustainable city assessment. *IEEE International Smart Cities Conference (ISC2)*, Casablanca, Morocco, pp 658–664, https://doi.org/10.1109/isc246665.2019.9071697

Angel S, Parent J, Civco D (2012) The Fragmentation of urban landscapes: global evidence of a key attribute of the spatial structures of cities, 1990–2000. Enviro Urban 24:249–283

Bai X (2007) Industrial ecology and the global impacts of cities. J Indust Ecol 11:1–6

Barles S (2009) Urban metabolism of Paris and its region. J Indust Ecol 13:898–913

Beninde J, Veith M, Hochkirch A (2015) Biodiversity in cities needs space: a meta-analysis of factors determining intra-urban biodiversity variation. Ecol Lett 18:581–592

Berto R (2014) The role of nature in coping with psycho-physiological stress: a literature review on restorativeness. Behav Sci 4:394–409

Bettencourt LM, Lobo J, Helbing D, Kuhnert C, West GB (2007) Growth, innovation, scaling, and the pace of life in cities. PNAS 104:7301–7306

Burton C, Rogerson JM (2017) The making of green infrastructure: the Klipriviersberg urban biodiversity corridor. Afr J Hosp, Tour Leis 6(3)

Carpenter S (ed) (2004) Growing green guide: a guide to green roofs, walls and facades in Melbourne and Victoria, Australia. State of Victoria, Melbourne

Caryl FM, Thomson K, Van Der Ree R (2013) Permeability of the urban matrix to arboreal gliding mammals: Sugar gliders in Melbourne, Australia. Aust Ecol 38:609–616

Childers DL, Bois P, Hartnett HE, McPhearson T, Metson GS, Sanchez CA (2019) Urban ecological infrastructure: an inclusive concept for the non-built urban environment. Elem Sci Anth 7:46

Chong KY, Yee AT, Yeo CK (2010) Biodiversity: linking Singapore's fragmented habitats. Nat 465:289

Cilliers SS, Breed CA, Cilliers EJ, Lategan LG (2021) Urban ecological planning and design in the Global South. In: Shackleton CM, Cilliers SS, Davoren E, du Toit MJ (eds) Urban ecology in the Global South. Springer, Cham, pp 365–401

Currie PK, Musango JK (2016) Assimilating urban metabolism into sustainability discourse and practice: African urbanisation. J Indust Ecol 21:1262–1276
Davoren E, Shackleton CM (2021) Urban ecosystem disservices in the Global South. In: Shackleton CM, Cilliers SS, Davoren E, du Toit MJ (eds) Urban ecology in the Global South. Springer, Cham, pp 265–292
Díaz S, Demissew S, Carabias J, Joly C, Lonsdale M, Ash N et al (2015) The IPBES conceptual framework: connecting nature and people. Curr Opin Enviro Sustain 14:1–16
Díaz S, Pascual U, Stenseke M, Martín-López B, Watson RT, Molnár Z, Hill R, Chan KMA, Baste IA, Brauman KA, Polasky S, Church A, Lonsdale M, Larigauderie A, Leadley PW, van Oudenhoven APE, van der Plaat F, Schröter M, Lavorel S, Aumeeruddy-Thomas Y, Bukvareva E, Davies K, Demissew S, Erpul G, Failler P, Guerra CA, Hewitt CL, Keune H, Lindley S, Shirayama Y (2018) Assessing nature's contributions to people. Sci 359(6373):270–272
Digital Matatu (2015) Available Online: http://www.digitalmatatus.com/
Dijst M, Worrell E, Böcker L, Brunner P, Davoudi S, Geertman S, Harmsen R et al (2018) Exploring urban metabolism: towards an interdisciplinary perspective. Res Conserv Recycl 132:190–203
Dobbs C, Vasquez A, Olave P, Olave M (2021) Cultural urban ecosystem services. In: Shackleton CM, Cilliers SS, Davoren E, du Toit MJ (eds) Urban ecology in the Global South. Springer, Cham, pp 245–264
Drescher AW, Isendahl C, Caridad Cruz M, Karg H, Menakanit A (2021) Urban and peri-urban agriculture in the Global South. In: Shackleton CM, Cilliers, SS, Davoren E, du Toit MJ (eds) Urban ecology in the Global South. Springer, Cham, pp 293–324
Edwards F, Dixon J (2016) Hum of the hive: negotiating conflict between humans and honeybee towards an ecological city. Soc Anim 24:535–555
Egerer M, Fouch N, Anderson EC, Clarke M (2020) Socio-ecological connectivity differs in magnitude and direction across urban landscapes. Sci Rep 10:4252
Escobedo FJ (2021) Understanding urban regulating ecosystem services in the Global South. In: Shackleton CM, Cilliers SS, Davoren E, du Toit MJ (eds) Urban ecology in the Global South. Springer, Cham, pp 227–244
FAO (2020) Food outlets and green spaces in Ilala and Kinondoni Districts, Dar es Salaam. FAO, Rome
Fernández JE (2014) Urban metabolism of the Global South. In: Parnell S, Oldfield S (eds) The Routledge handbook on cities of the Global South. Routledge, New York, pp 597–612
Forman RT (1995) Land mosaics: the ecology of landscapes and regions. Cambridge University Press, Cambridge
Grimm NB, Foster D, Groffman P, Grove MJ, Hopkinson CS, Nadelhoffer KJ, Pataki DE, Peters DP (2008a) The changing landscape: ecosystem responses to urbanisation and pollution across climatic and social gradients. Front Ecol Enviro 6:264–272
Godfrey L, Oelofse S (2017) Historical review of waste management and recycling in South Africa. Resources 6:1–11
Grimm N, Stanley S, Golubiewski N, Redman C, Wu J, Bai X, Briggs J (2008b) Global change and the ecology of cities. Sci 319:756
Grimm N, Schindler S (2018) Nature of cities and nature in cities: prospects for conservation and design of urban nature in human habitat. In: Lele S, Brondizio ES, Byrne J, Mace GM, Martinez-Alier J (eds) Rethinking environmentalism: linking justice, sustainability, and diversity. MIT Press, Cambridge, MA, pp 99–125
Himes A, Muraca B (2018) Relational values: the key to pluralistic valuation of ecosystem services. Curr Opin Enviro Sustain 35:1–7
Huchzermeyer M (2010) Pounding at the tip of the iceberg: the dominant politics of informal settlement eradication in South Africa. Politikon 37:129–148
Kennedy C, Cuddihy J, Engel-Yan J (2007) The changing metabolism of cities. J Indust Ecol 11:43–59

Klain SC, Olmsted P, Chan KM, Satterfield T (2017) Relational values resonate broadly and differently than intrinsic or instrumental values, or the New Ecological Paradigm. PLoS ONE 12(8):e0183962

Labuschagne P, Zulch B (2016) Green rooftop systems: a South African perspective. Energy Procedia 96:710–716

Mitchell M, Bennett E, Andrew G (2013) Linking landscape connectivity and ecosystem service provision: current knowledge and research gaps. Ecosys 16:894–908

Mostafavi M, Doherty G (2016) Ecological urbanism. Lars Muller Productions, Zurich

Muchadenyika D (2015) Slum upgrading and inclusive municipal governance in Harare, Zimbabwe: new perspectives for the urban poor. Habitat Intn 48:1–10

Muratet A, Lorrillière R, Clergeau P, Fontaine C (2013) Evaluation of landscape connectivity at community level using satellite derived NDVI. Landsc Ecol 28:95–105

Myers G (2021) Urbanisation in the Global South. In: Shackleton CM, Cilliers SS, Davoren E, du Toit MJ (eds) Urban ecology in the Global South. Springer, Cham, pp 27–49

Newell JP, Cousins JJ (2015) The boundaries of urban metabolism: towards a political—industrial ecology. Prog Hum Geog 39:702–728

Nor AN, Corstanje R, Harris JA, Grafius DR, Siriwardena GM (2017) Ecological connectivity networks in rapidly expanding cities. Heliyon 3:e00325

O'Farrell P, Anderson P, Le Maitre D, Holmes P (2012) Insights and opportunities offered by a rapid ecosystem service assessment in promoting a conservation agenda in an urban biodiversity hotspot. Ecol Soc 17(3):27

O'Farrell P, Anderson P, Culwick C, Currie P, Kavonic J, McClure A, Ngenda G, Sinnott E, Sitas N, Washbourne C-L, Audouin M, Blanchard R, Egoh B, Goodness J, Kotzee I, Sanya T, Stafford W, Wong G (2019) Towards resilient African cities: shared challenges and opportunities towards the retention and maintenance of ecological infrastructure. Glob Sustain 2:e19

Ohly H, White MP, Wheeler BW, Bethel A, Ukoumunne OC, Nikolaou V, Garside R (2016) Attention restoration theory: A systematic review of the attention restoration potential of exposure to natural environments. J Toxicol Enviro Heal B 19:305–343

Onodugo VA, Ezeadichie NH (2019). Future planning of Global South cities with inclusive informal economic growth in perspective. In Urban Design. IntechOpen. https://doi.org/10.5772/intechopen.89145

Parnell S, Pieterse E (2014) Africa's urban revolution. Zed Books, London

Pauleit S, Vasquéz A, Maruthaveeran S, Liu L, Cilliers SS (2021) Urban green infrastructure in the Global South. In: Shackleton CM, Cilliers SS, Davoren E, du Toit MJ (eds) Urban ecology in the Global South. Springer, Cham, pp 107–143

Peng J, Zhao H, Liu Y (2017) Urban ecological corridors construction: a review. Acta Ecol Sinica 37:23–30

Pieterse E (2009) Exploratory notes on African urbanism. In: 3rd European Conference on African Studies, 1–15. African Centre for Cities, University of Cape Town, Cape Town

Pieterse E, Simone AM (2017) New urban worlds: inhabiting dissonant times. Wiley, Hoboken

SDI (2016) Our practices for change. https://knowyourcity.info/our-practices-for-change/. Accessed February 13, 2020

Shackleton CM (2021) Ecosystem provisioning services in Global South cities. In: Shackleton CM, Cilliers SS, Davoren E, du Toit MJ (eds) Urban ecology in the Global South. Springer, Cham, pp 203–226

Shackleton CM, Ruwanza S, Sinasson Sanni GK, Bennett S, De Lacy P, Modipa R, Mtati N, Sachikonye M, Thondhlana G (2016) Unpacking Pandora's Box: understanding and categorising ecosystem disservices for environmental management and human wellbeing. Ecosys 19:587–600

Shi H, Shi T, Yang Z, Wang F, Wang C (2018) Effect of roads on ecological corridors used for wildlife movement in a natural heritage site. Sustain 10(8):2725

Simon D, Goodness J, Lwasa S, Puppim de Oliveira JA, Macedo LV, Kavonic J, Hermansson Török E, Elmqvist T (2021) Urban governance of and for urban green and blue infrastructure. In:

Shackleton CM, Cilliers SS, Davoren E, du Toit MJ (eds) Urban ecology in the Global South. Springer, Cham, pp 403–428

SLF (2016) Formalising informal micro-enterprises. Sustainable Livelihoods Foundation, June 8. http://livelihoods.org.za/projects/fime#lightbox[group-316]/5/

Smit S, Musango JK, Kovacic Z, Brent AC (2017) Conceptualising slum in an urban African context. Cities 62:107–119

Soulé ME (1985) What is conservation biology? Bioscience 35:727–734

Swilling M, Annecke E (2012) Just transitions: explorations of sustainability in an unfair world. UCT Press, Cape Town

Turok I (2013) Securing the resurgence of African cities. Local Econ 28(2):142–157

United Nations Department of Economic and Social Affairs (UNDESA) (2011) World population prospects 2011: the 2011 revision. UNDESA, New York

UN Population Division (2009) World Population Prospects: The 2008 Revision Population Database. New York: United Nations, available from http://esa.un.org/unpd/wpp2008/index.htm

UN Population Division (2009) World Population Prospects: The 2008 Revision Population Database. New York: United Nations. Available from http://esa.un.org/unpd/wpp2008/index

van Rensburg NJ, Matheri AN, Meyer J (2019) Bridging the digital divide in an African smart city. IEEE International Smart Cities Conference (ISC2), Casablanca, pp 518–526

Vergnes A, Kerbiriou C, Clergeau P (2013) Ecological corridors also operate in an urban matrix: a test case with garden shrews. Urb Ecosys 16:511–525

Wiggering H, Müller K, Werner A, Helming K (2003) The concept of multifunctionality in sustainable land development. In: Helming K, Wiggering H (eds) Sustainable development of multifunctional landscapes. Springer, Heidelberg, pp 3–18

Urban Ecological Planning and Design in the Global South

Sarel S. Cilliers, Christina A. Breed, E. Juanee Cilliers, and Louis G. Lategan

Abstract Tremendous increases in urbanisation predicted for the Global South compel urban planners and designers to be more innovative and ecological wisdom is needed for urban plans and designs to be relevant across different scales, communities and regions. Scientific emphasis on sustainable planning and design originated in the Global North, but is gaining increased consideration in the Global South. This chapter investigates to what extent research areas in ecological planning and design, such as (1) ecosystem services, (2) impact of natural disasters, (3) ecological restoration and rehabilitation and (4) ability of urban residents to link knowledge to action, have been dealt with in the Global South. Differences with the Global North in research needs are reflected upon and challenges in the application of ecological planning and design in the Global South are discussed. These challenges include aspects such as increased urban sprawl, land invasion and ecological degradation, weak environmental support from local government, lack of ecological planning initiatives and incentives and investments in ecosystems and their services, ineffective policy and legislation and narrow research approaches and agendas focusing too much on the Global North. Lastly, recommendations are made of research foci and methods to address these challenges drawing from Global South examples.

S. S. Cilliers (✉)
Unit for Environmental Sciences and Management, North-West University, Potchefstroom, South Africa
e-mail: sarel.cilliers@nwu.ac.za

C. A. Breed
Department of Architecture, Faculty of Engineering, Built Environment and Information Technology, University of Pretoria, Private Bag X20, Hatfield 0028, South Africa

E. J. Cilliers · L. G. Lategan
Unit for Environmental Sciences and Management, School of Geo- and Spatial Sciences: Urban and Regional Planning, North-West University, Private Bag X6001, Potchefstroom 2520, South Africa

E. J. Cilliers
Faculty of Design, Architecture and Building, University of Technology Sydney, PO Box 123, Broadway, NSW 2007, Australia

C. M. Shackleton et al. (eds.), *Urban Ecology in the Global South*, Cities and Nature,
https://doi.org/10.1007/978-3-030-67650-6_15

Keywords Ecological design · Ecological planning · Landscape design · Sustainable planning · Innovative planning

1 Introduction

> Balancing urban growth and development interests against the need to safeguard natural resources and socially equitable open space – forests, fields, allotment gardens and parks – constitutes a classic urban planning issue. (Erixon et al. 2013)

In light of the tremendous increase in urbanisation predicted for most countries in the Global South the "classic urban planning issue" captured in the quotation above is growing in pressure, compelling urban planners and designers to be more innovative. Steiner (2016, 2018) argued that urban planners and designers need ecological wisdom for their plans and designs to be relevant across different scales, communities and regions in the complex contemporary world. This implies flexibility, a response to the specific circumstances, and demand for a multiple solution process (Palazzo and Steiner 2011). Traditional ecological knowledge (TEK) and indigenous knowledge systems (IKS) do not differentiate between knowledge and wisdom, but rather emphasise ecological literacy that is attained through constant environmental observation, close perception and experience (Liao and Chan 2016; Salazar and Jalabert 2016). Wisdom manifests as applied place-specific knowledge, mediated by values towards a greater good (Liao and Chan 2016).

Scientific emphasis on sustainable planning and design, which include ecological aspects, originated in the Global North (Yigitcanlar and Kamruzzaman 2015; Geneletti et al. 2017), but has gained increasing consideration in the Global South (e.g. Swilling 2006; Chen et al. 2008; Cilliers et al. 2014; Yigitcanlar and Kamruzzaman 2015). Similarly, the concept of green infrastructure (GI) planning, first recognised in the work of landscape designers such as Olmsted and Vaux (Sinnett et al. 2015), focusing on a deliberate network of multifunctional and integrated green spaces (Benedict and McMahon 2006; Pauleit et al. 2021), was developed in the Global North (e.g. Lovell and Taylor 2013; Pauleit et al. 2019). However, there has been incremental uptake of GI planning in Global South countries (Frischenbruder and Pellegrino 2006; du Toit et al. 2018; Lindley et al. 2018; Pauleit et al. 2021) aiming to address sustainable urban development. Resilience theory has recently offered scientists a complementary approach to sustainability, in which nature (environmental aspects) forms the basis for society to adapt to foreseen inevitable change while maintaining system functioning (Erixon et al. 2013; Sitas et al. 2021). In practice, sustainable and resilient cities need to follow an adaptive planning and design approach in which various stakeholders (transdisciplinarity) contribute to realising specific ecosystem services (Ahern et al. 2014) and considering the trade-offs in a transparent way (O'Farrell et al. 2012). This implies cyclical processes of human interaction and feedback with the environment (Zipperer et al. 2011), a process well known to traditional Global South communities (Chen and Wu 2009; Liao and Chan 2016; Dieleman 2017).

Academic literature often links the concepts of urban ecological design and planning. Planners focus on the "bigger picture", mainly on a regional scale that ideally leads to the development of more detailed plans and set guidelines for the design at a local scale. Design engages with the relationships between and distribution of elements in the landscape, along with the fulfilment of local expectations (Frischenbruder and Pellegrino 2006; Palazzo and Steiner 2011; Armitage et al. 2014). The first publications on ecological design strived for planning and design that could reduce destructive environmental impacts through integration with living processes (McHarg 1969; Van der Ryn and Cowan 1996; Thompson and Steiner 1997). Ecological design considers ecological functioning, preservation and generation of resources, and a more resilient approach (Rottle and Yocom 2010). Design research experiments, using scientific and local knowledge, and following landscape ecological principles, have become critical (Nassauer and Opdam 2008; Felson and Pickett 2005).

Steiner (2014) identified four key areas of research that need to be enhanced to address the significant impact of urbanisation, including (1) ecosystem services, (2) the impact of natural disasters and how GI can mitigate their effects, (3) how urban green spaces can be restored or rehabilitated and (4) the ability of urban residents to link knowledge to action. Steiner (2014) mainly uses examples from the Global North and therefore, the aim of this chapter is to determine to what extent these proposed research areas have been addressed in the Global South. Additionally we reflect on some differences between the Global North and Global South in terms of research needs, and lastly we discuss challenges in the application of ecological planning and design in the Global South.

We divide the Global South into three main areas, South and Central America (Latin America—LA), Asia, and Sub-Saharan Africa (SSA), that we used together with the following keywords: "urban" AND "ecological design" OR "ecological planning" in searches of Scopus and Web of Science (from 1990 to 2019). Search queries also included the search terms "urban" AND "environmental/ sustainable/ landscape" AND "design" OR "planning". We included all studies rendered by the search that were conducted about or inside Latin/South America/Caribbean/Central America/Mexico as part of the Global South. We excluded Hong Kong, Singapore, Korea and Japan from the Asian studies. We considered only articles written in English. The focus in this chapter is on the four main key research areas in ecological planning and design (Steiner 2014) with examples from all three regions in the Global South.

2 Main Research Areas in Ecological Planning and Design

2.1 Attention Given to Ecosystem Services

The concept of ecosystem services (ES) that refers to the benefits that people derive from ecosystems (Díaz et al. 2015) is now widely studied. The implementation of the ES concept in GI planning in the Global South was described by Pauleit et al. (2021). The question we are addressing in this section is: How is the ES concept adopted across cultures? (Steiner 2014).

Dobbs et al. (2019) indicated a deficiency of publications on urban ES in LA countries. They emphasise that LA cities can be socio-ecologically unique, which implies different drivers of ES than in the Global North, and that the context and demand need to be deliberated during valuation of ES (Dobbs et al. 2019). In a thematic review of ES research Vihervaara et al. (2010) found that only 17% of studies and 20% of authors were from of Africa. According to du Toit et al. (2018) only 23% of SSA countries have published works on urban ES, with a clear bias towards South Africa (54% of the studies) and therefore predicting limited application of the ES concept in planning and design in SSA cities. Vihervaara et al. (2010) also found that 46% of studies and 27% of authors on ES were based in Asia and the Pacific. Research on ES, and particularly ES values, has been conducted in China since the 1990s (Long et al. 2014).

As in LA, the local socio-economic status influences the perceptions and beliefs of urban residents in South Africa in terms of the value of urban GI and ES (Adegun 2018). Poorer residents focus more on provisioning ES and wealthier residents more on regulating ES, but these preferences also change over time (du Toit et al. 2018). It is therefore, important that planners and decision-makers understand the spatial and temporal dynamics of ES and reflect them in strategic plans (du Toit et al. 2018). By far the majority of LA studies on ES focus on regulating services, such as climate and flood regulation, with few studies considering supporting services. Quite a number documented provisioning services, illustrating a variety of context-specific values e.g. Schwab et al. (2018). Du Toit et al. (2018) found similar results in SSA and emphasise the lack of studies on cultural and supporting ES. In a South African study on the different ES discussed in green space projects in three industry magazines aimed at urban planners, designers and managers, it was evident that cultural and regulating ES are valued higher in urban landscape design than are supporting and provisioning ES (Breed et al. 2015). Markedly influenced by "socio-economic and historic-political realities" in South Africa the study illustrates the important role of planners and designers in influencing ES outcomes (Breed et al. 2015). For example, in many cities in SSA the needs of the poor for ES were not addressed, mainly due to a lack of collaboration between planners, policymakers and residents and often "exogenous planning processes" imported from the Global North (Bolay 2015).

Fortunately, there are examples of SSA cities where attempts are made to address the ES needs of all citizens in urban planning and design. With a long history of urban open space planning, Durban was the first city in South Africa to apply a Metropolitan

Open Space System (MOSS) approach in city planning (Boon et al. 2016). The Durban MOSS initially focused on biodiversity, but since 1994 the focus shifted towards ES and their values (Roberts et al. 2005). However, proposals for planning and development in Durban still threaten the provision of some ES, such as carbon storage, water yield, sediment retention, nutrient retention and flood attenuation (Davids et al. 2018). The MOSS approach partially addresses these threats as it incorporates ES into decision-making processes at local government level, but some ES hotspots (Davids et al. 2016) are outside the protected and managed MOSS areas and need to be taken into consideration (Davids et al. 2018).

In China, the deliberate focus on ecology in development and planning since the 1990s has spurred rigorous research on and practical application of urban ES projects. Examples include the Sponge City (see Box 1) (Chan et al. 2018; Nguyen et al. 2019), various Eco-Cities (Lin 2014; Liu and Jensen 2018; Ng 2019) and Eco-Parks (Chen and Wu 2009). Across East Asia, the eco-city is promoted as an innovative urban policy to advance sustainable urbanisation and realign the post-industrial urban economy. Ecological urbanism, respective technological and urban design strategies, the relationship between eco-city building and local economic development, and the roles played by different stakeholders, were all identified as crucial considerations in the quest towards Asian eco-cities (Lin 2014). As a result, awareness of ES is high amongst scientists, policymakers, and a proportion of Asian society, although the underlying ecological roles of biodiversity are still debated (Thomas et al. 2004). Despite high awareness, the level of local participation remains limited.

Box 1: The Sponge City approach in Chinese cities

The Sponge City concept is premised on urban water resourcing and ecological water management. It allows for more efficient use of water resources through engineering and green infrastructure to control urban runoff and storm water storage during extreme wet events (Chan et al. 2018) and recycle storm water for use during droughts (Nguyen et al. 2019). The approach aims to increase the urban land area able to absorb surface water discharge by 20% and, retain or reuse 80% of storm water by 2030 by improving permeation, retention, storage, purification and drainage (Chan et al. 2018; Nguyen et al. 2019). Sponge Cities resemble international examples like the USA's Low Impact Development (LID); the UK's Sustainable Urban Drainage Systems (SuDS) and Blue-Green Cities (BGC); Australia's Water Sensitive Urban Design (WSUD) and New Zealand's Low Impact Development Urban Design (LIDUD) (Wang et al. 2017; Chan et al. 2018). The Sponge City is based on: (1) improving efficient control of urban peak runoff; (2) temporarily storing, recycling and purifying storm water; (3) upgrading traditional drainage systems utilising flood-resistant infrastructure like underground water storage; (4) increasing existing drainage protection standards to offset peak discharges and limit storm water; (5) integrating natural water bodies; (6) promoting multi-functionality by enhancing ES and 7) adding artificial water bodies and green spaces as amenities (Chan et al. 2018). Infrastructure includes bio-swales, rain gardens, pervious paving and green roofs to simulate natural hydrological responses and absorb storm water via soil infiltration, storm water retention, storage, purification, groundwater recharge and improved runoff quality (Everett et al. 2015; Qin et al. 2013; Chan et al. 2018). The Chinese National Government is spearheading funding for Sponge City pilots in thirty cities, including Beijing, Shanghai, Shenzhen and Tianjin (Chan et al. 2018).

The Sponge City initiative presents the potential to engage more effectively with land use and spatial planning and improve urban ecosystem diversity and social wellbeing (Chan et al. 2018). However, several barriers hinder Sponge City development (Nguyen et al. (2019). Low impact development is still applied narrowly in China given limited domestic research and that Chinese expertise is still rooted in centralised drainage systems with limited experience regarding green materials or guidance and training on holistic water management strategies. China often relies on outdated simulations and insufficient data for innovative planning and design. The initial stages of Sponge City development have relied on technocratic engineering-orientated approaches. Fragmented, top-down governance structures and capacity, unrealistic time frames and limited project evaluation procedures further risk poor design and project quality (Jiang et al. 2018; Zevenbergen et al. 2018). Physical obstacles, including intensifying urban pluvial flooding and droughts exacerbated by climate change; water and soil pollution due to urbanisation and industrialisation; changing soil conditions resulting from poorly planned urban expansion and the unparalleled conversion of previously pervious surfaces to concrete; a continuous loss of aquatic ecosystems and green spaces and local geographical variations in weather and landscape, for example soil-freeze and thaw, hinder broad implementation of Sponge City strategies. Such obstacles challenge Sponge City initiatives based on conditions and approaches that evolved over time in the Global North and are now applied with haste in the world's biggest developing country (Jiang et al. 2017; Zevenbergen et al. 2018). More research is needed on the financial viability of Sponge City projects, clarifying benefits for social wellbeing, return on investment, value of and for public–private partnerships and the role of regional organisations to encourage broader acceptance and support (Nguyen et al. 2019).

References

Chan FKS, Griffiths JA, Higgitt D, Xu S, Zhu F, Tang YT, Xu Y, Thorne, C.R. 2018. "Sponge City" in China—a breakthrough of planning and flood risk management in the urban context. Land Use Pol 76:772–778.

Everett G, Lamond J, Morzillo AT, Chan FKS, Matsler AM (2016, April). Sustainable drainage systems: helping people live with water. In Proceedings of the Institution of Civil Engineers-Water Management (Vol. 169, No. 2, pp. 94–104). Thomas Telford Ltd

Jiang Y, Zevenbergen C, Ma Y (2018) Urban pluvial flooding and stormwater management: A contemporary review of China's challenges and "sponge cities" strategy. Environ Sci & Policy 80:132–143. Liu L, Jensen MB (2018) Green infrastructure for sustainable urban water management: Practices of five forerunner cities. Cities 74:126–133.

Jiang Y, Zevenbergen C, Fu D (2017) Understanding the challenges for the governance of China's "sponge cities" initiative to sustainably manage urban stormwater and flooding. Nat Hazards 89(1): 521–529.

Nguyen TT, Ngo HH, Guo W, Wang XC, Ren N, Li G, Ding J, Liang H (2019). Implementation of a specific urban water management-Sponge City. Sci Total Environ 652:147–162.

Qin HP, Li ZX, Fu G (2013) The effects of low impact development on urban flooding under different rainfall characteristics. J Environ Manag 129:577–585.

Wang Y, Sun M, Song B (2017) Public perceptions of and willingness to pay for sponge city initiatives in China. Resour, Conserv Recycl 122:11–20.

Zevenbergen C, Fu D, Pathirana A (2018). Transitioning to sponge cities: challenges and opportunities to address urban water problems in China.

Balancing human activity and the environment is firmly rooted in Chinese philosophy, exemplified by the "unity of man with nature" philosophy, "peach blossom spring" ideal, "world-in-a-pot" model, and Feng Shui theory (Chen and Wu 2009). This philosophy denotes that all human activities should be integrated with natural patterns and processes to achieve harmony between humans and nature (Chen and Wu 2009). A more deliberate focus on balance, conservation and the values potentially delivered by nature, and ecosystems by extension, entered the national research and development arenas with the resurgence of the ecological civilisation concept. This concept insists that traditional Chinese philosophy valued "ecological harmony" and that such values should be revived for a green future informed by ecologically sustainable modes of resource extraction, production and trade, exercised by environmentally conscious, responsible citizens (Hansen et al. 2018).

Mapping and valuation of ES are not common in Global South cities. A rapid assessment and mapping of certain ES in Cape Town, South Africa, has shown how overlying spatial models of different ES could potentially be used to determine trade-offs between urban development and conservation (O'Farrell et al. 2012). It was, however, also argued that biodiversity as a rationale for conservation should sometimes override ES in a biodiversity hot spot. Protecting only multifunctional areas that provide various ES could jeopardise the conservation of biodiversity. Schäffler and Swilling (2013) have emphasised the importance of ES valuations for urban GI planning that must be relevant to the unique conditions of the Global South. Some studies on ES in South Africa attempted to determine the monetary value of the ES provided by urban ecosystems (Roberts et al. 2005; Cilliers and Cilliers 2015; de Wit et al. 2012). Although there is no indication that these efforts have led to direct changes in urban planning and design, they did sensitise policymakers and planners to the importance of urban green spaces in Durban (Roberts et al. 2005) and better collaboration between different municipal departments in Cape Town on environmental issues (de Wit et al. 2012).

City planning in the Global South must take local value articulation aspects into consideration (Ernstson 2013). For example, pine plantations in Cape Town, South Africa, are highly valued by some for recreational purposes (cultural ES) but rejected by others due to their high water-use and displacing habitat of local endemic vegetation (Ernstson 2013). In LA ES supply and demand are usually quantified using neoclassical economic methods, sociology and other qualitative methods that measure people's perception of ES (Dobbs et al. 2019). Despite the advances in the supply–demand research, it is rarely implemented by decision-makers in LA countries (Juanita et al. 2019). There is a need for context-specific information on how to promote, educate and apply ES, along with ES management and planning (Dobbs et al. 2019).

2.2 *Adaptation of Settlements for Natural Disasters*

Rapid and informal urbanisation can increase vulnerability to natural disasters (Aguilar and Santos 2011; Bertilsson et al. 2019; Dobbs et al. 2019). Dobbs et al. (2019) argue that in Global South regions inequities, rapid urbanisation and climate change mainly affect the adaptive capacity of socio-political and ecological systems. Most urban expansion in LA occurs towards floodplains and lower mountain slopes, frequently occupied by low income groups following unplanned growth. Informal settlements can often cause vegetation clearing on slopes and in riverbeds, thus increasing vulnerability to natural disasters (Dobbs et al. 2019). Bertilsson et al. (2019) therefore conclude that natural disaster avoidance becomes unpredictable, and as a consequence people need to learn how to live with disasters. Exacerbated by climate change, adversities such as flooding severely effect poor urban residents in the Global South (Douglas et al. 2008; Carmin et al. 2012; Bertilsson et al. 2019). Poor city planning and the lack of infrastructure along rivers have been stated as the main causes of flooding in five large African cities in Ghana, Uganda, Nigeria, Mozambique and Kenya (Douglas et al. 2008). China's distinctive and diverse geographical and climatic features, presenting great plains and deserts, hills and mountains, sunken basins, large rivers, deltas and an extensive coastline, as well as temperatures ranging from tropical to subarctic, make it a disaster-prone country with an extensive history of risk (Zheng et al. 2018).

Although resilience as the basis for adaptation planning in cities has increasingly been investigated (Steiner 2014) in both Global North and Global South countries

(Sitas et al. 2021), it is not widely or deliberately implemented in the Global South. In a review on studies from 2000 to 2013 of how global urban planning and design literature has addressed climate adaptation, it was found that only 4% of the studies were from the Global South (Dhar and Khirfan 2017). Although most studies have addressed climate change mitigation and social learning and vulnerability assessments, few have addressed adaptation and specific planning and design issues and interventions (Dhar and Khirfan 2017). Despite a marked increase in studies since 2013, relatively few are from the Global South. The question we are addressing in this section is: how are the concepts of resilience and GI advanced or implemented in ecological planning and design?

There is evidence that some LA cities are conserving, planning and implementing GI that could provide coping strategies against natural disaster management (Pauleit et al. 2021). With a long urban tradition embedded since pre-Hispanic cultures in LA, public and green spaces have traditionally evolved as a mechanism of social inclusion and interaction (Hermida et al. 2017). Hermida et al. (2017) illustrate that treatment of rivers and riverbanks are the main elements of both environmental and social resilience of cities, the latter due to the social ties created by green space activities. Many Global South nations have shown a desire for advancement and Global North ideals. For example, Brazil became synonymous with modernism in the twentieth century (Stephan 2000), and more recently pursued North American suburban lifestyles resulting in urban sprawl (Lara 2011). Global North influences in Puerto Rico include a reduced pool of food and ornamental species in the commercial nursery industry (Vila-Ruiz et al. 2014), which in turn has impacted the quality and diversity of GI implementation. Conversely, British Garden City town and neighbourhood designs that prioritise street trees and large parks, have persisted and resemble contemporary eco-cities—advancing GI planning and design with educational, recreational and aesthetic principles (Macedo 2011; Rego 2011; Gallanter 2012) and protecting cities against heat island effects and flooding. Where many cities in the Global North need to retrofit, due to the rapid expansion rates in the Global South, protecting existing GI and integrating it in new development is important (Dobbs et al. 2019). In Argentina (Gallanter 2012) and Brazil (Macedo 2011) the Garden City legacy has led to the safeguarding of green areas and native vegetation.

In Ecuador, the national plan for Buen Vivir calls for spatial planning to guarantee territorial and global environmental sustainability, and to increase people's safety by minimising the impact of natural hazards such as floods (Serra-Llobet and Hermida 2017. The application of the Global North concept of greenways in Brazil has been successfully implemented through federal laws, addressing regional planning and site scale design to improve the urban environmental quality (Frischenbruder and Pellegrino 2006). In Chile, Rojas et al. (2019) argue for the planned conservation of urban wetlands as a regulatory service against flooding. The general lack of connection between adaptive strategies and design disciplines is addressed by Dhar and Khirfan (2016). The authors combine experiential and expert knowledge with the participatory methods of urban planning to provide climate change adaptation in vulnerable coastal communities. The local community preferred soft adaptations or small-scale engineering interventions due to their reversibility and their minimal environmental impact. The underlying ecological design concepts can be useful to other coastal areas once the input of local communities is obtained (Dhar and Khirfan 2016).

Byrne et al. (2015) suggest that through extensive promotion by both science and policy, in Asia there is growing interest and support for GI and the part it plays in disaster risk reduction and management. Urban flooding in China has received significant attention following extreme weather events caused by climate change and urbanisation (Wang et al. 2017). A 2014 study found that 641 out of 654 Chinese cities suffer from frequent floods, with at least 130 experiencing flooding nearly every year. In 2012 flash floods in Beijing impacted more than 1.6 million people, while in 2016, massive floods in southern and northern China affected 60 million people, killing 300 and causing the evacuation of a further 500,000. Since 2010, flood disasters as the result of excessive rainfall caused damage of more than US $ 15 billion annually (Jiang et al., 2018). The Sponge City initiative (see Box 1) places a focus on disaster prevention (Wang et al. 2018; Nguyen et al. 2019) and mitigation through proactive urban water management to address future climatic extremes such as floods and droughts (Chan et al. 2018), although climate adaptation does not form part of this approach.

According to Borie et al. (2019) there are different reasons why resilience planning and strategies are rarely implemented in climate adaptation plans in the Global South. One reason is the lack of GIS technology to collect and analyse data, e.g. Nairobi, Kenya (Borie et al. 2019). Another reason is a top-down approach, as in Cape Town where a lack of engagement failed to address the expectations of the poor (Rodina 2019). Due to recent (2017–2019) severe droughts and water shortages, Cape Town has realised the importance of developing a water resilient agenda. In interviewing several experts involved in planning or policy-making, Rodina (2019) revealed two clear opposing trends in terms of "resilience-building strategies". The first is a focus on ecosystems (eco-hydrology approach, aligned with Water Sensitive Design, see Sect. 2.3a) proposing that the quality of urban watersheds, connectivity between different green areas, health of the urban ecosystem and the provision of water services to the community must improve and align with the social differentiation and participatory planning with locals (Rodina 2019). The second, an engineering focus, considers technical aspects and makes use of past hydrological data to predict future trends (Rodina 2019). An integration of these two approaches is needed in Cape Town to break down the silos in which water supply is managed in the city (Rodina 2019). According to Carden et al. (2016) a community of practice needs to develop where different stakeholders can learn from each other. Important concepts in resilience thinking such as co-production of knowledge, network building across scales and sectors, developing the capacity of stakeholders and stimulating innovative problem solving need to be applied to develop water-sensitive cities (Carden et al. 2016).

Several cities in the Global South have developed climate adaptation plans but often struggle to implement them. In Tanzania a national adaptation plan funded by the European Union was developed that assisted local governments to develop their own adaptation plans (Macchi et al. 2013). In Dar Es Salaam existing plans were changed to mainstream adaptation. This led to a better understanding of urban sprawl and the vulnerability to climate change in coastal areas and the conservation of water resources by developing methods to decrease groundwater salinisation (Macchi et al. 2013). Roy et al. (2018) indicated that although there are some examples of targeted greening in the city of Dar Es Salaam that will increase adaptation to climate change, residents of informal settlements are usually omitted from these measures. As one of the global leaders in climate adaptation planning, the city of Durban, South Africa, has developed and implemented municipal adaptation plans for water, health and disaster management and has also attempted to increase the adaptive capacity of local communities through several community pilot projects in terms of food security and reforestation (Carmin et al. 2012; Roberts et al. 2012).

2.3 *Ecological Renewal of Urban Areas*

With the increasing urban populations globally there is great potential for the "renewal and regeneration" of cities (Steiner 2014). In this section, we focus on planning and design of ecological infrastructure that conserves water, ensures co-living with nature, creates basic safety and wellbeing in urban areas, and address ES equity and environmental justice applications.

a. How are cities planned and designed to conserve water?

In Southeast Asia most governments have adopted policies for controlling land-use in upland watersheds, but the extent to which users and residents are involved in planning differ between various regions and cities (Lebel and Daniel 2009). China's Sponge City initiative espouses aims that directly address the conservation of water (see Box 1). In LA, a few studies have focused on water conservation. For example, de la Barrera et al. (2016) consider GI planning to include urban vegetation that is well-adapted to the regional climatic conditions and thereby reduce the need for watering.

Research on planning and design for urban conservation of water in SSA is biased towards South Africa. In the past, planning for water supply and demand was not linked in South Africa (Rohr and Fourie 2014), but was addressed in 2013 by new planning legislation, SPLUMA (Spatial Planning and Land Use Management Act) (Rohr and Fourie 2014). The legislation called for urban water cycle management in a more sustainable way. A Water Sensitive Design framework was developed for South Africa with the notion to advance water-sensitive cities, following an Australian model (Armitage et al. 2014). Water management becomes a prime objective in the design of urban areas, but being context-sensitive, different tools, tactics and ways to transfer the knowledge are needed in developing countries (Fisher-Jeffes et al. 2017a). Social acceptance and capacity building around this concept is important due to the high inequality in South Africa, and therefore the concept of "learning alliances" is introduced through which different stakeholders are informed and trained to enable the establishment of a community of practice to implement Water Sensitive Design (Carden et al. 2016).

Although Water Sensitive Design and its values are not yet acknowledged over a large scale in spatial planning decisions in South Africa, there are examples where it has been applied (Carden et al. 2018). The Cape Town Spatial Water Framework emphasises the importance of spatial planning in interventions in water management issues at different scales (Cameron and Katzschner 2017). The city has made great strides in focusing on the local context in water challenges (e.g. wastewater treatment, flood risk and water scarcity), implementing important policies in terms of the management of urban storm water impacts and floodplain and river corridor management (Madonsela et al. 2019). In the Liesbeek River catchment in Cape Town various stakeholders from public and local government were involved in the development of a framework to implement the Water Sensitive Design strategies—the Liesbeek Life Plan (Carden et al. 2016). Some Water Sensitive Design strategies were implemented in the Liesbeek catchment to alleviate flooding, improve the quality as well as the aesthetics of the river and increase biodiversity (Carden et al. 2016, 2018). Strategies include, for example, rainwater, storm water and grey water harvesting and sustainable urban drainage systems (SuDS) (Carden et al. 2018). Fisher-Jeffes et al. (2017b) argued that further research is needed on the advantages of storm water harvesting and social perceptions on willingness to harvest storm water across socio-economic groups.

Another example of SuDS is from the city of Johannesburg, where conventional storm water management techniques have been replaced by SuDS (Carden et al. 2016; Mguni 2015). However, in Johannesburg SuDS is implemented in a top-down manner with no social networks or learning alliances (Mguni 2015), which is common to developed countries. In a SWOT analysis to determine the potential value of SuDS replacing traditional storm water management systems in SSA, more threats (e.g. low priority of storm water management, lack of data, high maintenance costs, open flooded areas have safety concerns) than opportunities were identified, indicating risks of failure (Mguni et al. 2016). However, if the context of each city is considered, SuDS may still be applied successfully in SSA cities if local communities are involved in innovative SuDS experiments, increasing opportunities for learning and collaboration (Mguni et al. 2016).

b. Are cities planned and designed for co-living with nature?

In many Global South countries, land use policy often overrides environmental regulations due to social pressures (Aguilar and Santos 2011). In LA biodiversity protection in urban areas is gaining more interest due to the ES they provide, yet the assessment, in spatially and temporally explicit terms, of landscape-change factors threatening biodiversity has not gained much attention (Rojas et al. 2013a, b). Methods are strongly constrained by the lack of detailed biodiversity data (Rojas et al. 2013a). However, a number of countries are progressively incorporating a Strategic Environmental Assessment rationale via Environmental Impact Assessments (environmental frameworks and environmental mainstreaming in policies and plans) (Rojas et al. 2013b). Environmental Impact Assessment evaluations often fail to balance isolated impacts produced by a project with the overall territorial model (Rojas et al. 2013b). A useful planning tool advocated by Rojas et al. (2013b) is the Land and Suitability Index for urban development that combines three main sub-indices, i.e. (1) the vulnerability of the biosphere, lithosphere and hydrosphere to potential impacts; (2) the natural heritage value of the target area and (3) its contribution to terrestrial ecological connectivity.

One major cause of environmental degradation in the Global South and North is the decline of green open spaces that reduce potential ES values. Research indicates that almost all Global South cities in Asia have inadequate and poor quality greenspace, associated with poor social conditions, economic and environmental deterioration (Said and Mansor 2011). Where urban green spaces exist, many are sterile open spaces that are vacant or underutilised. In addition, many existing green spaces are not valued as assets or heritage and are therefore compromised in favour of other land-uses (Said and Mansor 2011). In LA, Salazar and Jalabert (2016) argue that to address problems with ecological urban design one needs to understand how contemporary city inhabitants perceive and use their local landscapes (Salazar and Jalabert 2016). Junior and Santos (2017) illustrate the importance of perception in an urban Protected Area in Brazil in which problems such as crime and a lack of infrastructure have prevented feelings of belonging and nature appreciation amongst residents. Junior and Santos (2017) argue that the participation of surrounding residents in the management of PAs is critical for GI conservation (see Sect. 2.4.1).

Due to competition with other land-uses, Jakarta reduced green space provision from 37% to 14% in their Regional Plan of Area Arrangement (Said and Mansor 2011). A lack of awareness and sense of civic mindedness further contributed to the degradation of existing green spaces in several Asian countries (Said and Mansor 2011). In Bangkok, research quantifying GI based on the total amount of green space, found that per capita park space was approximately 1.8 m^2. A master plan proposed increasing the per capita park area to 4 m^2 within the next 25 years (Thaiutsa et al. 2008). There have also been plans to redevelop some of the city's informal settlements with increased access to greenery. Concerns relating to GI in Kuala Lumpur included a lack of proper planning, implementation, management and reinforcement in terms of low standards of maintenance and lack of human resources and budgetary allocations, a lack of awareness, skills, knowledge and expertise (Said and Mansor 2011).

Future urban expansion is predicted in China's high biodiversity areas leading to habitat fragmentation and destruction (Wang et al. 2009; Güneralp et al. 2015), especially along the coast and the country's deltas (He et al. 2014). Hangzhou is unique in China for its initiatives to restore lost green space (Wu et al. 2012; Wolch et al. 2014). The city is acknowledged as a Garden City, an official designation bestowed on cities meeting national standards for forest cover, green spaces and parks. Hangzhou is celebrated for its tree-lined streets, scenic West Lake National Park and China's first urban wetland park (Wolch et al. 2014) and its efforts to harness ecological, economic and social functions from urban green spaces which include underutilised areas under freeways and on former industrial sites (Chen and Wu 2009; Wolch et al. 2014). China's Tianjin eco-city master plan's design is evaluated by a Key Performance Indicator system of 26 qualitative and quantitative indicators (Caprotti et al. 2015) based on local and international standards and Leadership in Energy and Environmental Design (Lin 2014). The Key Performance Indicator system prescribes an endemic species quota in vegetation composition and cover, through an urban plant community with native characteristics adapted to the local environment, but was not met in 2015 (Caprotti et al. 2015).

In SSA there is also a large loss of urban green space due to urban expansion into flood prone areas, especially riverbanks, floodplains and wetlands (Abo-El-Wafa et al. 2018; Douglas et al. 2008; Herslund et al. 2018). The importance of urban green spaces is well understood in SSA but often not implemented in planning due to other more pressing issues, land use conflicts and a lack of adequate policy support, to name a few (Herslund et al. 2018; Cilliers 2019; Bobbins and Culwick 2015). Although green areas are included in planning of cities in SSA they are often the left-over spaces remaining after planning for other land-uses (Cilliers 2019; Chishaleshale et al. 2015), are often illegally occupied (Lategan and Cilliers 2016) or legally rezoned for other land-uses; the latter issues are also prominent in LA (Frischenbruder and Pellegrino 2006). Not much research has been done in terms of green space planning to conserve biodiversity in urban areas in most SSA cities. However, South Africa is an exception in SSA, especially in large cities such as Cape Town, Durban and Johannesburg, the first two situated in global biodiversity hot spots. See Box 2 for a discussion of the systematic conservation planning approach that prioritises natural ecosystems areas inside city boundaries. Although urban planners in South Africa include conservation plans in land use planning frameworks, they rarely acknowledge green areas as part of a GI network and that they can provide services on par with conventional infrastructure (Bobbins and Culwick 2015).

Box 2: Systematic conservation planning in South African cities

Conservation of natural areas in South Africa promoted by the National Environmental Management and Biodiversity act (NEMBA) has been strongly influenced by systematic conservation planning (SCP). Each province has a systematic conservation plan, some more refined than others (Department of Environmental Affairs 2016). SCP is an important part of conservation biology and is used worldwide to prioritise areas for protection. This approach is built upon three principles, (1) conserve a representative sample of species and habitats, (2) protect ecological and evolutionary processes allowing biodiversity to persist and (3) set quantitative conservation targets identifying priority conservation areas for each habitat type (Driver et al. 2003). In South Africa, vegetation units are used as one of the basic data layers for SCP as they represent biodiversity in general.

Provincial systematic conservation plans need to be refined and put into context on a local level. Cape Town, which is situated in a global biodiversity hotspot, is the first South African city that developed a systematic conservation plan (Holmes et al. 2012). Two other comprehensive systematic conservation plans for SA cities include Durban (McLean et al., 2016) and the Gauteng province (including Johannesburg and Pretoria—Pfab et al. 2017).

SCP consists of two phases, the first is a systematic conservation assessment in which current protected areas, other critical biodiversity areas (CBAs) and other important conservation areas are mapped, based on data of biodiversity, vegetation types, plant species distribution, rare and endangered species, etc. (Holmes et al. 2012; Pfab et al. 2017). The second phase of a SCP is the implementation phase. Driver et al. (2003) suggest that during the assessment phase an operational framework consisting of different "key ingredients" should be followed from the start to ensure implementation. Important aspects included are involvement of implementation agencies (e.g. conservation agencies, municipalities, NGOs) in the assessment team, addressing the needs and interest of all stakeholders involved (bottom-up approach) and ensuring that maps are meaningful to implementation agencies (Driver et al. 2003). In Cape Town different implementation strategies such as spatial planning tools, CEPA (communication, education and public awareness), biodiversity impact assessments, and acquisition of land to be conserved are followed (Holmes et al. 2012). As land is expensive in the city, biodiversity stewardship partnerships between local government and land-owners have been established in Cape Town in which private land is secured for biodiversity and land-owners get tax rebates and management advice (Holmes et al. 2012).

Where systematic conservation plans exist, they form part of spatial development frameworks which are legally binding in South Africa but are mainly implemented in larger cities with dedicated environmental units. These plans are also important "decision support tools" in biodiversity studies for environmental impact assessments (Pfab et al. 2017). Systematic conservation plans are often not used in small and medium-sized cities due to lack of capacity and sensitivity about conservation issues on local government level, shifting the responsibility to environmental consultants (Wilhelm-Rechmann and Cowling 2013). It is, however, still the land use planners that need to make the decisions regarding land use change, but it seems the newer generation of planners is less sensitive to biodiversity conservation needs (Wilhelm-Rechmann and Cowling 2013).

Although SCP contributes to conservation of natural areas in and around large cities in South Africa, none of the other urban green spaces forming part of the urban green infrastructure are protected in this approach. SCP does, however, form a solid basis for the development of a green infrastructure plan for those cities.

References

Department of Environmental Affairs (2016) National protected areas expansion strategy for South Africa 2016. Department of Environmental Affairs, Pretoria

Driver A, Cowling RM, Maze K (2003) Planning for living Landscapes: perspectives and lessons from South Africa. www.cepf.net/documents/living.landscapes.pdf. Accessed 29 Mar 2017

Holmes PM, Rebelo AG, Dorse C, Wood J (2012) Can Cape Town's unique biodiversity be saved? balancing conservation imperatives and development needs. Ecol Soc 17(2): 28

McLean CT, Ground LE, Boon RG, Roberts DC, Govender N, McInnes A (2016) Durban's systematic conservation assessment. EThekwini Municipality, Environmental Planning and Climate Protection Department, Durban, http://www.durban.gov.za/City_Services/development_planning_management/environmentalplanning_climate_protection/Publications/Documents/DurbansSystematicConservation.pdf. Accessed 26 Apr 2019

Pfab MF, Compaan PC, Whittington-Jones CA, Engelbrecht I, Dumalisile L, Mills L, West SD, Muller PJ, Masterson GP, Nevhutalu LS, Holness SD, Hoare DB (2017) The Gauteng conservation plan: planning for biodiversity in a rapidly urbanising province. Bothalia 47: a2182

Wilhelm-Rechmann A, Cowling RM (2013) Local land-use planning and the role of conservation: an example analysing opportunities. S Afr J Sci 109: #0026

c. How does ecological planning and design improve safety and wellbeing in urban areas?

Little specific research has been done on this aspect in LA and SSA, although one of the main aims of the development of climate adaptation plans for cities in SSA, which have been discussed, is to improve the safety and wellbeing of the residents (Roberts et al. 2012; Macchi et al. 2013). Better water management using Water Sensitive Design principles could improve the quantity and quality of water resources in South Africa and the quality of life of residents, especially those living in informal settlements (Carden et al. 2016).

Said and Mansor (2011) confirm that GI can promote public health and environmental sustainability in Southeast Asian cities, but also identify challenges in terms of GI implementation to be addressed in urban planning and design (Said and Mansor 2011). Studies in Taiping, Malaysia, suggested that the characteristics and experiences of the urban green network resulted in progressive physical, cognitive and social functioning of urban residents, hence advancing wellbeing (Mansor and Said 2008). In a study exploring the impacts of several neighbourhood characteristics on subjective wellbeing in Beijing, only proximity to an urban park was shown to exert a significant impact (Dong and Qin 2017). Research by Wang et al. (2009) in China stated that urban green space is critical for promoting physical activity and improving public health. Beijing, therefore envisions that no resident should be more than 500 m from a green space, encouraging green space development and urban tree planting to establish green belts along main ring roads and to green one million m^2 of rooftop space. Such strategies are also employed to address poor air quality concerns related to high levels of nitrogen dioxide, sulphur dioxide and suspended particulate matter and an increased number of sandstorms in the city (Asian Green Cities Index 2012).

d. How are cities planned and designed to address equity and environmental justice implications?

As indicated earlier, in the Global South the urban poor are often relegated to environmentally vulnerable locations, often residing in low-lying areas, along the banks of rivers or lakes, on steep slopes (Aguilar and Santos 2011; Dobbs et al. 2019) or in the proximity of waste dump-sites. Urban sprawl, through rapid urban growth that is typical in the Global South (Myers 2021), results in high inequality in terms of public services accessibility and other socio-economic variables (Wei et al. 2017). Vulnerabilities are further exacerbated by the impacts of climate change (Dobbs et al. 2019). Furthermore, unequal power relations and perverse incentive structures often shape development and risk reduction decisions in ways that allocate benefits to elites and emerging middle classes, while disproportionately allocating ecological and social costs to the urban poor and marginalised (Garschagen and Marks 2019). In a systematic literature review on environmental inequality in terms of access of urban residents from different socio-economic groups, most of the studies were done in Asia (70%, 52% of which in China), 17% in Latin America and 13% in Africa (mainly South Africa) (Rigolon et al. 2018). The high number of studies in China could be due to its large population and its higher education sector growth (Rigolon et al. 2018). Comparisons between the results from this study and a similar study in the Global North have indicated similar patterns in terms of green space quantity and quality, being higher in more affluent areas, but not in terms of proximity in green space. In the Global South poorer communities and households are further from green spaces than more affluent one compared to the Global North (Rigolon et al. 2018). The vulnerable that face multiple environmental injustices warrant more attention in the Global South.

The planning and design of urban areas have a significant effect on the health and wellbeing and therefore also the health equity of residents which have been well-studied in cities in the Global North (Smit et al. 2011). Smit et al. (2011) suggest, a "research and action agenda" on urban planning and design and health in low- and middle-income countries for interventions to improve equity, based on sound research.

Serra-Llobet and Hermida (2017) explain that Ecuador's new constitution recognises "rights of nature" and peoples' right to benefit from the environment and natural resources that enhance the "Quality of Life". Similarly positive, the new constitution in Brazil, 1988, has prompted the planning and design implementation of greenways in many Brazilian cities (Frischenbruder and Pellegrino 2006), which besides promoting wellbeing, also has potential to reduce crime. For example, Escobedo et al. (2018) in Bogota indicated that increased tree density and size were associated with lower homicide occurrences. In LA the concept of sustainability often involves social justice and equity, a better quality of life for all (Sagaris 2014; Centner 2009). However, in spite of greater awareness, equity and environmental justice are rarely addressed in LA, since policies and federal requirements are often neither mandatory nor implemented (Dobbs et al. 2019). Aguilar and Santos (2011) illustrate how urban policy does not support new innovative schemes to address housing, basic services and land tenure security in informal settlements, along with conservation of the local environment. They demonstrate environmental injustice in how policies are currently implemented in urban peripheral areas. In Brazil, Macedo and Haddad (2016) found that though spatial equity has improved in Curitiba over the last 10 years, parks are not yet equitably distributed to working-class families. There are no provisions in the legislation enacting policies that guarantee an equitable spatial distribution of green open space. Although Curitiba is privileged in Brazil in terms of park space, with 27 parks and 16 wooded areas, comprising roughly 5.3% of the urban area, there are entire neighbourhoods in the city with no park within walking distance to residents.

Environmental justice is an emergent problem in China (Wolch et al. 2014), as eastern China is prioritised in development focus above the central and western regions, resulting in eastern cities and municipalities historically enjoying better environmental quality compared to their western counterparts where environmental hazards have been more concentrated. Although disparities have been decreasing, the west still bears a disproportionate environmental burden, while the east reaps more benefits (Zhang et al. 2016).

Even seemingly environmentally centred development programmes, such as the Chinese eco-city, may contribute to environmental injustices and add to environmental degradation if not done thoughtfully and based on sound environmental understandings (Caprotti et al. 2015; Liu 2018; Hansen et al. 2018). Given the high standards and financial stability required to commit resources towards clean-up programmes, meet the standards for a green economy, circular economy, renewable energy consumption, wastewater treatment, reforestation and city beautification for accreditation as an eco-city by the Chinese government, it is almost impossible for poor communities to comply. This is evidenced by accredited cities being some of the wealthiest in China (Liu 2018). Caprotti et al. (2015) commented on the striking contrast between the high-tech new urban environments in Tianjin Eco-City and the temporary dwellings in the periphery for the impoverished workers who build these eco-cities (Liu 2018). Access to and engagement with urban greenery has also been related to socio-demographic characteristics (Lv et al. 2011). Wang et al. (2017) identified public support for Sponge City initiatives as being dependent on occupation, income level and especially level of education. Those working in the public sector and those with higher incomes and more advanced levels of education were more likely to support Sponge Cities (Wang et al. 2017). These findings confirm that certain socio-demographic characteristics may influence support for eco-initiatives and that social variables must be considered when developing and generating support for such projects through participatory processes. King et al. (2003) however noted the paucity of local participation and top-down approaches that are being employed in three case studies in Asia. Various cultural or ethnic minority groups were not given the opportunity to participate in planning processes. Although the need was recognised, the implementation was not successful (King et al. 2003; Dai et al. 2018).

Several studies from SSA cities have indicated environmental injustice through inequality in terms of the distribution of urban green spaces and therefore also the ES that they provide (Ernstson 2013; Roy et al. 2018). Due to the inherently unjust political regimes in South Africa prior to 1994, the emergence of urban environmental justice is relatively recent with residual effects visible in the spatial fabric of contemporary towns and cities (McConnachie et al. 2008; Landman 2019; Venter et al. 2020). Since democracy there has been a marked increase in development of parks in previously marginalised areas (Stoffberg et al. 2012), but are dwarfed by the high rates of urban growth and in-migration and consequently the proportional areas are actually declining (Venter et al. 2020). The Lynedoch Eco Village is the first ecologically designed urban area in South Africa with the intention to mix different social groups and indicate that a balance between growth, equity and sustainability is possible (Swilling and Annecke 2006). Unfortunately, there were also some practical challenges in terms of empowerment opportunities, contributions to levy costs and uneven representation of different ethnic groups in this project (Lotter 2017).

As the provision of ES is linked with several social and political processes, Ernstson (2013) suggested a framework to analyse the environmental justice of ES in urban areas in the Global South consisting of three analytical aspects, namely the generation, distribution (spatial and temporal) and articulation (argumentation for protecting land) of ES. The first two analyses can be done at city level, but the last one deals with the place-specific struggles and opposed ways of city planning (Ernstson 2013). Use of this framework will address two often overlooked contemporary challenges in natural resource management, namely that ES are linked to socio-economic aspects and the danger that decisions on ES trade-offs will be driven by economic incentives (Ernstson 2013).

2.4 Link Knowledge to Action to Affect Positive Change

The importance of human initiative for greater urban ecological integrity goes beyond valuing nature and its benefits for stewardship, care and management, it involves the spread and uptake of relevant and workable ideas. Several Global South studies emphasised the prospect of policy to affect change (Eakin et al. 2010; Aguilar and Santos 2011; Rojas et al. 2013a; Kubiszewski et al. 2016; Rohr and Fourie 2014; Dobbs et al. 2019), in this section we will not discuss this in detail but rather focus on how ecological planning and design ideas spread.

2.4.1 Citizen Engagement in Planning and Design

Several case studies in LA made use of people as informants and included them in the design or planning process. The value of citizen engagement in GI management is illustrated by Junior and Santos (2017) in a protected urban forest fragment in Brazil. Interviews with residents indicated that crime, no basic services, conflicts in forest use and lack of communication with residents, are obstacles to the effectiveness of conservation actions and affected feelings of belongingness, nature appreciation and biodiversity conservation (Junior and Santos 2017). Interviews with management staff showed that their suggestions for improvement were complementary to those of residents. Junior and Santos (2017) recommend that the residents be allowed to participate as allies in the management of GI, sharing their needs and expectations and call for the inclusion of indigenous knowledge systems (IKS) in planning and management by combining life experiences with technical knowledge.

Widianingsih and Morrel (2007) state that Indonesia, as most of Southeast Asia, is slowly moving towards increased engagement between state and society. Although political structures have democratised, local-level civic engagement is yet to increase. Shatkin (2008) referred to a city developed around civic engagement as a "public city" and noted that such settlements have proven to be more resilient due to retained public responsibility for the built environment and greater public sector capacity. In contrast, Dai et al. (2018) highlight a lack of public participation as a major weakness in the Chinese Sponge City approach (Box 1). China presents a fragmented government administration system that lacks cooperation between related functions and agencies (Jiang et al. 2017; Li et al. 2017; Nguyen et al. 2019), offering limited opportunities for participation and collaboration between ministries, local governments and public and private sectors.

There are several studies from SSA discussing the importance of partnerships between different stakeholders contributing to GI planning (Lindley et al. 2018); for example, the Water Sensitive Design approach in Cape Town discussed under 2.3a. Partnerships usually include stakeholders from different municipal departments and natural and social scientists (Abo-El-Wafa et al. 2018; Herslund et al. 2018), but citizens rarely participate (Ruwanza and Shackleton 2016). Bolay (2015) argued that urban planning in African cities should begin from a "participatory diagnosis" of the demographic, spatial, infrastructural, economic, social and environmental situation of a specific city. An interesting example of citizen engagement is one from Cape Town where a community forum intervened in the development of a shopping mall in a wetland by developing their own landscape design plan that included restoration of the natural vegetation, more recreation areas and a community park (Aalto and Ernstson 2017). Professional landscape designers were only later involved but made use of a co-creation design process. Aalto and Ernstson (2017) suggest that planners and designers should still be involved but must take a "humble approach" through developing strong alliances with citizen groups to develop bottom-up approaches in design.

2.4.2 Environmental Perception and Design

Environmental perception surfaced in the mid-60 s, but has since been supressed by functionalist and productivist visions in LA (Salazar and Jalabert 2016). Junior and Santos (2017) argue for the importance of "bio-physiological perceptions" formed through memories, experiences, preferences, interpretations, attitudes and expectations. They maintain that local relationships with the environment as part of group identity and values and including IKS, must surface as part of programmes and planning strategies. Salazar and Jalabert (2016) agree with this viewpoint and illustrate, through qualitative fieldwork that makes use of semi-structured interviews, how urban inhabitants in Villarrica, Chile, lament having a high level of disconnect with the surrounding landscape. They argue that the task of ecological design is to generate a more deep-seated ecological literacy amongst urban inhabitants. This can be achieved by putting planning and design emphasis on facilitating an experiential closeness with the surrounding landscape through perception and use (Salazar and Jalabert 2016).

The importance of environmental perceptions of different stakeholders in urban green areas have been highlighted in several South African studies, for example community and ecosystem-based adaptation in Durban (Roberts et al. 2012) and the development of systematic conservation plans as in Cape Town (Holmes et al. 2012, Box 2).

In the Asian Global South it is recognised that successful broad application of initiatives, such as the Sponge City (Box 1), depends on public acceptance and support (Wang et al. 2017). Yet, despite the programme's expansive application, a large percentage of the Chinese public are not familiar with the Sponge City concept (Wang et al. 2017) and several obstacles still hamper meaningful stakeholder engagement, public participation and public–private partnerships to support Sponge City initiatives (Wang et al. 2017, 2018; Nguyen et al. 2019).

2.4.3 Linking Local and Scientific Knowledge

A recurrent challenge for planning in Asia is obtaining adequate information about ES at scales relevant to decision-making (Turner and Daily 2008). Local knowledge is often crucial but only available if planning agencies allow opportunity for meaningful local participation (Lebel and Daniel 2009). Any meaningful engagement with resilience building in Southeast Asia's small and mid-sized cities needs to start from a vulnerability perspective if it is to bring about sustainable and equitable risk reduction (Garschagen and Marks 2019). Such approaches should draw on the experiences and capacities of local communities to arrive at plausible local response strategies as part of broader integrated risk reduction strategies (Lebel 2013). Lebel (2013) notes that the strengths and limitations of local knowledge, as with science, should be understood to realise opportunities to link different ways of knowing towards local adaption strategies that are sensitive to specific ecosystems and social conditions. In addition, it is also imperative that existing fervour for green alternatives be capitalised on. In Dehli, India, the local government is disseminating the green agenda with a bottom-up approach by taking advantage of existing attitudes in support of urban environmental sustainability, through "eco-clubs". These eco-clubs inform students about environmental campaigns and engage them in a range of projects, such as tree planting, water conservation, establishing nature trails and minimising waste (Asian Green City Index 2012).

The involvement of scientists and universities in GI planning, design and implementation is recommended in LA by Frischenbruder and Pellegrino (2006), and techniques for local and expert knowledge integration for management, planning and design by Junior and Santos (2017), Salazar and Jalabert (2016), and Dhar and Khirfan (2016). Schwab et al. (2018) recommend context-specific ethnographic examination of how spaces are locally used and valued by different actors. Sagaris (2014) recommends small group discussions for authentic dialogue to build new consensuses. Participation must anticipate more than "educating" the community about a project's benefits; it must truly alter the final product and develop civil society (Sagaris 2018). In South Africa, there are several examples of partnerships between local municipalities and universities, e.g. the science-action partnership involving the eThekwini municipality. One aim of this partnership is to ensure that different ways of decision-making are developed through implementation-driven research (e.g. Cockburn et al. 2016).

2.4.4 Disseminating Knowledge in Future Research

Many authors see the improvements in future research processes as a way to disseminate knowledge. Aspects articulated in studies include data availability, access and quality, and improved methods for a greater spectrum of data sources to be integrated (Frischenbruder and Pellegrino 2006; Escobedo et al. 2018; Dobbs et al. 2019). There is a need for increased funding to address gaps in ES and planning research, education, and institutional capacity in urban areas in the Global South (Bolay 2015; du Toit et al. 2018; Dobbs et al. 2019). There is also a need to develop interdisciplinary research communities in different institutional contexts and areas of knowledge, principally those related to environmental challenges (Frischenbruder and Pellegrino 2006; Cockburn et al. 2016). The application of knowledge to planning and design and the training of designers and planners are also specifically articulated in LA (Dobbs et al. 2019) and South Africa (Wilhelm-Rechmann and Cowling 2013).

Not only are there obstacles in terms of domestic application of eco-initiatives such as the Sponge City in China (see Box 1), but other limitations may also impede the dissemination of valuable research and design approaches. For example, many publications dealing with the Chinese experience in eco-planning and development are published in Chinese language journals, which inhibited international dissemination (Wolch et al. 2014; Luo and Hyland 2019). The effectiveness of translation of scientific text remains contested, given potential inaccuracies in content, vocabulary, grammar and spelling (Luo and Hyland 2019). Language biases also constrain acknowledgement of all publications (Dobbs et al. 2019) and a consistent scientific dialogue on ecological planning and design in LA.

3 Discussion

3.1 Differences with Global North in Terms of Research Focus in Ecological Planning and Design

Although there are several structural, social, ecological and economic differences between cities in the Global North and Global South as highlighted in other chapters in this book, in this section we will briefly highlight some of the main differences in terms of ecological planning and design of cities where research is urgently required, namely the reality and effects of socio-economic inequity, projected urban population increases and climatic conditions exacerbating ecosystem disservices. These differences link the research areas discussed in Sect. 2 together.

Although research on ES in urban planning and design in the Global South is ongoing and has developed over the last decade, it has not yet reached the level of importance it has in the Global North. The socio-economic inequity experienced in many Global South cities tends to weigh more heavily in the supply and demand for ES than in the Global North and need to be addressed through planning and design research. Poor communities are often housed in ecologically vulnerable locations (wetlands, steep slopes, low-lying areas) (Douglas et al. 2008; Jenkins et al 2010; Aguilar and Santos 2011) and further away from urban green spaces (Macedo and Haddad 2016; Rigolon et al. 2018) that hamper ES benefits to them. Even in China environmental injustice is an emerging problem in terms of the high standards for accreditation for eco-cities that poor communities can't comply with (Caprotti et al. 2015) and public support for Sponge City initiatives which remains largely reserved for affluent communities (Wang et al. 2017). The socio-economic inequity in the Global South is starkly different from the Global North where effective public institutions and governance can maximise the provision of ES via well-established land use and conservation policies which are effectively implemented and which differ greatly from ineffective public policies and issues of poor transparency in many cities in the Global South (Dobbs et al. 2019).

Major urban population increases are predicted in the Global South in comparison with the Global North (United Nations 2018; Myers 2021). This will exacerbate existing urban conditions in Global South cities characterised by peripheral growth and sprawl, rural-urban migration, strained basic services, socio-economic inequity and ecological degradation, amongst other influences (Kubiszewski et al. 2016; Dobbs et al. 2019; Landman 2019) creating unique research and development challenges (Shackleton et al. 2021). Additionally most population growth will occur in small and medium-sized cities with segregated urbanisation patterns, leading to significant loss in remnant natural ecosystems in peripheral areas (Chan et al. 2018;

Güneralp et al. 2018; Dobbs et al. 2019). Although protection of natural resources through planning and design is also a challenge in cities in the Global North, they rarely experience the same priority challenges created by these unique urban and socio-economic conditions in the Global South.

In many cities in the Global South human health and wellbeing is compromised by climatic factors. The tropical climate and large areas prone to natural disasters exacerbated by climate change have a negative effect on health and wellbeing while the large biodiversity, besides its many unrecognised benefits, also entail some ecosystem disservices (e.g. disease vectors, safety, wildlife and insect nuisances, allergens, thermal comfort) (Escobedo et al. 2018; Dobbs et al. 2019). More research emphasis needs to be placed on urban ecological planning and design to illustrate how it can improve human health and wellbeing specifically in Global South cities (Aguilar and Santos 2011; Said and Mansor 2011; Carden et al. 2016) especially in light of the lack or limited importance currently given to GI (du Toit et al. 2018; Dobbs et al. 2019). This is an exercise that provides huge challenges, as expounded in Sect. 3.2. Although climate adaptation planning is not widely practised in Global South cities (Dhar and Khirfan 2017), South African cities, such as Durban, are regarded as one of the early adaptors and leaders in climate resilience research in the Global South (Carmin et al. 2012). These case studies could serve well as best practice examples for other cities.

3.2 Challenges in the Global South in Terms of Implementation of Ecological Planning and Design

In terms of the four research areas that are important for the development of best practice guidelines for ecological planning and design in the Global South, there are several implementation challenges. Some of these challenges are summarised in Table 1 and research projects to address these challenges are recommended. General research challenges mentioned in Sect. 2.4.4, such as access and quality of data, funding for research, education and institutional capacity are not discussed in detail. Transdisciplinary research is often suggested in Table 1 and it has become a common term over the past decades (Cilliers et al. 2014), but uneven power relations remain a significant challenge for its successful implementation. Due to the prevalence of social inequities in the Global South that lead to disparities in income and education and result in unequal levels of power (Dobbs et al. 2019; Landman 2019), transdisciplinary research is of extreme importance to make sure that all voices are heard and different ways of knowing are acknowledged when considering solutions to planning and design problems (Ernstson 2013).

Table 1 Challenges in the Global South that will influence the implementation of ecological planning and design and examples of research foci and methods that could advance these themes

Challenges	Unique considerations	Examples of research foci and methods to advance these themes
Increased urbanisation and sprawl	Unplanned sprawl, slum-like peripheral developments on steep slopes, low-lying areas, wetlands, areas with a lack of basic infrastructure and services—equity, health and wellbeing risks to inhabitants and environmental degradation (Aguilar 2008; Douglas et al. 2008; Abo-El-Wafa et al. 2018; Dobbs et al. 2019)	– Building a community of practice of different stakeholders through long-term research projects (co-production of knowledge, network building, innovative thinking) (Inzulza-Contardo and Moran-Figueroa 2018; Carden et al. 2016), as suggested for implementation of water-sensitive planning and design in South Africa (Carden et al. 2016)
Land invasion and ecological degradation	Loss of urban open space, natural areas and critical/sensitive biodiversity areas (e.g. presence of biodiversity hot spots)—resulting in biodiversity loss and low GI cover (Aguilar and Santos 2011; Said and Mansor 2011; Holmes et al. 2012)	– Transdisciplinary research on fragmented natural areas in urban and peri-urban areas, e.g. systematic conservation planning in SA (e.g. McLean et al. 2016; Pfab et al. 2017) – Qualitative social research on reasons and results behind ineffective land policy (Aguilar and Santos 2011) – Case studies on best practice guidelines for GI through garden cities examples in China, Argentina and Brazil that could capture GI benefits for other Global South cities (Gallanter 2012; Wu et al. 2012; Wolch et al. 2014)
Weak environmental support from local government	Many urban problems emerge from incapacity of local governments to manage rapid urban growth, to provide public goods for all social groups, to not neglect environmental protection in favour of market growth and to show adequate political and technical capacity (Dobbs et al. 2019; Rodina 2019)	– Transdisciplinary research on participatory planning and design following a bottom-up approach leading to strong alliances between local government and citizen groups in support of environmental and societal benefits—examples from Cape Town, South Africa (Aalto and Ernstson 2017) and Santiago, Chile (Sagaris 2014)—Qualitative social research on reasons and results behind ineffective land policy resulting in environmental injustice (Aguilar and Santos 2011), including governance issues and reasons behind it (Junior and Santos 2017)

(continued)

Table 1 (continued)

Challenges	Unique considerations	Examples of research foci and methods to advance these themes
Lack of planning initiatives and incentives for GI and ecological design	Agencies and professional disciplines operate in silos; lack of planning coordination between the urban areas subject to planning regulations; lack of strong local leadership, lack of national directives to encourage implementation and lack of local ordinances to make green infrastructure easier to implement (Frischenbruder and Pellegrino 2006; Roberts et al. 2012; Serra-Llobet and Hermida 2017)	– Case studies on how research on ecological planning and design can be made more relevant for decision-making and implementation on local government level. E.g.—universities-local governments partnerships in South Africa and India to develop implementation-driven projects (Asian Green City Index 2012, Cockburn et al. 2016) – Case study and experimental research for GI planning, design and implementation guidelines from best practice (Frischenbruder and Pellegrino 2006; Felson and Picket 2005) – Qualitative social research on indigenous knowledge systems (IKS) to include local know how in planning and management combined with technical knowledge (Dhar and Khirfan 2016; Junior and Santos 2017)
General lack of investment in ecosystems and their services due to economic market models, social pressures and weak governance	Local governments rarely invest in urban ecosystem restoration—basic and necessary social and economic programmes are prioritised, such as access to housing (social pressures), health and sanitation, while investing in green spaces and ES provision considered less important. Pressure for growth and development further disfavour public health investments in green spaces (Aguilar and Santos 2011; Schäffler and Swilling 2013; Dobbs et al. 2019)	– Transdisciplinary research on community-based ecosystem-based adaptation (Roberts et al 2012; Dhar and Khirfan 2016) – Case study research that demonstrates the health and wellbeing benefits of GI (Bobbins and Culwick 2015)
Ineffective policy and legislation create poor transparency and silo effects in addressing urban challenges	Complex peri-urban and peripheral areas are not seen and regulated as part of the city (Aguilar 2008; Eakin et al. 2010)	– Transdisciplinary studies on the integration of ecosystem and engineering approaches in planning and design (Rodina 2019). – Policy and framework gap research (Aguilar and Santos 2011; Dobbs et al. 2019)

(continued)

Table 1 (continued)

Challenges	Unique considerations	Examples of research foci and methods to advance these themes
Unique biodiversity, climatic, socio-political and economic aspects	Tropical environmental conditions and steep topography in various Global South countries influence productivity and ecosystem structure—supply and demand (Dobbs et al. 2019). Social and ecological inequity and environmental injustice (Wolch et al. 2014; Ernstson 2013; Macedo and Haddad 2016; Roy et al. 2018; Dobbs et al. 2019). High biodiversity and tropical climates often result in a need to address EDS (Escobedo et al. 2018)	– Transdisciplinary research on demand and supply of ES amongst different socio-economic groups, and the spatial and temporal dynamics thereof (Ernstson 2013), and interventions to improve equity (Smit et al. 2011). – Research on the development of learning alliances to get social acceptance and capacity building of all stakeholders on the implementation of "new" approaches in planning and design (Carden et al. 2016) – Redefining civic inclusion and participation by means of case study research in design and planning (Frischenbruder and Pellegrino 2006)
Research challenges, narrow approaches and agendas in urban ES and GI	Often too much focus on examples from Global North in terms of ES and GI instead of providing context-specific ES information, case studies, instruments and guidelines that can easily be integrated into decision-making and context relevant policies (Lara 2011; Bolay 2015; Dobbs et al. 2019). A lack of funding, data and data sharing (Dobbs et al. 2019)	– Research on using different techniques in mapping and valuation of ES provided by GI of different cities (including small and medium-sized cities) acknowledging the social-ecological uniqueness and demonstrating trade-offs in a transparent manner (O'Farrell et al. 2012; de Wit et al. 2012)
Vulnerability to climate change and natural disasters	In Global South regions inequities, rapid urbanisation and climate change mainly affect the adaptive capacity of socio-political and ecological systems (Dobbs et al. 2019). Informal settlements and unplanned growth further increase vulnerability to natural disasters, while climate change will affect these vulnerable populations most (Dobbs et al. 2019)	– Case study research to integrate local and expert knowledge for planning and design to formulate ecologically based adaptations that are context specific to vulnerable areas (Dhar and Khirfan 2016; Bertilsson et al 2019)

4 Conclusions

From this review on ecological planning and design in the Global South, we are able to conclude that Global South researchers have identified the context-specific inability of neoliberalism to assist the changes required for equity and environmental sustainability (Centner 2009; Dieleman 2017; Dobbs et al. 2019). Arguments given for LA also applicable to the entire Global South include the notion that political will and orientation could assist ecological planning and design through various means. These include regulation of market conditions, enforcement of land use norms, adaptive urban governance at local and city level (Eakin et al. 2010; Dobbs et al. 2019), provision of land and housing alternatives for poorer groups (Aguilar 2008), alternative productive activities such as agriculture and recreation activities, more participation of local residents to co-design and implement policies, and socially constructed consensus for effective urban development (Jenkins et al. 2010). The following aspects summarise the importance of context and local aspects in ecological planning and design in Global South cities.

4.1 A Focus on Local Relationships, Values and Perceptions

Current research and ecological design approaches in the Global South are increasingly focusing on uncovering local values, examples and ways of knowing (Ernstson 2013; Salazar and Jalabert 2016; Junior and Santos 2017). This is because international models have often been unsuccessfully applied to local contexts because they tend to oversimplify complex local realities (Schwab et al. 2018). Global South researchers have found that social values ascribed to land becomes part of local identity and intrinsically motivate local involvement in GI design, planning and management projects (Carden et al. 2016; Junior and Santos 2017; Nadal et al. 2018). Additionally, reverting to local terms and values in research and then finding commonalities with transnational models and concepts has proven to be more successful than top-down approaches (Mguni 2015; Schwab et al. 2018). Researchers further emphasise the value of capturing social perceptions in planning and design to create activities and spaces where citizens can connect with their environment and feel a part of it (Salazar and Jalabert 2016; Aalto and Ernstson 2017).

4.2 Global North Influences on Planning and Design

The historic British Garden City model has had positive consequences in Argentina and Brazil in terms of liveable green neighbourhoods that exist today (Macedo 2011; Gallanter 2012). Chinese garden cities and the Sponge City concept (see Box 1) are more recent examples of an Asian interpretation of ecologically minded strategies from the Global North (Wang et al. 2017). Its implementation across China presents new case studies to investigate such strategies in an Asian context. Although the local interpretation of Global North influences is worth expanding on in ecological planning and design, these influences don't always have positive results in the Global South. In Brazil, Global North modernist ideals eroded the local identify and has led to urban planning problems such as gated communities, unsustainable sprawl and alienation, as also seen in many other Global South countries, such as South Africa (Landman 2019). In the same way, international research metaphors (Dobbs et al.

2019) and transnational models are often blindly accepted and become more influential on planning than "translocal" dynamics (Schwab et al. 2018). The importance of own identity and the adaption of models and metaphors to local circumstances have become recognised in Global South research, design and planning. One example is water-sensitive design in South Africa, though based on an Australian model, different tools, tactics and ways of knowledge transfer have been used (Fisher-Jeffes et al. 2017a).

4.3 *Local Instruments to Protect Biodiversity and Ecosystem Services*

The Asian experience illustrates that site configurations, microclimate conditions and environmental perception could lead to appropriate design guidelines and planning practices (Xue et al. 2017). However, the importance of context-based planning is reinforced in Asia, similar to other Global South areas (Güneralp et al. 2018), increasingly dealing with informality (Myers 2021). In agreement with Revell (2010), we should question the current western models of development that are rooted within the Global North. We should reflect on how appropriate these models really are for analysing and developing cities of the Global South (Roy 2005). The informality that is part of most Global South cities presents an opportunity to rethink sustainability and how sustainable and resilient cities are contextualised through ecological planning and design. It is important to celebrate local examples of policies and actions that protect ecological integrity, such as Brazil's heritage of an internationally renowned landscape designer, Roberto Burle Marx who was a pioneer of tropical environmentalism and whose work remains highly influential today (Stephan 2000). China's Tianjin eco-city master plan prescribes greenery to develop based on an urban plant community with "native characteristics" adapted to the local ecology (Caprotti et al. 2015). In Cape Town, South Africa, the unique natural vegetation (fynbos) is protected on private land due to biodiversity stewardship partnerships between local government and land-owners as part of implementation of systematic conservation plans (Holmes et al. 2012) (see Box 2).

Urban designers and planners are not often in the position to make amends where political will and market conditions are unsupportive of decisions for greater socio-ecological good or urban sustainability. Furthermore, policy enforcement alone is not a guarantee for best practice (Frischenbruder and Pellegrino 2006). This chapter has presented examples of avenues for ecological design and planning to advance in the Global South. Firstly, there are examples and more potential for citizen engagement in planning and design with prospects to redefine civic inclusion and participation (Sagaris 2014, 2018). There is potential to mend and strengthen human-nature relationships through ecological design and planning processes by capturing, understanding and building on local environmental perceptions (Salazar and Jalabert 2016; Junior and Santos 2017). Lastly, case study research offers the potential to illustrate context-specific planning and design guidelines and exemplify best practice possibilities, which are required to improve planning, design and implementation (Said and Mansor 2011; Dhar and Khirfan 2016; Fisher-Jeffes et al. 2017a).

In conclusion, the best practice examples that are available through case study research in the Global South lead us to conclude that local challenges should be addressed by acknowledging differences and concentrating on unique conditions for enriched ecological planning and design in the Global South. Further research on the unique drivers, values and environmental perceptions in the Global South could assist best practice in urban planning and design. Place-specific knowledge, reconciled with local needs and values for greater socio-ecological good (Cilliers 2015; Liao and Chan 2016) might render wisdom in the ecological planning and design practices of the Global South. This wisdom could generate urban conditions that can engender value connections with surrounding landscapes (Salazar and Jalabert 2016) and in the long run improve living conditions, wellbeing, conservation and sustainability.

References

Aalto HE, Ernstson H (2017) Of plants, high lines and horses: civic groups and designers in the relational articulation of values of urban natures. Landsc Urb Plan 157:309–321

Abo-El-Wafa H, Yeshitela K, Pauleit S (2018) The use of urban spatial scenario design model as a strategic planning tool for Addis Ababa. Landsc Urb Plan 180:308–318

Adegun OB (2018) Resident's relationships with green infrastructure in Cosmo City, Johannesburg. J Urbanis 11:329–346

Aguilar AG (2008) Peri-urbanisation, illegal settlements and environmental impact in Mexico City. Cities 25:133–145

Aguilar AG, Santos C (2011) Informal settlements' needs and environmental conservation in Mexico City: an unsolved challenge for land-use policy. Land Use Pol 28:649–662

Ahern J, Cilliers SS, Niemelä J (2014) The concept of ecosystem services in adaptive urban planning and design: a framework for supporting innovation. Landsc Urb Plan 125:254–259

Armitage N, Fisher-Jeffes L, Carden K, Winter K, Naidoo V, Spiegel A, Mauck B, Coulson D (2014) Water sensitive urban design (WSUD) for South Africa: framework and guidelines. Water Research Commission (WRC) Report No. TT 588/14, Pretoria

Asian Green Cities Index (2012) Asian green city index: assessing the environmental performance of Asia's major cities. http://sg.siemens.com/city_of_the_future/_docs/Asian-Green-City-Index.pdf. Accessed 17 Aug 2019

Benedict MA, McMahon ET (2006) Green infrastructure: linking landscapes and communities. Island Press, Washington, DC

Bertilsson L, Wiklund K, de Moura Tebaldi I, Rezende OM, Veról AP, Miguez MG (2019) Urban flood resilience: a multi-criteria index to integrate flood resilience into urban planning. J Hydrol 573:970–982

Boon R, Cockburn J, Douwes E, Govender N, Ground L, Mclean C, Roberts D, Rouget M, Slotow R (2016) Managing a threatened savanna ecosystem (KwaZulu-Natal Sandstone Sourveld) in an urban biodiversity hotspot, Durban, South Africa. Bothalia 46:a2112

Breed CA, Cilliers SS, Fisher RC (2015) Role of landscape designers in promoting balanced approach to green infrastructure. J Urb Plan Dev 141:A5014003

Bobbins K, Culwick C (2015) Green growth transitions through a green infrastructure approach at the local government level: case study for the Gauteng city region. J Pub Admin 50:32–49

Bolay J-C (2015) Urban planning in Africa: which alternative for poor cities? The case of Koudougou in Burkina Faso. Curr Urb Stud 3:413–431
Borie M, Pelling M, Ziervogel G, Hyams K (2019) Mapping narratives of urban resilience in the Global South. Glob Enviro Change 54:203–213
Byrne JA, Lo AY, Jianjun Y (2015) Residents' understanding of the role of green infrastructure for climate change adaptation in Hangzhou, China. Landsc Urb Plan 138:132–143
Cameron R, Katzschner T (2017) Every last drop: the role of spatial planning in enhancing integrated urban water management in the City of Cape Town. S Afr Geog J 99:196–216
Caprotti F, Springer C, Harmer N (2015) 'Eco' for Whom? Envisioning eco-urbanism in the Sino-Singapore Tianjin eco-city, China. Intn J Urb Reg Res 39:495–517
Carden K, Ellis D, Armitage NP (2016) Water sensitive cities in South Africa: developing a community of practice. Proceedings of the 3rd Intn Conf on Design, Construction, Maintenance, Monitoring and Control of Urban Water Systems (UW 2016) www.witconferences.com. WIT Transactions on The Built Environment 165
Carden K, Armitage N, Fisher-Jeffes L, Winter K, Mauck B, Sanya T, Bhikha P, Mallett G, Kanyerere T, Gxokwe S (2018) Challenges and opportunities for implementing water sensitive design in South Africa. WRC Project report K5/2412, Pretoria
Carmin J, Anguelovski I, Roberts DC (2012) Urban climate adaptation in the Global South: planning in an emerging policy domain. J Plan Educ Res 32:18–32
Centner R (2009) Conflictive sustainability landscapes: the neoliberal quagmire of urban environmental planning in Buenos Aires. Local Enviro 14:173–192
Chan FK, Griffiths JA, Higgitt D, Xu S, Zhu F, Tang YT, Xu Y, Thorne CR (2018) "Sponge city" in China: a breakthrough of planning and flood risk management in the urban context. Land Use Pol 76:772–778
Chen X, Wu J (2009) Sustainable landscape architecture: implications of the Chinese philosophy of "unity of man with nature" and beyond. Landsc Ecol 24:1015–1026
Chen H, Jia B, Lau SS (2008) Sustainable urban form for Chinese compact cities: challenges of a rapid urbanised economy. Habitat Intn 32:28–40
Chishaleshale M, Shackleton CM, Gambiza J, Gumbo D (2015) The prevalence of planning and management frameworks for trees and green spaces in urban areas of South Africa. Urban For Urban Green 14:817–825
Cilliers EJ (2019) Reflecting on green infrastructure and spatial planning in Africa: the complexities, perceptions, and way forward. Sustain 11:455
Cilliers EJ, Cilliers SS (2015) From green to gold: a South African example of valuing urban green spaces in some residential areas in Potchefstroom. Town Region Plan 67:1–12
Cilliers SS, du Toit MJ, Cilliers EJ, Drewes E, Retief F (2014) Sustainable urban landscapes: South African perspectives on transdisciplinary possibilities. Landsc Urb Plan 125:260–270
Cilliers L (2015) Evaluating the spatial and environmental benefits of green space: an international and local comparison on rural areas. Agric for Fish 4:45–51
Cockburn J, Rouget M, Slotow R, Roberts D, Boon R, Douwes E, O'Donoghue S, Downs CT, Mukherjee S, Musakwa W, Mutanga O, Mwabvu T, Odindi J, Odindo A, Procheş S, Ramdhani S, Ray-Mukherjee J, Sershen D, Schoeman MC, Smit AJ, Wale E, Willows-Munro S (2016) How to build science-action partnerships for local land-use planning and management: lessons from Durban, South Africa. Ecol Soc 21:28
Dai L, van Rijswick HF, Driessen PP, Keessen AM (2018) Governance of the sponge city programme in China with Wuhan as a case study. Intn J Water Res Dev 34:578–596
Davids R, Rouget M, Boon R, Roberts D (2016) Identifying ecosystem service hotspots for environmental management in Durban, South Africa. Bothalia 46(i2):2118
Davids R, Rouget M, Boon R, Roberts D (2018) Spatial analyses of threats to ecosystem service hotspots in Greater Durban, South Africa. PeerJ 6:e5723
de la Barrera F, Reyes-Paecke S, Banzhaf E (2016) Indicators for green spaces in contrasting urban settings. Ecol Indic 62:212–219

de Wit M, van Zyl H, Crookes D, Blignaut J, Jayiya T, Goiset V, Mahumani B (2012) Including the economic value of well-functioning urban ecosystems in financial decisions: evidence from a process in Cape Town. Ecosys Serv 2:38–44
Dhar TK, Khirfan L (2016) Community-based adaptation through ecological design: lessons from Negril, Jamaica. J Urb Design 21:234–255
Dhar TK, Khirfan L (2017) Climate change adaptation in the urban planning and design research: missing links and research agenda. J Enviro Plan Mgmt 60:602–627
Díaz S, Demissew S, Carabias J, Joly C, Lonsdale M, Ash N, Larigauderie A, Adhikari JR, Arico S, Báldi A et al (2015) The IPBES conceptual framework: connecting nature and people. Curr Opin Enviro Sustain 14:1–16
Dieleman H (2017) Urban agriculture in Mexico City: balancing between ecological, economic, social and symbolic value. J Clean Prod 163:S156–S163
Dobbs C, Escobedo FJ, Clerici N, de la Barrera F, Eleuterio AA, MacGregor-Fors I, Reyes-Paecke S, Vásquez A, Zea Camaño JD, Hernández HJ (2019) Urban ecosystem services in Latin America: mismatch between global concepts and regional realities? Urb Ecosys 22:173–187
Dong H, Qin B (2017) Exploring the link between neighborhood environment and mental wellbeing: a case study in Beijing, China. Landsc Urb Plan 164:71–80
Douglas I, Alam K, Maghenda M, McDonnell Y, Mclean L, Campbell J (2008) Unjust waters: climate change, flooding and the urban poor in Africa. Enviro Urbanis 20:187–205
du Toit MJ, Cilliers SS, Dallimer M, Goddard M, Guenat S, Cornelius SF (2018) Urban green infrastructure and ecosystem services in sub-Saharan Africa. Landsc Urb Plan 180:240–261
Eakin H, Lerner AM, Murtinho F (2010) Adaptive capacity in evolving peri-urban spaces: responses to flood risk in the Upper Lerma River Valley, Mexico. Glob Enviro Change 20:14–22
Erixon H, Borgström S, Andersson E (2013) Challenging dichotomies: exploring resilience as an integrative and operative conceptual framework for large-scale urban green structures. Plan Theory Prac 14:349–372
Ernstson H (2013) The social production of ecosystem services: a framework for studying environmental justice and ecological complexity in urbanised landscapes. Landsc Urb Plan 109:7–17
Escobedo FJ, Clerici N, Staudhammer CL, Feged-Rivadeneira A, Bohorquez JC, Tovar G (2018) Trees and crime in Bogota, Colombia: is the link an ecosystem disservice or service? Land Use Pol 78:583–592
Everett G, Lamond J, Morzillo AT, Chan FK, Matsler AM (2016) April. Sustainable drainage systems: helping people live with water. Proceedings of the Institution of Civil Engineers-Water Management 169:94–104
Felson AJ, Pickett ST (2005) Designed experiments: new approaches to studying urban ecosystems. Front Ecol Enviro 3:549–566
Fisher-Jeffes L, Carden K, Armitage N (2017a) A water sensitive urban design framework for South Africa. Town Region Plan 71
Fisher-Jeffes L, Carden K, Armitage NP, Winter K (2017b). Stormwater harvesting: improving water security in South Africa's urban areas. S Afr J Sci 113: #2016-0153
Frischenbruder MT, Pellegrino P (2006) Using greenways to reclaim nature in Brazilian cities. Landsc Urb Plan 76:67–78
Gallanter E (2012) Ciudad Jardín Lomas del Palomar: deriving ecocity design lessons from a garden city. Plan Persp 27:297–307
Garschagen M, Marks D (2019) Towards an agenda for profound urban climate resilience in Southeast Asia. In: Daniere A, Garschagen M (eds) urban climate resilience in Southeast Asia. Springer, Cham, pp 223–228
Geneletti D, La Rosa D, Spyra M, Cortinovis C (2017) A review of approaches and challenges for sustainable planning in urban peripheries. Landsc Urb Plan 165:231–243
Güneralp B, Perlstein AS, Seto KC (2015) Balancing urban growth and ecological conservation: a challenge for planning and governance in China. Ambio 44:532–543

Güneralp B, Lwasa S, Masundire H, Parnell S, Seto KC (2018) Urbanisation in Africa: challenges and opportunities for conservation. Enviro Res Lett 13:015002
Hansen MH, Li H, Svarverud R (2018) Ecological civilisation: interpreting the Chinese past, projecting the global future. Glob Enviro Change 53:195–203
He C, Liu Z, Tian J, Ma Q (2014) Urban expansion dynamics and natural habitat loss in China: a multi-scale landscape perspective. Glob Change Biol 20:2886–2902
Hermida MA, Neira M, Cabrera-Jara N, Osorio P (2017) Resilience in Latin American cities: behaviour vs. space quality in the riverbanks of the Tomebamba River. Procedia Engin 198:467–481
Herslund L, Backhaus A, Fryd O, Jørgensen G, Liu L, Mguni Mkupasi M, Workalemahu L, Yeshitela K (2018) Challenges and opportunities for developing water resilient green cities in Addis Ababa and Dar es Salaam: in search of champions and paths for urban transition. Landsc Urb Plan 180:319–327
Holmes PM, Rebelo AG, Dorse C, Wood J (2012) Can Cape Town's unique biodiversity be saved? Balancing conservation imperatives and development needs. Ecol Soc 17:28
Inzulza-Contardo J, Moran-Figueroa P (2018) Towards an analytical framework based on the principles of civic design: the case of post-earthquake Talca, Chile 2010–2016. Cities 72:356–368
Jenkins P, Abiko A, Frediani AA, Moraes O (2010) Urban upgrading interventions and engaging residents in fuzzy management: case studies from Novos Alagados, Salvador, Brazil. Habitat Intn 34:105–114
Jiang Y, Zevenbergen C, Fu D (2017) Understanding the challenges for the governance of China's "sponge cities" initiative to sustainably manage urban storm water and flooding. Nat Haz 89:521–529
Jiang Y, Zevenbergen C, Ma Y (2018) Urban pluvial flooding and stormwater management: a contemporary review of China's challenges and "sponge cities" strategy. Environ Sci Pol 80:132–143
Juanita AD, Ignacio P, Jorgelina GA, Cecilia AS, Carlos M, Francisco N (2019) Assessing the effects of past and future land cover changes in ecosystem services, disservices and biodiversity: a case study in Barranquilla Metropolitan Area (BMA), Colombia. Ecosys Serv 37:100915
Junior VS, Santos BA (2017) Using environmental perception and local knowledge to improve the effectiveness of an Urban Park in Northeast Brazil. Ethnobiol Conserv 6:1–24
King P, Annandale D, Bailey J (2003) Integrated economic and environmental planning in Asia: a review of progress and proposals for policy reform. Progr Plan 59:233–315
Kubiszewski I, Anderson SJ, Costanza R, Sutton PC (2016) The future of ecosystem services in Asia and the Pacific. Asia Pacif Pol Stud 3:389–404
Landman K (2019) Evolving public space in South Africa: towards regenerative space in the post-apartheid city. Routledge, New York
Lara FL (2011) New (sub)urbanism and old inequalities in Brazilian gated communities. J Urb Design 16:369–380
Lategan L, Cilliers J (2016) Considering urban green space and informal backyard rentals in South Africa: disproving the compensation hypothesis. Town Region Plan 69:1–16
Lebel L, Daniel R (2009) The governance of ecosystem services from tropical upland watersheds. Curr Opin Enviro Sustain 1:61–68
Lebel L (2013) Local knowledge and adaptation to climate change in natural resource-based societies of the Asia-Pacific. Mitigat Adapt Strat Glob Change 18:1057–1076
Li H, Ding L, Ren M, Li C, Wang H (2017) Sponge city construction in China: a survey the challenges and opportunities. Water 9:594
Liao K-H, Chan JK (2016) What is ecological wisdom and how does it relate to ecological knowledge? Landsc Urb Plan 155:111–113
Lin Z (2014) Constructing utopias: China's emerging eco-cities. ARCC Conference Repository, https://doi.org/10.17831/rep:arcc%y310

Lindley S, Pauleit S, Yeshitela K, Cilliers SS, Shackleton C (2018) Rethinking urban green infrastructure and ecosystem services from the perspective of sub-Saharan African cities. Landsc Urb Plan 180:328–338
Liu L (2018) A sustainability index with attention to environmental justice for eco-city classification and assessment. Ecol indic 85:904–914
Liu L, Jensen MB (2018) Green infrastructure for sustainable urban water management: practices of five forerunner cities. Cities 74:126–133
Long H, Liu Y, Hou X, Li T, Li Y (2014) Effects of land use transitions due to rapid urbanisation on ecosystem services: implications for urban planning in the new developing area of China. Habitat Intn 44:536–544
Lotter A (2017) Meeting halfway: lessons in social sustainability from the Lynedoch EcoVillage. http://ourfuturecities.co/2017/09/meeting-halfway-lessons-in-social-sustainability-from-the-lynedoch-ecovillage/. Accessed 11 Apr 2020
Lovell ST, Taylor JR (2013) Supplying urban ecosystem services through multifunctional green infrastructure in the United States. Landsc Ecol 28:1447–1463
Luo N, Hyland K (2019) "I won't publish in Chinese now": publishing, translation and the non-English speaking academic. J English Acad Purp 39:37–47
Lv J, Liu Q, Ren Y, Gong T, Wang S, Li L (2011) Socio-demographic association of multiple modifiable lifestyle risk factors and their clustering in a representative urban population of adults: a cross-sectional study in Hangzhou, China. Intn J Behav Nutrit Phys Act 8:40
Macchi S, Ricci L, Congedo L, Faldi G (2013) Adapting to climate change in coastal Dar es Salaam. AESOP-ACSP Joint Congress, 15–19 July 2013, Dublin. http://www.planning4adaptation.org/Docs/newsInfoMaterial/01-2014/Paper-AESOP-Track3-MACCHI.pdf
Macedo J, Haddad MA (2016) Equitable distribution of open space: Using spatial analysis to evaluate urban parks in Curitiba, Brazil. Enviro Plan B: Plan Design 43:1096–1117
Macedo J (2011) Maringá: a British Garden City in the tropics. Cities 28:347–359
Madonsela B, Koop S, Van Leeuwen K, Carden K (2019) Evaluation of water governance processes required to transition towards water sensitive urban design: an indicator assessment approach for the City of Cape Town. Water 11:292
Mansor M, Said I (2008) Green infrastructure network as social spaces for wellbeing of residents in Taiping. J Alam Bina, Malaysia, p 11
McConnachie MM, Shackleton CM, McGregor GK (2008) The extent of public green space and alien plant species in 10 small towns of the Sub-Tropical Thicket Biome, South Africa. Urban For Urban Green 7:1–13
McHarg I (1969) Design with nature. The Natural History Press, New York
McLean CT, Ground LE, Boon RG, Roberts DC, Govender N, McInnes A (2016) Durban's systematic conservation assessment. EThekwini Municipality, Environmental Planning and Climate Protection Department, Durban, South Africa. http://www.durban.gov.za/City_Services/development_planning_management/environmentalplanning_climate_protection/Publications/Documents/DurbansSystematicConservation.pdf. Accessed 26 Apr 2019
Mguni P (2015) Sustainability transitions in the developing world: exploring the potential for integrating sustainable urban drainage systems in Sub-Saharan cities. PhD thesis, University of Copenhagen, Copenhagen
Mguni P, Herslund L, Jensen MB (2016) Sustainable urban drainage systems: examining the potential for green infrastructure-based stormwater management for Sub-Saharan cities. Nat Haz 82:S241–S257
Myers G (2021) Urbanisation in the Global South. In: Shackleton CM, Cilliers SS, Davoren E, du Toit MJ (eds) Urban ecology in the Global South. Springer, Cham, pp 27–49
Nadal A, Cerón-Palma I, García-Gómez C, Pérez-Sánchez M, Rodríguez-Labajos B, Cuerva E, Josa A, Rieradevall J (2018) Social perception of urban agriculture in Latin-America. a case study in Mexican social housing. Land Use Pol 76:719–734
Nassauer JI, Opdam P (2008) Design in science: extending the landscape ecology paradigm. Landsc Ecol 23:633–644

Ng MK (2019) Governing green urbanism: the case of Shenzhen, China. J Urb Aff 41:64–82
Nguyen TT, Ngo HH, Guo W, Wang XC, Ren N, Li G, Ding J, Liang H (2019) Implementation of a specific urban water management: sponge city. Sci Total Enviro 652:147–162
O'Farrell PJ, Andersen PM, Le Maitre DC, Holmes PM (2012) Insights and opportunities offered by a rapid ecosystem service assessment in promoting a conservation agenda in an urban biodiversity hotspot. Ecol Soc 17:27
Palazzo D, Steiner F (2011) Urban ecological design: a process for regenerative places. Island Press, Washington DC
Pauleit S, Ambrose-Oji B, Andersson E, Anton B, Buijs A, Haase D, Elands B, Hansen R, Kowarik I, Kronenberg J, Mattijssen T, Stahl Olafsson A, Ral IE, van der Jagt AP, Konijnendijk van den Bosch C (2019) Advancing urban green infrastructure in Europe: outcomes and reflections from the GREEN SURGE project. Urban For Urban Green 40:4–16
Pauleit S, Vasquéz A, Maruthaveeran S, Liu L, Cilliers SS (2021) Urban green infrastructure in the Global South. In: Shackleton CM, Cilliers SS, Davoren E, du Toit MJ (eds) Urban ecology in the Global South. Springer, Cham, pp 107–143
Pfab MF, Compaan PC, Whittington-Jones CA, Engelbrecht I, Dumalisile L, Mills L, West SD, Muller PJ, Masterson GP, Nevhutalu LS, Holness SD, Hoare DB (2017) The Gauteng conservation plan: planning for biodiversity in a rapidly urbanising province. Bothalia 47:a2182
Qin HP, Li ZX, Fu G (2013) The effects of low impact development on urban flooding under different rainfall characteristics. J Enviro Mgmt 129:577–585
Rego RL (2011) A tropical enterprise: British planning ideas in a private settlement in Brazil. Plan Persp 26:261–282
Revell K (2010) Working with informality: increasing resilience in cities of the Global South. Proceedings of 46th ISOCARP Congress, Nairobi, p 1–13
Rigolon A, Browning MH, Lee K, Shin S (2018) Access to urban green space in cities of the Global South: a systematic literature review. Urb Sci 2:67
Roberts DC, Boon R, Croucamp P, Mander M (2005) Resource economics as a tool for open space planning in Durban, South Africa. In: Tryzna T (ed) The urban imperative, urban outreach strategies for protected areas agencies. California Institute of Public Affairs, Sacramento, pp 44–48
Roberts D, Boon R, Diederichs N, Douwes E, Govender N, McInnes A, Mclean C, O'Donoghue S, Spires M (2012) Exploring ecosystem-based adaptation in Durban, South Africa: "Learning-by-doing" at the local government coal face. Enviro Urbanis 24:167–195
Rodina L (2019) Planning for water resilience: competing agenda's among Cape Town's planners and water managers. Enviro Sci Pol 99:10–16
Rohr HE, Fourie W (2014) Securing water sustainability through innovative planning. WIT Trans Ecol Enviro 191:140241
Rojas C, Pino J, Basnou C, Vivanco M (2013a) Assessing land-use and -cover changes in relation to geographic factors and urban planning in the metropolitan area of Concepción (Chile): implications for biodiversity conservation. Appl Geog 39:93–103
Rojas C, Pino J, Jaque E (2013b) Strategic environmental assessment in Latin America: a methodological proposal for urban planning in the Metropolitan Area of Concepción (Chile). Land Use Pol 30:519–527
Rojas C, Munizaga J, Rojas O, Martínez C, Pino J (2019) Urban development versus wetland loss in a coastal Latin American city: lessons for sustainable land use planning. Land Use Pol 80:47–56
Rottle N, Yocom K (2010) Landscape architecture 02: ecological design. AVA Publishing, Lausanne
Roy A (2005) Urban informality: toward an epistemology of planning. J Am Plan Assoc 71:147–158
Roy M, Shemdoe R, Hulme D, Mwageni N, Gough A (2018) Climate change and declining levels of green structures: life in informal settlements of Dar es Salaam, Tanzania. Landsc Urb Plan 180:282–293
Ruwanza S, Shackleton CM (2016) Incorporation of environmental issues in South Africa's municipal Integrated Development Plans. Intn J Sustain Dev World Ecol 23:28–39

Sagaris L (2014) Citizen participation for sustainable transport: the case of "Living City" in Santiago, Chile (1997–2012). J Transp Geog 41:74–83
Sagaris L (2018) Citizen participation for sustainable transport: lessons for change from Santiago and Temuco, Chile. Res Transp Econ 69:402–410
Said I, Mansor M (2011) green infrastructure in cities and towns in Southeast Asian countries: quest for research. In: 2nd Intn Seminar on Sustainable Development, Jakarta, Indonesia. http://www.medinanet.org/index.php/articles/uplanning/211-green-infrastructure-in-cities-and-towns-insoutheastasian-countries-quest-for-research. Accessed 12 Nov 2019
Salazar G, Jalabert D (2016) Towards a landscape ecosophy: interpreting how the Villarrica-Pucón urban system inhabitants in the Araucanía region of Chile perceive and relate with the dynamics of Landscape. Urbe 8:28–41
Schäffler A, Swilling M (2013) Valuing green infrastructure in an urban environment under pressure: the Johannesburg case. Ecol Econ 86:246–257
Schwab E, Caputo S, Hernández-García J (2018) Urban agriculture: models-in-circulation from a critical transnational perspective. Landsc Urb Plan 170:15–23
Serra-Llobet A, Hermida MA (2017) Opportunities for green infrastructure under Ecuador's new legal framework. Landsc Urb Plan 159:1–4
Shackleton CM, Cilliers SS, du Toit MJ, Davoren E (2021) The need for an urban ecology of the Global South. In: Shackleton CM, Cilliers SS, Davoren E, du Toit MJ (eds) Urban ecology in the Global South. Springer, Cham, pp 1–26
Shatkin G (2008) The city and the bottom line: urban megaprojects and the privatisation of planning in Southeast Asia. Enviro Plan A 40:383–401
Sinnett D, Smith N, Burgess S (eds) (2015) Handbook on green infrastructure: planning, design and implementation. Edward Elgar, Cheltenham
Sitas N, Selomane O, Hamann M, Gajjar SP (2021) Towards equitable urban resilience in the Global South within a context of planning and management. In: Shackleton CM, Cilliers SS, Davoren E, du Toit MJ (eds) Urban ecology in the Global South. Springer, Cham, pp 325–345
Smit W, Hancock T, Kumaresen J, Santos-Burgoa C, Sánchez-Kobashi Meneses R, Friel S (2011) Toward a research and action agenda on urban planning/design and health equity in cities in low and middle-income countries. J Urb Health 88:5
Steiner F (2014) Frontiers in urban ecological design and planning research. Landsc Urb Plan 125:304–311
Steiner F (2016) The application of ecological knowledge requires a pursuit of wisdom. Landsc Urb Plan 155:108–110
Steiner F (2018) The ecological wisdom of plan-making. J Urb Mgmt 7:124–130
Stephan NL (2000) Tropical modernism: designing the tropical landscape. Singapore J Trop Geog 21:76–91
Stoffberg H, Hindes C, Muller L (eds) (2012) South African landscape architecture: a compendium. Unisa Press, Pretorria
Swilling M (2006) Sustainability and infrastructure planning in South Africa: a Cape Town case study. Enviro Urbanis 18:23–50
Swilling M, Annecke E (2006) Building sustainable neighbourhoods in South Africa: learning from the Lynedoch case. Enviro Urbanis 18:315–332
Thaiutsa B, Puangchit L, Kjelgren R, Arunpraparut W (2008) Urban green space, street tree and heritage large tree assessment in Bangkok, Thailand. Urban For Urban Green 7:219–229
Thomas PT, Thomas DE, Van Noordwijk M (2004) Environmental services and land use change in Southeast Asia: from recognition to regulation or reward? Agric Ecosys Enviro 104:229–244
Thompson GF, Steiner FR (1997) Ecological design and planning. Wiley, New York
Turner RK, Daily GC (2008) The ecosystem services framework and natural capital conservation. Enviro Res Econ 39:25–35
United Nations, Department of Economic and Social Affairs, Population Division (UN/DESA/PD) (2018) World urbanization prospects: the 2018 revision. Retrieved from: https://population.un.org/wup/Download/

Van der Ryn S, Cowan S (1996) Ecological design. Island Press, Washington DC
Venter ZS, Shackleton CM, van Staden F, Selomane O, Masterson VA (2020) Green apartheid: urban green space remains unequally distributed across income and race geographies in South Africa. Landsc Urb Plan 203:103889
Vihervaara P, Rönkä M, Walls M (2010) Trends in ecosystem service research: early steps and current drivers. Ambio 39:314–324
Vila-Ruiz CP, Meléndez-Ackerman E, Santiago-Bartolomei R, Garcia-Montiel D, Lastra L, Figuerola CE, Fumero-Caban J (2014) Plant species richness and abundance in residential yards across a tropical watershed: implications for urban sustainability. Ecol Soc 19:22
Wang H, Mei C, Liu JH, Shao WW (2018) A new strategy for integrated urban water management in China: Sponge city. Sci China Techn Sci 61:317–329
Wang S, Li J, Wu D, Liu J Zhang K, Wang R (2009) The strategic ecological impact assessment of urban development policies: a case study of Rizhao City, China. Stoch Enviro Res Risk Assess 23:1169–1180
Wang Y, Sun M, Song B (2017) Public perceptions of and willingness to pay for sponge city initiatives in China. Res Conserv Recycl 122:11–20
Wei C, Padgham M, Barona PC, Blaschke T (2017) Scale-free relationships between social and landscape factors in urban systems. Sustain 9:84
Widianingsih I, Morrell E (2007) Participatory planning in Indonesia: seeking a new path to democracy. Pol Stud 28:1–15
Wilhelm-Rechmann A, Cowling RM (2013) Local land-use planning and the role of conservation: An example analysing opportunities. S Afr J Sci 109: #0026
Wolch JR, Byrne J, Newell JP (2014) Urban green space, public health, and environmental justice: the challenge of making cities 'just green enough'. Landsc Urb Plan 125:234–244
Wu WT, Zhao HY, Ren Y, Tian MM, Shen D (2012) Investigation on shading effect of urban green space in summer in Hangzhou. Adv Mat Res 424:916–919
Xue F, Gou Z, Lau SS (2017) Green open space in high-dense Asian cities: site configurations, microclimates and users' perceptions. Sustain Cities Soc 34:114–125
Yigitcanlar T, Kamruzzaman M (2015) Planning, development and management of sustainable cities: a commentary from the guest editors. Sustain 7:14677–14688
Zhang X, Zhao K, Jennings ET Jr (2016) Empirical evidence and determinants of region-based environmental injustice in China: does environmental public service level make a difference? Soc Sci Quart 97:1082–1095
Zheng Y, Xin-Lu XI, Chen-Zhen LI, Mou WA, Xiao-Jia HE (2018) Development as adaptation: Framing and measuring urban resilience in Beijing. Adv Clim Change Res 9:234–242
Zipperer WC, Morse WC, Johnson-Gaither C (2011) Linking social and ecological systems. In: Niemelä J (ed) Urban ecology: patterns, processes and applications. Oxford University Press, New York, pp 298–308

Urban Governance of and for Urban Green and Blue Infrastructure

David Simon, Julie Goodness, Shuaib Lwasa, José Antônio Puppim de Oliveira, Laura V. Macedo, Jess Kavonic, Ellika Hermansson Török, and Thomas Elmqvist

Abstract To capture the many complexities, we adopt a broad approach to urban governance, encompassing the diverse combinations of formal, informal and/or customary/traditional institutions and practices in urban areas of the Global South. The broad arguments are illustrated with appropriate examples and boxed case studies to illustrate important dimensions of diversity but also the scope for generalisation. In many contexts, inclusion of urban ecology, biodiversity, and green–blue infrastructure within urban governance is quite novel, thus presenting challenges to often rigid and outdated systems in times of unprecedented change. Hence, the chapter addresses key aspects needing change, including guidelines and examples of how this has been and can be achieved. A comprehensive and holistic approach is vital to provide a logical context for prioritisation and integration. This will facilitate joined-up action to achieve multiple co-benefits through targeted interventions rather than a

D. Simon (✉)
Royal Holloway, University of London, Egham, UK
e-mail: d.simon@rhul.ac.uk

J. Goodness
University of Cape Town, Cape Town, South Africa
e-mail: julie.goodness@uct.ac.za

S. Lwasa
Makerere University, Kampala, Uganda

J. A. Puppim de Oliveira · L. V. Macedo
Getúlio Vargas Foundation, Rio de Janeiro, Brazil
e-mail: jose.puppim@fgv.br

J. Kavonic
ICLEI Africa, Cape Town, South Africa
e-mail: jessica.kavonic@iclei.org

E. Hermansson Török
SwedBio, Stockholm, Sweden
e-mail: ellika.hermansson.torok@su.se

T. Elmqvist
Stockholm Resilience Centre, Stockholm, Sweden
e-mail: Thomas.Elmqvist@su.se

C. M. Shackleton et al. (eds.), *Urban Ecology in the Global South*, Cities and Nature,
https://doi.org/10.1007/978-3-030-67650-6_16

scatter gun approach. Novel approaches that prioritise transdisciplinary co-design or co-production over conventional adversarial and top-down expert-led mechanisms have considerable potential in this regard. These are examined across relevant spatial scales, from key global agreements and conventions, to national initiatives, local authorities and the potential of transnational municipal networks.

Keywords Biodiversity · Co-benefits · Green–blue infrastructure · Transdisciplinary co-design · Transnational municipal networks · Urban governance

1 Introduction

Broadly defined, urban governance encompasses the diverse combinations of formal, informal and customary/traditional institutions and practices that exist and operate in, or partly in, areas defined as urban (Rudd et al. 2018). The precise mixture varies from locality to locality and often changes over time in a single place. We consciously focus here on governance, which comprises both the institutions of government and the accumulated set of procedures and practices through which they operate, including non-governmental stakeholders (e.g. Rosenau 1995). Particularly in Global South contexts and in relation to environmental resources, this is essential to embrace the complexities, overlaps and perhaps gaps among and between the many different institutions of different origins and how they operate.

With a multiplicity of policies, both endogenous and exogenous, there is evidence of mixed urban governance models, including state-controlled, top-down systems, autonomous and semi-autonomous decentralised governance and more recently cooperative-type models, as in Kigali (Rwanda) (Goodfellow 2013; Goodfellow and Titeca 2012). Reality is often complex, however. For example, Kampala (Uganda) exemplifies both a semi-autonomous decentralised system and a co-operative model. In a nutshell, each city presents a unique urban governance system reflecting the prevailing political and social dispensations but often with strong historical (including colonial) legacies.

From this, it follows that urban governance for addressing important new issues, environmental change, sustainability and environmental resources, of which blue–green infrastructure is a subset, can be complex. Indeed, the inclusion of urban ecology within urban governance arrangements has only recently started to receive attention (e.g. Wilkinson et al. 2013), with long-term evidence of the roles of urban agriculture and more recently urban natural assets and riparian management (Lwasa et al. 2014). Overlapping, ambiguous or even contradictory jurisdictions among the respective sets of institutions and their procedures complicate the universal issues of the adequacy of existing laws, rules, customary practices and societal norms and how well they are observed and enforced (Wilkinson et al. 2013). One simple illustration will suffice here since other detailed examples are given later in the chapter. Land, a fundamental environmental resource that is centrally relevant to the subject of this chapter, is often highly contested because of its competing use, exchange and symbolic values as perceived by various stakeholders.

Within urban areas, and especially in rapidly growing centres typical of the Global South that have spread outward to embrace previously rural areas, land in different parts of a city may fall under the jurisdiction of formal local authorities and/or statutory bodies as well as various peri-urban or rural local authorities and traditional authorities, all practising various forms of statutory and customary law and regulations. In some countries, such as Eritrea, Ethiopia, Mozambique and Tanzania, land has been nationalised and is technically vested in the state. In practice, however, many forms of de facto if not de jure individualisation or privatisation occur, while urban local authorities may manage land within their boundaries on behalf of the central state. Control over water bodies and even subterranean water resources may also be complex, linked to different formal or customary institutions and practices.

Such complexities present very real problems in the context of efforts to provide coherent and integrated urban leadership and (co-)management across a city, let alone a city region (International Expert Panel 2018). Yet, the latter is often the most appropriate spatial scale for governance of ecosystem-based and hydrological resources (see Sitas et al. 2021). The limitations of conventional institutions and procedures are therefore being increasingly widely recognised, and more appropriate approaches are being experimented with (see Box 1 on the Urban Natural Assets for Africa (UNA) programme). In terms of formal, statutory institutions, one common approach is the establishment of a strategic metropolitan or city regional authority with responsibility for one or more functions well suited to that scale, often including urban planning, land, environment (including air quality) and water resources. Such bodies are given defined powers, which differ according to whether the establishment occurs through top-down processes by central or regional government or via bottom-up processes by constituent local authorities or other bodies. Even so, however, there can be contestations over jurisdiction and the division of powers if the enabling legislation or regulations are not unambiguous, logical and widely acceptable.

Box 1: Urban Natural Assets for Africa Programme (UNA)

With the unprecedented rate of urbanisation in Africa, cities in the region face severe urban challenges, such as inadequate planning, poor infrastructure, and a lack of financial, human and knowledge resources (Goodness and Anderson 2013). These challenges are exacerbated by the exponential loss of natural assets currently being experienced on the continent, which in turn impacts on the resilience of cities. The Urban Natural Assets for Africa (UNA) programme, implemented from 2014 to 2020 in eight cities and six countries on the African continent, specifically seeks to improve human well-being and build climate resilience through integrating nature-based solutions into land-use planning and decision-making processes.

The UNA programme, guided by international targets and processes, provides a model of effective nature and climate action at the local level, where methods that prioritise and support shifts in natural asset management are tailored to the local context in line with good international practice (e.g. Martens and Carvalho 2017). Ensuring that international climate change and biodiversity processes are applied and made relevant at the local level is key to the design and implementation of the programme. Valuable experiences and co-produced new knowledge generated feed into regional and international processes by ICLEI and local partners, making the bridging of scales a two-way process.

ICLEI Africa and UNA are therefore at the forefront of influencing urban governance landscapes of the UNA cities and countries. In particular, local governments in sub-Saharan Africa are exposed to new methods for integrating urban natural assets into climate planning, and vice versa. The UNA programme grounds activities in both a country's National Biodiversity Strategy and Action Plan (NBSAP) as well as its National Determined Contribution (NDC) to reduce emissions in terms of the 2015 Paris Agreement, influencing the review of these from a local government perspective.

A key learning to date has been that producing information and searching for 'entry points' in decision-making into which climate change or natural asset information can be inserted is not adequate enough to facilitate change in urban governance of Southern African cities, especially when considering the nuanced complexity surrounding decision-making, as well as financial and political influences of cities in the Global South. Instead, for climate change and natural asset information to have relevance in Southern Africa, a fundamental shift in planning and decision-making is required by cities; one that incorporates ongoing learning, knowledge construction and reflection.

The UNA programme has witnessed that prioritising these key 'softer' considerations when integrating nature into decision-making is essential to embracing the complexities, overlaps and gaps between horizontal and vertical governance spheres as well as between the many different institutions working on natural asset management. In addition, these knowledge generation and learning spaces effectively promote sustained relationship-building which in turn subtly shifts the way stakeholders work together, thereby generating entry points' for nature mainstreaming (ICLEI CBC 2019d). This is an important insight as mainstreaming or urban governance processes that stem from the Global North conditions cannot be easily applied. The UNA programme has adapted these approaches to better suit the complexity of the Global South with more innovative methods being applied.

These include innovative games and interactive exercises that equip participants with new tools and understandings, and address everything from process, procedures and practices to language discrepancies (ICLEI CBC 2019e). The importance of using innovative games for capacity building lies in their ability to level the power dynamics in a room. This is essential for improved governance. This was clearly seen when land-use planners and environmental officers from Malawi's Lilongwe City Council came together to discuss how to better embed natural asset considerations in city planning processes (ICLEI CBC 2019b). This was a significant development in itself as cross-departmental meetings that aimed to link planners and environmentalists had never been held previously. While the outputs themselves are important, the emergent discussions and ideas, perhaps, are even more valuable (ICLEI CBC 2019f). The journey that planners and environmental officers embarked on together facilitated a mind-set shift that led to the prioritisation of 'planning with nature' and encouraged a completely different way of thinking about natural systems. Much of the success of all the UNA projects has hinged on the energy invested in building relationships with key stakeholders (ICLEI CBC 2019d). Strong relationships are the foundation of improved urban governance, and hence, a significant part of the UNA programme methodology is devoted to fostering connections based on trust. The UNA programme makes relationship-building a priority by (a) working with the same city over extended periods, (b) building on existing work and (c) facilitating engagement spaces and interactive learning activities that bring different stakeholders together to jointly reflect and problem-solve (ICLEI CBC 2019f; ICLEI CBC 2019 g). Working with and investing in a suite of 'champions' in the UNA project cities has been instrumental in taking forward the nature-climate agenda in urban areas.

The nuanced complexity surrounding decision-making as well as financial and political influences cannot be underestimated when working on urban ecological governance in Global South cities. Lower levels of decentralisation in many African countries means that key decisions and associated budgets affecting cities often reside with national government. In addition, a large proportion of funding for public infrastructure comes from international loans to national governments, which then constrains decision-making. Coupled with lower human resource capacity compared with cities in the Global North (Smit 2018; Resnick et al. 2019), this further complicates the entanglement of institutional structures, financial flows and access to information—encapsulating the challenges to effective MLG examined in the main text.

A key emergent lesson from UNA was that both formal and informal governance structures exist (ICLEI CBC 2019c). The former are supported by policies and legal frameworks, and can be easily mapped; the latter are based on unspoken agreements and invisible power dynamics, and are therefore more complex to grasp. UNA prioritises understanding both the official (formal) and unofficial (informal) decision-making processes in African cities, and how they play out in conjunction with each other. This understanding has been essential in ensuring that relationships are built with the right contact people, has improved communication, avoided conflicts, and co-created entry points for real change.

The UNA programme is unique in that it embraces processes that are highly reflective and adaptive. The programme has found that flexibility and a willingness to pause, think and adapt to what is happening on the ground is essential in order to be responsive to city needs (ICLEI CBC 2019a). This is worth reflecting on in the context of global development cooperation being increasingly more focused on results-based management (RBM) and short term delivery of results that may not always favour co-designed projects applying adaptive management.

References

Goodness J, Anderson PM (2013) Local assessment of Cape Town: navigating the management complexities of urbanisation, biodiversity, and ecosystem services in the Cape Floristic Region. In: Elmqvist, T. Fragkias M, Goodness J, Güneralp B, Marcotullio PJ, McDonald RI, Parnell S, Schewenius M, Sendstad M, Seto KC, Wilkinson C (eds) Urbanisation, biodiversity and ecosystem services: challenges and opportunities: a global assessment. Springer, Dordrecht, pp 461–484

ICLEI CBC (2019a) Handbook 1: Key pillars of the UNA programme approach. In: The value of urban natural assets when planning for resilient Africa cities: consideration and decision-making processes. Cape Town, South Africa: ICLEI CBC.

ICLEI CBC (2019b) Handbook 2: Alternative approaches to planning with nature: Prioritisation mapping in Malawi. In: The value of urban natural assets when planning for resilient Africa cities: consideration and decision-making processes. Cape Town, South Africa: ICLEI CBC.

ICLEI CBC (2019c) Handbook 4: Working with formal and informal governance structures: ICLEI Africa's learnings. In: The value of urban natural assets when planning for resilient Africa cities: consideration and decision-making processes. Cape Town, South Africa: ICLEI CBC.

ICLEI CBC (2019d) Handbook 7: The importance of relationship building: Fostering trust through time investment and innovative interactive exercises. In: The value of urban natural assets when planning for resilient Africa cities: consideration and decision-making processes. Cape Town, South Africa: ICLEI CBC.

ICLEI CBC (2019e). Handbook 8: Capacitating cities: Creating spaces for reflection and effective dialogue in sub-Saharan Africa. In: The value of urban natural assets when planning for resilient Africa cities: consideration and decision-making processes. Cape Town, South Africa: ICLEI CBC.

ICLEI CBC (2019f) Handbook 9: Prioritising processes: Bridging the divide between different sectors and disciplines in sub-Saharan Africa. In: The value of urban natural assets when planning for resilient Africa cities: consideration and decision-making processes. Cape Town, South Africa: ICLEI CBC.

ICLEI CBC (2019 g) Handbook 11: Multi-level engagement: Improving national and local working relationships through dialogues in Uganda. In: The value of urban natural assets when planning for resilient Africa cities: consideration and decision-making processes. Cape Town, South Africa: ICLEI CBC.

Martens ML, Carvalho MM (2017) Key factors of sustainability in project management context: a survey exploring the project managers' perspective. Intn J Proj Mgmt 35: 1084–1102.

Resnick D, Siame G, Mulambia P, Ndhlovu D, Shicilenge B, Sivasubramanian B (2019) Deepening decentralisation in Zambia: political economy constraints and opportunities for reform. International Food Policy Research Institute Discussion Paper 1893. https://www.ifpri.org/publication/deepening-decentralisation-zambia-political-economy-constraints-and-opportunities-reform. Accessed Oct 2019.

Smit W (2018) Urban governance in Africa: an overview. Intn Dev Pol 10: 55–77.

Increasingly, too, the limitations of conventional adversarial approaches to governance of heterogeneous populations have led to a loss of effectiveness and legitimacy, eventually stimulating experimentation with a wide variety of co-design, co-creation and co-production approaches. What these all have in common is bringing together different stakeholder groups, many of whom have long conflictual histories, to work out a mutually acceptable solution via one or more deliberative processes. Underpinning them is the belief that by engaging together in the entire process of co-producing services, research, planning or budgeting, for instance, shared understandings and the realisation will develop that what divides them is usually less than what they have in common (e.g. Durose and Richardson 2016; Durose et al. 2018; Patel et al. 2015; Polk 2015; Simon et al. 2018a, 2020; Trencher et al. 2014; Watson 2014). In this sense, they go beyond longer-established participatory methods such as participatory municipal budgeting and community-based natural resource management.

The remainder of this chapter provides a comprehensive overview of governance issues in relation to urban ecology with particular reference to green and blue infrastructure (GBI) in the Global South. Following a brief overview of how thinking about the relationship between people and the natural environment within urban areas has evolved over time, attention shifts to governance arrangements in relation to urban ecology and biodiversity, nature-based solutions (NBS) and GBI. We reveal the limitations of conventional approaches and how changing theoretical conceptualisations and challenges of rapid urbanisation on the one hand, and climate and environmental change on the other, have driven institutional and policy initiatives to embed such thinking into good practice at all scales from the urban to the global. Key global agreements, from the Convention on Biodiversity to the New Urban Agenda are surveyed, before attention shifts to urban collaborations across national borders by means of transnational municipal networks (TMNs). The final section examines practical experiments in biodiversity conservation and GBI, illustrated by means of boxes containing detailed case studies drawn from several cities in Africa, Brazil, Thailand and China.

2 Tracing the Evolution of Ecological Thought About Biodiversity and Ecosystem Services as Well as Urban Green and Blue Infrastructure: From Nature *in* Cities, to Nature *of* Cities, to Nature *for* Cities

Approaches to governance of and for GBI have paralleled the evolution of ecological thought on how nature has been conceptualised in cities (e.g. Wilkinson et al. 2013). This has developed from an impression of cities as places devoid and depleted of nature, to places comprising important interlinked networks of green and blue infrastructure that are key for human well-being (McDonnell et al. 2009; Elmqvist et al. 2013). Historically, this is rooted in an understanding of city life as distinct and separate from that in rural areas (McDonnell 2011). Moreover, according to this view, cities by definition comprise dense conglomerations of people, built infrastructure, and industrial enterprise, and hence are epicentres of pollution bereft of natural elements (Campbell 1996). This also hinges upon cultural notions, derived from modernist town planning precepts, of the only legitimate nature being found in rural and 'wilderness' areas (the latter involving land and natural resources that are supposedly pristine or unaffected by human beings [Cronon 1996]), and therefore located outside of urban areas.

Gradually, this progressed to a conception of nature *in* cities (Pickett 1997) and acknowledgement of the authenticity of urban nature through deeming urban ecosystems as legitimate subjects of ecological study (McDonnell 2011; Grimm et al. 2000; Grimm et al. 2008). Earlier studies documented the ecological make-up of particular spaces (e.g. remnant land or green spaces), or detailed the compositions of specific types of organisms in urban areas, such as plants in European cities (Sukopp 2008). This further progressed to a view of the nature *of* cities (Pickett 1997; Grimm et al. 2000; Grimm et al. 2008; Pickett and MacGregor-Fors 2016), which proceeds beyond viewing individual patches of ecology within the city to instead considering the full array and interplay of the entire integrated urban mosaic. Several global studies, including the Millennium Ecosystem Assessment (MEA) and the Economics of Ecosystems and Biodiversity (TEEB), helped to usher in and embed this understanding, also providing authenticity and tools for assessments of the nature of cities (MEA 2003; Kumar 2010).

Finally, there has been a paradigm shift to the idea of nature *for* cities (Pickett and MacGregor-Fors 2016; Childers et al. 2015; McPhearson et al. 2016), which recognises the social side of the social-ecological systems of cities, and has an aim of "linking ecological science with civic processes" (Pickett and MacGregor-Fors 2016: 5). Such civic processes include governance and decision-making, as in the 'nature for cities' approach elaborated upon below. This adds social justice, equity, and stewardship components to the study and governance of urban ecosystems. In addition, the development of the field of political ecology has also raised the profile of the social side of urban social-ecological systems through highlighting and advocating for the social justice implications of ecological issues (Swyngedouw and Heynen 2003; Heynen 2013). Overall, in Global South contexts, the attention to a holistic patchwork collection of green and blue infrastructure elements in a city, as well as the social justice issues that are crucially and inextricably linked with these elements, are particularly important and salient themes. Governance of this patchwork of spaces requires negotiation and management brokered through the diversity of actors responsible for them. Derkzen et al. (2017) provide examples in south India, while Nagendra's (2016) important monograph analyses in detail the changing fortunes of GBI in Bengaluru, an erstwhile imperial garden city and recently India's rapidly growing information technology metropolis.

3 Governance Approaches to Urban Ecology and Biodiversity

3.1 From Urban Areas as Islands to Parts of Effective Multilevel Governance

Underpinning the arguments so far is an understanding of the importance of integrated governance of urban areas as social-ecological (also called socio-natural) systems, comprising both natural and anthropogenic components (Elmqvist et al. 2013). The latter constitute the built environment, which has often spread at the expense of the former, itself progressively reduced in scale and even ecological integrity over time. In terms of 'natural' environment, open space systems often comprise a mosaic of isolated green–blue 'islands', perhaps linked by corridors, that constitute a mixture of remnant, partially natural areas and anthropogenic green space designed as part of the built environment. Examples range from sports fields to parks and gardens; green roofs and walls; and reservoirs, rivers and canals with different degrees of shoreline greenery and habitats. Trying to make them more than the sum of their parts requires a strategic overview and governance approach (Nagendra 2016; Demir and Baylan 2019; Li et al. 2020).

3.1.1 From Urban Local Authorities as Discrete Planning Units to Functional or Eco-Bio-Urban Regions

By analogy with the previous paragraph, at the regional or city regional scale, it is important to understand and govern urban areas not as isolated 'islands' delimited by their municipal boundaries or natural barriers like a river, but as part of larger biophysical and social-ecological systems. Only by so doing can the integrity of natural or partially natural systems be secured. Since many such systems and urban regions nowadays straddle national political boundaries (as do pollution, environmental change and trade in endangered species), international conventions and treaties play an important role in fostering the necessary collaboration. Because the signatories to and parties responsible for international agreements are national governments, this point illustrates the need for effective multilevel governance embracing these levels as well as regional and local government institutions.

Multilevel governance (MLG) has become a well-established concept to address the limitations of state authority in global environmental politics since the early 1990s (see Marks 1992; Hooghe and Marks 1997; Piattoni 2010). In some instances, such as international environmental regimes, no overall authority has the power to reach an agreement or implement the policy unilaterally. Also, in many domestic public policies, governments need support from a broad range of stakeholders to implement policies effectively and efficiently. As acknowledged later by scholarship investigating climate governance (Bulkeley and Betsill 2005; Betsill and Bulkeley 2006; Corfee-Morlot et al. 2009; Puppim de Oliveira 2009; Leck and Simon 2013; Bäckstrand et al. 2017), in order to address the complexities of global policy interactions, MLG includes vertical relations between states and subnational (i.e. regional and local) governments, as well as horizontal transboundary linkages with and between other local authorities and also non-state actors (NSAs) such as businesses, transnational networks and civil society organisations (CSOs).

3.1.2 The Role of International Conventions, Treaties and Environmental Agendas (CITES, CBD, Agenda 2030 and the SDGs, and the New Urban Agenda)

The most important international agreements in relation to the subject of this chapter are the suite adopted by the UN and its member states in 2015 and 2016. Together they constitute the global commitment to sustainable development until 2030 or 2040 (the end dates vary), and all explicitly acknowledge the central importance of subnational entities (regional and local authorities) to achievement of their objectives, as the Convention on Biodiversity had done already in 2010. In other words, as elaborated in the previous paragraph, effective multilevel governance will be essential (Rudd et al. 2018). These agreements are the Sendai Framework for Disaster Risk Reduction (2015), Agenda 2030, which incorporates the 17 Sustainable Development Goals (SDGs) as monitoring and evaluation framework (2015), The Paris Agreement of the UN Framework Convention on Climate Change (2015), and the New Urban Agenda (NUA) (2016).

Inevitably, these agreements are broad in nature and often lack specific means of implementation and monitoring. For instance, the SDGs were initially intended to form the monitoring and evaluation framework for the very aspirational and progressive NUA but that formal link did not gain the requisite support from member states. UN-led efforts are remedying the situation by means of new tools and reporting procedures. This search for operational synergy encapsulates the essential strategic point that the effect and effectiveness of these agreements will be maximised through their coherent implementation as a package, since their complementarities far outweigh some ambiguities and potential or real trade-offs. Hence, for instance, The Sendai Framework recognises the well-being and development implications of disaster risk, while the urban SDG (Goal 11) has targets and indicators addressing disaster risk reduction (DRR) and climate change. The following paragraphs identify the most relevant provisions of the respective agreements in the context of urban ecology, especially in the Global South, where urban resilience is often low and vulnerability high as a result of a complex mixture of factors (e.g. Pelling 2003). These are bound up with income distributions, higher levels of absolute and relative poverty, and inadequate governmental resources, capacities and sometimes political will to address effectively the hazards and risks from disasters, extreme events and changing environmental conditions. For a survey of the evolution of approaches to sustainability and resilience, see Simon et al. (2018b).

In keeping with current good practice, the Sendai Framework (UNISDR 2015) is underpinned by the understanding of the importance of environmental integrity to DRR, in urban areas as elsewhere. Investment in DRR is linked explicitly to enhanced resilience. Particular urban examples include coastal mangroves and dune vegetation as barriers against storm surges and coastal inundation, and wetlands and riverbanks to slow water flow, absorb excess run-off and act as sediment traps. Similarly, urban tree cover and other elements of urban green infrastructure play roles in soil moisture retention and shade provision, thereby reducing heat island effects. These examples address both extreme events and the ongoing, insidious effects of changing prevailing conditions.

Agenda 2030 (UN 2015) embodies a holistic perspective on sustainable development to which environmental, social and economic components are integral. Put simply, development that continues to undermine environmental integrity and resources is unsustainable. Despite inevitable limitations and some trade-offs, the 17 SDGs constitute the most comprehensive effort to date to capture the diversity and holistic nature of sustainable development. No fewer than eight of the 17 goals have particular environmental emphases, with urban as well as societal relevance (Table 1).

Table 1 SDGs with particular environmental and ecological relevance

6	Clean water and sanitation
7	Affordable and clean energy
11	Sustainable cities and communities
12	Responsible consumption and production
13	Climate action
14	Life below water
15	Life on land
17	Partnerships for the goals

The symbolic as well as substantive importance of having an explicitly urban goal, complemented by urban-related indicators in 11 other goals, reaffirms the understanding that cities do not exist in isolation from the territories and societies of which they form part and also cannot be sustainable without wider societal sustainability (Seto et al. 2012a).

There are several targets in Goal 11 with explicit environmental and ecological content:

- 11.4: strengthen protection for the world's cultural and natural heritage;
- 11.5: significantly reduce the number of deaths, people affected direct economic losses from disasters, including water-related disasters, while protecting the poor and vulnerable;
- 11.6: reduce adverse per capita environmental impact of cities, especially air quality and municipal and other waste management;
- 11.7: provide universal access to safe, inclusive and accessible green and public spaces, in particular for women, children, older persons and those with disabilities;
- 11a: support positive economic, social and environmental links between urban, peri-urban and rural areas by strengthening national and regional development planning, and
- 11b: substantially increase the number of cities and human settlements adopting and implementing integrated policies and plans towards inclusion, resource efficiency, mitigation and adaptation to climate change, resilience to disasters, and holistic disaster risk management at all levels (UN 2019).

Although as with all UN processes, it is national governments that report annually to the UN on progress against the SDGs, they require inputs from regional and local authorities to do so if those reports are to contain more than centrally available data. That said, meaningful engagement by urban governments will inevitably require localisation of the SDGs in locally appropriate ways (SDSN 2016); in consequence, how this is being carried out varies widely within and between countries (Valencia et al. 2019).

Finally, the NUA (UN 2017) contains many key paragraphs and provisions relevant to this chapter. Chronologically, it makes explicit reference to protecting the environment (paragraph 5) and affirms that implementation "contributes to the implementation and localisation of the 2030 Agenda for Sustainable Development in an integrated manner" (paragraph 9). The essential desired characteristics that cities and human settlements should fulfil set out in paragraph 13 include:

- "(a) Fulfil their social function, including the social and ecological function of land …";
- "(b) … prioritise safe, inclusive, accessible, green and quality public open spaces that are friendly …";
- "(g) adopt and implement disaster risk reduction and management, reduce vulnerability, build resilience and responsiveness to natural and human-made hazards and foster mitigation of and adaptation to climate change"; and

- "(h) protect, conserve, restore and promote their ecosystems, water, natural habitats and biodiversity, minimise their environmental impact and change to sustainable consumption and production patterns".

Furthermore, one of the three guiding principles to achieve the NUA's vision is "(c) Ensure environmental sustainability by promoting clean energy and sustainable use of land and resources in urban development, by protecting ecosystems and biodiversity, including adopting healthy lifestyles in harmony with nature, by promoting sustainable consumption and production patterns, by building urban resilience, by reducing disaster risks and by mitigating and adapting to climate change" (paragraph 14). Many of the same issues and principles are repeated in subsequent paragraphs in various permutations, thereby affirming the considerable importance attached to them as part of the holistic framing. This latter also includes a commitment to compact urban design (paragraph 52) as a means to maximise efficiency and effectiveness of use of land and other natural environmental resources. All these issues are then elaborated further in the section on environmentally sustainable and resilient urban development (paragraphs 63-80) and the environment is linked to other sectors and levels of governance in the section on implementation (paragraphs 85–92).

Convention on Biodiversity and Convention on the Trade in International Species

Two earlier international conventions require mention, despite not having an explicit urban focus. These are the Convention on Biodiversity (CBD) and Convention on the Trade in International Species (CITES). The goal of the CITES is to ensure that international trade in specimens of wild animals and plants does not threaten their survival. The CITES has implications for specimens of, and products derived from, endangered species originating and deriving from or moving through urban areas; this is crucial as expanding urban areas may be home to key biodiversity hotspots (Seto et al. 2012b). The CBD, first opened for signature at the United Nations World Conference on Environment and Development (the Rio Earth Summit) in 1992, aims to facilitate the conservation and sustainable use of biological diversity across the globe.

While principally international in focus, the CBD has particular implications for urban areas through a specific expanded focus on cities and subnational governments. The first decision on cities and subnational governments (Decision X/22) was adopted at the Convention of the Parties (COP) 9 in 2008, and the *Plan of Action on Subnational Governments, Cities and Other Local Authorities for Biodiversity (2011–2020)* was adopted at COP 10 in Nagoya in 2010. Furthermore, the COP 12 in Hyderabad in 2012 gave support for a Global Partnership associated with local governments, invited Parties of the convention to develop and support tools and initiatives that facilitate subnational implementation, and provided suggestions on how to help bring national strategies and plans into local contexts. In addition, a

Biodiversity Summit for Cities and Subnational governments has been held at the CBD COP meetings each year since 2010.

Intergovernmental Panel on Climate Change and Intergovernmental Science-Policy Platform on Biodiversity and Ecosystem Services

Two intergovernmental panel platforms initiated by the UN also have implications for governance of urban blue and green infrastructure. First, the Intergovernmental Panel on Climate Change (IPCC) Working Group II has developed specific report sections on urban areas and climate change. Second, the Intergovernmental Science-Policy Platform on Biodiversity and Ecosystem Services (IPBES), established by UN member States in 2012, aims to strengthen the science-policy interface for biodiversity and ecosystem services in order to ensure the conservation and sustainable use of biodiversity, long-term human well-being and sustainable development. IPBES has produced scholarly synthesis information that can be used by urban areas in management and governance.

3.2 Transnational Cooperation for Urban Biodiversity Protection Experiments (TMNs, City Paradiplomacy and Innovation)

3.2.1 Transnational Municipal Networks

Transnational cooperation among cities, understood here as interactions between local governments (municipal or metropolitan), is not a new phenomenon. However, in the MLG framework, transnational municipal networks have evolved rapidly from facilitators of cooperation among peers to become collective agents in their own right vis-à-vis the arena of international negotiations. These networks have also increased in scope and size since the 1990s, fostered by UN agencies. An assessment of city networking by the University College London's (UCL´s) City Leadership Initiative investigated 168 networks, of which 29% focus on environment as the main issue; other issues include poverty, peace and culture (Acuto and Rayner 2016). Since then, network activities focusing on climate have increased, fostered by developments in the wake of the Paris Agreement.

Transnational municipal networks (TMNs) have been key drivers in the participation of cities in climate governance, the overwhelming majority being originally from Europe and North America (See Betsill and Bulkeley 2006; Gordon and Acuto 2014; Fischer et al. 2015; Hickmann 2016; Bansard et al. 2017). Nonetheless, engagement of cities from the Global South with TMNs has been on the rise, leveraged by projects funded by multi- or bilateral cooperation initiatives (Macedo and Jacobi 2019). Countries from the Global South have been increasingly involved in world politics as significant stakeholders, setting an agenda for cooperation determined by their own priorities, both within North–South and South–South partnerships. This process is replicated at the subnational level, orchestrated mostly by national governments and supranational agents, and supported by TMNs. Cities in these countries are prioritising issues that concern their interests, such as water scarcity, biodiversity loss, adaptation to climate change and food security.

As a membership-based network, ICLEI has been instrumental in expanding the urban climate agenda to include these issues, having established several programmes and projects such as the Sustainable Urban Resilient Water for Africa (SURe-Water4Africa), IAdapt(India), Resilient Cities (global), Integrated Subnational Action for Biodiversity (INTERACT-Bio)(Brazil, India, Tanzania), Local Protected Areas (South America), the Urban Natural Assets for Africa (UNA) programme (see Box 1), Local Action for Biodiversity (LAB) (Global) and the global ICLEI-RUAF Cityfood network. Another global-scale local authority membership organisation, United Cities and Local Governments (UCLG), has also embraced the global agendas for sustainable development, increasingly in collaboration with ICLEI.

Other global city networks of note in the environmental, DRR and climate change context are the C40 network of major cities tackling climate change (with a membership now approaching 100) and the 100 Resilient Cities (100RC) network, which had a specific urban resilience focus over its 10-year lifespan to 2019, funded by the Rockefeller Foundation. Several other networks or platforms promote urban biodiversity protection, including Biophilic Cities, The Nature of Cities, the Urban Biodiversity Hub, and the CitieswithNature, a knowledge sharing initiative within ICLEI's Cities Biodiversity Centre, with support from its partners, The Nature Conservancy (TNC) and the International Union for Conservation of Nature (IUCN). CitieswithNature is now the CBD-endorsed official platform for local and subnational governments to connect and report on their ambitions and commitments to the Action Agenda for Nature and People, and the Post-2020 Global Biodiversity Framework.

3.2.2 Experiments in Biodiversity Protection and Green and Blue Infrastructure

Biodiversity protection had so far been primarily of regional or national concern, but the linkages between climate change and biodiversity loss become more salient as the contribution to climate change from land use, land-use change and forestry (LULUCF) in the Global South increases. Tropical biodiversity hotspot countries, such as Brazil and Indonesia, are significant emitters, their main sources being land-use change, agriculture and deforestation, as reported in platforms such as the UNFCCC´s Climate Watch coordinated by WRI, and the Global Carbon Project (GCP). Cities have been important agents in climate governance and socio-technical experimentation, contributing to the transition towards sustainable development since the onset of the UNFCCC´s implementation process, and even before that, by engaging in actions established by Agenda 21 in terms of the World Conference on Environment and Development in 1987. Under Agenda 21, cities have participated in initiatives that foster innovation and experiments at local level promoted by TMNs, many of which focused on climate action since the early 1990s (Keck and Sikkink 1998; Bulkeley and Betsill 2013). Here, again, the climate agenda has been driven by high income countries with mandatory greenhouse gas (GHG) reduction targets, mostly oriented towards mitigation policies and measures related to energy emissions (Hickmann 2016). The same applies to TMNs that in turn concentrated local climate action on reducing emissions from transport and energy efficiency (Bulkeley and Castán Broto 2013; Barbi and Macedo 2019). In the area of biodiversity, TMNs have been a prominent player since the 2000s. During the CBD COP of 2006 in Curitiba, cities launched the Global Initiative on Cities and Biodiversity supported by the CBD Secretariat, led by ICLEI and other partners, including a network of researchers called URBIO. In the following CBD COPs, parallel summits on cities and biodiversity were hosted in the same cities as the COP. Over time, the network expanded and the Biodiversity Summit for Cities and Subnational Governments became an important part of the CBD COP meetings. It is noteworthy that ICLEI's Africa secretariat (ICLEI Africa) was the leading catalyst of the network, thus bringing many cities of the Global South to discuss and act on biodiversity in the city.

As cities from the Global South gained more prominence in TMNs, so did their agenda on biodiversity and climate action, which included greening urban spaces, increasing the number of parks and tree planting. Furthermore, taking into account social issues in addressing global environmental concerns, such as poverty alleviation and food security, is a key feature of local action in the Global South. In the case of São Paulo, for instance, a municipal climate law passed in 2009 had a strong focus on urban greening as an adaptation strategy; establishing linear parks and increasing green areas and tree planting to restore degraded land in favelas (slum areas) turned out much more successful than other recommended activities such as investments in clean and accessible urban public transport (Macedo 2017). African cities have also been at the forefront of nature-based solution (NBS) implementation, supported by programmes such as UNA (Boxes 1 and 2), while Brazilian and Chinese cities have also been very active in TMNs and in the implementation of biodiversity friendly initiatives (see Boxes 4 and 5).

Box 2: Kampala's approach to governing green and blue infrastructure

The current urban environmental management system in the administrative structure of Kampala (Uganda) is largely oriented towards public health management. Thus, the focus is on sanitation, water management, greening and landscape management. Only recently has urban governance of city ecological resources and functions started to be recognised and mainstreamed into the urban planning and development plans, let alone routine city management functions. A systems approach to city green, blue and grey infrastructure is one of the models for urban ecological management to enhance the connections between these types of infrastructure. Kampala Capital City Authority (KCCA) worked with ICLEI Africa under the Urban Natural Assets: Rivers for Life (UNA Rivers) initiative to initiate a process of entrenching management of urban natural assets in KCCA's routine management functions. KCCA recognised that natural assets play an important role in livability of the city. Thus, protecting the natural assets and enhancing their functionality has become a priority action area for the city. It was also recognised that, given the city's spatial spread and jurisdiction covering a surface area of 195 km^2, it was appropriate to plan with urban natural assets at a city-regional scale of the Greater Kampala Metropolitan Area (GKMA). Although the GKMA comprises five districts (Wakiso, Kampala, Mpigi, Buikwe and Mukono), this natural asset mapping and planning task focused on the adjacent municipalities to Kampala city, spanning an area of about 839 km^2. The objective was to map the assets and analyse hot spots so that these data could then be integrated into spatial planning.

Through a combination of participatory Geographic Information System, desktop reviews and field surveys, a range of natural asset classes was developed from satellite imagery, including broad canopy, forest plantation, grasslands, hedge green, swamp forest, swamp papyrus, woodlot, built up land, and water bodies. These data enabled spatial hotspot analysis to identify intervention points and mainstreaming into spatial planning. The findings indicate that the city-region has undergone tremendous ecosystem change in the last 20 years. Urban development is the main driving force for built up areas, extraction of materials for construction and urban infrastructure. Land tenure is an underlying factor as it strongly influences degradation of natural assets in Kampala through the relations of production governing use of the land under different tenure systems. The stock of natural assets within KCCA is estimated to have land patches totalling 120 km^2 (Table 2).

The city-region illustrates intact and protected areas but highly threatened by the expanding urban infrastructure and residential development. This limits ecosystem services provision but there is potential for tourism. The categories of broad canopy, grasslands, woodlots, swamp papyrus and hedge green patches registered losses while forest plantation and swamp forests gained. KCCA apparently has limited zones that can offer habitat for birds with numbers adequate enough to attract meaningful tourism but the GKMA still has sizeable patches for this and other functions. The lakefront and habitat can support birds, although there are limited wetlands at the lakefront habitat and conditions to attract birds that fluctuate seasonally (Kenworthy 2006).

Reference

Kenworthy JR (2006) The Eco-City: ten key transport and planning dimensions for sustainable city development. Enviro Urbanis 18: 67–85

Table 2 Natural assets within the KCCA

Natural Asset	Description	Area km^2	Percentage
Broad Canopy	A combination of evergreen broad canopy trees and deciduous trees	23.1	19.1
Forest Plantation	Plantations of differing sizes manly eucalyptus, pinewood and Cyprus spp	9.8	8.1
Grasslands	Colonies of Herbaceous spp that transitioned between evergreen native spp	14.4	12.2
Hedge Green	Largely non-native spp used for ornamental value. A mix of shrubs and broad canopy trees	12.8	11.0
Swamp Forest	Lowland broad canopy evergreen vegetation with remnants of naïve spp	15.9	13.2
Swamp Papyrus	An aquatic flowering plant in shallow water bodies	24.6	20.3
Woodlot	Scattered broad canopy trees in scrapped landscapes mixed with planted non-native spp	16.1	13.1
Lake Victoria	The biggest lake in Uganda. The captured section borders Kampala City	3.5	2.9
Other water bodies	Other water bodies scattered across Kampala other than Lake Victoria	0.2	0.1
Total area km^2		120.4	100

The establishment of concepts linking the natural and anthropogenic environments (cities and other human settlements) to global concerns was a key development to strengthen urban biodiversity protection. By addressing the multi-scalar dimension of climate change and biodiversity through conceptualising integrated approaches, scholars and practitioners can investigate and tackle challenges such as adaptation and urban liveability through NBS that include GBI and consider ecosystem services.

From its long pedigree in urban planning and design, landscape design and ecology, exemplified by seminal work by Geddes (1915) and McHarg (1969), urban GBI is fast becoming a growing trend in the urban climate action debate as an alternative to more complex and expensive grey infrastructure-based measures such as drainage systems and reservoirs, to deal with adaptation. In light of global environmental problems that require urban responses, GBI, consisting of local-scale interventions such as street trees, greenways, gardens, parks and wetlands, is often viewed as a more beneficial approach to urban management. However, the contribution of these measures to greenhouse gas (GHG) emissions reduction or biodiversity protection is still poorly quantified. Practitioners and many scholars highlight the positive outcomes of promoting GBI and other nature-based solutions (Benedict and McMahon 2002; Kabisch et al. 2016; Andersson et al. 2019). GBIs are supporting a shift to healthier, more amenable cities (see, e.g. the case of Bangkok (Thailand) in Box 3 and Pauleit et al. (2021) and Cilliers et al. (2021) for different approaches in the Global South. Nonetheless, urban greening and the use of green infrastructure (GI) is not without controversy. Some studies question whether they are mostly used as 'green branding' or 'greenwashing' by politicians to enhance competitiveness or as marketing tools by developers, contending that they also contribute to gentrification and aggravate inequalities in the spatial distribution of the city (Cho 2010; Friedmann 2010; Kabisch et al. 2016; Anguelovski et al. 2019).

Box 3: Governance beyond government for climate mitigation in Bangkok (Thailand)

Bangkok, the capital of Thailand since 1782, reached a population of over 8.8 million inhabitants by 2018 in the city alone, and has become an economic hub in Asia. Throughout the twentieth century, the city's development accelerated, attracting migrants from across the country and beyond, resulting in haphazard urbanisation and aggravated infrastructure issues. Coupled with its environmental characteristics, rapid urban growth is threatening the well-being of its inhabitants. Being situated between 1 metre below and 2 metres above mean sea level, many of the low-lying areas are particularly vulnerable. The whole city has been prone to flooding, and groundwater pumping is causing land subsidence across the city, aggravating risks to water and land-use management. Bangkok is said to be sinking about 1–2 cm a year. So far, the Bangkok Metropolitan Administration (BMA), responsible for the management of the city of Bangkok, has addressed flooding with grey infrastructure, such as the underground tunnels to drain excessive water in the city area directly into the Chao Praya River for quick discharge.

Bangkok was built on soft layers of clay in the low-lying plains of the Chao Phraya River, and has developed an effective canal management system as part of urban planning and the transport network since the early 1800 s. Once known as the Venice of the East, the city still has several canals that also form part of the drainage system. However, these waterways, along with orchards, trees and green areas which helped to detain floodwater effectively, have been replaced with continuously expanding impervious surfaces, such as housing and ground transport networks. Furthermore, their functionality has been compromised due to excessive garbage dumping, particularly by residents of settlements along the canals. Green space has also been lost to urbanisation, leaving the city with one of the lowest percentages of any major capital in Asia. According to the Economist Intelligence Unit's Asian Green City Index 2011, while urban citizens in Asia enjoyed 39 m^2 of green space each on average, Bangkok had 3.3 m^2 per inhabitant. In 2014, Bangkok began planning a resilience-building strategy to address these issues, within the 100 Resilient Cities initiative led by the Rockefeller Foundation. The strategy includes green and blue infrastructure as a solution for flooding in the coming years.

In 2012, the BMA initiated the Bangkok 2032 Project in collaboration with Chulalongkorn University. It was a visionary planning process for Bangkok City, involving local stakeholders from private sector, academia and civil society. That same year, Chulalongkorn University launched a competition for the design of a park in the campus area to celebrate its centennial. Chulalongkorn University Centenary Park, inaugurated and opened to the public in 2017, was the first blue and green infrastructure project designed to address climate change and mitigate flooding. Its 5 ha area integrated to the neighbourhood, includes a 1.3 km green avenue with many ecological functions, such as collecting and treating water, decreasing flood risks, reducing the urban heat island, and promoting active mobility. Features include landscape modelled on rice terraces acting as retention ponds, artificial catch basins and tanks, pedestrian and cycling lanes, and recreational equipment. Visitors can ride water treatment stationary bikes to create movement and introduce more oxygen into the ponds to prevent stagnation. This project has spurred similar initiatives in Bangkok and other Thai cities as a nature-based solution to address flooding, stormwater management and air pollution in the urban environment.

The use of NBS has taken different paths in cities of the North and South, although there are various overlaps and no clear division. Whereas urban design in the former has often adopted sophisticated and manicured approaches to integrating NBS, city managers and technicians in the Global South have generally been more concerned about increasing green areas as a measure to tackle flooding and mudslides, promote social justice and to improve liveability and health, particularly in large cities riddled with pollution, insufficient public spaces, besides land tenure and housing issues. Bilateral initiatives are also endeavouring to assist Southern countries such as Brazil to embed principles of good planning and governance practice to achieve multiple co-benefits, such as use of NBS to promote urban residents' senses of well-being, climate change adaptation and thereby overall urban resilience (Herzog and Rozado 2019).

Nevertheless, the perception about greening is not always positive in these countries. For instance, when implemented in poor areas and informal settlements, experiments in India, Ghana and Kenya demonstrated that restoration of GI to address flooding is often met with social resistance and political contestation (Tauhid and Zawani 2018), since they may involve displacing occupants of public areas, disrupting communities and leaving families homeless and unprotected. This is increasingly being recognised and addressed, as many countries, such as Brazil, China, Colombia, Ghana and South Africa, have been developing and implementing multilevel governance strategies that include legal and institutional frameworks contemplating synergies between biodiversity protection and urban planning (see Boxes 4 and 5). The observed trend suggests a convergence of agendas at national and subnational levels (Puppim de Oliveira et al. 2011; Leck and Simon 2013, 2019; Chen 2015; Vásquez-Muñoz 2016; Dhakal and Chevalier 2017; Cilliers 2019). Although these strategies focus predominantly on health, safety and economic issues, and need to be perfected (Zhang et al. 2020; Ahmed and Puppim de Oliveira 2017), they do bring about benefits such as improving urban air and water quality, reducing disaster risks, and recovering degraded land (Brondizio et al. 2019). Interestingly, a key insight from the UNA programme (see Box 1) is that African decision-makers have been using the concept of NBS for many years, but now that they (and GBI) have been established internationally as key concepts, city officials' confidence seems to have improved, knowing that they are on an accepted path for developing solutions.

Box 4: Urban Governance of Green and Blue Infrastructure in Brazilian Cities

Brazilian cities have adopted various strategies to use green and blue infrastructure for providing different ecosystem services to their citizens. Several initiatives have been key to the success of urban biodiversity conservation, supported by multilevel governance strategies through integrating different policies and departments in planning and decision-making, as well as involving non-governmental stakeholders since early stages of planning and implementation. In this box, the initiatives in three cities of different sizes are highlighted.

Curitiba´s success as a sustainable city can be attributed to continuing public support. Environmental conservation has been a cornerstone of Curitiba's urban planning since the 1970 s, but most of its 30 parks were created or upgraded between 1992 and 2003. In 2007, Curitiba launched Biocidade (BioCity), a programme that places biodiversity protection at the centre of urban planning; it takes environmental conservation to the next level by including water management and urban agriculture and promoting awareness about native fauna and flora. BioCity´s key strategy establishes conservation units and linear parks that include restoration of the main water basins. It also implements projects to reintroduce and expand the use of ornamental indigenous plant species within the city, promoting familiarity with the region's native flora. Despite a threefold increase in population since 1970, the city´s ratio of green space grew from 0.5 m^2 to 58.1 m^2 per capita in 2010. Green and blue infrastructure including recreational equipment helped the municipality to engage users in conservation and avoid illegal occupation, while protecting biodiversity and improving the city´s resilience. The Curitiba case demonstrates that committed and forward-thinking leadership fostering nature-based strategies can make a difference in moving toward urban sustainability.

As a large megalopolis with increasing risk of water supply shortages, São Paulo has a long-standing legal and institutional framework for environmental protection, coordinated with policies that address social vulnerability, and more recently climate change. Between 2003 and 2012, the environmental department spearheaded several initiatives that prioritised green areas and parks, conservation units and private natural reserves. The city contains large fragments of Atlantic Forest (covering 27% of its territory) that are under constant threat by urban sprawl. The Sapé Favela redevelopment project is an example of this approach. A densely populated slum covering 120,000 m^2 of land along the Sapé stream, with precarious housing, no sanitation infrastructure, urban services or green space was transformed into a safe urbanised neighbourhood. A partnership between São Paulo city and state government restored the watershed, including water supply and sewage collection, waste collection, new housing and recreational green spaces along the stream banks and surrounding areas. These measures reduced flooding, landslides and proliferation of waterborne diseases. The GBI provided several urban ecosystem services to the community, such as stormwater drainage, better air and water quality, adequate housing and public recreation, thus contributing towards climate adaptation and urban resilience.

Sorocaba is a medium-sized city, centre of a metropolitan area within a conurbation including 174 municipalities, among which is São Paulo´s metropolitan region. It is an industrial hub, with high quality education and a robust economy. There are 33 dispersed public areas designated as parks, mostly smaller recreational green areas formed by land that was either expropriated or granted as compensation for environmental impacts of development projects. Established by law, many lack specifications such as objectives or management plans addressing biodiversity conservation. In 2018, the city issued a norm on urban afforestation mandating new plots and constructions to include trees in their sidewalks. As an implementation strategy that also addresses social inclusion, the environmental department promotes tree planting rallies involving municipal staff, residents and NGOs. Members of an association of ex-convicts and their families participate by preparing recipient areas, while seedlings are produced by convicts of a state prison located in the city. Despite criticism about poor monitoring and maintenance in public areas, as well as the use of inadequate tree species in many sites, the plantation initiative engages NGOs, communities and students, and provides a way to connect people to nature.

References

Cardoso-Leite E, Faria LC, Capelo FF, Tonello KC, Castello AC (2014) Composição florística da arborização urbana de Sorocaba/SP, Brasil, Revista da Sociedade Brasileira de Arborização Urbana REVSBAU, Piracicaba—SP 9(1):133–150

Curitiba Municipality. Website: https://www.curitiba.pr.gov.br/conteudo/sobre-areas-verdes/123. Accessed 27 June 2019

ICLEI (2008) Local action on biodiversity case studies: Curitiba. Online http://old.iclei.org/fileadmin/PUBLICATIONS/Case_Stories/6._LAB/ICLEI_LABCaseStory_Curitiba_2008.PDF. Accessed 27 June 2019

Macedo LS (2017) Participação das Cidades Brasileiras na governança multinível das mudanças climáticas. Doctoral thesis, University of São Paulo, São Paulo

Mota ML, Sola E, Kaline FM (2016) Categorização da infraestrutura verde do município de Sorocaba (SP) para criação de um sistema municipal integrando espaços livres e áreas protegidas. Revista Brasileira de Ciências Ambientais 122–140

PINI-Infraestrutura magazine PINI online (2019) Website: infraestruturaurbana17.pini.com.br/solucoes-tecnicas/30/favela-do-sape-recuperacao-das-margens-de-corrego-ocupadas-294289-1.aspx. Accessed 29 June 2019

PMS (Prefeitura Municipal de Sorocaba) (2019) https://leismunicipais.com.br/a1/sp/s/sorocaba/lei-ordinaria/2018/1182/11815/lei-ordinaria-n-11815-2018-institui-o-espaco-arvore-e-da-outras-providencias. Accessed 27 June 2019

PMSP (Prefeitura Municipal de São Paulo) (2019) www.prefeitura.sp.gov.br/cidade/secretarias/meio_ambiente/parques/index.php?p=144010

PMSP (Prefeitura Municipal de São Paulo) (2019) www.prefeitura.sp.gov.br/cidade/secretarias/meio_ambiente/unid_de_conservacao/index.php?p=3339. Accessed 27 June 2019

Box 5: Collaborative design for green infrastructure in China

In the past decade, China has implemented many urban ecosystem restoration initiatives, and has a national policy establishing the Sponge City Program with guidelines for cities to use green infrastructure (GI) in addressing urban water management issues, such as pollution, flooding, runoff and stormwater drainage. And although GI design has been widely applied throughout China, there are few examples that respond to site-specific problems in a multidisciplinary approach (Luan et al. 2017).

Xianyang is a historic city, with over 5 million inhabitants, part of the metropolitan area of Xi-An, Shaanxi province capital, approximately 900 km (560 miles) south of Beijing. It is situated in the Wei River floodplain, the largest and one of the main tributaries of the Yellow River. The whole basin has suffered urban encroachment in the past decades, both upstream and downstream of Xi-An, and few natural areas still remain. In one of the sections outside Xianyang city, the government commissioned the firm Yifang Ecoscape to design the Weiliu Wetland Park (WWP), an area of approximately 125 ha. along 3.2 km of the riverbank.

The designers adopted a collaborative, multidisciplinary approach combining research, engineering technology and landscape architecture to deliver the Weiliu Wetland Park in an area that was rapidly degrading and being reclaimed for housing and urban agriculture. The main goal was to restore the local ecosystem and implement a series of green and blue infrastructure strategies including adaptive flood control, stormwater management, water quality improvement, wastewater reuse and biodiversity restoration. Techniques for flood protection, biodiversity restoration and habitat protection include willow mattress revetments, gabions and grassed slopes (Chen and Shao 2019). Furthermore, the design team was committed to providing urban responses to climate change impacts and improving the city´s resilience. The park´s management includes a partnership with Peking University to assess its environmental and social impacts, enabling its improvement and replication. Besides its functional aspects, it is a much-appreciated recreational area for inhabitants of and visitors to Xianyang since May 2017, when it was inaugurated. The WWP is considered a model for site-scale design using GBI, also addressing increasing vulnerability to climate change. Yifang Ecoscape won the Best of the Best Landscape Design Award in the Eighth edition of IDEA-KING International Landscape Planning & Design Competition, in 2018.

References

Chen J, Shao W (2019) Restoring Wei River Floodplain: a resilient design for Weiliu Wetland Park. Available athttp://landezine.com/index.php/2019/01/weiliu-wetland-park-by-yifang-ecoscape/?unapproved=171563&moderation-hash=c713227e48f04287b99c6a3e65eb7283#comment-171563. Accessed 27 Apr 2020

Luan B, Wang X, Yueyan Jin Y, Chai M, Hu C (2017) Collaborative design of site-scale green infrastructure: a case study on the ecological restoration design of Weiliu Wetland Park in Xianyang. Landsc Archit Front 5:26–43

As concerns about urban vulnerability due to climate change increase, TMNs are stepping into support Southern cities to step up. They play an important role in introducing the global dimension to the local environment by highlighting the nexus between climate change and biodiversity loss and urban liveability, thereby contributing towards more integrated multilevel governance.

4 Conclusions

As elsewhere, there have been profound shifts in thinking about human–environment relationships within urban areas of the Global South over recent decades. Often influenced by global initiatives and conventions, these have revolutionised policy and practice geared to promoting urban sustainability and resilience in the context of urban restructuring, climate and broader environmental change, and overall human well-being. Of course, the diversity of conditions within and between countries means that there can be no single blueprint and all such broad principles and guidelines of good practice need to be adapted to local circumstances. Many problems and challenges arise before and during implementation, so that outcomes are not guaranteed.

The multitude of useful mechanisms and approaches for governing GBI surveyed in this chapter demonstrates the need for governance actions across multiple spatial scales (e.g. international, regional, and local city, with attention and respect to ecosystem and watershed boundaries) as well as multiple institutions (e.g. city governments, community-based organisations, and private enterprise) to promote resilient landscapes. In the diverse Global South contexts, these can provide constructive opportunities to produce creative and innovative governance interventions, especially in situations where formal government institutions may play a weaker or, in extreme cases, even minor role in governance. Linked to broader issues of vulnerability and weak resilience discussed above, this often reflects inadequate capacity in terms of finance and human resources) and where informality may be a pervasive characteristic of cities.

To this end, the detailed case studies have demonstrated innovative ways forward that are being developed in ways that are locally appropriate in the very different contexts of Kampala, Bangkok, Brazilian and Chinese cities. The UNA Rivers project examined in Box 1 has many lessons applicable not just in sub-Saharan Africa but also in other parts of the world because of its comparative approach, using collaborative and co-production methods that bring multiple stakeholder groups together to explore the challenges and work out appropriate solutions. The approaches can all contribute to reshaping existing towns and cities, as well as the appropriate design of new urban components and entire future urban areas that are not only more ecologically sustainable and resilient but also more enjoyable and social places to live.

References

Acuto M, Rayner S (2016) City networks: breaking gridlocks or forging (new) lock-ins? Intn Affairs 92:1147–1166

Ahmed A, Puppim de Oliveira JA (2017) Integration of biodiversity in urban planning instruments in developing countries: the case of Kumasi Metropolitan Assembly, Ghana. J Enviro Plan Mgmt 60:1741–1764

Andersson E, Langemeyer J, Borgström S, McPhearson T, Haase D, Kronenberg J, Barton DN, Davis M, Naumann S, Röschel L, Barö F (2019) Enabling green and blue infrastructure to improve contributions to human well-being and equity in urban systems. Bioscience 69:566–574

Anguelovski I, Irazábal-Zurita C, Connolly JJ (2019) Grabbed urban landscapes: socio-spatial tensions in green infrastructure planning in Medellín. Intn Urb Reg Res 43:133–156

Bäckstrand K, Kuyper J, Linnér W, Lövbrand E (2017) Non-state actors in global climate governance: from Copenhagen to Paris and beyond. Enviro Polit 26:561–579

Bansard JS, Pattberg PH, Widerberg O (2017) Cities to the rescue? Assessing the performance of transnational municipal networks in global climate governance. Intn Enviro Agree Polit Law Econ 17:229–246

Barbi F, Macedo LS (2019) Transnational municipal networks and cities in climate governance: experiments in Brazil. In: Van der Heijden J, Bulkeley H, Certoma C (eds) Urban climate politics. Cambridge University Press, Cambridge, pp 59–79

Benedict MA, McMahon ET (2002) Green infrastructure: smart conservation for the 21st century. Renew Res J 20:12–17

Betsill M, Bulkeley H (2006) Cities and the multilevel governance of global climate change. Glob Govern 12:141–159
Brondizio ES, Settele J, Díaz S, Ngo HT (eds) (2019) Global assessment report on biodiversity and ecosystem services of the intergovernmental science-policy platform on biodiversity and ecosystem services. IPBES Secretariat, Bonn
Bulkeley H, Betsill M (2005) Rethinking sustainable cities: multilevel governance and the urban politics of climate change. Enviro Polit 14:42–63
Bulkeley H, Castán Broto V (2013) Government by experiment? Global cities and the governing of climate change. Trans Inst Brit Geog 38:361–375
Bulkeley H, Betsill M (2013) Revisiting the urban politics of climate change. Enviro Polit 22:136–154
Chen WY (2015) The role of urban green infrastructure in offsetting carbon emissions in 35 major Chinese cities: a nationwide estimate. Cities 44:112–120
Campbell S (1996) Green cities, growing cities, just cities? Urban planning and the contradictions of sustainable development. J Am Plan Assoc 62:296–312
Childers DL, Cadenasso ML, Grove MJ, Marshall V, McGrath B, Pickett ST (2015) An ecology for cities: a transformational nexus of design and ecology to advance climate change resilience and urban sustainability. Sustain 7:3774–3791
Cho MR (2010) The politics of urban nature restoration: the case of Cheonggyecheon restoration in Seoul, Korea. Intn Dev Plan Rev 32:145–165
Cilliers EJ (2019) Reflecting on green infrastructure and spatial planning in Africa: the complexities, perceptions, and way forward. Sustain 11:455
Cilliers SS, Breed C, Cilliers EJ, Lategan LG (2021) Urban ecological planning and design in the Global South. In: Shackleton CM, Cilliers SS, Davoren E, du Toit MJ (eds) Urban ecology in the Global South. Springer, Cham, pp 365–401
Corfee-Morlot J, Kamal-Chaoui L, Donovan MG, Cochran I, Robert A, Teasdale PJ (2009) Cities, climate change and multilevel governance. OECD Environmental Working Papers N° 14, OECD Publishing, Brussels
Cronon W (1996) The trouble with wilderness getting back to the wrong nature. Enviro Hist 1:7–28
Demir A, Baylan E (2019) The determination of green infrastructures components of Van City center and its near surroundings. Intn J Sci Tech Res 5:328–343
Derkzen, ML, Nagendra H, Van Teeffelen A, Purushotham A, Verburg PH (2017) Shifts in ecosystem services in deprived urban areas: understanding people's responses and consequences for well-being. Ecol Soc 22(1) art 51
Dhakal KP, Chevalier LR (2017) Managing Urban stormwater for urban sustainability: barriers and policy solutions for green infrastructure application. J Enviro Mgmt 203:171–181
Durose C, Richardson L (2016) Designing public policy for co-production: theory, practice and change. Policy Press, Bristol
Durose C, Richardson L, Perry B (2018) Craft metrics to value co-production. Nat 562:32–33
Elmqvist T, Fragkias M, Goodness J, Güneralp B, Marcotullio PJ, McDonald RI, Parnell S, Schewenius M, Sendstad M, Seto KC, Wilkinson C (eds) (2013) Urbanisation, biodiversity and ecosystem services: challenges and opportunities: a global assessment. Springer, Dordrecht
Fischer K, Dellas E, Schreiber F, Acuto M, London D, Tänzler D, Carius A (2015). Urbanisation and climate diplomacy: the stake of cities in global climate governance. Climate Diplomacy Series. Berlin: Adelphi. Online https://www.adelphi.de/en/system/files/mediathek/bilder/Urbanization%20and%20Climate%20Diplomacy.pdf
Friedmann J (2010) Place and place-making in cities: a global perspective. Plan Theory Prac 11:149–165
Geddes P (1915) Cities in Evolution: an introduction to the town planning movement and to the study of civics. London: Williams and Norgate
Goodfellow T (2013) Urban planning in Africa and the politics of implementation: contrasting patterns of state intervention in Kampala and Kigali. In: Arlt V, Marcamo E, Obrist B (eds) In Living the city in Africa: processes of invention and intervention. LIT Verlag, Basel, pp 45–62

Goodfellow T, Titeca K (2012) Presidential intervention and the changing 'politics of survival' in Kampala's informal economy. Cities 29:264–270

Gordon D, Acuto M (2014) If cities are the solution, what are the problems? The promise and perils of urban climate leadership. In: Johnson C, Toly N, Schroeder H (eds) The urban climate challenge: rethinking the role of cities in the global climate regime. Routledge, London, pp 65–81

Grimm N, Faeth SH, Golubiewski NE, Redman C, Wu J, Bai X, Briggs JM (2008) Global change and the ecology of cities. Sci 314(5864):756–760

Grimm NB, Grove JM, Pickett ST, Redman C (2000) Integrated approaches to long-term studies of urban ecological systems. Bioscience 50:571–584

Herzog CP, Rozado CA (2019) The EU—Brazil sector dialogue on nature-based solutions: contribution to a Brazilian roadmap on nature-based solutions for resilient cities. Directorate-General for Research and Innovation, European Commission, Brussels. https://oppla.eu/sites/default/files/docs/EU-Brazil-NBS-dialogue-2409-light.pdf

Heynen N (2013) Urban political ecology I: the urban century. Progr Hum Geog 38:1–7

Hickmann T (2016) Rethinking authority in global climate governance: how transnational climate initiatives relate to the international climate regime. Routledge, London

Hooghe L, Marks G (1997) Contending models of governance in the European Union. In: Cafruny A, Lankowski C (eds) Europe's ambiguous unity: conflict and consensus in the post-Maastricht era. Lynne Rienner, Boulder

International Expert Panel on Science and the Future of Cities (2018) Science and the Future of Cities, London and Melbourne

Kabisch N, Frantzeskaki N, Pauleit S, Naumann S, Davis M, Artmann M, Haase D, Knapp S, Korn H, Stadler J (2016) Nature-based solutions to climate change mitigation and adaptation in urban areas: perspectives on indicators, knowledge gaps, barriers, and opportunities for action. Ecol Soc 21:39

Keck ME, Sikkink K (1998) Activists beyond borders: advocacy networks in international politics. Cornell University Press, Ithaca and London

Kumar P (2010) The economics of ecosystems and biodiversity ecological and economic foundations. Earthscan, London

Leck H, Simon D (2013) Fostering multiscalar collaboration and co-operation for effective governance of climate change adaptation. Urb Stud 50:1221–1238

Li L, Van Eetvelde V, Cheng X, Uyttenhove P (2020) Assessing stormwater runoff reduction capacity of existing green infrastructure in the city of Ghent. Intn J Sustain Dev World Ecol. 27:749–761

Lwasa S, Mugagga F, Wahab B, Simon D, Connors J, Griffith C (2014) Urban and peri-urban agriculture and forestry: transcending poverty alleviation to climate change mitigation and adaptation. Urb Clim 7:92–106

Macedo LS (2017) Participação das Cidades Brasileiras na governança multinível das mudanças climáticas. Doctoral thesis University of São Paulo, São Paulo

Macedo LS, Jacobi PR (2019) Subnational politics of the urban age: evidence from Brazil on integrating global climate goals in the municipal agenda. In: Acuto M, Setzer J, Robin E, Ljungkvist K (eds) The politics of an urban age collection. Palgrave Communications, online publication, https://www.nature.com/articles/s41599-019-0225-x

Marks G (1992) Structural policy in the European community. In: Sbragia A (ed) Euro-politics, institutions and policymaking in the "New" European Community. The Brookings Institution, Washington, DC, pp 191–225

McDonnell MJ (2011) The history of urban ecology: an ecologist's perspective. In: Niemelä J (ed) Urban ecology: patterns, processes, and applications. Oxford University Press, New York, pp 5–13

McDonnell MJ, Hahs AK, Breuste JH (eds) (2009) Ecology of cities and towns: a comparative approach. Cambridge University Press, Cambridge

McHarg I (1969) Design with Nature. Garden City, New York: Natural History Press

McPhearson T, Pickett ST, Grimm NB, Niemela J, Alberti M, Elmqvist T, Weber C, Haase D, Breuste J, Qureshi S (2016) Advancing urban ecology toward a science of cities. Bioscience 66:198–212
MEA (2003) Millennium ecosystem assessment: ecosystems and human well-being: a framework for assessment. Island Press, Washington, DC
Nagendra H (2016) Nature in the city: Bengaluru in the past, present and future. Oxford University Press, New Delhi
Patel Z, Greyling S, Parnell S, Pirie G (2015) Co-producing urban knowledge: experimenting with alternatives to 'best practice' for Cape Town, South Africa. Intn Dev Plan Rev 37:187–203
Pauleit S, Vasquéz A, Maruthaveeran S, Liu L, Cilliers SS (2021) Urban green infrastructure in the Global South. In: Shackleton CM, Cilliers SS, Davoren E, du Toit MJ (eds) Urban ecology in the Global South. Springer, Cham, pp 107–143
Pelling M (2003) The vulnerability of cities: natural disasters and social resilience. Routledge, London
Piattoni S (2010) The theory of multi-level governance: conceptual, empirical and normative challenges. Oxford University Press, Oxford
Pickett ST, MacGregor-Fors I (2016) Evolution and future of urban ecological science: ecology in, of, and for the city. Ecosys Heal Sustain 2:1–16
Pickett ST (1997) Integrated urban ecosystem research. Urb Ecosys 1:183–184
Polk M (2015) Transdisciplinary co-production: designing and testing a transdisciplinary research framework for societal problem solving. Futures 65:110–122
Puppim de Oliveira JA, Balaban O, Doll CN, Moreno-Peñaranda R, Gasparatos A, Iossifova D, Suwa A (2011) Cities and biodiversity: perspectives and governance challenges for implementing the convention on biological diversity (CBD) at the city level. Biol Conserv 144:1302–1313
Puppim de Oliveira JA (2009) The implementation of climate change related policies at the subnational level: an analysis of three countries. Habitat Intn 33:253–259
Rosenau JN (1995) Governance in the twenty-first century. Glob Govern 1:13–43
Rudd A, Simon D, Cardama M, Birch E, Revi A (2018) The UN, the urban sustainable development goal, and the new urban agenda. In: Elmqvist T, Bai X, Frantzeskaki N, Griffith C, Maddox D, McPhearson T, Parnell S, Romero-Lankao P, Simon D, Watkins M (eds) Urban planet. Cambridge University Press, Cambridge, pp 180–196
SDSN (2016) Getting started with the SDGs in cities: a guide for stakeholders. Sustainable Development Solutions Network, New York
Seto KC, Boone CG, Branislav O, Fragkias M, Haase D, Langanke T, Marcotulio P, Munro D, Reenberg A, Simon D (2012a) Urban land teleconnections and sustainability. Proc Nat Acad Sci 109:7687–7692
Seto KC, Güneralp B, Hutyra LR (2012) Global forecasts of urban expansion to 2030 and direct impacts on biodiversity and carbon pools. Proc Nat Acad Sci 109:16083–16088
Simon D, Palmer H, Smit W, Riise J, Valencia S (2018a) The Challenges of transdisciplinary co-production: from unilocal to comparative research. Enviro Urbanis 30:481–500
Simon D, Griffith C, Nagendra H (2018b) Rethinking urban sustainability and resilience. In: Elmqvist T, Bai X, Frantzeskaki N, Griffith C, Maddox D, McPhearson T, Parnell S, Romero-Lankao P, Simon D, Watkins M (eds) Urban Planet. Cambridge University Press, Cambridge, pp 149–162
Simon D, Palmer H, Riise J (eds) (2020) Comparative Urban research from theory to practice: co-production for sustainability. Bristol University Press, Bristol
Sitas N, Selomane O, Hamann M, Gajjar SP (2021) Towards equitable urban resilience in the Global South within a context of planning and management. In: Shackleton CM, Cilliers SS, Davoren E, du Toit MJ (eds) Urban ecology in the Global South. Springer, Cham, pp 325–345
Sukopp H (2008) On the early history of urban ecology in Europe. In: Marzluff JM, Shulenberger E, Endlicher W, Alberti M, Bradley G, Ryan C, ZumBrunnen C, Simon U (eds) Urban ecology: an international perspective on the interaction between humans and nature, Boston, MA: Springer, Boston MA, pp 79–97

Swyngedouw E, Heynen NC (2003) Urban political ecology justice and the politics of scale. Antipode 35:898–918

Tauhid FA, Zawani H (2018) Mitigating climate change related floods in urban poor areas: green infrastructure approach. J Region City Plan 29:98–112

The Nature Conservancy (TNC) (2018) Nature in the urban century. The Nature Conservancy, London

Trencher G, Bai X, Evans J, McCormick K, Yarime M (2014) University partnerships for co-designing and co-producing urban sustainability. Glob Enviro Change 28:153–165

Valencia S, Simon D, Croese S, Davidson A, Diprose K, Aishwarya KS, Nesprias J, Nordqvist J, Oloko M, Sharma T, Buck NT, Versace I (2019) Adapting the sustainable development goals and the new urban agenda to the city level: initial reflections from a comparative research project. Intn J Urb Sustain Dev 11:4–23

Vásquez-Muñoz J (2016) Agenda verde: propuestas de investigación en servicios ecosistémicos para Medellín. In: Mejía MA (ed) Naturaleza urbana. Instituto de Investigación de Recursos Biológicos, Alexander von Humboldt, Bogotá, pp 112–115

Watson V (2014) Co-production and collaboration in planning: the difference. Plan Theory Pract 15:62–76

Wilkinson C, Sendstad M, Parnell S, Schewenius M (2013) Urban governance of biodiversity and ecosystem services. In: Elmqvist T, Fragkias M, Goodness J, Güneralp B, Marcotullio PJ, McDonald RI, Parnell S, Schewenius M, Sendstad M, Seto KC, Wilkinson C (eds) Urbanisation, biodiversity and ecosystem services: challenges and opportunities: a global assessment. Springer, Dordrecht, pp 539–587

UN (2015) Transforming our world: the 2030 agenda for sustainable development. A/RES/70/1. United Nations, New York

UN (2019) Global indicator framework for the Sustainable Development Goals and targets of the 2030 Agenda for Sustainable Development, after the 2019 refinement. United Nations Statistical Office, New York. Online: https://unstats.un.org/sdgs/indicators/GlobalIndicatorFrameworkafter2019refinement_Eng.pdf

UNISDR (2015) Sendai framework for disaster risk reduction: 2015–2030. United Nations Office for Disaster Risk Reduction, Nairobi

United Nations (UN) (2017) The New Urban Agenda. https://habitat3.org/the-new-urban-agenda/

Zhang F, Calvin Chung CK, Yin Z (2020) Green infrastructure for China's new urbanisation: a case study of greenway development in Maanshan. Urb Stud 57: 508–524

Advancing Urban Ecology in the Global South: Emerging Themes and Future Research Directions

Marié J. du Toit, Charlie M. Shackleton, Sarel S. Cilliers, and Elandrie Davoren

Abstract The rapid growth in urban ecological research and application has been led by countries of the Global North, particularly Europe and the USA, albeit not restricted to them. However, this belies that most urban growth is currently in the Global South, with the differential set to increase in the future. Thus, there is an imbalance between where the bulk of urban ecological research and framings are developed and where the majority of urban citizens globally live, work and seek to meet their aspirations. This chapter illustrates how this book addressed the identified research gaps in the Global South. We also highlight dominant recurring themes in Global South research discourse and importantly, where future research efforts over the next decade should be focussed. Eight themes are considered, namely: inequality; informality; urban–rural links; small and medium-sized towns and cities; urban green infrastructure, biodiversity and ecosystem services; understanding and accommodating multiple worldviews of urban nature; human health and urban nature; and specific research approaches.

Keywords Colonialism · Ecosystem services · Environmental justice · Governance · Informality · Urban green infrastructure

M. J. du Toit (✉) · S. S. Cilliers
Unit for Environmental Sciences and Management, North-West University, Potchefstroom, South Africa

S. S. Cilliers
e-mail: sarel.cilliers@nwu.ac.za

C. M. Shackleton · E. Davoren
Department of Environmental Science, Rhodes University, Makhanda 6140, South Africa
e-mail: c.shackleton@ru.ac.za

E. Davoren
e-mail: e.davoren@ru.ac.za

C. M. Shackleton et al. (eds.), *Urban Ecology in the Global South*, Cities and Nature,
https://doi.org/10.1007/978-3-030-67650-6_17

1 Introduction

Urban ecology is a growing research and applied discipline guiding urban planning, management, sustainability and resilience (Gill et al 2008; Wu 2014; Cilliers et al. 2021). Consequently, urban ecological understandings, principles and applications have significant relevance and expression in the quality of life and well-being of urban citizens around the world. With more people now living in urban areas than rural ones, the relevance of urban ecology cannot be underestimated.

Although urban ecology has been studied and debated for several decades (Rebele 1994; Pickett et al. 2001; Shochat et al. 2006; Ramalho and Hobbs 2012; Wu 2014; McPhearson et al. 2016) it is only in the last two decades, as rates of urban population growth have escalated, that it has come to the fore, transmuting from the periphery of urban planning and sustainability debates to the centre. This has seen the simultaneous evolution of broad-scale conceptions of urban ecology moving from ecology "in" the city, to ecology "of" the city, and more latterly ecology "for" the city (Wu 2014; Pickett et al. 2016). This has been premised on the gradual integration of the previously separate biophysical and social components of urban systems and thinking into more holistic framings as social-ecological systems and recently as social-ecological-technical systems (McPhearson et al. 2016).

The rapid growth in urban ecological research and application has been led by countries of the Global North (GN), particularly Europe and the USA (Shackleton et al. 2021), albeit not restricted to them. However, this belies that most urban growth is currently in the Global South (GS), with the differential set to increase in the future. Thus, there is an imbalance between where the bulk of urban ecological research and framings are developed and where the majority of urban citizens globally live, work and seek to meet their aspirations. At an overarching level, this does not matter because urban ecology is a global discipline that seeks principles that have global relevance and application. But at a finer scale, context is all-important in any ecological investigation and management; meaning that if urban ecology is to have relevance and compatible application in the GS, it must be able to accommodate GS contexts and dynamics. If it does not, at best it risks becoming redundant, and at worst, misguiding urban development and management to the detriment of urban sustainability and resilience, and human well-being in the GS.

In accepting that context is an essential consideration of social-ecological form and dynamics, then appreciation of the characteristics of GS towns and cities is fundamental if urban ecology is to have relevance in the GS. Principles, framings and applications of urban ecology must incorporate understandings of and from GS contexts (McHale et al. 2013; Nagendra et al. 2018). Whilst there is as much variation within the GS contexts as there is between the GS and the GN, many of the fundamental differences have a significant bearing on urban social-ecological dynamics. Core contextual differences have been introduced in the Chapter "The need for an Urban Ecology of the Global South" of this book (Shackleton et al. 2021), including rates of urban population growth being far higher than were ever experienced in the GN; pervasive poverty; large informal economies, housing and governance;

relatively weak central governance systems; rapid land-use change; extensive urban agriculture and livestock; and many more. Most of these have been elaborated further to some degree in the other chapters of this book. It is for these reasons that we brought together the various authors to compile this book on urban ecology in the GS with the purpose to: (1) review and present coherent, current and scientifically sound information and case studies on the urban ecology of cities and towns in the GS; (2) consider how, where and why urban ecology in the GS may differ or parallel that of the GN and the implications of that; (3) offer a GS contribution in a field that is predominantly shaped by and reported from the GN, and (4) through each of these, advance knowledge of and stimulate scholarship around urban ecology in the GS.

Having addressed these four aims through the preceding 16 chapters of this book, in this final chapter we seek to (1) illuminate ecology "in", "of" and "for" in urban areas of the GS as discussed in this book, (2) foreground some of the core and iterative themes emerging from the preceding chapters, and (3) identify research gaps aimed at advancing urban ecology research in the GS.

2 Ecology "in", "of" and "for" in Urban Areas of the Global South

As was elucidated in the first chapter of this book (Shackleton et al. 2021), different approaches have globally been followed in urban ecological studies. Ecology "in" cities, by studying the patterns and processes of urban plant (Anderson et al. 2021) and urban animal diversity (Reynolds et al. 2021), is still a more common approach in the GS than the ecology "of" cities approach, which studies cities as entire ecosystems. Although there are similar urban biodiversity patterns to GN cities, the steep socio-economic gradients characterised by high levels of poverty and inequality in GS cities results in other unique patterns that are often overlooked, mainly due to lack of long-term monitoring of urban biodiversity (Reynolds et al. 2021). The influence of social aspects on urban biodiversity have been studied in GS urban areas especially in terms of plant diversity patterns (Anderson et al. 2021), but few studies on urban wildlife have included socio-economic aspects as drivers of biodiversity (Reynolds et al. 2021).

Most of the chapters in this book stressed, however, the importance of studying cities in the GS using interdisciplinary and transdisciplinary approaches, where cities are studied as entire ecosystems in which social and ecological aspects need to be integrated, but also acknowledge the failure of applying the ecology "of" cities approach in many GS cities (Cilliers et al. 2021). From the different routes that ecologists take to understand the ecology of cities, the ecosystem services approach, although not as well-developed as in the GN, is valuable as indicated in several chapters (Davoren and Shackleton 2021; Dobbs et al. 2021; Drescher et al. 2021; Escobedo 2021; Shackleton 2021). It was, however, explained that the unique socio-economic context of GS cities does challenge the provisioning, management and perceptions of the value of ecosystem services, and addressing these issues in research and implementation should be more context-specific (Shackleton et al. 2021) instead of just applying principles developed in the GN. There is, for example, a higher demand for a variety of provisioning ecosystems services in GS than in GN cities such as wild foods, energy, medicine, material for construction and tools (Shackleton 2021).

An important aspect of the ecology "for" cities approach is to analyse cities as complex social-ecological systems in which the ecological and social aspects are coupled, enabling co-produced research in which all relevant stakeholders are involved (see Shackleton et al. 2021). According to McPhearson et al. (2016) using the "complex adaptive systems" model will increase understanding of the complexity of cities as they can be regarded as "interconnected and unpredictable, consisting of modular subsystems that can be redundant and are capable of resilience". It is also believed that more emphasis needs to be placed on technological aspects in the SES approach incorporating engineering and design disciplines in urban ecology, creating a social-ecological-technical systems (SETS) framework (McPhearson et al. 2016). A social-ecological systems perspective contributes to the understanding of how cities may deal with changes, uncertainties and surprises to improve urban sustainability and urban resilience. Examples of collaboration and stakeholder involvement from the GS include attempts to address the drought crisis in Cape Town, South Africa, and the influences of the Asian tsunami in Chennai, India (Sitas et al. 2021). The importance of developing sustainable and resilient cities is mentioned in most chapters in this book.

The resilience of cities can further be promoted by increasing social, ecological and social-ecological connectivity that will enhance the provision of ecosystem services and nature's contributions to people (Diaz et al. 2018) provided by urban green infrastructure (Maciejewski et al. 2021; Pauleit et al. 2021). The ecology "for" cities approach also focuses on the importance of merging urban ecology with design because "most of the environmental impacts of any project are manifest at the point of design" according to Childers et al. (2015). Cilliers et al. (2021) discussed specific avenues for ecological planning and design in GS cities highlighting citizen engagement, an increase in human–nature relationships and case study research focusing on addressing local relationships, values and perceptions. Novel approaches prioritising transdisciplinary co-design or co-production in urban governance, instead of traditional top-down and expert-led approaches, have extreme value in GS cities, especially where the capacity of central government institutions may be constrained and if they are focusing on multiple spatial scales and multiple institutions (Simon et al. 2021).

One of the most important meta-principles in urban ecology is consideration of heterogeneity at multiple spatial scales within complex social-ecological systems (Pickett and Cadenasso 2017). This importance is confirmed in several chapters in this book also acknowledging that the scale of analysis varies widely from fine-scale studies in parks and gardens (household scale) to entire towns and cities (local scale) and even on regional (landscape) and global scales. Biodiversity studies have been conducted at various scales with an emphasis on local, and it is often found that heterogeneity of habitats on a local scale may increase biodiversity, but heterogeneity at a landscape scale may have a negative effect on biodiversity (e.g. amphibian diversity) (Reynolds et al. 2021). The influence of ecosystem services also varies in terms of the scale of analysis; for example at local scales the importance of livelihood benefits (specifically provisioning services) is often the focus, whilst regulating services such as nutrient cycling and carbon sequestration tend to be considered at coarser scales (Anderson et al. 2021; Dobbs et al. 2021; Escobedo 2021; Shackleton 2021). The interaction between local and coarser scale factors determining the delivery of different provisioning (including urban agriculture) and cultural ecosystem services, especially in terms of the effect of different cultures and traditions, is quite complex and needs to be taken into consideration (Dobbs et al. 2021; Drescher et al. 2021; Shackleton 2021). Equally important, but less often studied than ecosystem services, is the effect of various ecosystem disservices, firstly at local scales because local knowledge and perceptions on ecosystem disservices are crucial in management plans and policies, but interventions at coarser scales are also necessary (Davoren and Shackleton 2021). Although urban green infrastructure (UGI) is characterised in different ways in GS cities, there are good examples of both top-down planning of urban green spaces at coarse scales as well as citizen-led initiatives at a local scales (Pauleit et al. 2021), such as tree planting and other activities in urban green spaces, gardens and wetlands, which are often viewed as more beneficial approaches to urban management (Simon et al. 2021).

The importance of multiple scales is also specifically mentioned in the four chapters on urban planning and management in the of GS (Cilliers et al. 2021; Maciejewski et al. 2021; Pauleit et al. 2021; Sitas et al. 2021). Addressing urban resilience, social-ecological connectivity, ecological planning and design and urban governance in GS cities should emphasise local-scale challenges informing and governing challenges at coarser scales (Cilliers et al. 2021; Maciejewski et al. 2021; Simon et al. 2021; Sitas et al. 2021). Sitas et al. (2021) discussed several global initiatives, that also extended to GS cities, which are strengthening institutional capacities but also establishing research-action networks to build resilience at different scales. In terms of spatial planning, the Cape Town Spatial Water Framework emphasises the importance of intervening in water management to increase water security at different scales (Cilliers et al. 2021), namely the "household (building), neighbourhood, district and metropolitan scales" (Cameron and Katzschner 2017). From a governance point of view, Simon et al. (2021) discussed several case studies from GS cities of governing the UGI demonstrating the importance for governance over different scales.

3 Advancing Urban Ecology in the Global South: Building on Previously Identified Gaps

The dominance of GN research in various aspects of urban planning, ecology, forestry and agriculture is well documented (e.g. Shackleton 2012; McHale et al. 2013; Lindley et al. 2018; Nagendra et al. 2018; Rigolon et al. 2018). Simultaneously, with highlighting the mal-distribution of urban ecology research and framings in favour of the GN, these authors also identified important research gaps or proposed research themes critical for advancing GS understandings and framings. Through this book, we attempted to advance the research in the GS by addressing some of these previously identified research gaps. In Table 1 we indicate to what degree this book has dealt with these gaps. We do so by scoring the contribution on a range from 0 to 3, with 0 meaning no or negligible contribution and 3 denoting a significant contribution to addressing that research need. Obviously, allocating the highest number to a research gap does not mean that this book provides all the answers and full coverage of the identified gap, but rather that this book directly addressed the gap and advanced the relevant research and science.

The most significant contribution of this book towards addressing previously identified research gaps is to illuminate the differences in contexts and drivers in the GS generally relative to the GN, as well as specific regions within the GS (Nagendra et al. 2018). This book, through multiple local examples, indicates the diversity of contexts found in the GS and the importance of incorporating context into decision-making and planning for successful local outcomes.

Moderate contributions are made through updated research in the GS on urban biodiversity, UGI and the associated ecosystem services. Furthermore, the colonial and urbanisation history and current situation in the GS are elucidated, along with discussions on related governance and stakeholder issues. The book also highlights examples of the successful use of local knowledge and innovative solutions from bottom-up, local actions. There is only one previously identified gap in Table 1 against which this book offers no significant inputs, which was the identification of specific urban social-ecological system typologies (Lindley et al. 2018) and focussing on rural areas that will become future urban areas (McHale et al. 2013).

From these prior identified gaps in Table 1 and the current book, we have identified further emerging themes from the book as a whole (Sect. 4), as well as priority research opportunities and needs to advance urban ecology in the GS in the medium term.

4 Emerging Themes Regarding Urban Ecology in the Global South

Throughout the book, several themes recurred across different chapters and hold general realities regarding urban dynamics in the GS and hence core aspects for any urban ecology research and frameworks. The ten themes, namely: rapid change;

Table 1 List of Global South research gaps identified by previous authors and the level to which they were explored in this book. The scale is from 0–3, 0 = not addressed, 1 = slightly, 2 = moderately, and 3 = directly addressed

Research need	Extent to which covered in this book
1. There is a knowledge gap on the differences in contexts and drivers in different Global South regions (Nagendra et al. 2018)	3
2. Is the current understanding of urbanisation developed in the Global North relevant in Africa and the rest of the Global South with its significantly higher percentage of informal settlements (McHale et al. 2013)?	2
3. Need for contextual and locally based knowledge to address diverse urban issues. Available information on the urban South is limited; there are large blind spots in critical locations such as sub-Saharan Africa and the Amazon (Lindley et al. 2018; Nagendra 2018; Nagendra et al. 2018)	2
4. There is a paucity of socioecological studies of urbanisation in developing nations (McHale et al. 2013). More empirical evidence is needed to inform decision-making and advance the development of theory on the social and ecological dimensions of ecosystem services and urban green infrastructure in Africa (Lindley et al. 2018)	2
5. Lack of information on innovation in Global South cities (Nagendra et al. 2018)	2
6. A better understanding of institutional diversity and their impact on urban sustainability (Nagendra et al. 2018)	2
7. More knowledge is needed on the potential contributions of urban forestry to poverty alleviation in the Global South. The extent and impact of harvesting and usage of consumptive goods from green spaces and trees in urban environments are practically unknown (Shackleton 2012)	2
8. Need a "new type of urban science to better reflect the actual plurality of contexts from the north and south" (Nagendra et al. 2018) The Western bias in sustainability suppresses progress, what works in the Global North often fail in Global South settings (Nagendra 2018)	1
9. Urban sustainability research needs a systems approach for evaluating potential benefits and trade-offs (Nagendra et al. 2018)	1

(continued)

Table 1 (continued)

Research need	Extent to which covered in this book
10. Need to address impacts of human livelihoods on the provisioning of ecosystem services and environmental degradation with a specific focus on local sustainability and issues such as excessive consumption, inequity and social injustice. (McHale et al. 2013; Nagendra 2018)	1
11. "How communities reshape traditional approaches to grapple with twenty-first-century challenges, how they address gender and caste inequities, and how philosophies and faiths influence people's attitudes to nature" (Nagendra 2018) There is a lack of research on social elements, especially on how "strong cultural and spiritual affiliations with nature can be recognised and protected" (Lindley et al. 2018)	1
12. More research needs to be integrated on GI governance (Lindley et al. 2018)	1
13. Identification of specific urban social-ecological system typologies that represent heterogeneous urban zones (Lindley et al. 2018), including how dense rural settlements converge with African cities—rural is the new urban (McHale et al. 2013)	0

informality; vulnerability; legacies of colonialism; weak or constrained planning and implementation of policies, plans and regulations; connectivity; partnership and local knowledge and action; biological invasions; environmental injustice; and context matters; are discussed below.

4.1 Rapid Change

The urban population of the GS has seen exponential growth over the last few decades, coupled with rapid rates of urbanisation (Mehring et al. 2021; Myers 2021; Shackleton et al. 2021), which has led to rapid and often unplanned urban sprawl, and as a result, the creation of slums, shantytowns and urban villages (Myers 2021). Urban areas are the most intensely altered areas on Earth, and the history of the landscape can have a profound impact on currently observed environments (du Toit et al. 2021). For example, legacies of the "apartheid" era's racial segregation and town planning in South African cities not only manifests as separated areas but also in disparities in the quantity and quality of UGI (McConnachie and Shackleton 2010; Venter et al. 2020).

The rapid rate of change in GS countries has gone, in most cases, hand in hand with a lack of sufficient and effective governance and poor service delivery in urban settlements (Myers 2021; Escobedo 2021; Cilliers et al. 2021), which means that the demand on local urban nature and biodiversity is greater (Anderson et al. 2021; Shackleton 2021). However, there are some cities in the GS that have produced models of effective governance and service delivery, such as Curitiba (Brazil) which is usually credited with developing the world's first Bus Rapid Transport system in 1974 (Myers 2021). Indeed, whilst the rapid pace of change can strain many urban systems, it also reduces the extent of path dependency, potentially allowing greater opportunities for innovation and positive changes at various scales.

4.2 *Informality*

A defining feature across much of the GS is the informal nature of many patterns and processes (Myers 2021). It is perhaps most noticeable in terms of the nature and magnitude of the informal economy and the informal housing (in terms of location and structures) that large proportions of the population call home. However, it may also refer to land-use patterns, such as urban agriculture in interstitial spaces, water sources and use, sanitation systems, regulation of urban bylaws, and management of UGI. By definition, informality mocks the rigid and predictable that characterise urban form and dynamics in the GN.

Informality in the economy means that urban or national authorities have limited direct knowledge or statistics of the nature of the economy, the number of people engaged and the benefits that they receive. Consequently, government institutions have less influence on what is carried out and where and have only limited means of collecting the legislated taxes and levies. Thus, revenues to urban authorities are well below what would be expected in the GN, which compromises the ability of central authorities to carry out certain functions and offer basic services. A large segment of the economy is comprised of small businesses, frequently unregistered and that may not abide by all applicable labour, environmental or planning regulations. The same can be said of urban housing, with approximately one-third of urban citizens in the GS living in slums (UN-Habitat 2003). Slum-dwellers have heightened vulnerability to natural disasters, communicable diseases and the consequences of insufficient basic services (Martinez et al. 2008). They may also be more divorced from particular ecosystem services (such as recreational parks, secure sources of potable water), raising the issue of environmental justice.

However, informality should not be viewed or interpreted as a total lack of structure, systems or chaos, but rather a number of systems and decision-making processes that operate in parallel at a local level, usually with no, or only limited, "official" authorisation. This has marked implications for the study and application of urban ecology, because in many countries (other than perhaps China) the power of the central "formal" authority is considerably diluted, thereby requiring a much more bottom-up, and inclusive approach than may be typical in the GN. Thus, all facets of urban ecology research or action need to engage at smaller scales and be able to accommodate far higher levels of uncertainty, paradoxically a vital principle in planning for resilience (Sitas et al. 2021), which also provide space for local level novelties and innovation.

The importance of informal green spaces as vital spaces in the urban matrix has also been highlighted in this book. Such spaces are used for housing (Myers 2021), collection of provisioning resources (Shackleton 2021) and urban agriculture (Drescher et al. 2021), whilst also providing regulating (Escobedo 2021) and cultural (Dobbs et al. 2021) ecosystem services, including connectivity (Maciejewski et al. 2021) and resilience (Sitas et al. 2021). However, whatever benefits are provided by these informal spaces, their very informality also places them at risk of sudden land-use change or appropriation by specific groups or elites.

4.3 Vulnerability

Closely allied to the pervasive informality in the GS is the recurring theme of human vulnerability. Human vulnerability is typically greater in the GS than the GN due to the interplay of two factors. The first is the greater frequency of natural disasters and ecosystem disservices (Davoren and Shackleton 2021), and economic shocks or institutional failures. The second is the generally lower capacity of institutions and households to cope due to lower financial means and poverty.

Human vulnerability is a complex concept because it integrates across different domains of endeavour and research disciplines. Even at a basic level, any analysis and planning response needs to understand the different components of vulnerability (Adger 2006) as (i) exposure to a specific type of shock, such as floods, financial losses or a fire, (ii) sensitivity and (iii) response capacity. However, all of these are also shaped by the historical and current political and power dynamics and balances operating in the system under scrutiny (Ribot 2014). This is further complicated by the necessity for different scales of analysis and action, from individuals to households, to neighbourhoods, and to city authorities and higher-level governance structures. For example, a particular neighbourhood may be at risk of periodic flooding, even though the city as a whole might not, or poor households will be at greater risk of price changes to access basic services than would be the more well-off households.

Various instances and dimensions of human vulnerability are raised by authors in this book, clearly affirming that it is a core reality of lives and livelihoods in many GS urban settings. Davoren and Shackleton (2021) argue that the higher frequency of ecosystem disservices exacerbates the vulnerability of urban citizens in the GS, especially the urban poor. In a positive framing, Drescher et al. (2021) describe the significant potential of urban agriculture to provide food security, especially in times of political or economic upheaval. Mehring et al. (2021) also argue that the strong links that many urban households in the GS have to rural areas foster household resilience and thereby reduce vulnerability, especially to economic shocks. The extensive use of provisioning ecosystem services in meeting household needs and as safety nets in times of crisis is highlighted by Shackleton (2021). Some may argue that the roots of the heightened poverty and hence vulnerability in the GS lie in the former colonial domination and segregation that have lasting legacies to this current day (du Toit et al. 2021). This will have further ramifications into the future with the intensification of global climate change, with GS countries being disproportionately vulnerable to its effects (Scoville-Simonds et al. 2020).

4.4 Environmental Justice

Rapid, unplanned and haphazard urban development contributes to environmental injustices experienced by many of the urban poor and marginalised communities (Myers 2021; du Toit et al. 2021; Mehring et al. 2021; Pauleit et al. 2021; Cilliers et al. 2021). They are often relegated to low-income settlements and slums in environmentally vulnerable locations, which frequently are situated in low-lying areas, along the banks of rivers or lakes, or on steep slopes with low vegetation cover, and vulnerable to industrial pollution and disasters, such as floods and landslides (Myers 2021; Pauleit et al. 2021; Davoren and Shackleton 2021; Cilliers et al. 2021). Rapidly expanding urban populations and limited resources can result in much lower levels of UGI in developing countries and in poorer urban communities (Venter et al. 2020). Furthermore, urban amenities are unevenly distributed and often biased against marginalised populations, leading to high levels of inequality and social exclusion in both GN and South countries, but these inequalities and resulting injustices are far more prevalent in the GS (Mehring et al. 2021; Pauleit et al. 2021; Cilliers et al. 2021).

As pointed out by du Toit et al. (2021), socio-demographic legacies can have significant impacts on recovery after disasters and highlight the environmental injustices experienced by all urban communities in both GN and GS. After disasters, vulnerable people are most likely to experience environmental injustice, for example, after Hurricane Katrina in New Orleans, the most vulnerable people were located in the most high-risk areas, where post-disaster recovery was much slower than high-income areas (du Toit et al. 2021). The restoration and development of UGI can also involve the displacement and relocation of vulnerable communities as a consequence of environmental injustice (Pauleit et al. 2021). Climate change will further exacerbate existing environmental injustices, as well as creating new ones (Myers 2021). Coastal cities are especially vulnerable since the "structural vulnerabilities" in these cities "constitutes a form of unfairness or injustice" (Myers 2021). However, there are initiatives and programmes that redress existing injustices and which highlight the role that the GS can play in driving the movement to a more sustainable and equitable world (du Toit et al. 2021; Pauleit et al. 2021).

4.5 Context Matters

According to Shackleton et al. (2021) "there are a multitude of local and national-scale contextual differences between the Global North and the Global South that limit or question the universal application of Global North perspectives and knowledge". Urban environments around the world are the result of their evolutionary and geographic history, and more contemporary influences of the social, cultural, economic, political and technical systems that shaped their construction (du Toit et al. 2021). These contextual differences need to be accounted for in urban ecological research, theory and application towards the development of an urban ecology that is more relevant for the GS, and when conjoined with understandings from the GN allow for the exploration and development of truly universal urban ecology principles and frameworks (Shackleton et al. 2021).

Thus, there is a need for context-specific information on how to promote, educate and apply various ecological concepts, such as ecosystem services, to urban design, planning, development and management (Cilliers et al. 2021). For example, water management is an important aspect of urban design, but recognising the importance of local context results in the need for different context-sensitive tools, tactics and means of knowledge transfer and co-learning. For example, through development of "learning alliances" in which different stakeholders are informed and empowered to establish communities of practice to implement Water Sensitive Design (Cilliers et al. 2021). The Cape Town Spatial Water Framework (SWF) emphasises the importance of spatial planning in interventions in water management issues on different scales (Cilliers et al. 2021). Similar arguments can be made around a whole host of attributes and dynamics that are more common in the GS than the GN, such as poverty, urban farming and livestock, climate change vulnerability, and so on (Shackleton et al. 2021). Context sets the broad- and fine-scale template that shapes what factors come into play, their relative magnitudes, and in what combinations and sequences (Shackleton et al. 2021).

Context is also important in identifying urban ecosystem disservices. What constitutes an urban ecosystem disservice, which ones are the most relevant to a particular urban population and how to adequately manage the disservices in different cities, are determined by that community's context, heterogeneity, perceptions, prior experiences and scale (Davoren and Shackleton 2021). Context refers to the environmental, economic, cultural, social and political realities of an urban community and the value and importance that they ascribe to both urban ecosystem services and disservices (Davoren and Shackleton 2021). The impact of ecosystem disservices on the urban poor will become more severe under future climate change scenarios.

4.6 Legacies of Colonialism

An important factor that influenced almost the entire GS is colonialism. The GS nearly divides the colonisers from the colonised (Shackleton et al. 2021). The impacts of colonialism pervade through urban form and development, governance, biodiversity, and societal perceptions and behaviour. Colonialism is responsible for the introduction of multitudes of alien species and subsequent invasiveness of many (Shackleton et al. 2021; du Toit et al. 2021; Anderson et al. 2021). Species were ostensibly imported to resemble the home countries of colonial settlers and to "improve" local biodiversity. However, despite the pointedly negative effects of many invasive species in new host countries, in some cities, such as Bangalore and Delhi, the colonial-era tree planting and city greening significantly improved the liveability of these cities (Anderson et al. 2021). Notwithstanding the often intentional use of exotic plants, some colonial-era parks still vitally complement the UGI of various urban areas and provide critical ecosystem services to residents.

Colonialism facilitated the focus of urban development on port cities due to an economic focus on resource extraction from the colonised countries (Myers 2021). The concentration of megacities on the coast also means that they are disproportionately vulnerable to climate change impacts and natural disasters such as flooding (Myers 2021; du Toit et al. 2021). Colonial cities also had a distinctive urban form with segregated residential areas based on racial grounds, purportedly for hygiene reasons (Myers 2021). These segregated residential areas also had unequally distributed green spaces and vegetation cover, which can result in distinct surface temperature differences between these neighbourhoods (du Toit et al. 2021). Infrastructure investments in colonial cities strongly reflect the urban hierarchy, which in many cases is visible to this day (Myers 2021). Moreover, inherited colonial governance structures plague many postcolonial governments (Myers 2021; Simon et al. 2021).

4.7 *Biological Invasions*

Biological invasions remain a major problem around the world with no signs of saturation (du Toit et al. 2021). Control of invasive species in the GS is often hindered by weak governance, inadequate capacity and funding constraints (Anderson et al. 2021; Reynolds et al. 2021). The greatest initial source of invasive species was colonial administrations, who imported and traded them for economic and ornamental value (du Toit et al. 2021; Anderson et al. 2021).

Invasive species can have both negative and positive effects. Invasive bird species can outcompete natives to dominate urban environments (Reynolds et al. 2021). Invasive species are a major cause of species extinctions, and they also drive biotic homogenisation (du Toit et al. 2021; Anderson et al. 2021). Moreover, suppression of valuable resource species by invasive species can have a detrimental effect on those that rely on resource species for their invaluable provisioning services. Urban mammals such as feral domestic cats, house mice and rats typically have large, negative impacts on biodiversity, human health and food security (Reynolds et al. 2021). Many alien tree species also have disservices which include allergenic pollen and invasive insect tree pests have killed thousands of urban trees (Reynolds et al. 2021; Davoren and Shackleton 2021).

Positive values include essential ecosystem services of regulation and provisioning; and cultural and heritage values especially linked to urban trees (du Toit et al. 2021; Anderson et al. 2021). Exotic tree species can increase bird species richness, and many raptor species use invasive tree species as nesting sites (Reynolds et al. 2021). Such positive values can make control and management of invasive species in urban areas highly contentious in some cities (Anderson et al. 2021).

4.8 Weak or Constrained Planning and Implementation of Policies, Plans and Regulations

Several of the chapters in this book refer to the challenges of planning for sustainability in GS cities. Some of these challenges include rapid rates of urban growth, unplanned sprawl, loss of UGI, weak support on environmental issues by local governments, lack of initiatives and incentives for ecological planning and design, to name a few (Cilliers et al. 2021). It was also indicated that some of these challenges had been addressed to a certain extent by focusing on local relationships, values and perceptions as GN models sometimes do not address the complex realities of GS cities (Cilliers et al. 2021; Myers 2021). Aspects in terms of planning and design that specifically need to be addressed include, (1) a focus on participatory planning and bottom-up initiatives at neighbourhood scales, (2) local projects focusing on community-based, ecosystem-based adaptation and design, (3) issues of environmental and social injustice in planning and design, (4) integration of local and expert knowledge for planning and design that are context-specific, (5) successful transmission of research findings through partnerships between local governments and research entities to inform decision-making, and (6) household food and nutrition security (Cilliers et al. 2021; Drescher et al. 2021; Mehring et al. 2021; Pauleit et al. 2021). The sponge city concept of China focusing on "urban water resourcing and ecological water management" and adaptive planning for climate-resilient cities and systematic conservation planning in South Africa, have been discussed as important ecological planning approaches that the rest of world could adapt or emulate (Cilliers et al. 2021; Pauleit et al. 2021; Sitas et al. 2021).

Although a failure in formal governance of UGI is a reality in many GS cities, as noted in several chapters in this book, Simon et al. (2021) highlighted that each city has its own unique and complex governance system, mainly defined by historical legacies, often including racial segregation (du Toit et al. 2021) that need to be considered. Furthermore, there is a need for governance actions across multiple spatial scales (international, regional and local city) as well as multiple institutions (city governments, community-based organisations, and private enterprise) (Simon et al. 2021). These could be accommodated by polycentric governance systems that are flexible enough to deal with changing contexts, but also aim for collective management in a decentralised way (Sitas et al. 2021). Simon et al. (2021) provided a number of success stories from GS cities where strong partnerships based on trust have improved governance (Urban Natural Assets for Africa Programme—UNA) and where committed leadership on local government level contributed to the use of nature-based aspects in addressing sustainability (Curitiba, Brazil) and also mentioned the potential of informality providing "creative and innovative governance interventions".

4.9 Connectivity

The important role of connectivity in the sustainability and resilience of urban areas is generally acknowledged in urban ecology (Andersson 2006; Gaston et al 2013). In cities as complex social-ecological systems, the focus should be placed on connectivity between humans and biophysical components (social-ecological connectivity) but it depends greatly on connectivity between humans (social connectivity) and between natural components inside and around cities (ecological connectivity) (Maciejewski et al. 2021). In urban studies, connectivity is more often described by "links, networks and intertwined economies" of GN metropolises and GS megacities are only mentioned from a GN perspective (Myers 2021). As an important principle in resilience, connectivity can lead to faster recovery after disturbances (Sitas et al. 2021). Future research in urban resilience should, however, not only focus on the advantages of increasing connectivity between cities in terms of governance, but also whether over-connected cities would not lead to specific social and ecological vulnerabilities (Elmqvist et al. 2019; Sitas et al. 2021).

Increasing connectivity between people (social connectivity) in the built infrastructure is a challenge in GS cities due to the high density of people influencing city structure, the high proportion of new migrants, and also the layout of communication networks. The informal economies of many GS cities pose several ecological and social challenges, including low or constrained service provision. Maciejewski et al. (2021) discussed "informal enterprises" such as "waste picking" which address the lack of certain services and which may "fulfil the role of connecting the Municipality's service value chain to the material value chain of recycling companies" if well planned and managed. Maciejewski et al. (2021) also mentioned two other challenges in increasing social connectivity in GS cities. Firstly, in upgrading informal settlements or relocating urban residents, the unique informal but often well-functioning economies and existing social interactions provided in these informal settlements are usually overlooked. Secondly, the value of information communication technologies has not fully been realised in GS cities as "educational systems, health systems, office work and finance systems have yet to adapt fully to the potentials offered by decentralised communication" (Maciejewski et al. 2021).

The connection between people and the biophysical environment is equally important in terms of the provision of ecosystem services by UGI to improve human health and well-being, and the opportunities and challenges have been dealt with by Shackleton (2021), Dobbs et al. (2021), and Escobedo (2021) in this book. Ecosystem services and nature's contributions to people frameworks, developed on the basis of worldviews, knowledge and understandings primarily from the GN, may have some relevance or application to GS cities (Maciejewski et al. 2021), but need to be used in the full knowledge of the contextual differences. For example, conceptions of nature and hence what it offers to or confers on local citizens may not fit typical framings of the flow of goods and services *from* nature *to* humans (Shackleton and Cocks 2021).

Ecological connectivity linking natural areas in and around cities increases the flow of ecosystem services and the functioning of ecological processes, especially in terms of the movement of animals and plant propagules. "Landscape connectivity is correlated to animal diversity", for example, a lack of connectivity decreases amphibian diversity in GS cities (Reynolds et al. 2021). Connectivity as a principle in UGI planning is well-developed in some GS cities, and although the focus is mainly on ecological connectivity, in providing dispersal corridors for species (Reynolds et al. 2021), there are examples where social connectivity is aimed for in terms of creating links for social upliftment of the poor and addressing the effects of climate change (Pauleit et al. 2021). Maciejewski et al. (2021) discussed the development of an ecological corridor in Johannesburg, South Africa, focusing on environmental, economic and social benefits. The reality is that we have limited knowledge on ecological connectivity in urban areas, especially in GS cities as only eight out of 174 studies on urban ecological connectivity investigated by La Point et al. (2015) were from GS cities. More studies are needed to quantify functional connectivity as it is vitally important for conservation and management of the urban green infrastructure.

4.10 Partnerships and Local Knowledge and Action

Effective practices and interventions in GS cities are strongly linked to successful partnerships with local stakeholders, bottom-up approaches and use of local knowledge. Pauleit et al. (2021) describe the successful revision of the Addis Ababa master plan through cooperation between local stakeholders. They also highlight the importance of City Lab initiatives in establishing partnerships between society, researchers and local municipalities to transmit research results into practice. Successful informal, citizen-led, small-scale initiatives illustrate the value and importance of community perspectives on UGI problems (Pauleit et al. 2021). Effective use of local provisioning services mostly relies on the local ecological knowledge of the users (Shackleton 2021). It is therefore essential that such knowledge should not be lost, and care should be taken to ensure that this knowledge is passed on. Moreover, researchers should acknowledge the value of local knowledge in potentially uncovering innovative and alternative ways to use UGI for livelihoods, economic, cultural and well-being activities.

The importance of context-specific, local solutions to urban issues is often the difference between failure and success. The Urban Natural Assets for Africa Programme (UNA) is one such example where the local context is explicitly addressed through tailor-made solutions that encourage co-production of knowledge to provide effective nature and climate action at the local level (Simon et al. 2021). However, we need more cooperation and knowledge-sharing amongst societies and institutions of the GS, with less dependence on the GN for knowledge. Researchers and authorities should share examples and experience from other GS countries in trying to solve pressing urban issues.

5 Research Needs for Advancing Urban Ecology in the Global South

This section presents some significant issues and themes that will need to be addressed in urban ecology research in the GS in the short- to medium-term. The eight themes are inequality; informality; urban–rural links; small and medium-sized towns and cities, urban green infrastructure, biodiversity and ecosystem services; understanding and accommodating multiple worldviews of urban nature; human health and urban nature; and specific research approaches. Listed with each theme are *illustrative* questions that require exploration across a range of GS contexts.

5.1 Inequality

According to Nijman and Wei (2020) "inequality has long been an important theme in the social sciences, and it has acquired renewed prominence in the last couple of decades", as is evident by the establishment of specialised research centres such as UCLA's Institute on Inequality and Democracy, Stanford's Center on Poverty and Inequality, the London School of Economics and Political Science's International Inequality Institute and the University of Amsterdam Centre for Inequality Studies. Urban inequality is multidimensional and highly complex (Nijman and Wei 2020). It encompasses all aspects of the city, such as economic inequality, education inequality, housing and residential inequalities, transportation inequity, spatial mismatch, environmental injustice, the digital divide, food deserts and unequal access to government services (Nijman and Wei 2020). Urban inequality has, therefore, also been identified as an emerging research theme by several authors in this book.

Myers (2021) highlighted the high degree of inequality and injustice that exists spatially in many GS cities in terms of the negative externalities of environmental change, including climate change. For example, Cape Town in South Africa is broadly divided between largely white neighbourhoods at higher elevations and mixed-heritage neighbourhoods that predominate in the low-lying Cape Flats (Myers 2021). Climate change makes the Cape Flats more susceptible to flooding, and at the same time, ironically, to chronic potable water shortages, whilst the predominantly white areas face less risk from rising sea-levels and have greater wealth to purchase water for domestic consumption from commercial sources (Myers 2021).

Both Mehring et al. (2021) and Pauleit et al. (2021) emphasised that poorly planned UGI can exacerbate social inequality and environmental injustice (Sect. 4.4) in GS cities. According to Rigolon et al. (2018), GN and GS cities experience similar inequalities in terms of green space quantity and quality, but GS cities also experience inequalities in terms of proximity to urban green spaces. For example, in Mexico City UGI of higher quantity and quality is mostly located close to conservation zones, leaving other areas of the city without access to the ecosystem services provided by UGI (Mehring et al. 2021). According to du Toit et al. (2021), approaching development with too narrow a focus can result in legacies of inequitable access to nature and biodiversity.

Inequality in terms of infrastructure for water and sanitation and waste treatment is also a major issue in many GS cities. Nagendra et al. (2018) argue that they are the biggest problems for urban sustainability in the GS, which is a sentiment shared by many of the authors in this book who highlighted such inequalities experienced by the urban poor living in slums and informal settlements in GS cities (Escobedo 2021; Maciejewski et al. 2021; Sitas et al. 2021). As Shackleton (2021) points out "inadequate water supplies or quality compromise what food can be cooked, along with general hygiene and sanitation, and increases the risk of water-borne diseases".

Illustrative research questions pertaining to inequality in GS cities

- How does inequality shape the use and needs of urban space generally and UGI specifically?
- How can UGI development help alleviate urban inequality and minimise environmental injustice in GS cities?
- How can the voices of the "voiceless" be included to plan more resilient and sustainable cities?
- What are the specific reasons for inequality in terms of green space provision in GS cities, and how can that knowledge assist urban planners, managers and policymakers in addressing environmental justice in GS cities?
- To what extent does informal green space mitigate against environmental and ecosystem services inequality?
- What planning and governance approaches exacerbate or reduce inequality?
- How does the rapid development of digital technologies contribute to urban inequality?

5.2 *Informality*

Informality in the GS can be best summed up by Myers' (2021) interpretation of the term as *normality,* highlighting its ubiquity in cities of the GS. He emphasises that "Global South urbanisation is inseparable from an engagement with informality, especially regarding employment, housing and everyday life, as convoluted and multi-faceted as informality may be". Featuring as one of the emergent themes in the GS, much of this book highlights the importance of directly acknowledging the role of informality. Moreover, of also explicitly incorporating it into urban decision-making and planning, and not the common solution of trying to demolish it in a misguided bid to eradicate poverty in the pursuit of a world-class city vision (Sheppard et al. 2020). Informal settlements and slums are often portrayed in negative terms, even by the UN-HABITAT, which Arfvidsson et al. (2017) caution can lead to the perception that residents of informal settlements lack dignity. However, in some urban areas such as Addis Ababa, informal areas account for most of the cities' economic activity (Arfvidsson et al. 2017). Similarly, informal green spaces are often neglected in urban ecology research and planning (Rupprecht et al. 2015), yet may provide vital spaces for citizens' physical and mental well-being and even livelihoods (Shackleton 2021).

As discussed in Sect. 4.2, the nature of informality means that urban and national authorities have limited knowledge and influence on informal spaces, informal economies and informal institutions that shape land and nature's contributions to people. Moreover, several paradoxes in informality exist that belies the official legal system's ability to control or tax it (van Gelder 2013). The lack of formal control means that research requires a much more inclusive, bottom-up approach, incorporating high levels of uncertainty, focussing and engaging at smaller scales. It also means that researchers frequently cannot rely on official statistics or the perspectives of those in formal authority, but rather must seek alternative sources of information and triangulate it across different sectors and role-players.

Illustrative research questions pertaining to informality

- What is the relationship between formal and informal areas of the city, and how does this affect urban governance?
- How can "invisible" informal areas be "formally" accommodated in city planning and decision-making?
- How do informal areas contribute to livelihood, ecological and city resilience?
- Is informality necessary to a few GS cities as alternative stable states, and if so how can it be accommodated?
- How to navigate the benefits of informal green spaces in areas at risk of sudden land-use change or appropriation?
- What informal rules and institutions shape access to and use of UGI and the ecosystem services it provides?

5.3 Urban–Rural Links

The GS is experiencing rapid rates of rural to urban migration, which is one contributor to the high rates of urban expansion (Myers 2021; Mehring et al. 2021). However, unlike much of the GN, not all this migration is uni-directional and permanent. There are massive flows of people, resources, knowledge and culture in both directions, and many urban households have kin in rural areas and visit them regularly or for specific occasions (Tacoli 2006). For example, Mehring et al. (2021) reveal how people in Senegal and Mali migrate from villages to urban areas during the dry season and return to their villages for the harvest season, whilst Cocks and Wiersum (2014) describe how urban Xhosa in South Africa visit rural family and communities for specific ceremonies and rituals. Seasonal and circular migration over longer periods are common throughout the GS. These visits to rural family and landscapes contribute to their sense of identity, well-being and culture. These connections provide mutual support mechanisms to both urban and rural members of extended families and communities. For example, in Windhoek (Namibia) food insecurity is lower amongst urban households that have rural links (Frayne 2004).

The often strong urban–rural links in GS contexts have considerable implications for all aspects of urban ecology, because not all urban citizens are entirely reliant on the urban ecosystem for their needs and well-being. This means that studies and understanding of urban livelihoods, vulnerability, resilience, relationships with nature and more, need to acknowledge the connections and contributions from rural areas. The same applies with respect to the study of rural livelihoods, economies and culture, which also receive regular injections or contributions from urban relatives and activities. These connections do not play out solely at the household scale, but also pertain to the flow of resources, knowledge and experiences at large scales in the form of teleconnections (e.g. Güneralp et al. 2013).

Illustrative research questions pertaining to urban–rural links in GS cities

- How do urban–rural links shape the demand for urban ecosystem services?
- Do urban–rural links increase household and city resilience?
- How do urban–rural links contribute to or maintain local ecological and agricultural knowledge?
- Do the urban-rural links mitigate against the "extinction of experience" with respect to biodiversity?
- How does urbanisation impact rural–urban connectivity, agriculture and food production systems on a local and regional scale?
- What agricultural land-use patterns influence rural-urban connectivity, and how can this knowledge be utilised to formulate relevant policy responses to rapid urbanisation?
- What broader-scale teleconnections between urban and rural areas shape urban sustainability?

5.4 Small and Medium-Sized Towns and Cities

The process of urbanisation results in massive changes in land use, biodiversity, access to services, consumption patterns, and human health and well-being to name but a few. However, there is very little understanding of how the positive and negative effects of urbanisation scale with town or city size. Towns and cities of less than half a million people currently house more than 50% of the world's urban population (World Bank 2009), and close to 45% live in urban centres of less than 100,000 people (Bell and Jayne 2009). Collectively, small towns and cities cover 3–7 times more land area than do large cities and megacities (Łopucki and Kitowski 2017). Moreover, most of the new built-up urban areas and future urban growth are expected to occur in small and medium-sized cities (Secretariat of the Convention on Biological Diversity 2012). Despite these patterns, most dimensions of urban studies focus on larger cities and economic hubs (Parnell and Walawege 2011), and smaller centres are overlooked (Bell and Jayne 2009). This also applies to the majority of studies on urban biodiversity and ecology, leading Łopucki and Kitowski (2017) to state "the basic generalisations and laws of urban ecology have been formulated on the basis of large city studies".

This is likely to offer skewed or incomplete understandings of urban ecological patterns and processes because some properties or dynamics only emerge as cities attain particular sizes. For example, the urban heat island effect, alteration of precipitation patterns, critical mass of qualified urban ecologists/foresters/agriculturalists in urban governance institutions, development and implementation of climate change adaptation plans, and so on. Most small towns and cities are socially, economically and ecologically very different to large cities and megacities. A few examples include that they are generally less compact thereby leaving more space for UGI and biodiversity, they have higher core to periphery ratios which has implications for connectivity and access to nature, often lower per capita incomes across city scales and lower pollution levels. These, and many more, mean that urban ecology is likely to be very different in smaller cities and towns and hence requires greater research, understanding and consideration in urban policies and programmes.

Illustrative research questions relating to urban ecology and towns/city size in the GS

- How do patterns of biodiversity change with increasing urban size?
- At what size do particular emergent properties manifest?
- Are levels of vulnerability to ecosystem disservices higher and capacities to respond lower in small and medium-sized urban centres?
- Is access to UGI and urban nature higher in small towns and cities than in large ones?
- How does a declining core to periphery ratio affect species exchanges?
- How does town/city size influence resilience?

5.5 Urban Green Infrastructure, Biodiversity and Ecosystem Services

Shackleton et al. (2021) highlighted the disparities between the GN and GS in terms of UGI provision and development, perceptions and valuation of biodiversity, and ecosystem services and disservices. Disparities within GS cities can also be seen in the quantity and quality of UGI due to the legacy of colonialism (du Toit et al. 2021) (and apartheid in the case of South Africa [Venter et al. 2020]). The importance of UGI has been highlighted throughout this book. Urban green infrastructure encompasses the entire diversity of green and blue spaces, the green on built and grey infrastructure (i.e. green roofs, walls and facades) and the ecosystem services and disservices that these spaces provide (Pauleit et al. 2021).

Future research needs to ensure that basic principles of UGI planning, such as multifunctionality, social-ecological connectivity, green-grey integration and social inclusion, are understood to address the specific situations of the urban areas in the GS, prioritising risk management, and dealing with poverty and informality (Maciejewski et al. 2021; Mehring et al. 2021; Pauleit et al. 2021). Collective knowledge of UGI, as well as urban ecology, could greatly benefit from the development of inventories and identification of case studies on urban plant and animal diversity patterns and processes in GS urban areas and the filters determining them (climate, biogeography and land use; human facilitation; urban form and history and socio-economic and cultural factors). This could indicate that GN generated biodiversity conservation solutions are often not applicable in the GS context and that novel, context-specific, solutions will be needed to secure urban biodiversity in the GS.

These solutions should embrace the high utility of the local natural environment to many urban citizens in the GS and the effect of climate change on city sustainability and resilience in the GS (Anderson et al. 2021; Reynolds et al. 2021). There is a need to move beyond spatial studies of urban animal and plant diversity patterns (e.g. along urban–rural gradients) and analyse the urbanisation impacts over time providing rigorous data from long-term studies in GS urban areas (Anderson et al. 2021; Reynolds et al. 2021).

Illustrative research questions pertaining to urban green infrastructure and biodiversity

- What types of UGI are present in a spatially explicit format in GS cities?
- What kind of ecosystem services and disservices do specific types of UGI provide and how can ecosystem disservices be minimised without compromising ecosystem services?
- What is the demand for ecosystem services along socio-economic gradients, and who benefits from them?
- What are the perceived and real values of ecosystem services in addressing inequalities in terms of UGI provision, and how can these values be quantified?
- How can UGI be used to mitigate climate change and develop resilient cities?
- What innovative solutions are there for UGI development, climate change mitigation and/or ecosystem disservice management?
- How does informality shape the supply and demand for ecosystem services?

5.6 Understanding and Accommodating Multiple Worldviews of Urban Nature

The predominance of GN research on the roles and values of urban nature is premised very much on an implicit philosophy, or worldview, dating back several centuries, that there is a separation between humans and nature. In this view, nature is there to be tamed, exploited and to provide services to humans, a worldview that is both anthropocentric and reductionist (Skandrani 2016) in comparison to many other worldviews. There are many different worldviews of what nature is and how humans relate to it or are part of it, besides the dominant Eurocentric/Christian view (Flint et al. 2013, and case examples in Cocks and Shackleton 2021). Indeed, for many cultural groups, and especially First Nations peoples, nature and what makes one human, are indivisible. As such nature, and the innate feelings and connections between humans and nature shape personal and collective identity, mental well-being, spirituality, sense of place and much more (Shackleton and Cocks 2021). Acknowledgement of the existence of multiple worldviews has deep consequences for what types of nature are provided and maintained in urban spaces, where, what sorts of activities people may wish to engage in when in urban nature, how compatible other types of activities might be with their worldviews, and the extent to which local citizens will participate in conservation. This has not been acknowledged and incorporated into research on the uses and benefits of urban green spaces and nature, in general, and consequently is totally overlooked in urban planning and design. This is an arena requiring urgent consideration across the GS, and even globally as cities of the GN become increasingly more multicultural.

Illustrative research questions pertaining to multiplicities of worldviews of nature and urban nature

- What are the different worldviews of nature?
- What sort of practices in and of urban nature do such worldviews require?
- How can they be accommodated in the planning and design of urban green spaces?
- What specific species, sites or landscapes hold spiritual or religious significance, and how can they be protected, restored or maintained?
- What can urban design, planning and governance learn from such worldviews?
- What are the human well-being implications of severing or preventing expression of such worldviews?
- How does a multiplicity of worldviews contribute to urban resilience and adaptability?

5.7 *Human Health and Urban Nature*

The significance of urban nature and UGI for human mental and physical well-being and quality of life has been well recognised (Hartig et al. 2014; Shanahan et al. 2016) and is now a major research theme in many countries, especially in the GN. Whilst the findings from the GN have relevance in the GS; they are premised on particular conceptions of what is health (and also on what is nature), especially mental health, and how that is affected by urban nature. Yet health, especially mental or psychological health, is culturally defined, mediated and treated (Shavitt et al. 2016; Capasso et al. 2018) and consequently, indicators and expressions of health and well-being must be culturally relevant and situated. Thus, to have relevance in the GS, any research into links between nature and health must include cultural constructions, indicators and definitions and health, especially mental health (although in some cultures there is a limited separation of physical and mental health, and the former is entirely dependent on the latter). Consequently, the relationships between urban nature and health may take different forms in different cultural settings.

Illustrative research questions pertaining to health and urban nature in GS cities

- What are local constructions and meanings of mental and physical well-being, and urban nature, across cultures?
- What culturally appropriate indicators are best able to represent the relationship between urban nature and human well-being and health?
- Do different measures of health reveal different needs of urban nature?

- How can urban planners and managers provide for a range of culturally mediated health needs through urban nature?
- How might cultural constructions of nature and/or health affect the types and nature of urban green infrastructure required?

5.8 Research Approaches

To adequately understand the GS context, research approaches need to be adapted and focussed on addressing multidisciplinary issues and largely ignored research gaps. New research needs to follow a transdisciplinary approach in terms of planning, design and management. Scientists need to build a community of practice, redefining civic participation and building learning alliances of different stakeholders (bottom-up approach) through long term research projects producing case study scenarios (co-production of knowledge, network building and innovative thinking). The bias on the predominant western view of urban ecosystems illustrated by a majority of GN studies must be addressed. Socioecological studies of urbanisation in developing nations remain rare (McHale et al. 2013). A "new type of urban science is needed to better reflect the actual plurality of contexts from the north and south" (Nagendra et al. 2018). Therefore, capacity building of GS researchers should be emphasised and addressed. Global South scientists need to be doing GS research informing GN scientists (Nagendra et al. 2018). Local expertise ought to be recognised and invited to contribute to research and frameworks.

Comparative studies need to be encouraged as well the establishment of City Labs to foster transdisciplinary engagement (Pauleit et al. 2021). Studies could also be designed to analyse the impacts of urbanisation over time, providing rigorous data from long-term studies in GS urban areas. Furthermore, despite the fact that most urban growth and the most pressing urban challenges are found in small and medium-sized GS cities, there is a lack of information and research on these settlements. The focus on enhancing urban sustainability needs to be a systems approach for evaluating potential benefits and trade-offs (Nagendra et al. 2018). Scientists should develop a research-action agenda for urban resilience in GS cities within the context of planning and management focusing on the conditions, constraints and dynamics of the GS as many urban "solutions" developed within the context of the GN, do not apply to GS cities (Nagendra 2018; Sitas et al. 2021). Lastly, there is a need for "creative planning" in unplanned contexts (Lindley et al. 2018).

Illustrative questions pertaining to research approaches

- In what ways can transdisciplinary approaches and understandings improve urban ecology research and management in the GS?
- What innovative research approaches are needed to better understand the unique and complex dynamics of GS towns and cities?
- How can citizen science and other participatory approaches be better used to enhance urban ecological research and management in GS cities?

- How do urban environments change over time, and what are the implications for urban ecosystem services and conservation?

6 Concluding Remarks

In April 2019, the National Geographic magazine released a special issue on cities, in which they asked a prestigious architecture and urban planning firm from the USA (Skidmore, Owings and Merrill) how they would design a city of the future? The firm envisioned that:

> the plan allows ecology to guide development. Water sources are protected and systems are designed to capture, treat, and reuse it. Energy is renewable, and the city becomes more liveable even as it becomes more densely populated. All waste becomes a resource. Food is grown locally and sustainably. High-speed rail improves mobility. The culture and heritage of the increasingly diverse population are publicly supported. The infrastructure is carbon-neutral, and the economy is largely automated and online. (National Geographic 2019)

In most of the GS and even in the GN, much of this vision remains a distant and ambitious future. Urbanisation in the GS is often treated as if it will "catch up" with the GN, whereas it most likely represents alternative states or trajectories (McHale et al. 2013). The future cities of the GS have yet to be fully designed and developed. Our future is an existing one, and it's up to current day researchers, policymakers, stakeholders and citizens of these future cities to help develop them and shape them into sustainable and resilient cities that address GS problems with uniquely GS solutions. As Nagendra (2018) laments, the disciplines of sustainability and conservation often highlight only failures, but next-generation researchers and citizens need cases of hope to oppose common narratives of doom and gloom. They need to know that success stories "can be found on their doorstep", and not only in the GN (Nagendra 2018). Initiatives such as the "Seeds of the Good Anthropocene" (Bennett et al. 2016) can provide catalyst in this respect. However, other than South Africa, the current database on "Seeds of a good Anthrpocene" have insufficient examples from the GS.

We do not wish less for the GS, in fact, we wish more… more research, more acknowledgment of its rightful place in global urban ecological science, more local, context-driven solutions, more innovative participatory action plans, and more action towards an equitable, inclusive, sustainable and resilient future.

References

Adger N (2006) Vulnerability. Glob Environ Change 16:268–281

Anderson PM, Potgieter LJ, Chan L, Cilliers SS, Nagendra H (2021) Urban plant diversity: understanding informing processes and emerging trends. In: Shackleton CM, Cilliers SS, Davoren E, du Toit MJ (eds) Urban ecology in the Global South. Springer, Cham, pp 145–168

Andersson E (2006) Urban landscapes and sustainable cities. Ecol Soc 11:34
Arfvidsson H, Simon D, Oloko M, Moodley N (2017) Engaging with and measuring informality in the proposed Urban Sustainable Development Goal. Afr Geogr Rev 36:100–114
Bell D, Jayne M (2009) Small cities: towards a research agenda. Intn J Urb Reg Res 33:683–699
Bennett EM, Solan M, Biggs R, McPhearson T, Norström AV, Olsson P, Pereira L, Peterson GD, Raudsepp-Hearne C, Biermann F, Carpenter RC, Ellis EC, Hichert T, Galaz V, Lahsen M, López BM, Niholas KA, Preiser R, Vince G, Vervoort JM, Xu J (2016) Bright spots: seeds of a good Anthrpocene. Front Ecol Enviro 14:441–448
Cameron R, Katzschner T (2017) Every last drop: the role of spatial planning in enhancing integrated urban water management in the City of Cape Town. S Afr Geogr J 99:196–216
Capasso R, Zurlo MC, Smith AP (2018) Ethnicity, work-related stress and subjective reports of health by migrant workers: a multi-dimensional model. Ethn Heal 23:174–193
Childers D, Cadenasso M, Grove J, Marshall V, Mcgrath B, Pickett S (2015) An ecology for cities: a transformational nexus of design and ecology to advance climate change resilience and urban sustainability. Sustain 7:3774–3791
Cilliers SS, Breed C, Cilliers EJ, Lategan LG (2021) Urban ecological planning and design in the Global South. In: Shackleton CM, Cilliers SS, Davoren E, du Toit MJ (eds) Urban ecology in the Global South. Springer, Cham, pp 365–401
Cocks ML, Wiersum F (2014) Reappraising the concept of biocultural diversity: a perspective from South Africa. Hum Ecol 42:727–737
Cocks ML, Shackleton CM (eds) (2021) Diversities of nature in cities: enriching belonging, wellbeing and bioculture. Earthscan/Routledge, London
Davoren E, Shackleton CM (2021) Urban ecosystem disservices in the Global South. In: Shackleton CM, Cilliers SS, Davoren E, du Toit MJ (eds) Urban ecology in the Global South. Springer, Cham, pp 265–292
Diaz S, Pacual U, Srenseke M, Martin-López B, Watson RT, Molnár Z, Hill R, Chan KM, Baste IA, Brauman KA, Polasky S, Church A, Lonsdale M, Larigauderie A, Leadley PW, van Oudenhoven AP, van der Plaat F, Schröter M, Lavorel S, Aumeeruddy-Thomas Y, Bukvareva E, Davies K, Demissew S, Eroul G, Failler P, Guerra CA, Hewitt CL, Keune H, Lindley S, Shirayama Y (2018) Assessing nature's contributions to people. Sci 359:270–272
Dobbs C, Vasquez A, Olave P, Olave M (2021) Cultural urban ecosystem services. In: Shackleton CM, Cilliers SS, Davoren E, du Toit MJ (eds) Urban ecology in the Global South. Springer, Cham, pp 245–264
Drescher AW, Isendahl C, Caridad Cruz M, Karg H, Menakanit A (2021) Urban and peri-urban agriculture in the Global South. In: Shackleton CM, Cilliers SS, Davoren E, du Toit MJ (eds) Urban ecology in the Global South. Springer, Cham, pp 293–324
du Toit MJ, Hahs AK, MacGregor-Fors I (2021) The effect of landscape history on the urban environment: past landscapes, present patterns. In: Shackleton CM, Cilliers SS, Davoren E, du Toit MJ (eds) Urban ecology in the Global South. Springer, Cham, pp 51–78
Elmqvist T, Andersson E, Frantzeskaki N, McPhearson T, Olsson P, Gaffney O, Takeuchi K, Folke C (2019) Sustainability and resilience for transformation in the urban century. Nat Sustain 2:267–273
Escobedo FJ (2021) Understanding urban regulating ecosystem services in the Global South. In: Shackleton CM, Cilliers SS, Davoren E, du Toit MJ (eds) Urban ecology in the Global South. Springer, Cham, pp 227–244
Flint CG, Kunze I, Muhar A, Yoshida Y, Penker M (2013) Exploring empirical typologies of human–nature relationships and linkages to the ecosystem services concept. Landsc Urb Plan 120:208–217
Frayne B (2004) Migration and urban survival strategies in Windhoek, Namibia. Geofor 35:489–505
Gaston KJ, Ávila-Jiménez ML, Edmondson JL (2013) Managing urban ecosystems for goods and service. J Appl Ecol 50:830–840

Gill SE, Handley JF, Ennos AR, Pauleit S, Theuray N, Lindley SJ (2008) Characterising the urban environment of UK cities and towns: a template for landscape planning. Landsc Urb Plan 87:210–222

Güneralp B, Seto KC, Ramachandran M (2013) Evidence of urban land teleconnections and impacts on hinterlands. Curr Opin Enviro Sust 5: 445–451

Hartig T, Mitchell R, De Vries S, Frumkin H (2014) Nature and health. Ann Rev Publ Heal 35:207–228

Łopucki R, Kitowski I (2017) How small cities effect the biodiversity of ground-dwelling mammals and the relevance of this knowledge in planning urban land expansion in terms of urban wildlife. Urb Ecosys 20:933–943

La Point S, Balkenhol N, Hale J, Sadler J, van der Ree R (2015) Ecological connectivity research in urban areas. Funct Ecol 29:868–878

Lindley S, Pauleit S, Yeshitela K, Cilliers S, Shackleton C (2018) Rethinking urban green infrastructure and ecosystem services from the perspective of sub-Saharan African cities. Landsc Urb Plan 180:328–338

Maciejewski K, Currie P, O'Farrell P (2021) Social-ecological connectivity in Global South cities. In: Shackleton CM, Cilliers, SS, Davoren E, du Toit MJ (eds) Urban ecology in the Global South. Springer, Cham, pp 347–364

Martinez J, Mboup G, Sliuzas R, Stein A (2008) Trends in urban and slum indicators across developing wold cities 1990–2003. Habitat Intn 32:86–108

McConnachie MM, Shackleton CM (2010) Public green space inequality in small towns in South Africa. Habit Intn 34:244–248

McHale MR, Bunn DN, Pickett ST, Twine WC (2013) Urban ecology in a developing world: why advanced socioecological theory needs Africa. Front Ecol Enviro 11:556–564

McPhearson T, Pickett ST, Grimm NB, Niemelä J, Alberti M, Elmqvist T, Weber C, Haase D, Breuste J, Qureshi S (2016) Advancing urban ecology toward a science of cities. Bioscience 66:198–212

Mehring M, Katharina Geitmann-Mügge K, Frick-Trzebitzky F, Hummel D (2021) Urban social ecology. In: Shackleton CM, Cilliers SS, Davoren E, du Toit MJ (eds) Urban ecology in the Global South. Springer, Cham, pp 79–105

Myers G (2021) Urbanisation in the Global South. In: Shackleton CM, Cilliers SS, Davoren E, du Toit MJ (eds) Urban ecology in the Global South. Springer, Cham, pp 27–49

Nagendra H (2018) The global south is rich in sustainability lessons that students deserve to hear. Nat 557:485–488

Nagendra H, Bai X, Brondizio ES, Lwasa S (2018) The urban south and the predicament of global sustainability. Nat Sustain 1:341–349

National Geographic (2019) Infographic: cities of the future. *National Geographic.* April ed. Washington, DC: National Geographic Partners, LLC

Nijman J, Wei YD (2020) Urban inequalities in the 21st century economy. Appl Geogr 117:102188

Parnell S, Walawege R (2011) Sub-Saharan African urbanisation and global environmental change. Glob Enviro Change 21S:S12–S20

Pauleit S, Vasquéz A, Maruthaveeran S, Liu L, Cilliers SS (2021) Urban green infrastructure in the Global South. In: Shackleton CM, Cilliers SS, Davoren E, du Toit MJ (eds) Urban ecology in the Global South. Springer, Cham, pp 107–143

Pickett ST, Cadanesso ML, Grove MJ, Nilon CH, Pouyat RV, Zipperer WC, Costanza R (2001) Urban ecological systems: linking terrestrial ecological, physical and socioeconomic components of metropolitan areas. Ann Rev Ecol Syst 32:127–157

Pickett ST, Cadenasso ML (2017) How many principles of urban ecology are there? Landsc Ecol 32:699–705

Pickett ST, Cadenasso ML, Childers DL, McDonnell MJ, Zhou W (2016) Evolution and future of urban ecological science in, of and for the city. Ecosys Health Sustain 2:e1229

Ramalho CE, Hobbs RJ (2012) Time for change: dynamic urban ecology. Trends Ecol Evol 27:179–188

Rebele F (1994) Urban ecology and special features of urban ecosystems. Glob Ecol Biogeogr Lett 4:173–187
Reynolds C, Byrne MJ, Chamberlain DE, Howes CG, Seymour CL, Sumasgutner P, Taylor PJ (2021) Urban animal diversity in the Global South. In: Shackleton CM, Cilliers SS, Davoren E, du Toit MJ (eds) Urban ecology in the Global South. Springer, Cham, pp 169–202
Ribot J (2014) Cause and response: vulnerability and climate in the Anthropocene. J Peasant Stud 41:667–705
Rigolon A, Browning MH, Lee K, Shin S (2018) Access to urban green space in cities of the Global South: a systematic literature review. Urb Sci 2:67
Rupprecht CD, Byrne JA, Garden JG, Hero J-M (2015) Informal urban green space: a trilingual systematic review of its role for biodiversity and trends in the literature. Urban For Urban Green 14:883–908
Scoville-Simonds M, Jamali H, Hufty M (2020) The hazards of mainstreaming: climate change adaptation politics in three dimensions. World Dev 125:104683
Secretariat of the Convention on Biological Diversity (2012) Cities and biodiversity outlook: executive summary, Montreal. Available at: https://www.cbd.int/authorities/doc/cbo-1/cbd-cbo1-summary-en-f-web.pdf. Accessed 7 Nov 2020
Shackleton CM (2012) Is there no urban forestry in the developing world? Sci Res Essays 7:3329–3335
Shackleton CM (2021) Provisioning services in Global South cities. In: Shackleton CM, Cilliers SS, Davoren E, du Toit MJ (eds) Urban ecology in the Global South. Springer, Cham, pp 203–226
Shackleton CM, Cocks ML (2021) Urban nature and biocultural realities. In: Cocks ML, Shackleton CM (eds) Diversities of nature in cities: enriching belonging, wellbeing and bioculture. Earthscan/Routledge, London, pp 1–24
Shackleton CM, Cilliers SS, du Toit MJ, Davoren E (2021) The need for an urban ecology of the Global South. In: Shackleton CM, Cilliers SS, Davoren E, du Toit MJ (eds) Urban ecology in the Global South. Springer, Cham, pp 1–26
Shanahan DF, Bush R, Gaston KJ, Lin BB, Dean J, Barber E, Fuller RA (2016) Health benefits from nature experiences depend on dose. Sci Rep 6:28551
Shavitt S, Cho YI, Johnson TP, Jiang D, Hollbrook A, Stavrakantonaki M (2016) Culture moderates the relation between perceived stress, social support, and mental and physical health. J. Cross-Cult Psych 47:956–980
Sheppard E, Sparks T, Leitner H (2020) World class aspirations, urban informality, and poverty politics: a North-South comparison. Antipode 52:393–407
Shochat E, Warren PS, Faeth SH, McIntyre NR, Hope D (2006) From patterns to emerging processes in mechanistic urban ecology. Trends Ecol Evol 21:186–191
Simon D, Goodness J, Lwasa S, Puppim de Oliveira JA, Macedo LV, Kavonic J, Hermansson Török E, Elmqvist T (2021) Urban governance of and for urban green and blue infrastructure. In: Shackleton CM, Cilliers SS, Davoren E, du Toit MJ (eds) Urban ecology in the Global South. Springer, Cham, pp 403–428
Sitas N, Selomane O, Hamann M, Gajjar SP (2021) Towards equitable urban resilience in the Global South within a context of planning and management. In: Shackleton CM, Cilliers SS, Davoren E, du Toit MJ (eds) Urban ecology in the Global South. Springer, Cham, pp 325–345
Skandrani Z (2016) Considering the socio-ecological co-construction of nature conceptions as a basis for urban environmental governance. J Geog Nat Disast 6:1000153
Tacoli C (2006) (ed) The Earthscan reader in urban-rural linkages. Earthscan, London
UN-Habitat (2003) The challenge of slums: global report on human settlements 2003. Earthscan, London
van Gelder J-L (2013) Paradoxes of urban housing informality in the developing world. Law Soc Rev 47:493–522
Venter ZS, Shackleton CM, Van Staden F, Selomane O, Masterson VA (2020) Green apartheid: urban green infrastructure remains unequally distributed across income and race geographies in South Africa. Landsc Urb Plan 203:103889

World Bank (2009) Urban and local government strategy: concept and issues note. Sustainable Development Network, World Bank, Washington DC
Wu J (2014) Urban ecology and sustainability: the-state-of-the-science and future directions. Landsc Urb Plan 125:209–221